ELECTROMANIPULATION
of **CELLS**

ELECTROMANIPULATION
of **CELLS**

Ulrich Zimmermann
G.A. Neil

CRC Press
Boca Raton   New York   London   Tokyo

Library of Congress Cataloging-in-Publication Data

Zimmermann, Ulrich, 1942-
    Electromanipulation of cells / U. Zimmermann and G.A. Neil.
       p.   cm.
    Includes bibliographical references and index.
    ISBN 0-8493-4476-X
    1. Cells--Electric properties. 2. Electrofusion. 3. Cytology-
-Methodology. I. Neil, G. A. (Garry A.) II. Title.
    QH645.Z56   1995
    574.87′6041--dc20   95-14982
                                                          CIP

No claim to original U.S. Government works
International Standard Book Number 0-8493-4476-X
Library of Congress Card Number 95-14982
Printed in the United States of America 1 2 3 4 5 6 7 8 9 0
Printed on acid-free paper

Dedicated to Gertraud Zimmermann

PREFACE

The past decade has seen extraordinary advances in the electromanipulation of cells, a technology with foundations extending backward many years and enormous future potential in many areas of biology, biophysics, biochemistry, microbiology, medicine, and biotechnology. Although the techniques and applications of cellular, and indeed subcellular, electromanipulation are obviously diverse, spanning genetic engineering of prokaryotic and eukaryotic cells, cell motion, and orientation in time-varying fields as well as biophysical modeling of cell membranes, a common element of principles links them all.

The impetus for this book arose from a long-standing collaboration between the editors and a perceived need to explore and review the techniques in sufficient detail to satisfy the requirements of both the biophysicist as well as researchers who need and want only to apply these powerful techniques to their research. No discipline can stand long without a solid foundation. We recognized that the understanding of the "first principles" will advance the field, drive the development of newer techniques, further the refinement of existing technology, and provide the best possible framework for specific applications. We have endeavored, therefore, to provide not merely a collection of papers, but rather a comprehensive, critical, and balanced overview of the foundations of cellular electromanipulation as well as the present state-of-the-art of basic research. Equally important is the discussion of current practical applications of electric field technologies based on these principles.

The book begins with the impact of electric field effects of high (Chapters 1–4) and moderate intensity (Chapters 4 and 5) on cell and artificial membranes as well as the "electric structure" of the cell (Chapter 6). The principles are illustrated by widely used techniques of mammalian and human cell fusion, electropermeabilization of various cell types, and advanced cellular electromanipulation. Where relevant, key historical references are discussed to put the field in proper perspective for the investigator lacking a comprehensive command of the literature. An attempt to provide an inclusive reference list for each chapter has been made as a resource for the investigator requiring an in-depth treatment of a specific topic. Finally, in Appendix A we have provided a compendium of "ready-to-use" protocols for those seeking rapid application to specific projects. These protocols should be partially useful to those with no previous experience of electromanipulation techniques who wish to try these methods for the first time, as well as to undergraduates (especially project students). These same investigators are, however, directed to the earlier chapters to allow optimization and refinement of the techniques according to their needs.

Although the virtual explosion of cellular electromanipulation in recent years has been driven to a large extent by technical developments, the interest by scientists from many different fields was the key driving factor. The "cross-fertilization" engendered by this multidisciplinary interest has stimulated the increasingly broad spectrum of applications (which appear to be nearly unlimited) and has also enriched the science immensely. The authors, and indeed the editors themselves, reflect this scientific diversity. While the text is aimed at researchers in biology, physiology, microbiology, immunology, biophysics, biochemistry, and semiconductor technology, the level is that appropriate to graduate students in these fields. Therefore, we hope that this book will prove valuable to researchers in specific aspects seeking a broader view and that it also can serve as a textbook for graduate students taking advanced courses or carrying out research in the field. The biologist may wish to skip the mathematical aspects in some of the chapters but should be able to follow the principles and appreciate the applications. It is our sincere hope that this book, in its own modest way, will encourage the development of new ideas emerging from the interface between biological and biophysical scientists.

We would like to thank all contributing authors for providing us with their experience in various areas. All of them have worked closely with us in planning their chapters and minimizing overlaps. We also would like to thank CRC, in particular Jeff Holtmeier and Paul Petralia for their efforts to bring to fruition a difficult, but highly fascinating project.

Furthermore, we wish to thank the staff at the University of Würzburg who assisted in many ways as the book developed. Special appreciation goes to Dr. Cholpon S. Djuzenova and Dr. Vladimir Sukhorukov, but also to Dr. Gerd Klöck, Dr. Reiner Schnettler, Dr. William Michael Arnold, Joachim Schmitt, Petra Gessner, Birgit Herrmann, Heike Schneider, Willibald Bauer, and Thomas Igerst. They provided solid constructive criticism and helped, among other things, in the development of the electrofusion and electroinjection protocols. The first editor is also deeply obliged to his secretaries, Sigrid Mönkmeyer and Astrid Thal. They have read, proofed, and corrected the many type-writing mistakes and checked all the references. They never lost their humor and their patience. The outstanding and unstinting help of all of them made the book possible.

The editors and authors wish to acknowledge the generous financial supports provided by various German Grant Foundations. The work described in Chapters 1 and 4, and Appendix A was supported by the Deutsche Forschungsgemeinschaft (Sonderforschungsbereich 176), by the German Space Agency (DARA), and by the Bundesministerium für Forschung und Technologie, Bonn; the work described in Chapters 3 and 5 was granted by the Deutsche Forschungsgemeinschaft (Sonderforschungsbereich 176), by the VDI/VDE-Technologiezentrum (Bundesministerium für Forschung und Technologie, Bonn), and by the Fraunhofer-Institut für Siliziumforschung, Berlin, respectively, and the work described in Chapter 6 was supported by the Deutsche Volkswagenstiftung, Hannover.

U. Zimmermann and G. A. Neil   *September 1995*

CONTRIBUTORS

Roland Benz
Lehrstuhl für Biotechnologie
Biozentrum der Universität Würzburg
Am Hubland
Würzburg, Germany

Günter Fuhr
Lehrstuhl für Membranphysiologie
Institut für Biologie
Humboldt-Universität zu Berlin
Berlin, Germany

Roland Glaser
Lehrstuhl für Experimentelle
  Biophysik
Institut für Biologie
Humboldt-Universität zu Berlin
Berlin, Germany

Karl-Heinz Klotz
Department of Biophysical Chemistry
Biocenter of the University of Basel
Basel, Switzerland

Stephen G. Shirley
Lehrstuhl für Membranphysiologie
Institut für Biologie
Humboldt-Universität zu Berlin
Berlin, Germany

Akira Taketo
Department of Biochemistry I
Fukui Medical School
Matsuoka, Fukui, Japan

Mathias Winterhalter
Department of Biophysical Chemistry
Biocenter of the University of Basel
Basel, Switzerland

Ulrich Zimmermann
Lehrstuhl für Biotechnologie
Biozentrum der Universität Würzburg
Am Hubland
Würzburg, Germany

CONTENTS

Chapter 1
The Effect of High Intensity Electric Field Pulses on Eukaryotic Cell
Membranes: Fundamentals and Applications 1
Ulrich Zimmermann

Chapter 2
Electrotransformation of Bacteria 107
Akira Taketo

Chapter 3
High-Intensity Field Effects on Artificial Lipid Bilayer Membranes:
Theoretical Considerations and Experimental Evidence 137
Mathias Winterhalter, Karl-Heinz Klotz, and Roland Benz

Chapter 4
Electrofusion of Cells: State of the Art and Future Directions 173
Ulrich Zimmermann

Chapter 5
Cell Motion in Time-Varying Fields: Principles and Potential 259
Günter Fuhr, Ulrich Zimmermann, and Stephen G. Shirley

Chapter 6
Electric Properties of the Membrane and the Cell Surface 329
Roland Glaser

Appendix A
Protocols for High-Efficiency Electroinjection of "Xenomolecules" and
Electrofusion of Cells 365
Ulrich Zimmermann

Index 375

Chapter 1

THE EFFECT OF HIGH INTENSITY ELECTRIC FIELD PULSES ON EUKARYOTIC CELL MEMBRANES: FUNDAMENTALS AND APPLICATIONS

Ulrich Zimmermann

CONTENTS

I. Introduction 2

II. Membrane Breakdown: Experimental Background 4
 A. Determination of the Membrane Breakdown Voltage 4
 B. Factors Affecting the Breakdown Voltage 7
 C. Calculation of the Breakdown Voltage from
 the External Electric Field Strength 11
 D. Symmetric and Asymmetric Breakdown 17

III. Field Exposure Time and Post-Pulse Processes 24
 A. Short Duration Field Pulses 24
 B. Long Duration Field Pulses 24
 C. Selection of the Optimal Pulse Duration Time 26
 D. Resealing 27
 E. Electrointernalization 29

IV. Wave Form of the Field Pulse 31

V. Compromising and Beneficial Effects of the Pulse Medium 33

VI. Mechanism of Membrane Breakdown 36

VII. Electropermeabilization in Basic and Applied Research 40

VIII. Potentials of Electropermeabilization in Diagnosis and Therapy 77

IX. Conclusions 79

References 80

0-8493-4476-X/96/$0.00+$.50
© 1996 by CRC Press Inc.

1

I. INTRODUCTION

Although the phenomena of electric disruption of the cell membrane (irreversible membrane breakdown)[1] and electric alignment of freely suspended cells (dielectrophoresis)[2-4] were described in the early part of the 20th century, the foundations of cellular electromanipulation techniques are based on the much more recent discovery of reversible membrane breakdown, i.e. electric field-induced membrane perturbations that can spontaneously reseal without lethal consequences for the cell.[5-7] This seminal finding led rapidly to the development of electropermeabilization and electrofusion of prokaryotic and eukaryotic cells.

Despite our relatively recent recognition of these powerful technologies it is quite possible that electropermeabilization of membranes and electrofusion of cells were employed by nature during the evolution of life. The Earth has always been exposed continuously to electromagnetic waves emitted from the sun and other sources. The frequency of the radiation that reaches the ground depends upon the ionization of the atmosphere. The present day atmosphere screens the Earth's surface from frequencies below approximately 10–30 MHz. However, at the time of the first appearance of life, it is presumed that the level of ionization was much lower than today. Thus, electromagnetic waves of lower frequencies could pass through the atmosphere and reach the Earth's surface.[8] In the frequency range between about 100 kHz to perhaps 50 MHz particles (including cells) are capable of positive or negative dielectrophoretic alignment (a prerequisite for electrofusion), provided that the electric fields in question are even slightly inhomogeneous. The presence of frequencies in this critical range seems virtually certain given the state of the early terrestrial environment.

High intensity, short duration electric fields sufficient to induce reversible membrane breakdown (which is associated with electropermeabilization of the membrane and uptake of "xenomolecules" and "xenoparticles" into living cells) are created by many natural phenomena including earthquakes, ice formation, and lightning (literature quoted in References 8, 9). It is interesting to note that prolonged and intense lightning is thought to have occurred during the early development of life on Earth. The voltage necessary for the reversible breakdown of a cell membrane could readily result from a lightning discharge or the direct effect of the currents radiating from a nearby lightning strike on the ground. The duration and strength of a few of these "natural" field discharges would certainly have been short enough to create reversible electric membrane breakdown.

The existence of naturally conductive veins of ore within the Earth's crust, often considerably longer than 50 cm, which could have acted as the necessary antennas for the electromagnetic waves and field pulses, adds plausibility to this hypothesis. Laboratory experiments in which ore-containing rocks and electric field pulses similar to those that might result from a lightning strike were used to induce alignment and electrofusion of cells and provide supportive experimental evidence.[10,11] Thus whole primordial cells might have been fused with wholesale exchange of genetic material by a natural electrofusion process. Such mechanisms support early evolution and the generation of biodiversity by saltatory rather than continual development (as has often been asserted).

Electropermeabilization of freely suspended cells also may have contributed to the development of life on Earth. According to the symbiotic theory,[12,13] mitochondria and chloroplasts were originally derived from free-living bacteria and blue-green algae, respectively.

Laboratory experiments have shown that proteins, DNA, and also organelles, small particles and whole cells, can be "electrotransferred" or "electro-internalized" very easily into host cells.[14-18] The electrointernalized particles are surrounded by plasmamembrane from the host cell. This could explain why mitochondria and chloroplasts have a double membrane. Research on the interaction of electric field effects with biological systems offers, therefore, not only interesting basic and industrial applications, but also fascinating insights that might contribute to our understanding of the evolution of life.

In the laboratory, carefully controlled reversible electric membrane breakdown is now a powerful and indispensable tool in basic research, plant breeding, biomedicine, and biotechnology. This technology is now widely accepted as the state of the art in a number of important areas of research. There are two major categories of electric cell membrane breakdown technology. The first of these, termed electropermeabilization or electroinjection,* focuses on the use of reversible membrane breakdown for the injection of soluble material including macromolecules such as nucleic acids or proteins and/or pharmaceutical agents. As I will discuss further below, this technology has contributed enormously to the studies of genetic recombination, antigen expression, and intracellular regulatory processes, as well as that of cytotoxicity.

The second of these technologies, electrofusion, is a more complex sequential electromanipulative process whereby entire cells may be melded for the production of hybrids or formation of giant cells and other applications. The process, described in detail in Chapter 4, is based upon electric field-mediated dielectrophoretic alignment of cells, reversible (and irreversible) membrane breakdown of two or more aligned cells leading to fusion, followed by selection of fused cells. Even though they require specialized equipment, these

---

* Several authors use the term "electroporation" in place of electropermeabilization or electroinjection. The use of this term is somewhat unfortunate since it implies the formation of transient channels or "electropores" in the membrane during and after the application of reversible electric membrane breakdown pulses. As discussed further in this chapter and elsewhere in this monograph, there is no convincing evidence supporting the existence of such "electropores" in cell membranes. While the concept of pore formation may provide a useful construct for conceptualizing certain processes or making certain transfer calculations, I believe the use of the term "electroporation" has led to considerable confusion in the literature. I cannot therefore condone the use of this term, preferring the more neutral "electroinjection", "electropermeabilization," or "electric breakdown." I have termed a related, but distinct, transfer method based upon vesicle uptake induced by electric field mediated membrane invagination, "electrointernalization."

techniques have gained acceptance because they are more controllable, reproducible, and efficient than corresponding chemical (or viral) methods. An explosion of new publications in the field attests to its growing importance (see Chapter 4).

In this chapter, I will focus on the processes involved in reversible electric membrane breakdown of freely suspended eukaryotic cells and on the injection of low and high molecular weight substances into these cells. Although several important differences exist between prokaryotic and eukaryotic cells, the fundamental processes of membrane breakdown of bacteria are comparable to those occurring in yeast, plant, and mammalian cells. Peculiarities of the prokaryotic cell membrane and wall, the specialization of techniques of prokaryotic genetic manipulation, and the profusion of literature in this field dictated the need for a separate chapter devoted to this topic (Chapter 2).

As in any vital scientific area, controversies and contradictions exist in the field of electromanipulation of cells. I have endeavored to provide both a useful review as well as a critical analysis in order to assist the reader in coming to grips with the burgeoning literature in the field. The results and information presented here should also be seen as an introduction to the various facets of electrofusion of dielectrophoretically aligned cells described in Chapter 4. The reader is further directed to previous reviews of the subject of electropermeabilization.[8,19-30]

II. MEMBRANE BREAKDOWN: EXPERIMENTAL BACKGROUND

A. DETERMINATION OF THE MEMBRANE BREAKDOWN VOLTAGE

In cells that are large enough for the introduction of microelectrodes, the electric breakdown can be measured directly by determining the membrane current-voltage characteristics[31-33] or by using the charge pulse technique[34-37] by analogy with the experiments on planar lipid bilayers (Chapter 3). Such experiments also yield information on the resealing time of the membrane after breakdown.[32,36] For smaller cells (diameter of 1–40 μm) or cells in suspension, reversible membrane breakdown can be demonstrated:

1. by measuring the size distribution of a cell population in a particle analyzer as a function of the strength of an externally applied field pulse;[5-7,19,20,38-45]
2. by studying the breakdown-induced uptake or release of radioactive isotopes[7,41-43,46-53] or of other indicator substances[6,20,54-69] in response to an external field (see below);
3. by microscopic observation of the distribution of chemotactic bacteria when critical, low-molecular weight substances are released from the electropermeabilized cell;[70,71]

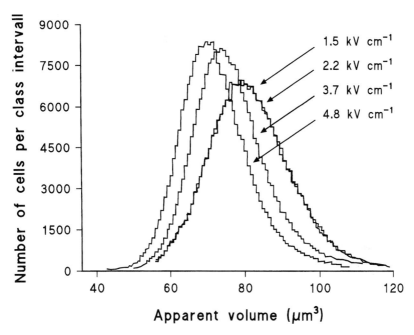

FIGURE 1. The apparent size distribution of a human erythrocyte suspension as a function of the field strength in the orifice of a hydrodynamically focusing particle analyser. Above the critical field strength leading to membrane breakdown, the size distribution becomes underestimated because of current flow through the permeabilized membrane and the highly conductive cell interior.

4. by determination of the spatial changes in the transmembrane voltage and conductance using voltage-sensitive (i.e., charge-shift potentiometric) fluorescent dyes[72-80] (see also Chapter 6);

5. through the use of patch clamp techniques,[81] special voltage clamp techniques,[82] or special planar electrode arrays for analysis of adherent cells.[83] These are discussed in detail below, with the exception of the planar electrode array which is discussed in Chapter 5.

In the particle analyzer, cells (suspended in highly conductive, physiological solutions) are forced to pass through an electric field (generated in a small diameter orifice) for a very short time (μsec range). Due to the high inherent membrane resistance of the cell, a voltage signal is induced which is proportional to the cell size, provided that the field strength used is below a critical threshold required for breakdown. At strengths exceeding this critical level, the cell size will be underestimated because the current (field) lines pass (in part) through the highly conductive cell interior (Figure 1). The degree of this underestimation depends upon the total conductivity of the cells. Therefore, this method is useful in measuring the breakdown voltage of the membrane but also allows an estimate of the internal conductivity of the cell,[38,84] a parameter which is otherwise difficult to measure.*

Estimation of the breakdown voltage from studies of the breakdown-induced exchange of (normally membrane-impermeable) substances between cell and environment in a discharge chamber is less accurate. However, the use of marker molecules for determination of the breakdown voltage does not require sophisticated equipment. In common protocols, a capacitor is charged to a known voltage by a high voltage power supply and discharged through a cell suspension placed between two (noble) metal plate electrodes in a chamber.[6,48,51,59,85-90] Once the plasmamembranes have been rendered electropermeable, the cytosolic components will exchange with those present in the suspension medium. The time for solute exchange and equilibration depends on many parameters, including the field conditions and media composition (strength and duration of the pulse, the homogeneity of the field, the conductivities of the medium and the cytosol,[67,91] etc.), the resistance of the cells to osmotic swelling after permeabilization (see below and also Footnote on page 11), the resealing properties of the membrane after pulse application (see below), the size of the indicator molecule,[54] the cell volume, and so on. Experimental variations and the use of cells with unusual structural and electrical properties might explain some of the many inconsistencies present in the literature with respect to the breakdown voltage deduced from discharge experiments (see, for example, References 64, 92). Generally speaking, small cells (such as platelets, yeast, bacteria, etc.) have a much more favorable surface-to-volume ratio and will, therefore, equilibrate faster than larger cells (such as plant protoplasts, mammalian cells, etc.), with the proviso that appropriate field conditions and media compositions were used.

Although it is difficult to estimate the accurate value of the breakdown voltage from discharge chamber experiments, this is the method of choice to manipulate the cytosolic content and to study the events after breakdown.[69] The solute exchange can be determined by removal of aliquots from the pulsed suspension, followed by microscopic examination (in the case of dyes) or by investigation of the cell pellet and/or the supernatant after centrifugation. Experiments under the microscope require a microslide onto which two or more parallel wires or strips have been fixed[21,22,30,93] at least a few millimeters apart in order to ensure a nearly homogeneous field. The nonuniform fields such as used by some authors[62,94-96] should be avoided for reasons given below and in Chapter 4. Electropermeabilization of the cell membranes can also be easily detected by uptake or release of appropriate fluorescent-labelled dyes. Continuous monitoring of the efflux- or influx-kinetics using flow cytometry or other fluorimetric devices may yield additional information about the diffusion and binding properties of the cytosol, about the resealing process of the

---

*   The relatively new technique of electrorotation described in detail in Chapter 5 offers a novel and sensitive means of measuring internal conductivity. Such measurements should become easier as this technology becomes more widely available.

electropermeabilized membrane, and about the ratio of viable to electro-killed cells (Figure 2). In fact, the technique of flow cytometry has recently been recognized as an excellent (quantitative) method for analysis of net solute exchange through electropermeabilized membranes.[67,68,97-110] Additional potential refinements for combined cell selection/electromanipulation of cells based upon the adaptation of flow cytometers equipped with field pulse instruments for selection, electroinjection, or killing (negative selection) of cells are foreseen in the near future.[111]

The release kinetics of intracellular compounds (organic acids and/or amino acids) in response to a breakdown pulse can also be detected by accumulation and retention of chemo- or aerotactic bacteria (e.g., *Pseudomonas aeruginosa*) close to the permeabilized membrane area[70,71] (Figure 3). This method is indirect but allows reasonable estimations of the breakdown voltage, the duration of membrane electropermeabilization, and information about the site(s) of breakdown. The spatial resolution of this technique, however, is only sufficient for cells with a diameter of >20 μm.

The use of voltage-sensitive fluorescent dyes in combination with pulse-laser fluorescence microscopy, with its sub-microsecond temporal as well as fine spatial resolution,[74-80] is a very powerful and highly sensitive technique to measure membrane charging quantitatively, the spatial distribution of the voltage generated across the membrane, and the breakdown time and voltage, as well as the spatial kinetics of the conductance increase and resealing of the membrane (i.e., parameters which can also be measured by impaled microelectrodes). The "disadvantage" of the fluorescence imaging method is that it is at present restricted to relatively large cells (e.g., sea urchin eggs) or to giant liposomes.

Sophisticated cell-attached patch clamp techniques have been used to study the dynamics of cell membrane permeabilization and breakdown.[81] Although such direct measurement techniques are influenced by variations in the morphology of the applied membrane patch with respect to surface area and location of ion channels, pumps, and other membrane structures, as well as by the membrane effects that might be induced by application of the patch itself, they are essentially in agreement with, and provide corroborative evidence of, the aforementioned direct and indirect techniques.

B. FACTORS AFFECTING THE BREAKDOWN VOLTAGE

The membrane breakdown voltage for most eukaryotic cells has been repeatedly found to be around 1 V (0.5 – 1.5 V) at room temperature independent of the methods used[5-7,20,27,30-33,36,40,42,58,70,74,78,80,81,84,98,109,112] (with the proviso that the pulse duration was very short; see below). Interestingly, no systematic differences in this relatively constant parameter can be demonstrated between cells of widely different species, phyla, and kingdoms. This is not surprising given the nearly constant structure and composition (lipid/protein ratio 1:1) of all eukaryotic cell membranes.

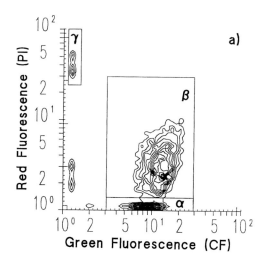

FIGURE 2. Flow cytometric determination of the ratio of viable to electro-killed cells by measuring the uptake of the (normally membrane-impermeable) propidium iodide (PI) through the electropermeabilized membranes of murine SP2/0-Ag14 myeloma cells which had been preloaded with the fluorescent dye carboxyfluorescein, CF. Before pulsing the cells were incubated in iso-osmolar phosphate buffer solution containing 6 μM carboxyfluorescein diacetate (CFDA, a non-fluorescent, membrane-permeable precursor of carboxyfluorescein) at 37°C for 1 min. Due to cytoplasmic esterase activity CFDA is transformed into the membrane-impermeable CF. Under these conditions of low-level staining, the majority of CF molecules seems to be bound to cytoplasmic macromolecules and/or organelles. The CF-loaded cells were incubated at a suspension density of 10^6 cells ml^{-1} in iso-osmolar solutions containing 30 mM KCl, 0.35 mM KH$_2$PO$_4$, 1.15 mM K$_2$HPO$_4$, 0.2% bovine serum albumin, 25 μg ml^{-1} PI and appropriate amounts of inositol. The pH was 7.2. Up to three breakdown pulses of 2 kV cm^{-1} field strength and 40 μs duration were applied at 20°C (interval between consecutive pulses 1 min). 5 min after pulsing the cells were analyzed in the flow cytometer. Electro-injected PI molecules bind to the nucleic acids and give red fluorescence. The PI fluorescence was calibrated *versus* the PI fluorescence of lysed cells (fmol per cell). Figure 2a shows a typical contour-plot cytogram (PI- *versus* CF-fluorescence). Several subpopulations can be distinguished: (α) intact living cells containing CF, but not PI. Such cells represent the majority of the cell population before pulsing: the plasmamembrane of these cells is impermeable to PI; (β) electropulsed, but living cells which are characterized by an unchanged CF- and a low-level of PI fluorescence. The low-level red fluorescence (despite the high PI-concentration in the pulse medium) demonstrates that the membranes of these cells resealed within a few minutes after the pulse(s); γ) "dead" cells which exhibit no CF-, but very high PI-fluorescence. This sub-population is also seen in the control cells and increases after field pulse application. The enhanced disappearance of the CF-fluorescence after pulsing in some part of the cells indicate loss of the cytoplasmic macromolecules to which CF was bound. A sub-population with intermediate CF-fluorescence could not be detected indicating that the release of the (bound) CF molecules in response to the pulse(s) occurred according to an "all-or-none" mechanism. Figures 2b–d show the evaluation of the cytograms (such as shown in Figure 2a): (b) viability of the cells as a function of the number of pulses; (c) the corresponding increase of the PI-content; and (d) the corresponding CF-fluorescence in the living cells. Each point represents the mean value ± SD from at least three independent measurements. It is interesting to note that the uptake of PI decreased with the increase in the number of pulses. Such a kinetic is expected if PI enters the cells by diffusion through the electropermeabilized membrane. (Data courtesy of Sukhorukov, V. L. and Djuzenova, C. S.)

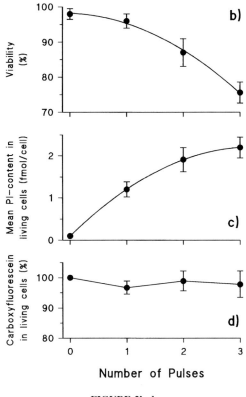

FIGURE 2b-d.

Incorporation of protein into biomembranes can be demonstrated to affect the membrane breakdown voltage moderately:

1. A reduction in the breakdown voltage of about 20% (i.e. 0.2 V) in developing chloroplasts has been reported.[113] Since the protein content of the chloroplasts is significantly higher than that of the etioplasts this finding suggests that breakdown occurs predominantly at the lipid/protein junctions.
2. Electrofusion experiments on isolated plant vacuoles have given evidence that the breakdown voltage of the tonoplast is somewhat higher than that of the plasmamembrane.[70] Vacuolar membranes usually contain less protein than cell membranes. Similarly, the finding that mitochondria can be fused at average field strengths of 2 kV cm⁻¹ despite their small size and their double membrane arrangement[114] might be partly a result of the high protein content leading to a considerable reduction of the breakdown voltage (see Chapter 4). In addition, electropermeabilization of prokaryotic membranes (lipid/protein ratio about 1:2) usually also needs less field strengths than theoretically expected (see Chapter 2).

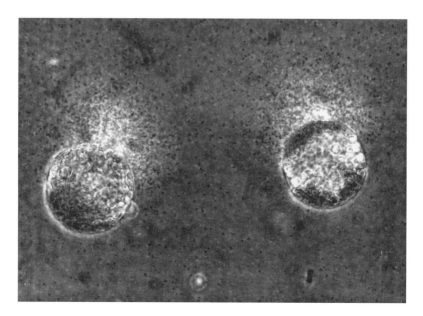

FIGURE 3. Demonstration of asymmetric release of intracellular solutes through the electropermeabilized membrane of oat protoplasts by using chemotactic bacteria (iso-osmolar sugar solution, a single field pulse of 600 V cm⁻¹ strength and 50 μs duration). Breakdown occurred initially in the hemisphere facing the anode. However, the movement of the bacteria rapidly changes the orientation of the cells relative to the electrodes. (From Mehrle, W., Hampp, R., and Zimmermann, U., *Biochim. Biophys. Acta,* 978, 267, 1989. With permission.)

3. Pre-treatment of cells with pronase before pulsing and fusion stabilizes biomembranes against intense field pulses.[115,116] The effect of proteolytic enzymes on membranes is interpreted in terms of the emergence of protein-free lipid domains following the degradation of membrane proteins (Chapter 4). It is also well-known that pharmacological agents including local anesthetics can influence the breakdown voltage.[45,117,118] The extent to which other membrane-stabilizing agents (e.g. Ca^{2+}-ions) or changes in membrane fluidity influence the breakdown voltage has not been investigated.

Other factors may also affect the membrane breakdown voltage. For example, it has been suggested that the breakdown voltage depends upon the cell cycle.[20,109] The mechanism underlying the latter observation has not yet been elucidated but may be due to changes of cytoskeletal attachment, membrane fluidity changes, or alterations in microvillous membranes (see below).

The breakdown voltage is greatly influenced by temperature.[20,25-27,30,32,36] At low temperatures (around 4°C) the breakdown voltage doubles to about 2 V. At higher temperatures (37°C) it decreases to about 0.5 V.[32] The dependence of the breakdown voltage on temperature seemed to be smaller when long pulses (900 μs) were used.[119] Similarly, the breakdown voltage also depends strongly on the duration of the field pulse, according to measurements using impaled microelectrodes inserted into giant algal cells.[34] This result is in agreement with analogous findings on planar lipid bilayer membranes (Chapter 3). A decrease in breakdown voltage as the pulse duration increases (above approximately 10 – 20 μs) presumably also occurs for smaller cells exposed to an external field pulse in a discharge chamber.[54,58,64] With the benefit of hindsight, the finding that the critical field strength for breakdown decreases with pulse duration as a hyperbolic function may be — at least partly — due to the pulse-length dependence of the breakdown voltage.

Other physical factors may also exert a profound influence on membrane breakdown. For example, precompression of the cell membranes by hydrostatic pressure[120-122] and pressure gradients (turgor pressure)[48,81,123-126] significantly decreases the membrane breakdown threshold voltage.* The relatively low breakdown voltage of the membrane barrier of the peat-bog alga *Eremosphaera viridis* (recorded directly using the charge pulse technique[37]) may also be due to the unusually high turgor pressure (about 0.8 MPa) in cells of this species. In addition, the inverse relationship between the breakdown voltage and turgor pressure may also be one of the reasons why the field strengths required for membrane breakdown of bacterial cells are significantly lower than theoretically expected (see below and Chapter 2).

C. CALCULATION OF THE BREAKDOWN VOLTAGE FROM THE EXTERNAL ELECTRIC FIELD STRENGTH

An externally applied electric field pulse induces a potential difference across the membrane capacitor due to charge movement in the cytosol and the external medium followed by charge accumulation at the membrane interfaces. The generated membrane potential difference, V_g, is superimposed upon the resting transmembrane potential difference, V_m. Breakdown occurs if the total membrane potential difference ($V_c = V_g + V_m$) exceeds the breakdown voltage at a given critical electric field strength, E_c.

Membrane potentials (including the breakdown voltage) induced by external field pulses are usually derived from a Laplacian treatment considering a cell surrounded by a perfectly insulating membrane (Figure 4). When we assume that the stationary membrane potential is reached (i.e., that the charging

---

* Chinese hamster ovary cells, which can maintain their volume constant over a large osmotic pressure range (presumably due to a very strong cytoskeleton network), did not show a dependence on the threshold field strength required for electropermeabilization. This result cannot be used as an argument that the breakdown voltage is not pressure dependent[127] because the hydrostatic pressure within the cells is only transiently built up by instant water uptake in response to hypo-osmolar conditions within a few seconds. Then, the number of intracellular osmotically active particles is regulated either by metabolic processes or by ion efflux until the original hydrostatic pressure difference is reached. The biphasic osmotic response of wall-less cells under hypo-osmolar stress and the resulted maintenance of the volume is well-studied and understood.[128]

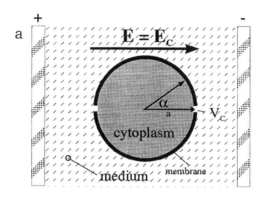

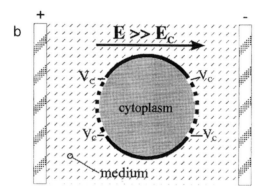

FIGURE 4. Schematic diagram of the induction of symmetric (a, b) and asymmetric (c) membrane breakdown in a spherical cell of radius, a, exposed to a stationary external field (see also Figure 4a, Chapter 3 and Discussion in Chapter 5). Because of the angle-dependence of the generated potential the breakdown voltage is first reached in field direction, i.e., $E = E_c$ (a), and then for higher (supercritical) field strengths, $E > E_c$, at membrane sites oriented at a certain angle, α, to the field direction (b). Membrane breakdown is illustrated schematically by the formation of "uniform pores". If the resting transmembrane potential, V_m, (or more accurately the intrinsic membrane field) is taken into account an asymmetric breakdown is expected (c). For the case that the cell interior is negatively charged with respect to the external medium, the generated membrane potential, V_g, is parallel to the transmembrane potential difference in the hemisphere facing the anode, but antiparallel in the hemisphere facing the cathode.

process of the membrane capacitor is over), the membrane potential of a freely suspended, spherical cell with a smooth surface leading to breakdown is usually calculated using the following expression:[6,72-74,78-80,87,129-133]

$$V_c = 1.5\, E_c\, a \cos \alpha \pm V_m \qquad\qquad (1)$$

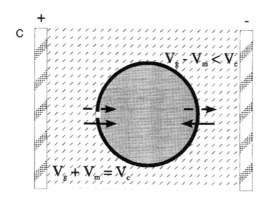

FIGURE 4c.

where:

E_c = critical external electric field (V cm^{-1})
V_c = membrane (breakdown) potential of a spherical cell (V)
V_m = resting transmembrane potential difference (V)
a = the cell radius
α = the angle between the membrane site and the field direction

The values of the breakdown voltage of cell membranes of different species mentioned in the foregoing section were calculated by using Equation 1. Equation 1 is also used in electropermeabilization experiments for calculation of the critical field strength resulting in breakdown by assuming that the breakdown voltage is around 1 V at room temperature or 2 V at 4°C.

However, in contrast to lipid vesicles (see Chapter 3) a biological cell has a finite specific membrane conductance per unit area, G_m. For example, measurements on giant algal cells (using intra- and extracellular electrodes) have shown[37,134] that G_m is of the order of $10^{-3} - 10^{-4}$ S cm^{-2}.

In those cases where the membrane conductance cannot be neglected, the following equation can be derived:[6]

$$V_c = \frac{\dfrac{3}{2+\rho_e/\rho_m} E_C \left[a_1 - a_2 + \dfrac{1-\rho_m/\rho_i}{2+\rho_m/\rho_i} a_1^3 \left(\dfrac{1}{a_1^2} - \dfrac{1}{a_2^2} \right) \right] \cos\alpha}{1+2\left(\dfrac{1-\rho_e/\rho_m}{2+\rho_e/\rho_m} \right)\left(\dfrac{1-\rho_m/\rho_i}{2+\rho_m/\rho_i} \right)\left(\dfrac{a_1}{a_2} \right)^3} \pm V_m \qquad (2)$$

where a_1 = inner radius and a_2 = outer radius of the cell ($a_2 - a_1$ = membrane thickness), ρ_i = internal specific resistivity of the cell (Ω cm), ρ_e = external specific resistivity of the medium (Ω cm), and ρ_m = membrane specific resistivity (Ω cm). If the membrane thickness is much smaller than the radius of the cell (i.e., $a_1 = a_2 = a$), Equation 2 goes over into Equation 3:[132]

$$V_c = \frac{1.5\, E_c a \cos\alpha}{1+\lambda} \pm V_m \qquad (3)$$

where $\lambda = G_m\, a\, (0.5\, \rho_e + \rho_i)$.

In the derivation of Equations 1 to 3 several assumptions have been made including:

1. that a biological cell can be modelled as a homogeneously conducting interior (cytosol) surrounded by an external shell (membrane);
2. that the field is homogeneous;
3. that V_g is superimposed linearly upon V_m;
4. that V_m is not changed by the external field;
5. that surface admittance and space charge effects do not play a role.

Condition 1 is a first order approximation, but may not be completely valid if the equations are applied to plant cells having two membranes in series or to mammalian cells possessing a large nucleus (see Chapter 6).

Condition 2 is fulfilled if a plate capacitor with sufficient spacing between two planar electrodes (about $0.5 - 1$ cm) and a suspension density of about 10^6 cells per ml or less are used. If the cell density is too high,[135,136] or if wire electrodes with a separation smaller than about 1 cm are used,[94,95] the electric field may become inhomogeneous. This, in turn, may result in the exposure of some cells in the suspension to field strengths which differ considerably from that expected theoretically. Under these conditions the calculation of the breakdown voltage by using Equations 1, 2, or 3 (and the elucidation of the electropermeabilization process) are rather unpredictable. Thus some cells are permeabilized reversibly, whereas others fail to reach the threshold of reversible membrane breakdown or are lysed. In addition, dielectrophoresis can occur, promoting the fusion of cells which receive appropriate breakdown field strength (see also Chapter 4).

Conditions 3 and 4 are certainly not valid (see below and Chapter 6). However, because of the lack of information, these assumptions seem reasonable given the state-of-the-art.

Condition 5 seems to be valid under the conditions used so far in electropermeabilization and in electrofusion (see Chapter 4). Contributions from the surface admittance and from space charges to Equations 1 to 3 will considerably lower the calculated breakdown voltage. Such effects are theoretically expected for cells suspended in poorly low conductive solutions[137,138] (as used frequently in electrorotation studies, see Figure 5 and Chapter 5) and may also play a role in bacterial transformation where almost nonconductive pulse media are used (Chapter 2).

Inspection of the three equations shows that Equation 1 can be used for calculation of the membrane voltage (including the breakdown voltage) of biological cells with a finite specific conductance per unit area, G_m, provided that $\lambda < 1$. Because ρ_i is of the order of 100 Ω cm for many mammalian cells[84] (and less for plant and algal cells), λ depends mainly on G_m and ρ_e for a given cell radius. Plots of λ *versus* ρ_e for typical values of G_m are given in Figure 5 thereby assuming that the cell radius is about 10 μm = 10^{-3} cm.* It is obvious that λ is significantly lower than 0.1 (and therefore negligible) for phosphate buffer solution (PBS) and conductive pulse media which are usually (but not always) used in electropermeabilization studies of eukaryotic cells (for media composition, see below). As indicated by Figure 5, λ can also be neglected when weakly conductive solutions for electrofusion are used (see Chapter 4). Furthermore, it is evident from Figure 5 that λ is very small even when relatively high G_m-values of the order of 10 mS cm^{-2} are assumed as deduced from electrorotation experiments.[139] As outlined in Chapter 5, cell electrorotation yields G_m-values that include the contribution from surface admittance to the overall conductance of the membrane. In light of these results we can conclude that the membrane conductance and surface admittance do not significantly influence the membrane potential generated by an external field under electropermeabilization (as well as electrofusion) conditions as assumed by some authors.[137,138] Their conclusion that (reversible) breakdown occurs at considerably lower membrane voltages (about 200 mV) because of membrane conductance and surface admittance effects is completely misleading (see also Footnote on page 23) and also inconsistent with the experimental results obtained using intracellular microelectrodes, voltage-sensitive dyes, or patch clamp technique (see above).

Thus, although Equations 2 and 3 are generally valid for calculation of the breakdown voltage from the external field strength, Equation 1 can be used with confidence provided that conductive solutions (as is the case, for example, when using a particle analyzer for determination of the breakdown voltage) are used and the problems of the contribution of the resting membrane potential and other conditions mentioned above are kept in mind.

Despite the fact that uncertainties in the calculation of the true breakdown voltage from discharge experiments remain, inspection of Equations 1 to 3 reveals a number of practically useful points. The first of these is that the generated membrane breakdown voltage, V_c, depends on the radius, a, of the

---

* Note that the specific membrane conductance per unit area is not independent of the external conductivity. G_m usually increases with increasing external conductivity.

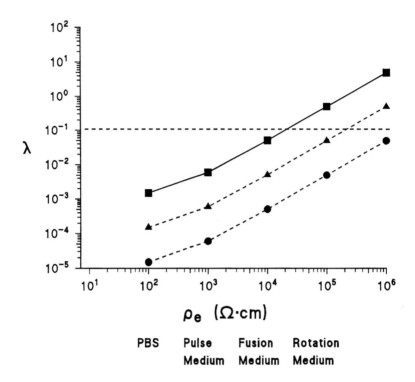

FIGURE 5.   The dependence of the parameter λ in Equation 3 on the resistivity of the medium, ρ_e, for typical values of the specific membrane conductance per unit area, G_m (squares: $G_m = 10^{-2}$ S cm^{-2}; triangles: $G_m = 10^{-3}$ S cm^{-2}; circles: $G_m = 10^{-4}$ S cm^{-2}). Data were calculated for a cell with a radius a = 10 μm and a cytosolic resistivity, $\rho_i = 10^2$ Ω cm.

cells.* This means that even though there is little difference in the electric effects on the membrane per se, higher field strengths are required for the membrane breakdown of smaller *versus* larger cells (for experimental proof, see References 43, 143). Thus a field strength of about 15 – 20 kV cm^{-1} is required for bacteria and platelets whereas about 1 kV cm^{-1} is all that is needed for the much larger plant protoplasts (radius up to 50 μm).[23] Owing to the radius-dependence of the generated voltage the organelles will, generally

---

* Some early workers[140] failed to take into account the radius dependence of the generated membrane voltage. Thus they erroneously assumed that in the presence of a 20 kV cm^{-1} pulse, only 20 mV would be generated across a 10 nm cell membrane. This potential is expected during certain cell physiological events, including neurosecretion, but would not be associated with membrane breakdown. When the data are properly calculated, the true transmembrane voltage generated by such a pulse corresponds to the breakdown voltage.[131] Although some authors later realized that it *could* occur under these circumstances, membrane breakdown was discounted[141,142] in favor of one of several other electrically mediated mechanisms. This was because it was widely believed that membrane breakdown was always irreversible. The irrefutable demonstration of reversible membrane breakdown[5-7,21,31-33,41,42,46,54] under appropriate conditions helped to settle this controversy.

speaking, not be affected by the breakdown of the plasma membrane.[21,87] Breakdown of the membrane of chloroplasts, mitochondria, and nuclei, as well as of the vacuolar membrane, in plant cells depend on their size and on the conductivity in the cytosol and in the organelle concerned.

As discussed above, the radius-dependence of the generated membrane potential difference may sometimes be offset by a very low breakdown voltage due to a high protein content within the membrane or to high temperature and/or pressure.

The angular dependence of the generated potential is also of considerable importance. In contrast to application of a field pulse via impaled microelectrodes, the potential depends on the angle, α, between a given membrane site and the field vector (as specified in Figure 4). In other words, the generated potential is maximal at membrane sites oriented in the field direction and zero at sites perpendicular to the field direction (for experimental proof, see References 72–80).

The radius- and angle dependence of the generated voltage dictate the advantage of selecting cell suspensions with a narrow size distribution for electropermeabilization applications in order to avoid much cell lysis or low yields of reversibly electropermeabilized cells. Another disadvantage of lysing large numbers of cells is the release of intracellular enzymes and ions which can have detrimental effects on remaining viable cells in a field-treated suspension. In addition, the field strength may decay uncontrollably as a consequence of this release during prolonged (ms duration) pulses (see discussion below and Chapter 2). These effects are probably amplified in the presence of irreversible membrane breakdown. Unselected cell populations (e.g., cell cultures or plant protoplast preparations) usually show a broad and nonoptimal size distribution.[144] Narrow cell-distributions can be obtained by preselection using fluorescence activated flow cytometry (FACS)[101,102,105] or free flow electrophoresis.[145] These techniques have as yet almost completely unrealized applications for the efficient separation of lysed cells and cell debris after field pulse application (see also above). Although access to these cell selection technologies is increasingly common, they are expensive, not always available, and somewhat labor-intensive. Substantial improvement and automation along these lines are likely and will be welcome.

D. SYMMETRIC AND ASYMMETRIC BREAKDOWN

The concept of asymmetric membrane breakdown is critical to the understanding of modern cell electromanipulation although our knowledge of this effect is very rudimentary. Providing that the resting membrane potential difference, V_m, is negligible in relation to the superimposed potential difference, Equations 1–3 predict symmetric membrane permeabilization in the two hemispheres of a cell exposed to a breakdown pulse (Figures 4a and 4b). However, if the transmembrane potential, V_m, is sufficiently high, asymmetric breakdown of only one hemisphere will be expected as illustrated in

Figure 4c. If we assume that V_m is directed towards the interior of the negatively charged cell, the vector of the field generated across the membrane is parallel to the transmembrane field vector in the hemisphere facing the anode, but antiparallel in the hemisphere facing the cathode. Thus membrane breakdown will occur in the anodic hemisphere at external field strengths lower than that necessary to induce breakdown in the membrane facing the cathode.

Experimentally, this so-called "asymmetric breakdown" can be easily detected and followed using fluorescent dyes that have been "preloaded" into the cell or added to the suspension medium prior to field application, as well as by the chemotactic bacteria assay. Microphotometric or microfluorimetric determination of the uptake of fluorescent dyes (Figure 6) or accumulation of chemotactic bacteria around the electropermeabilized area (Figure 3) allows measurement of the duration of the high permeability state in the anode-facing hemisphere and allows monitoring of the resealing process. Asymmetric breakdown of the anode-facing hemisphere was detected in chloroplast "blebs,"[146] in sea urchin eggs,[147] and in plant[60,70,71] and mammalian cells.[63,71,148,149] It has also been observed in dielectrophoretically aligned plant protoplasts[70,71] (see also Chapter 4). Isolated plant vacuoles,[70] which are positively charged with respect to the suspension medium, exhibited asymmetric breakdown in the cathode-facing hemisphere. The application of a second breakdown pulse of opposite polarity (after membrane resealing) resulted in breakdown of the other cellular hemisphere (now facing the former anode).

In contrast, in erythrocyte ghosts, breakdown occurred preferentially in the cathode-facing hemisphere,[150] even though these vesicles should not exhibit a membrane potential difference. In large unilamellar liposomes,[76,79] in which the membrane potential difference should also be zero, two large holes with a diameter of several micrometers were created on the anode side; however, holes on the cathode side or on both sides were sometimes observed.*

For mouse myeloma cells it was recently shown[151] that the asymmetry of the breakdown depended on the osmolality of the pulse medium (Figure 7). In the majority of cells, field-induced uptake of propidium iodide was observed through the anodic side when the cells were suspended in iso-osmolar pulse medium (field conditions: a single pulse of 1.5 kV cm⁻¹ strength and 40 μs duration; medium composition: 30 mM KCl, 1.15 mM phosphate buffer and appropriate amounts of inositol; conductivity 3 mS cm⁻¹, osmolality 300 mOsmol). In contrast, asymmetric breakdown occurred in the cathodic hemisphere when the cells were exposed to hypo-osmolar stress (150 mOsmol solutions). However, for mouse L-cells (which have a specific membrane capacity of 2 μF cm⁻² as measured by electrorotation, see Chapter 5) the breakdown was on the cathodic side in iso-osmolar and on the anodic side in hypo-osmolar solutions. In poorly conductive solutions (0.3–1.4 mS cm⁻¹) breakdown was symmetric both under iso- and hypo-osmolar conditions. Similar evidence for the complexity of the asymmetric breakdown phenomenon has

---

* The phenomena observed with liposomes may be unrelated to those observed with cells and may be due to mechanical stress resulting from elongation.[76]

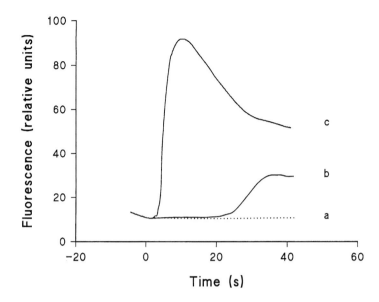

FIGURE 6.   Microfluorimetric determination of the uptake of ethidium bromide into oat proto-plasts subjected to a field pulse of 2 kV cm^{-1} strength and 20 µs duration. Curve a: untreated cells; curve b: fluorescence increase in the cell hemisphere facing the cathode; curve c: the corresponding increase of the fluorescent signal in the hemisphere facing the anode. (From Mehrle, W., Zimmermann, U., and Hampp, R., *FEBS Lett.*, 185, 89, 1985. With permission.)

been given by submicrosecond imaging of transmembrane potentials in sea urchin eggs.[74-80] Just after breakdown (which probably occurred within about 10–50 ns[33,36,80,152]) a conductance increase was observed — as expected from the membrane potential difference — in the anode-facing hemisphere followed by a somewhat delayed conductance increase in the other hemisphere (Figure 8). After about 2 µs the permeability change of the cathode-facing hemisphere surpassed that on the positive side. Consistent with this, the subsequent incor-poration of dye or Ca^{2+} (taking many seconds, Plate 1*) occurred almost exclusively on the cathode-facing side.[80]

This shows that predicting the conditions under which asymmetric or sym-metric breakdown occur is not always straightforward. Part of the "osmotic" data may be explained if we assume that the direction of the transmembrane potential depends more or less on the medium osmolality. At least for plant cells and protoplasts it is known[153] that the magnitude and the direction of the transmembrane potential are dependent on membrane tension and thus on osmotically induced expansion. Similar phenomena may occur in mammalian cells. The findings with sea urchin eggs also suggest that asymmetric structural elements of the membrane may be involved, e.g. asymmetric localization of Ca^{2+}-activated channels in the other hemisphere.[80] Reversible electric break-down experiments on the squid axon membrane have shown[36] that membrane

* Plate 1 follows page 276.

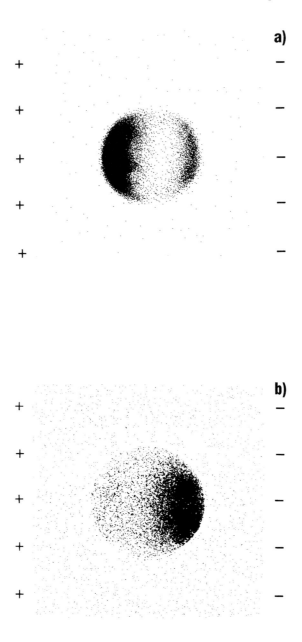

FIGURE 7. Asymmetric uptake of propidium iodide (black spots) through the membranes of SP2/0-Ag14 myeloma cells after application of a field pulse of 1.5 kV cm^{-1} strength and 40 µs duration. The cells were pulsed (a) in iso-osmolar (300 mOsmol) and (b) in hypo-osmolar (150 mOsmol) solutions. Measurements were performed with a Confocal Laser Scanning microscope. Photographs were made about 40–60 s after field application. (Sukhorukov, V. L., Frank, H., Richter, E., Djuzenova, C. S., Fuhr, G., and Zimmermann, U., unpublished data.)

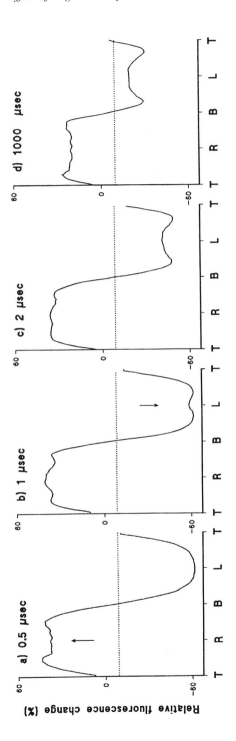

FIGURE 8. Time course of electropermeabilization (membrane conductance changes) in the two hemispheres of a sea urchin egg monitored by imaging the membrane potential at a submicrosecond resolution with the voltage-sensitive fluorescent dye RH292. Electropermeabilization was induced by application of a single, rectangular pulse of 400 V cm^{-1} field strength and 400 μs duration to the egg in Ca^{2+}-free sea water (solid line). The field strength was about 3-fold higher than required for breakdown of the membrane sites oriented in field direction (see Equation 1). The dotted line represents the control experiment in which a sub-critical field (67 V cm^{-1} field strength) was applied. T = top, B = bottom, R = right (hemisphere facing the anode, α = 0°) and L = left (hemisphere facing the cathode, α = 0°) of the image. Note that the amplitude of the dotted line remains practically constant up to 1 ms, whereas the amplitude of the solid line decreases during and after the pulse because of membrane breakdown. Breakdown occurs apparently first in the hemisphere facing the anode (as monitored about 0.5 μs after field application, arrow in (a) and then in the other hemisphere (about 1 μs after field application, arrow in (b). Thereafter, the conductance increased steadily. However, the change in conductance proceeds faster on the cathode than on the anode side. (From Hibino, M., Itoh, H., and Kinosita, K., *Biophys. J.*, 64, 1789, 1993. With permission.)

voltages close to the breakdown voltage open the Na^+-channels in a few microseconds but do not produce a decrease of the time constant of K^+-activation large enough to cause the opening of a significant percentage of channels in a period of about 10 μs.

A recent theoretical analysis of the asymmetric breakdown phenomenon[154] (based on a pore model[155,156]) has yielded some information which may be helpful in elucidating the underlying complex (and not well understood) processes of asymmetric breakdown. In agreement with Equations 1–3, the resting potential causes breakdown to occur first in the hemisphere facing the anode (if the cell interior is negatively charged). This happens even though the difference between the total membrane potential in the two hemispheres may be small. When the initial increase in conductivity of the anodic hemisphere is not too high, the permeabilized area will increase further. In the opposite case, breakdown in the other hemisphere can occur due to the rapid increase in the transmembrane potential on the cathodic side. This effect depends on the field strength and pulse duration (see also References 31–33, 35).

Theoretically, conditions can be envisaged (depending on the pulse parameters and the cell size) where the field-generated "pores" are larger, but the field-affected area is smaller on the cathodic than on the anodic side. In this case, conductance measurements will reveal a symmetric breakdown, whereas diffusion measurements with large marker molecules will indicate an asymmetric breakdown. If the diameter of the marker molecules is low enough to pass the "pores" generated in the hemisphere facing the positively charged electrode, an anodic asymmetric breakdown is recorded (because the area occupied by the "pores" is larger). Otherwise, if the diameter of the marker molecules is larger than the diameter of the anodic "pores" a cathodic asymmetric breakdown will be observed. This explains nicely why accumulation of chemotactic bacteria (due to the field-induced release of very small intracellular molecules) always occurred on the side which was predicted from the direction of the transmembrane potential. Among other things, the theory also shows the importance of the conductivity of the external pulse medium (see Equations 2 and 3).

A major drawback of these and other[137,138] considerations is, unfortunately, that the profile and magnitude of the intrinsic (resting) electric field within the membrane is not taken into account. The intrinsic electric field (and not the transmembrane potential difference as suggested by Equations 1–3) is the critical, but much less easily measured, factor in this circumstance. It is markedly influenced by the cell's surface potential which, in turn, is dependent upon the ionic concentration and the presence of multi-valent ions in the suspension medium. In addition, the profile and magnitude of the intrinsic field in any given area may change during the charging process of the membrane in an external field[137,157] as discussed above. Electric field-induced perturbations of the membrane structure in one hemisphere may also instantaneously affect lipid/protein domains in other membrane areas due to the high lipid mobility (see below) or due to a subsequent breakdown in the other hemisphere.

Although there now seems little doubt that the membrane potential and the intrinsic electric field play a significant role in asymmetric breakdown, this remains somewhat controversial for some authors.[158-160] In one study, human and rabbit erythrocyte ghosts loaded with FITC-dextran (MW = 10 kDa) or with NBD-glucosamine (MW = 342 Da) were electropulsed at either 6°C and 20°C. In agreement with the bulk of literature cited previously, it was found (using video microscopy) that the fluorescent dye always appeared outside the membranes as a transient cylindrical cloud directed toward the negative electrode, independent of the polarity of the pulse. These findings were interpreted to be consistent with symmetric membrane breakdown followed by an electro-osmotic-induced flow of water through the permeabilized membrane into the cell interior.[158-160] It was argued that the positive mobile counter-ions (which are present at high concentrations near the negatively charged pore walls) create an electro-osmotic-driven water flow toward the anode and carry the fluorescent probe.

Electrophoretic and electro-osmotic effects cannot *a priori* be excluded in these experiments. Unfortunately, the field strength used was 7 kV cm⁻¹ and therefore far above the critical field strength required for breakdown of ghost membranes (about 2–3 kV cm⁻¹). Additionally, the duration of the field pulse was long enough (1 ms) to induce pronounced current flow through the cells and salt shifts at the membrane interfaces. This could, in turn, lead to "punch through" effects[33] which can also lead to a reversible high-conductance state of the membrane in a manner completely different from that of reversible membrane breakdown.*

The cylindrical-shape of the fluorescent cloud asymmetrically released from the cell after electropermeabilization is apparently created by ejection of water through the permeabilized area owing to a slightly higher hydrostatic pressure within the cells, estimated to be about 2 mm H_2O column, and/or by the field-induced movement of the negatively charged cells in the direction of the anode, usually several μm in typical experiments of this type (see below). It is therefore difficult to accept electrophoresis as an explanation for these

---

* The "punch through" effect, originally observed in giant algal cells using microelectrode impalement[161-163] is observed with millisecond duration pulses and membrane potentials of 200–300 mV. Under these long pulse conditions the membrane cannot be charged to the breakdown voltage (1 V). In addition, differences in the transport numbers of cations and anions will, in general, result in considerable changes in salt concentrations at the membrane-solution interfaces and, in turn, in a current-induced volume flow, which may be mistaken for electro-osmosis[164,165] or for other effects.[166] Once established, the punch through phenomenon may have dire consequences for the cell as well as causing various artifacts that might confound experimental interpretation[138,167] (see also discussion in References 36, 168). This topic is also discussed in Chapter 6.

results, because such movements can only occur during pulse application.[79,169] Thus, while interesting, these experiments are at variance with a considerable body of evidence, including the prolonged asymmetric accumulation of chemotactic bacteria and atypical fluorescence exchange.

III. FIELD EXPOSURE TIME AND POST-PULSE PROCESSES

As noted above, intense electric field pulses may induce profound changes in the cell membrane, organelles, chromosomes, and DNA that are invariably lethal to the cell if the pulse duration, the number of pulses, the time interval between consecutive pulses, and/or the ambient temperature are not carefully controlled.[129,170-184]

A. SHORT DURATION FIELD PULSES

If the pulse length is in the range of a few µs, say 5–50 µs, thermal effects[185,186] can be excluded because heating of the cell suspension is negligible.[6,40] Salt shifts and "punch through" effects also do not occur. They are involved if the pulse lasts more than about 1 ms (see Footnote on page 23). When short pulses and appropriate field strengths for breakdown are used only **direct** and reversible membrane electric field effects are seen.[33,87] However, even with short duration pulses, if the field strength is too high (see Equations 1–3) the membrane breakdown can be irreversible due to the high membrane permeability and total collapse of solute gradients (see below). This is particularly evident at low temperatures because of slow membrane resealing.

Irreversible membrane breakdown can similarly occur with short duration pulses of appropriate field strength, if a series of pulses is applied with very short time intervals between, perhaps 1 s or less.[27] Under these conditions the membrane has insufficient time to recover its integrity between pulses and remains leaky, i.e., has not resealed when the next pulse is applied. Electric current can thus flow through the highly conductive cell interior producing profound cell damage. The time interval between consecutive pulses must, therefore, be adjusted to about 1 min (or longer at lower temperatures) in order to allow recovery of the membrane.[187,188] (In contrast, in electrofusion experiments consecutive pulses should be applied very rapidly in order to induce an irreversible breakdown in the contact zone of two adhering cells, see Chapter 4.)

The tolerance of the various electric field exposure regimes also depends upon the phase of the cell within the cell cycle. It has been demonstrated by flow cytometry that electropulsed cells in G_0/G_1 are more resistant to direct, short duration electric field exposure than cells in $S+G_2/M$.[109]

B. LONG DURATION FIELD PULSES

As noted above, pulse durations >100 µs produce usually complicated membrane effects at much lower membrane potential differences (see Footnote on page 23) and thereby cause adverse side effects. In addition, other detrimental effects can occur when long duration pulses are used, including electrophoresis of cellular[189] and of membrane components[190] (see also Chapter 6) as well as local heating effects in the cytosol and membrane resulting from current application to the suspension medium (see below).

It is tempting to postulate that heating effects (and not direct electric field effects) play a dominant role in electrotransfection of certain bacteria in the presence of long duration pulses (Chapter 2). Gram-negative bacteria have given much higher frequencies of transformation after electropermeabilization than their gram-positive cousins.[191,192] Local heating in the conductive porin pores of the outer wall of the gram-negative bacteria could explain the fundamental difference in "transformability" between these two classes of bacteria,[30] although there may be a number of other equally important factors as well.* Support for the assumption that DNA uptake in some bacterial strains is of thermal nature is also given by the finding that an alternating field as low as 100 V cm^{-1} was found to facilitate DNA entry into *Escherichia coli*.[196]

Of course, if there are sufficient input cells and adequate selection methods for viable cell recovery (e.g., DNA-electrotransfection of bacteria; see Chapter 2), such suboptimal experimental conditions might still yield satisfactory results. One is sometimes forced to resort to such methods when equipment capable of generating adequate field strengths is not available.

However, it should be noted that there are some interesting experimental conditions whereby the application of pulses of a few hundred microseconds or some milliseconds may also be beneficial for small eukaryotic cells. If exogenous material or particles are trapped within the membrane, current flowing through the membrane would be considerably diminished, thus shielding the cell from the side effects of long duration pulses. Trapping of material within the membrane is enhanced if electropulsing is performed at elevated temperatures (20°C to 37°C) in the presence of relatively high concentrations of exogenous proteins or other macromolecules.[14,188,197,198] This may explain why sometimes pulses of long duration can be used for electroinjection of antibodies, enzymes, and DNA into cells without severe lethal effects. We have termed this process "electrointernalization" or "electroinsertion" (see below).

Generally speaking, however, application of ms-duration pulses is not recommended, particularly for electroinjection of low-molecular weight substances into eukaryotic cells. This is not only because of the previously cited reasons, but also because of the lack of predictability and reproducibility of the results.

---

* As outlined in this chapter and in Chapter 2, turgor pressure may play a very important role in electrotransformation of bacteria (and of walled plant cells[193]), although the effect of pressure on the breakdown voltage may be partly offset by surface admittance effects. First, the breakdown voltage decreases with increasing turgor pressure. Second, it is well-known from studies on plant cells that excess internal hydrostatic pressure leads to local bursting of the cell wall. This process is completely reversible if the pressure is released.[194,195] Such effects can be envisaged to be different for gram-negative and gram-positive bacterial cells.

C. SELECTION OF THE OPTIMAL PULSE DURATION TIME

The optimal field exposure time is dictated by the time required for charging of the membrane and by the fact that breakdown occurs in nanoseconds[33,36,80,152] once the breakdown voltage has been reached.

The membrane capacitor is not charged instantaneously to the (stationary) voltage given in Equations 1–3. The relaxation time of the exponential charging process is given by Equation 4.[21,25,27,199-201]

$$\frac{1}{\tau} = \frac{1}{R_m C_m} + \frac{1}{a \ C_m (\rho_i + 0.5 \rho_e)} \tag{4}$$

where:

τ = relaxation time
R_m = specific membrane resistivity (= $1/G_m$)
C_m = specific membrane capacity

The $R_m C_m$ term in Equation 4 can normally be neglected provided that the medium conductivity is not too low (see above). Thus, for practical purposes, Equation 5 can be used:

$$\tau = a \ C_m (\rho_i + 0.5 \rho_e) \tag{5}$$

Using voltage-sensitive fluorescent dyes in combination with pulse-laser fluorescence microscopy (see above) agreement between the theoretical predictions of Equation 5 (and Equation 1) and the experimental data were found,[74-80] supporting the premise that over the range of conditions likely to be encountered in biological systems this equation provides helpful guidelines for optimizing electric field parameters for reversible electric breakdown of the membrane for all purposes (except electrotransformation of bacteria and electropermeabilization of eukaryotic cells when electrointernalization is involved).

Because C_m and ρ_i are nearly constant for a given cell (to a first approximation, see above), the charging time of the membrane in the presence of an external field depends mainly on the radius, a, and the external specific resistivity, ρ_e. Thus, the larger the radius of the cell and the higher the specific resistivity of the external pulse medium, the longer the pulse duration time must be selected for the stationary voltage to be reached. For example, according to Equation 5 the relaxation times for cells with a radius of 10 μm to 20 μm are calculated to be in the range of 0.5 to 5 μs (assuming that $C_m = 1$ μF cm^{-2} and $\rho_i = 100 \ \Omega$ cm). The stationary membrane voltage given by Equations 1–3 is reached approximately after about 5 times the relaxation time. Therefore, the pulse duration time for reversible membrane breakdown and electroinjection of exogeneous material should not exceed about 15–25 μs in order to avoid adverse side effects. Because of the radius dependence of the charging process, very large cells (such as oocytes[27,48,79]) must accordingly be treated with pulses of long duration (100 μs to a few ms). They are seemingly "resistant" to short duration pulses because of the slow charging process of the membrane.

Since the relaxation time increases with decreasing electrolyte concentration, slightly longer pulse durations must be applied under conditions used for electrofusion, that is in weakly conductive solutions (see Chapter 4). Furthermore, in studies of ionic-strength modulation of field-induced membrane permeabilization,[202] use of constant pulse duration conditions requires that the changes in membrane charging time that result from changes of the conductivity of the pulse medium are recognized and accounted for. Otherwise the interpretation of the results may be misleading.

D. RESEALING

The resealing process is temperature-dependent.[20-27,30,32,42,46,47,51,93,203-206] For example, complete resealing may require perhaps 30 min or more at 4°C. This feature of breakdown may be used to enhance uptake by **diffusion** of soluble exogeneous molecules or for the release of cytosolic contents. Conversely, resealing increases dramatically at temperatures >20°C for plant protoplasts and >30°C for yeast and mammalian cells, respectively. These temperature effects reflect active membrane processes that are optimal at temperatures best suited for the function of enzymes in these cell types.

The temperature-dependence of membrane resealing is easily measured by microelectrodes, patch clamp, exclusion of normally membrane-impermeable dyes (such as trypan blue, Phloxin B, or propidium iodide), or by accumulation of chemotactic bacteria (see above). At optimal temperatures, the prebreakdown state of the membrane (i.e., low-conductance and low-permeability) is usually regained within seconds by plant vacuoles or liposomes, and after 10–30 min by eukaryotic cells.

The restoration of the original high resistance and low permeability of the membrane does not necessarily mean that the original membrane structure and assembly is completely restored. Both of these parameters are largely determined by the lipid bilayer. Electropermeabilized, symmetric artificial lipid bilayer membranes[20] and liposomes[78,79,207] reseal in a few μs because of the high lateral mobility within the lipid bilayer (of the order of 0.1 to 1 ms,[208] see also Chapter 3). Resealing of the lipid bilayer in the membrane of mammalian cells may be slightly slower because of the presence of cholesterol and/or other components, which reduce the lateral diffusion coefficients of the lipids, but is still very rapid (in the range of seconds[21]).

In biomembranes one must take the membrane assembly into consideration (see Chapter 6). In many cells, phospholipids are distributed asymmetrically between the two monolayers. In the plasmamembranes of eukaryotes, aminophospholipid asymmetry appears to be due to the ATP-dependent aminophospholipid translocase activity that counter-balances the spontaneous lipid randomization by flip-flop, which is normally very low (see also Chapter 6).[209] Interactions of the acidic phospholipids with the membrane skeleton may also contribute to the maintenance of phospholipid asymmetry.[210] Electric breakdown results in enhanced transbilayer mobility, which then results in a partial disorder of the bilayer lasting for at least 1 hour.[210,211] Recent investigations on red blood cells suggest that the exposure of the acidic phospholipids at the outer cell surface was more likely to be due to a direct effect of the electric field pulses on plasmamembrane structure than to secondary effects (such as action of endogenous proteinases on the membrane skeleton) because the effects were seen within 6 min after field pulse application.[211] In addition, contact-angle measurements on pulsed plant protoplasts have shown an increase in the hydrophobicity of the interface[212] suggesting sustained partial membrane dis-aggregation. ^{31}P-NMR (nuclear magnetic resonance) studies of the organiza-tion of plasmamembrane phospholipids in Chinese hamster ovary cells have demonstrated a transient change in the orientation of the polar head groups during the permeable state leading presumably to a weakening of the hydration layer.[213] Furthermore, high-intensity, supercritical field pulses can cause en-hanced lipid-peroxidation observed for plant membranes.[214]

It is similarly likely that these subtle post-electropulse effects in the lipid domain are responsible for the prolonged "fusogenic" properties of pulsed erythrocyte ghost cells.[215*]

Furthermore, additional "repair" of field-induced conformational changes in membrane proteins, reorganization of ion channels, pumps, enzymes, recep-tors, and surface antigens, also appears to require minutes to hours. Such effects may be due, in part, to field-induced redox reactions of SH-groups in certain membrane proteins.[216,217]

There is additional evidence that some surface proteins may be influenced by the effect of electric fields on cell membranes. For example,[109] antibody staining of surface immunoglobulin is substantially reduced in lipopolysaccha-ride (LPS)-stimulated B lymphoblasts exposed to harsh pulse conditions (4–5 kV cm^{-1}, 3 pulses of 5 µs duration, at 4°C) as compared with controls. The addition of antibody to surface immunoglobulin prior to exposure to the field pulse prevents this reduction in staining. This observation suggests several

---

*   However, it must be noted that the "long-lived" fusogenic state had a much shorter life-time (up to 5 min[215]) than the change in phospholipid asymmetry.[211] Since cell alignment was performed in an alternating field of 60 Hz frequency, formation of electrolysis products which may interact with the membranes (e.g. by lipid-peroxidation) under creation of a "fusogenic state" cannot be excluded. In addition, electrophoresis of cells and membrane components can occur at these low frequencies (Chapters 4 and 6). Toxic products can be also produced by oxidation of chloride in the pulse medium to chlorine or hypochlorite.[124,172] Release of toxic heavy-metal ions from the electrodes which were made of solder-tinned copper hook-up wires in that work[215] must also be excluded before conclusions can be drawn about direct field effects on the membrane properties.

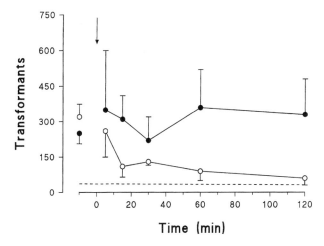

Time (min)

FIGURE 9. Electrotransformation of protoplasts of the yeast strain AH 215 of *Saccharomyces cerevisiae* (10^9 cell ml^{-1}) with supercoiled plasmid DNA pADH 040–2 (10 µg ml^{-1}) in iso-osmolar pulse medium. Three field pulses of 10 kV cm^{-1} strength and 10 µs-duration were applied at intervals of 2 min at 4°C (filled circles) and 20°C (open circles). DNA was added either *before* or at intervals of 5 to 120 min *after* field application (arrow). Clones were counted 2 to 3 weeks after transfer of the pulsed cells to regeneration and selection agar. Dashed line: number of transformants in control experiments in which DNA was added without field treatment. Note that a temperature of 4°C during pulsing resulted in significantly higher transformant yields than room temperature, presumably because the slow membrane resealing allowed more DNA molecules to be electroinserted within the membrane. This apparently favors a long-term electrointernalization process when the cells are returned to room temperature. (Data courtesy of Schnettler, R.)

possibilities, including irreversible conformational changes of the surface immunoglobulin or internalization/extrusion of the protein. Another possibility is the field-mediated loss of membrane material with its embedded surface protein.

E. ELECTROINTERNALIZATION

As shown for mammalian cells,[15,103,106,218,219] yeast protoplasts,[15,30] and liposomes,[220] components of the membrane can be endocytosed after breakdown when the pulsed suspension is kept at room or elevated temperature (see also Chapter 4). Under some conditions plasmalemma internalization continues over a few hours, during which time exogeneous DNA and proteins are incorporated into membrane vesicles (Figures 9 and 10). This effect provides clear evidence that permeabilized biomembranes remain disturbed long after the membrane resistance and permeability have normalized (or nearly so). This electrointernalization may be easily detected using fluorescence-labelled compounds.[15,103,106,218,219] When taken up by diffusion, fluorescein-labelled compounds are evenly distributed throughout the cell, whereas with electrointernalization fluorescent patches appear within the cell or adhere to the inner cell membrane (Figure 10). Field-induced membrane trapping and

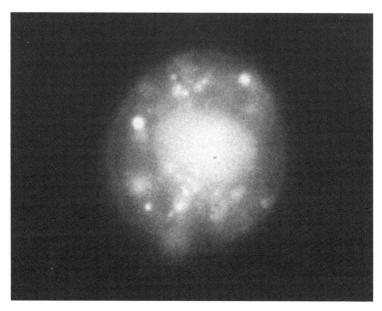

FIGURE 10. Field-induced, vesicular uptake of FITC-labelled bovine serum albumin in mouse L-cells (2×10^6 cells ml^{-1}, 90 mOsmol-solution). The membranes were electropermeabilized by application of a single pulse of 4.55 kV cm^{-1} and 140 µs duration at room temperature. The fluorescence micrograph was taken 100 min after transfer to iso-osmolar solution. (Photograph courtesy of Klöck, G. and Schnettler, R.)

internalization can also be easily visualized using latex particles of various sizes[14,221,222] (Figure 11). Electron microscopy shows that, depending upon the radius, the applied field strength, and the pulse duration, such particles are either stuck within the previously permeabilized membrane areas (thereby shielding the interior against millisecond pulses as mentioned above) or are found in the cell interior surrounded by small layers of membrane material. Therefore, it is possible that incorporation of "xenomolecules" (exogeneous material) into the membrane ("electroinsertion") triggers electrointernalization. The beneficial effect of albumin on the viability rate of pulsed cells (both in electrofusion and electroinjection)[27,29,30,100,223] may also be due to electroinsertion of these molecules into the electropermeabilized membrane, thus preventing high current density in the perturbed areas of the membrane.

Electric field-induced exocytosis (of intracellular organelles and macromolecules) has also been reported.[16] This phenomenon is strongly dependent upon the extracellular Ca^{2+}. It is very likely that membrane internalization is also involved in this effect.

Electrointernalization is, then, a unique mechanism, *distinct from diffusion*: efficient uptake of macromolecules can be obtained without causing cellular and membrane compromise (see above).

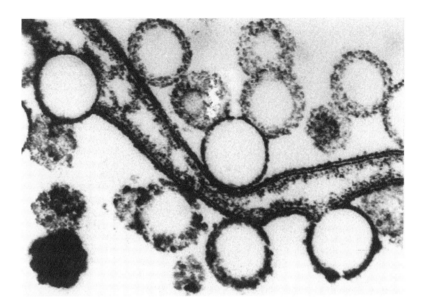

FIGURE 11.   Electron microscopy of latex spheres of 0.176 μm in diameter electroinserted into the membranes of human erythrocyte ghost cells after application of a field pulse of 16 kV cm⁻¹ and 40 μs duration. (From Vienken, J., Jeltsch, E., and Zimmermann, U., *Cytobiologie,* 17, 182, 1978. With permission.)

IV. WAVE FORM OF THE FIELD PULSE

Most published work on electropermeabilization has been based upon the use of one or more exponentially decaying field pulses (provided by the discharge of a capacitor). However, there is also a considerable body of literature using rectangular pulses[57,62,70,93,97,224-228] (see also Tables 1 and 2 below). A rectangular pulse of an amplitude comparable to the peak value of the exponentially decaying field pulse has the same effect on membrane breakdown because of the rapidity of the breakdown process.[19,62,229] The membrane cannot sustain a voltage difference greater than the critical value of about 1 V even for the brief period of membrane charging.[31-33,79] However, the total current that flows through the permeabilized cell is substantially less when the field strength decreases during pulse application (as compared to a direct current pulse). Under some circumstances the perturbations created in the membrane by current flow may be beneficial for promoting uptake of macromolecules.[230] Since alternative, physically better grounded protocols using exponentially decaying pulses are available, such conditions should be avoided.[231]

The time constant of the exponentially decaying field pulse is defined as the pulse duration time[54] and is determined by the capacity value of the storage capacitor and by the resistance setting of the instrument, with the proviso that the resistance of the instrument is well below the resistance of the cell suspension ($R = 2 - 20\ \Omega$).*

Unfortunately, these very important parameters have been ignored by many investigators. In many cases (see Chapter 2), only the capacity values are given, without the ratio of the resistances of the cell suspension and the instrument. Data about the electrode spacing are also missing in many publications. Without such critical information, it is very difficult to interpret the data or to reproduce the experiment, particularly if a different apparatus is used.

Membrane breakdown in suspended or dielectrophoretically aligned cells can also be induced by application of rotating[232,233] (see Chapter 5) and sine wave (oscillating) fields of appropriate frequency, peak amplitude, and duration.[21,25,63,169,234] When an alternating (sinoidal) electric field is applied (see Chapters 4 and 5), the generated membrane potential is given by the following equation for a spherical cell with a smooth surface:[21,200,201,235]

$$V_g = \frac{1.5aE\cos\alpha}{\sqrt{1+(2\pi f\tau)^2}} \tag{6}$$

where:

V_g = generated membrane potential (V)
τ = the relaxation time of the charging process (Equation 5)
f = frequency of the external field

Below the "cut-off" frequency $f_c = 1/2\pi\tau$ the denominator approaches 1 and Equation 6 reduces to Equation 1 (if the resting membrane potential and the term λ are neglected in the considerations). Above the so-called "cut-off" frequency f_c, the denominator becomes $2\pi f_c\tau$ and the generated membrane voltage is then given by Equation 7:

$$V_g = \frac{1.5E\cos\alpha}{2\pi fC_m(\rho_i+0.5\rho_e)} \tag{7}$$

Equation 7 states that the generated voltage across the membrane is independent of the cell radius, but decreases linearly with further increases in the

---

*   In most of the instruments the resistance of the discharge chamber (cuvette, electrodes plus cell suspension) is arranged in parallel to the resistance of the instrument. If the resistance of the discharge chamber is well below the resistance of the instrument, the discharge is controlled solely by the parallel instrument resistance. In the case of resistances of the order of 2 to 20 Ω, highly conductive (physiological) solutions can be used. In instruments, in which the resistance is in the range of 200 to 1000 Ω, only non-conductive (unphysiological) solutions can be used. Increase of the conductivity of the solution during pulsing (due to leaky cells) can considerably change the discharge time.

frequency of the alternating field (for experimental proof, see Reference 169). Higher external field strengths are, therefore, required to reach the breakdown voltage of the membrane. For $\tau = 0.15$ μs ($a = 10$ μm, ρ_i, $\rho_e = 100$ Ω cm corresponding to PBS solution), f_c is about 1 MHz and increases with increasing conductivity of the pulse medium.[235] Above f_c the plasmamembrane becomes electrically completely transparent and, in turn, the organelles are exposed fully to the electric field and become polarized. This is particularly so for the relatively large-sized nucleus in mammalian cells, thereby allowing for intracellular dielectrophoresis of nuclei where desired.[235]

Even though the radius-independence of the generated potential is at first glance appealing from a physical standpoint, electropermeabilization with an oscillating field of a frequency above f_c is not recommended (nor has it been reported) because of the very high external field strengths that are required.

The use of sine wave fields below f_c results in breakdown during each half-cycle (provided that the critical field strength is exceeded). This is, therefore, equivalent to the application of repetitive pulses without sufficient time interval between consecutive pulses. If the membrane does not reseal during each half-cycle, high current will flow through the cell interior. As discussed elsewhere in this chapter, such conditions are not favorable for the cell.

Some authors have recently reported the use of DC-shifted fields with a frequency of 100 kHz for electroinjection of macromolecules into cells (and also for electrofusion) in buffered solutions.[236-238] In contrast to statements by these authors, the generated membrane potential is still radius-dependent as given by Equations 1–3. Although this method succeeded, one must surmise that the mechanism producing the observed electropermeabilization and electrofusion may have been current-induced heat shock in the cells and in the membrane.

DC-shifted wave forms may be interesting from the physical point of view. However, for most applications described to date, exponentially decaying field pulses are very efficient and well-tolerated by the cell when appropriate medium conditions are used. There would appear to be no practical need for the use of DC-shifted wave forms at present.

V. COMPROMISING AND BENEFICIAL EFFECTS OF THE PULSE MEDIUM

The concentration and nature of the pulse medium ingredients are very important. Because of the loss of the selective osmotic barrier in electropermeabilized cells, the K^+ and the Na^+ gradients will collapse (for a rigorous discussion, see Chapter 6). The time course and magnitude of the gradient collapse depend upon the duration of the high permeability state. A decrease in the intracellular K^+ concentration and concomitant increase of Na^+ can have detrimental effects on enzymatic functions and other biochemical processes. This crucial point has been overlooked in many publications cited in Tables 1 and 2 (see below). The collapse of the ion gradients can be diminished if K^+ instead of Na^+ is added to the pulse medium. The selection of high K^+- and low Na^+ levels to mimic the intracellular ionic environment may lead to a depolarization of the membrane potential[27,59,239] (see also Chapter 6). In this case, breakdown of the cells may be more symmetric than asymmetric which is, however, not always desirable. K^+ concentrations of about 30–70 mM are, therefore, normally recommended.[27,240,241] Higher concentrations (up to 140 mM)[59,85,205,242-252] can be used, but they may be toxic to the cells when exposure exceeds half an hour in such solutions.[253] For the same reason, solutions should be buffered by K^+-phosphates. Organic buffer substances (such as HEPES, TRIS, etc.) should be omitted because their uptake by the cell may lead to adverse side effects. Similarly, the presence of EDTA or EGTA in the external solution should also be avoided, even at very low concentrations.[27,55] For example, it was found[44,55,254] that EDTA shifts the onset of electrically induced hemolysis of erythrocytes to higher electric field strengths.* I believe that conflicting data in the literature on electric field-mediated molecule uptake are often attributable to the use of pulse media which contain organic buffer and/or chelating compounds. It is interesting to note that in about 40% of the work cited in Tables 1 and 2, EGTA, or EDTA and/or HEPES were added to the pulse medium despite the detrimental effects of incorporated EDTA on the intracellular pH, Ca^{2+}, and/or ATP level. Reports in the literature[256-258] on the stimulation of DNA synthesis, cell division, and plant regeneration after application of a breakdown pulse in the presence of 5 mM MES might be the result of the electric field-mediated uptake of this pH-buffering substance in the cytosol. The pulse duration in this experiment was 10 to 50 μs: too short to induce electric field interactions within the cells. In most of the articles the cell viability was not tested after pulsing. In a few reports, the authors tested the viability of the pulsed cells with trypan blue or Phloxin B assays after resealing. However, the information from this test is limited because membrane impermeability to dyes does not necessarily yield information about electric field-induced changes either to the membrane (electrointernalization) or within the cell (see above). A crucial test of viability is the subsequent measurement of the growth rate of the cells.

Media supplementation with divalent cations,[25-27,30,168,259-261] such as Ca^{2+}, Mg^{2+}, and/or Zn^{2+}, is beneficial for membrane stabilization of electropermeabilized cells when low concentrations of about 0.1 mM are used (in

---

* This effect can be used for the preparation of erythrocyte ghosts under iso-osmolar conditions.[44,254] The electrically generated ghost cells contain only about 1% haemoglobin (which is bound to the membrane). SDS-polyacrylamide electrophoresis of electrically generated ghost cells shows that the band 6 (corresponding to glyceraldehyde-3-phosphate dehydrogenase, G3PD) is absent. This indicates that this enzyme is bound artificially to the membrane during osmotic haemolysis under low ionic strength conditions. Erythrocyte ghosts depleted of G3PD can be used as specific high affinity adsorbents for rapid extraction and purification of this enzyme from different tissue sources ("bio-affinity" chromatography).[255]

gene transfer experiments). Higher concentrations may chelate ATP and may thus be detrimental. A decrease in the intracellular ATP level may also reduce lipid fluidity[262] and thus adversely effect membrane reorganization and resealing.[263] For a rigorous discussion of the influence of divalent cations on the electropermeabilization process the reader is referred to Chapter 4.

As mentioned above, acetate,[27-30,264] gluconate,[265] or glutamate[205,244-247] salts should be used if possible in order to avoid the formation of toxic electrolysis products when pulse trains are applied (see Footnote on page 28 and also discussion in Reference 87). The high-permeability state of the membrane is maintained for some time after pulsing, therefore return of permeabilized cells to nutrition medium may also be detrimental if the medium contains high concentrations of Na^+ or pH indicators such as phenol red.[23,27]

In general, pulse media should be comprised of ultrapure substances dissolved in double distilled water to assiduously avoid trace contaminants. These may be nontoxic to untreated cells, but can produce reactions in the electropermeabilized cells which are mistaken as field effects and can even lead to cell death.[23] Heavy-metal ions such as Pb^{2+} are particularly dangerous here because they are frequent contaminants of sugars and salts. Field pulses solubilize metals from the electrode plates. Metal ions can thus be introduced by the electrode material.[24,266] Caution must therefore be exercised when using commercial, disposable discharge chambers with aluminum plate electrodes.[267-280] Al^{3+} ions are very toxic and change membrane properties even in untreated cells by binding to outer membrane surface charges.[281] An interesting example of the adverse side effects of Al^{3+} on the biochemistry of electropermeabilized cells was published recently.[282] Electropermeabilized L-1210 cells showed evidence of increased Ca^{2+} release. However, careful investigation of the effects gave clear-cut evidence that incorporated intracellular Al^{3+} ions enhanced the conversion of inositol 1,3,4,5-tetraphosphate into inositol 1,4,5-triphosphate which induced the Ca^{2+} release. Conflicting data in the literature regarding the sensitivity of cells to field pulses of various intensity and duration can be, in many cases, easily attributed to dissolution of Al^{3+} from the electrode cuvette.*

The osmolality of the pulse medium is also a very important parameter that has an enormous effect on the number of cells that survive electropermeabilization. There are several reports[283-285] that hypo-osmolar conditions during electropermeabilization (and also electrofusion) should be avoided. However, compromising effects of hypo-osmolar media occur only if the field strength is not properly adjusted according to Equations 1–5, taking into account the increased size (due to water uptake) of cells in hypo-osmolar solutions. Proper reduction of the field strength in accordance with the increased radius results in high uptake rates of exogeneous molecules

---

* The field strength may also differ from that expected when Al^{3+} electrodes are used because of the significant voltage drop across the oxide layer on the electrode surface. Although more expensive, platinum or gold electrodes are therefore more advantageous.[24]

without any decrease in the viability rates[286] (see also Chapter 4). Best results (particularly for mammalian cells) are achieved (both for electro-permeabilization and electrofusion) if strongly hypo-osmolar solutions of 75 to 100 mOsmol are used.[286] The use of such strongly hypo-osmolar media has led to very reproducible electrotransformation (and electrofusion) conditions.[27] One of the more obvious advantages of this technique is that in general a single high intensity pulse is sufficient to induce highly efficient uptake rates.[286] In contrast, under iso-osmolar conditions a train of about three or more pulses must be applied[93,187,213,260,261,274,287-290] in order to obtain a comparable uptake of exogeneous compounds. Hypo-osmolar conditions also accelerate the resealing process of the membrane of field-treated cells resulting in more viable cells.[286] The physical basis for the beneficial effects of strongly hypo-osmolar solutions is given by the secondary processes induced after membrane breakdown: these are considered in the next section and in Chapter 4.

VI. MECHANISM OF MEMBRANE BREAKDOWN

In contrast to the reversible and irreversible electric breakdown of planar lipid bilayer membranes (Chapter 3), the mechanism of electric breakdown of cell membranes is poorly understood. Elucidation of the reaction steps that may precede or occur during electric breakdown of the cell membrane would require measurements of high temporal and spatial resolution. Measurements of this kind are not possible at present. In the early stage, field effects might include domain-induced budding[291-294] (see also below), field-induced "flip-flop",[295] migration of charged molecules within the membrane,[190] conformational changes of membrane components,[217] changes in curvature, etc.[291] In addition, membrane breakdown involves electromechanical coupling that is normally neglected in models based on the assumption of pore formation.[81] The application of an electric field obviously exerts additional stress on the membrane, either as a tensile and/or an electrocompressive force,[32,36,48,121,122,124,296] as surface waves,[297,298] or as a pressure within ion channels to expand them[299] (for a rigorous discussion see Chapter 3). One should also not forget that electro-mechanical compression of the membrane material, when sufficiently large, could, for instance, expose transmembrane conduction modules, which by virtue of their relative size to the membrane are normally buried in the membrane matrix.[124] When this occurs the electric conductance through the membrane would increase sharply and breakdown associated with permeabilization could result. Field-induced thinning of the membrane may also be accompanied by a decrease in the Born energy required for ion injection into the membrane.[36,152]

Any or all of these effects may lead to changes in the membrane structure without formation of distinct pores. Electron microscopic evidence of cracks or pores observed in freely suspended cells[300,301] after pulsing cannot be taken as proof of primary generation of so-called "electropores" (by analogy to lipid bilayer membranes).* Fixation or freezing of the samples requires milliseconds to seconds. This is rather long compared with the secondary process induced by the nanosecond-lasting breakdown event. It therefore seems very likely that the "membrane openings" and "cracks" seen in electron microscopic images of cells electropermeabilized in iso-osmolar solutions are most probably due to artifactual processes created during the freezing process by the loss of the osmotic barrier.[189] As mentioned above, the intracellular hydrostatic pressure head of a few millimeter H_2O column will rapidly eject water through the permeabilized area even under hypo-osmolar conditions. This event, requiring several milliseconds, will certainly result in lesions and perturbations in the membrane assembly which may be visualized as "imprints" by electron microscopy.

The hydrostatic pressure-driven water flow will be followed by a rapid net water flow toward the cell interior because of the colloid-osmotic pressure of the intracellular proteins.[55,295] Cell swelling will occur resulting in rupture of the cytoskeleton "scaffolding" which is anchored to certain membrane proteins in eukaryotic cells, normally rendering these molecules "immobile"[303,304] (see also Chapter 4).

That secondary osmotically triggered effects also play an important and complex role in the generation of "pore-like structures" in "normal" iso-osmolar pulse media is undisputed.[305,306] This becomes clear if the primary electric field effects in the membrane are studied separately from the osmotically-induced ones. This can be achieved by using iso-osmolar, "pulse-balanced" media containing appropriate concentrations of osmotically active substances of higher molecular weight (e.g., inulin, polypeptides, Ficoll, dextrans, etc.).[55] If the high permeability state of the membrane does not exceed a critical level, the membrane remains impermeable to these "balancing" exogenous compounds (i.e., the reflection coefficient $\sigma = 1$, see also Chapter 4) so that the cells are osmotically protected. This can be proven by electronically measuring the cell size, if appropriate equipment is available. Experiments have shown that under these conditions the field-induced permeability and structural changes of the membrane are different compared to "normal" iso-osmolar solutions.[151]

Another approach for separation of electrically and osmotically induced effects is the use of strongly hypo-osmolar electropermeabilization (and electrofusion) solutions. In contrast to the effects seen in iso-osmolar conditions, in strongly hypo-osmolar solutions water uptake and dissolution of the cytoskeleton (associated with a very slight increase in membrane permeability) occurs predominantly before field pulse application. Therefore, reversible breakdown leads only to a further, but slight, increase in cell size.

Electron micrographs as well as electrorotation measurements (see Chapter 5 and Figure 12) have revealed that, for cultured mammalian cells,

---

* The situation may be different for dielectrophoretically aligned cells[302] because the cells adhere in the breakdown zone together. This should have a stabilizing effect on pores generated by a breakdown pulse (see Chapter 4).

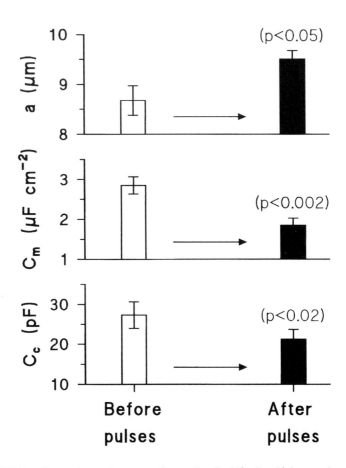

FIGURE 12. Changes in membrane area of mouse L-cells (10^6 cells ml^{-1}, iso-osmolar solution) after application of 3 pulses of 5 kV cm^{-1} strength and 40 μs duration at 20°C, followed by 15 min resealing in culture medium without phenol red at 37°C. Aphidicolin-synchronized cells, 3 h post-block, were used. Changes in membrane area were recorded by measuring the total capacity of the cell by means of the electrorotation method (see Chapter 5). The whole-cell capacitance, C_c, is proportional to the total membrane area (the membrane surface estimated from the radius of the cell, $4\pi a^2$, plus the area of the microvilli), whereas the "apparent" specific capacity, C_m, is the capacity per unit membrane area which increases with the density of the microvilli on the membrane. Electropermeabilization causes the cell radius to increase (due to colloid-osmotic swelling), whereas C_m and C_c decrease. The results indicate that the pulsing causes a decrease in the plasmamembrane villiation and partial loss (about 22%) of the plasmamembrane (electrointernalization and/or loss to the medium). Statistics were obtained by Student's t-test. (From Sukhorukov, V.L., Djuzenova, C.S., Arnold, W.M., and Zimmermann, U., *J. Membrane Biol.*, 142, 77, 1994. With permission.)

hypo-osmotically treated cells increase their membrane area rapidly by reducing their microvilli.[139] Below about 200 mOsmol the values of the whole-cell capacitance (as measured with electrorotation, see Chapter 5) indicate the progressive incorporation of internal or stored "extra" or "redundant" membrane material into the cell surface. This is immediately available.[139] During the resealing and regeneration processes this material is apparently removed by vesicle formation (i.e., by electrointernalization) as discussed above and in Chapter 4. Incorporation of membrane material derived from the microvilli is also observed in response to high-intensity field pulses under iso-osmolar conditions.[151] Controversial data published for Chinese hamster ovary cells[307] are most probably due to the resistance of these cells to osmotic swelling (see Footnote on page 11). Obviously, only cells lacking a cell wall or a tight cytoskeleton will demonstrate such changes in size and membrane area.

Under some circumstances field-treatment of cells can also be accompanied by the formation and growth of spherical and hemispherical protuberances of the cell membranes (so-called "blebs"). The occurrence of "blebs" is described for adherent cell lines given rectangular pulses in physiological media.[307,308] The formation of "blebs" is seen predominantly at the membrane sites subjected to the highest generated membrane potential difference (see Equations 1–3). The field-induced structural changes are nearly completely reversible at elevated temperatures associated with the formation of giant fused polykaryons. Supplementing the pulse media with inulin prevents "bleb" formation, indicating the osmotic nature of the processes involved.*

In light of the evidence given above it is quite clear to me that the structural changes of freely suspended cells measured after field treatment are due to both the application of electric field effects within the membrane, as well as to secondary osmotic processes.**

Carefully conducted permeability studies on field-treated human erythrocyte membranes have also shown[91,295] that there is no clear-cut evidence for the

---

* It is conceivable that the "bleb" formation is related to the domain-induced budding or invagination process known from lipid bilayers.[291-294] Theoretical considerations predict[291-294] that a flat or weakly curved domain becomes unstable at a certain limiting size and then undergoes a budding or invagination process. This instability is driven by the competition between the bending energy of the domain and the line tension of the domain edge. In lipid bilayers, the budding domain can rupture the membrane which then pinches off from the matrix.

** From the foregoing discussion, it is clear that membrane breakdown under hypo-osmolar conditions must be superior to iso-osmolar breakdown. Swelling and the associated dissolution of the cytoskeleton scaffolding within the cell and in the microvilli occur in a very reproducible manner because of the uniform uptake of water over the entire cell surface. In iso-osmolar solutions water will enter the cell predominantly through the unevenly distributed, permeabilized (and therefore low resistance) areas. Nonuniform cell swelling and cytoskeleton dissolution will more easily lead to rupture in those areas of "point defects" in contrast to hypo-osmolar conditions where the induced defects are presumed to be distributed evenly over the total surface area. Pre-swelling of cells in hypo-osmolar solutions also has the advantage that the cells and the nuclei assume a nearly spherical shape, and that plasmalemma and nuclear membrane undulations disappear. Equations 1–6 can, therefore, be more accurately applied than for cells in iso-osmolar solutions, which are usually somewhat irregular. The high dilution of the cytosol before field application may also be advantageous in electrotransfection, because predilution of endonucleases (which may otherwise degrade part of the entrapped DNA before incorporation into the nucleus) may cause them to be less likely to be active.

generation of "electropores." The "electroleaks" (generated in the erythrocyte membrane in response to an electric breakdown pulse) are weakly selective for small monovalent inorganic ions according to the sequences of their free solution mobilities. Sugars and organic anions were selected according to size and charge. Common properties of these electrically-induced defects and of leaks induced chemically (e.g., by adding diamide, periodate, t-butylhydroperoxide, etc.) in the erythrocyte membrane suggest close similarities in the molecular organization. The size selectivity of the "electroleaks" might suggest a "pore-like" structure.[295,309] In fact, when cylindrical "pores" were assumed, the dependence of the induced permeabilities on the permeants' radius could be fitted to the Renkin equation[310] which yields an apparent pore radius. Calculations have given an average (apparent) pore radius of at least 0.6 nm[295,309] in accordance with other studies[57] (but see also References 86, 205 and Chapter 3). However, the data could be also fitted by an alternative model[309] assuming non-Stokesian diffusion[311] through a network of intertwined hydrocarbon chains comparable to a soft polymer.

Furthermore, the previous discussion of electrointernalization has demonstrated that the uptake of DNA, organelles, and particles also cannot be used as an argument for the creation of large "electropores" as done by some authors.[305] The opening of pores large enough to accommodate "diffusion" of organelles and cells would certainly have lethal consequences for the cell.

It is undisputed that a description of the breakdown process in terms of formation of (hydrophilic) pores is a helpful construct for understanding the processes that may occur on the molecular level within the membrane of a field-treated, freely suspended cell.[21,306] However, it should be remembered that like other models, this construct is based upon unproven hypotheses. The concept of membrane breakdown does not, in any event, require the generation of cylindrical pores. Despite similarities between the breakdown of cell and lipid membranes[312] (see Chapter 3), the presence of the protein components (see discussion above and References 168, 295) and the asymmetry of the bilayer in a cell membrane introduce uncertainties regarding the location and the pattern of breakdown as well as the events that follow. Therefore, I believe that it is more accurate to speak about electroleaks[91,295] and not electropores[204,305,313] (see also Footnote on page 3).

VII. ELECTROPERMEABILIZATION IN BASIC AND APPLIED RESEARCH

The ability to introduce impermeable probes such as dyes, peptides, nucleosides, inhibitors, antibodies, enzymes, and other macromolecules into living cells through electropermeabilized membranes by diffusion and/or electrointernalization without compromising cellular and membrane functions has become an important approach for the study of regulatory genetic elements, cellular regulatory mechanisms, and biochemical reactions (see Tables 1 and 2).

The possibility of controlled release of soluble metabolites from the cytosol by electropermeabilization opens completely new avenues in basic research[50,59,69,86,206] and in biotechnology[17,544] such as (repetitive) electrorelease* of endogenous cytosolic contents (see Table 1).

Electropermeabilization appears to be a useful approach to perturb specific cellular processes by introduction of functional molecular species and probe molecules into the cytoplasm of viable cells (for details and References see Tables 1 and 2). Electrotransfection of cells with exogenous DNA or RNA is now recognized as the most efficient means of transforming many eukaryotic cells (including wall-less cells, walled cells, suspension cells, adherent cells, and embryonic cells). Transient and stable expression of electrically incorporated genes are well reported (for details and references see Table 2). Note that linear or supercoiled plasmid DNA is usually more efficient than circularized DNA for (co-)-transformation.

From the foregoing considerations we have seen that optimum electropermeabilization conditions can be precisely predicted, in most of the cases, provided that the electric field and medium parameters as well as the temperature during and after field pulse application have been carefully selected. Unfortunately, many researchers in the field of electropermeabilization continue to adopt an empirical approach to field pulse technology. Many commercial instruments cannot provide the required microsecond-duration field pulses of sufficiently high intensity to permit true reversible electric breakdown of the cell membrane. In an effort to compensate for these deficiencies, they must use millisecond-duration times and correspondingly lower field strengths. We have seen that for electrointernalization and electroinsertion applications (i.e., in the presence of high concentrations of exogenous macromolecules and/or at elevated temperatures) this may work quite well and may provide an efficient method for the uptake of exogenous macromolecules. However, generally speaking, the "beauty" of electropermeabilization (the predictability of the field conditions for cells of different origin and characteristics) is lost when these "long and low" field conditions are employed.

---

*   This new concept, which I have also termed "electromilking" of cells is based upon the intermittent application of reversible membrane breakdown pulses to cells in culture to facilitate the release of cell products (native or transfected). This technique may be used to increase the yield of secreted cell products over long periods of time or in place of "cell suicide" techniques for obtention of such products as monoclonal antibodies.

TABLE 1
Electroinjection of Exogeneous Material

Tissue/Species Tumor/Species/Site	Exogeneous material	Remarks/Objective	Ref.
	Plant Kingdom[a]		
Algae			
Chlamydomonas rheinhardtii	Serva Blue G	r.p.; cells; permeability; uptake	314
Higher Plants			
Avena sativa	ethidium bromide, berberine hemisulfate	r.p.; protoplasts; asymmetric uptake	60
Beta vulgaris	phenosafranine and/or [3H]-pABD1 plasmid DNA	r.p.; protoplasts, cells; no electrotransformation; uptake	93
Daucus carota	rhodaminyl lysine, phallotoxin (RLP)	e.p.; suspension culture cells; F-actin network	315
Dryopteris paleacea	[Ca^{2+}]-indicator fura-2	e.p.; single celled spores; [Ca^{2+}]$_i$	316
Lentil culinaris	anti-lipoxygenase-antibody	e.p.; protoplasts; inhibition of lipoxygenase activity	267
Lilium longiflorum	FITC-labelled dextrans, [Ca^{2+}]$_i$-indicator quin 2	e.p.; pollen; uptake	317
Nicotiana tabacum	fluorescein	e.p.; pollen; release	61
Phaseolus mungo	[Ca^{2+}]$_i$-indicator quin 2	e.p.; protoplasts; [Ca^{2+}]$_i$	318
Fungi			
Dictyostelium discoideum	[14C]-deoxyglucose (marker for permeabilisation), ATP, GTP-γ-S, ethidium bromide	r.p.; regulation of adenylate cyclase	264
	[3H]-inositol	e.p.; inositol cycle	319

Dictyostelium discoideum (strain NC4)	[Ca^{2+}]-EGTA	e.p.; regulation of guanylyl cyclase	320
Saccharomyces cerevisiae	FITC-dextran 70 kDa, propidium iodide	r.p.; intact cells; permeability; uptake	97, 98
Saccharomyces cerevisiae (wildtype and *pim* mutants)	antibody kt3g against PIP$_2$	r.p.; intact cells	321
Schizosaccharomyces pombe	FITC-dextran 70 kDa, propidium iodide	r.p. intact cells; permeability; uptake	98

Animal Kingdom

Echinodermata
sea urchin

eggs	[Ca^{2+}]$_e$	r.p.; membrane potential; asymmetric uptake	77, 79, 80, 147
eggs (*Echinus esculentus*)	[^{14}C]-mannitol, [^{23}Na$^+$]	e.p.; exocytosis	49
eggs (*Lytechinus pictus, Strongylocentrotus purpuratus*)	[^{32}P]-phosphate	e.p.; protein phosphorylation	265
eggs (*Lytechinus pictus, Strongylocentrotus purpuratus*)	DAPI, various metabolites	e.p.; uptake and release of metabolites	239

Vertebrate

Mammalian Cells[b]
Adipose tissue
rat

adipocytes	[^{14}C]-sucrose, [^{14}C]-EDTA, propidium iodide, insulin, GTP-γ-S, glucose-6-phosphate	e.p.; insulin signalling; fatty acid synthesis; lipolysis	274
	[Ca^{2+}]$_e$, insulin, spermin, spermidin	e.p.; signalling of insulin and polyamines	280
	protein kinase C inhibitor H-7, [^{14}C]-sucrose; cAMP	e.p.; insulin action; hexose transport; lipolysis	322
	protein kinase C (19–31) pseudosubstrate	e.p.; insulin action; hexose transport	323
	protein kinase C	e.p.; insulin action; hexose transport	324

TABLE 1 (continued)
Electroinjection of Exogeneous Material

Tissue/Species Tumor/Species/Site	Exogeneous material	Remarks/Objective	Ref.
Adrenal gland			
bovine			
adrenal chromaffine cells	[Ca²⁺]-EGTA	e.p.: horseradish peroxidase; exocytosis	325
	tetanus toxine, anti-tetanus toxine antibodies	e.p.; exocytosis; inactivation of intracellular toxin	326
adrenal medullary cells	guanine nucleotides	e.p.; exocytosis	327
	[Ca²⁺]-EGTA, noradrenalin, glucose analogous, LDH loss	e.p.; exocytosis; pore radius estimation	86
	[Ca²⁺]$_e$, [Mg²⁺]$_e$, phorbol ester, ATP	e.p.; exocytosis; endocytosis	50, 87, 328
chicken			
adrenal medullary cells	guanine nucleotides	e.p.; exocytosis	327
Amnion			
human			
AMA	MAbs 1E9H8, 22IID8B, anti-vimentine, trypan blue	e.p.: pulsing of cells in monolayer; uptake	96
	MAb C4C10, anti-vimentine	e.p.: pulsing of cells in monolayer; DNA synthesis	329
Carcinoma			
human			
breast			
MCF-7	antisense-myc oligonucleotides	e.p.: inhibition of cell adhesion	330
cervix			
HeLa	anti-asparagine synthetase, MAbs (3F3 and 2B4)	e.p.: incorporation; cellular metabolism	100

NHIK 3025	cis-dichloro-diammineplatinum(II)	e.p.; release; cytotoxicity *in vitro*	90
intestine, small			
HCT-8	[α-^{32}P]-dATP	e.p.; DNA synthesis	278
vulva			
A-431	MAbs 1E9H8, 22IID8B, anti-vimentine, trypan blue	e.p.; pulsing of cells in monolayer; cytoskeleton	96
Connective tissue			
human			
fibroblasts	human cytomegalovirus (HCMV AD169)	e.p.; synchronized infection	331
fibroblasts (EAR)	MAbs 1E9H8, 22IID8B, antivimentine, trypan blue	e.p.; pulsing of cells in monolayer; cytoskeleton	96
HT-5 fibroblasts	anti-asparagine synthetase MAbs (3F3 and 2B4)	e.p.; incorporation; cellular metabolism	100
mouse			
L-cells	FITC-BSA, trypan blue	e.p.; field-induced endocytosis; vesicles	15
	propidium iodide	e.p.; membrane incorporation; synchronized cells; electrorotation	67
L-cells (strain I-3; transfected with H-2K^{b})	ovalbumin	e.p.; antigen presentation	273
rat			
F111 fibroblasts (polyoma virus transformed)	chicken IgG	e.p.; uptake; pulsing of cells in monolayer;	332
Ear			
human			
fibroblasts (fetal skin)	MAb C4C10, anti-vimentine	e.p.; pulsing of cells in monolayer; DNA synthesis	329

TABLE 1 (continued)
Electroinjection of Exogeneous Material

Tissue/Species Tumor/Species/Site	Exogeneous material	Remarks/Objective	Ref.
Embryo / fetus, whole			
mouse			
10T 1/2 fibroblasts	FITC-dextrans 9, 41, 72, and 154 kDa	r.p; e.p.: pulsing of cells in monolayer; uptake	62
CB-17 fibroblasts	restriction enzymes RsaI, Sau3AI	e.p.: repair defects in SCID mice	333
C3H 10T 1/2 fibroblasts	restriction enzymes PstI, PvuII, XbaI	e.p.: malignant transformation	334
C3H fibroblasts	rhodamin-conjugated phalloidin, antibodies	r.f.p.: actin; nuclear proteins	237
NHIR fibroblasts	anti-peptide antibodies	e.p.: Glut 1 transporter in glucose transport	335
NIH3T3 fibroblasts	[α-^{32}P]-dATP ethidium bromide	e.p.; uptake; cell metabolism r.f.p.; r.p.: asymmetric uptake	278 63
NIH3T3-A11	*Phytolacca americana* (Pokeweed) antiviral protein	r.p.: cytotoxicity	289
Swiss 3T3 fibroblasts	bombesin, GTP analogous, PMA, β-phorbol-ester	e.p.: inositol phosphate generation	336
rat			
Rat-1 (Tk$^+$), Rat-3 (Tk$^-$)	FITC-dextran, recombinant thymidine kinase (rTk)	e.p.: phenotype conversion	337

Endothelium			
human			
endothelial cells (vein)	human cytomegalovirus (HCMV AD169)	r.p.; synchronized infection	331
Erythrocytes			
human			
erythrocyte ghosts	[14C]-sucrose, urease	e.p.; drug carrier systems	42
	latex particles	e.p.; electroinsertion; uptake; drug carrier	14, 222
erythrocytes	[3H]-methotrexate	e.p.; drug carrier systems	338
	FITC-dextran 70 kDa	r.p.; FACS; uptake studies	99
	FITC-dextran 70 kDa	e.p.; uptake studies	159
	magnetic particles	e.p.; electroinsertion; uptake; drug carrier	339
mouse			
erythrocyte ghosts	[3H]-methotrexate	e.p.; drug carrier systems	41, 46, 222, 340
rabbit			
erythrocytes	[14C]-sucrose	e.p.; drug carrier systems	112, 186
rat			
erythrocytes	FITC-dextran 70 kDa	e.p.; uptake studies	159
erythrocytes	latex particles	e.p.; electroinsertion; uptake; drug carrier	221
Gingiva			
human			
fibroblasts	FITC-BSA, propidium iodide	e.p.; field-induced endocytosis; FACS	219
	FITC-dextrans 4-150 kDa, BSA, mouse MAb IgG$_1$ to human vimentin, FITC-phalloidin, bovine pancreatic RNase A	e.p.; field-induced endocytosis; FACS; uptake	106

TABLE 1 (continued)
Electroinjection of Exogeneous Material

Tissue/Species Tumor/Species/Site	Exogeneous material	Remarks/Objective	Ref.
Glial tumor			
rat			
C6	lissamin-rhodamin-BSA	e.p.; FACS; electrointernalization	103
Hepatoma			
rat			
H4IIE	restriction enzymes EcoRI, PvuII	e.p.; chromosome damage; stress factors	241
Insuloma			
rat			
INS-1 (β-cell derived cell line)	[Ca²⁺]ᵢ, dyes, glycerol phosphate, ATP	e.p.; mitochondrial FAD-linked GPDH; regulation	341
Kidney			
monkey, African green			
BS-C1	MAbs 1E9H8, 22IID8B, anti-vimentine, trypan blue	e.p.; pulsing of cells in monolayer; uptake	96
CV-1	calcein, rhodamin-conjugated phalloidin and dextran (20 kDa)	r.f.p.; uptake; actin	237
mouse			
kidney cells	trypan blue	e.p.; protein synthesis; uptake and resealing kinetics	205
rat			
NRK cells	irradiated plasmid DNA	e.p.; repair of radiation damage	342

Leucocytes

human

polymorphonuclear leucocytes	FITC-phalloidin	r.p.; uptake; actin; phagocytosis	343

Leukemias & Lymphomas

human

10 F2 lymphoblastoid cell line (B-EBV$^+$)	Lucifer Yellow CH	r.p.; uptake; combined with electrotransfection	226
CEM lymphoblastoid leukemia cells	[α-^{32}P]-dATP	e.p.; DNA synthesis; cell metabolism	278
Daudi B-lymphoblastoid cell line	Lucifer Yellow CH	r.p.; uptake	226
HL-60 (promyelocytic leukemia cells)	[^{45}Ca^{2+}], ATP, inositol triphosphate	e.p.; [Ca^{2+}]$_i$-pools; differentiation	243
	deoxyribonucleoside triphosphates	e.p.; DNA synthesis	278
	nucleotides, nucleosides	e.p.; metabolism; toxicity	344
	GTP-γ-S, [Ca^{2+}]$_e$, staurosporine	e.p.; phospholipase A$_2$ regulation	345
	vanadate derivatives, TPA, GTP-γ-S	e.p.; phospholipase D activation	346
	Lucifer Yellow	r.p.; uptake; combined with electrotransfection	226
Jurkat	Tat protein from *E.coli*	e.p.; transcription transactivators	347
	Lucifer Yellow	r.p.; uptake; combined with electrotransfection	226
Jurkat ATCC cells	FITC-dextran 19, 71, 148 kDa, B-phycoerythrin	e.p.; co-loading; protein uptake; FACS	102
Jurkat-K^b	Lucifer Yellow, organic anion transport inhibitors	e.p.; dye retention; FACS; viability; function	108
Molt4 (T-lymphoblastoid cell line)	ovalbumin	e.p.; antigen presentation	273
	nucleotides ara-CTP, dCTP	e.p.; metabolism; toxicity	344

TABLE 1 (continued)
Electroinjection of Exogeneous Material

Tissue/Species Tumor/Species/Site	Exogeneous material	Remarks/Objective	Ref.
Raji F1 (pHAZE transfected lymphoblastoid cells)	restriction enzymes PvuI, PvuII, ClaI	e.p.: chromosome damage	348
T2-K^b lymphoblastoid cells	ovalbumin	e.p.: antigen presentation	273
XPA (lymphoblastoid)	DNA endonuclease complexes	e.p.: DNA repair	349
YT cells (transformed NK cells)	recombinant C3 exoenzyme from *Clostridium botulinum*	e.p.: cytotoxicity; *ras*-related *Rho* proteins	350
mouse			
3D0-54.8 T-cell hybridomas	Lucifer Yellow, organic anion transport inhibitors	e.p.: dye retention; viability; function; FACS	108
A20-2J B-cell lymphoma	Lucifer Yellow, organic anion transport inhibitors	e.p.: dye retention; viability; function; FACS	108
EL-4 lymphoma	ovalbumin	e.p.: antigen presentation	273
L1210 lymphoma	thapsigargin, Ins(2,4,5)P$_3$	e.p.: inositol triphosphate mediated Ca^{2+}-release	248
L5178Y lymphoma	anti-asparagine synthetase MAbs (3F3 and 2B4)	e.p.: incorporation; cellular metabolism	100
M12.B6 (mouse B-lymphoma fused to B cells)	ovalbumin	e.p.: antigen presentation	351
Liver			
rat			
hepatocytes	[^{14}C]-sucrose	e.p.: autophagy	51, 206
	[^{3}H]-raffinose, protein kinase inhibitors	e.p.: autophagy; cytoskeleton	352

Lung			
hamster, Chinese			
DC-3F fibroblasts	d[α]octothymidilate	r.p.; uptake	261
DC-3F (various sublines and clones)	Lucifer Yellow, anticancer drugs (methotrexate, bleomycin)	r.p.; cytotoxicity	289, 353
human			
airway epithelial cells	heparin, thapsigargin, chondroitin sulphate	65-kHz triangular wave; Ca^{2+}-signalling	238
IMR-90	rhodamin-conjugated phalloidin	r.f.p.; uptake; microfilaments	237
WI-38	lissamin-rhodamin-BSA	e.p.; FACS; electrointernalization	103
WI-38 VA-13 fibroblasts	horseradish peroxidase	r.p.; microstructure; adherent cells	83
Lymphocytes			
human			
CTL lines (cytotoxic T-cells; H-2K[b]-specific)	Lucifer Yellow, EGTA, protein kinase C pseudosubstrate peptide, [125I]-cytochrome C	e.p.; signalling pathways; expression of effector functions; FACS	272
CTL line	recombinant C3 exoenzyme from *Clostridium botulinum*	e.p.; cytotoxicity; *ras*-related *Rho* proteins	350
T-lymphocytes	Lucifer Yellow; organic anion transport inhibitors	e.p.; dye retention; FACS; viability; function	108
mouse			
B-cells	propidium iodide	e.p.; FACS; uptake studies	109
splenocytes	ovalbumin	e.p.; antigen presentation	273
thymocytes	ADP-ribosyltransferase from *C. botulinum*, [32P]-NAD	e.p.; inositol phosphate metabolism	354
Macrophages			
mouse			
macrophages	fluorescent pH indicator BCEFC, nucleotides, vanadate, NO_3^-	e.p.; efflux; phagosomal acidification; H^+-ATPase	355

TABLE 1 (continued)
Electroinjection of Exogeneous Material

Tissue/Species Tumor/Species/Site	Exogeneous material	Remarks/Objective	Ref.
rat			
macrophages (alveolar)	lissamin-rhodamin-BSA	e.p.; FACS; electrointernalization	103
macrophages (peritoneal)	$[Ca^{2+}]_e$, $[Mg^{2+}]_e$	e.p.; chemiluminescence	356
Melanoma			
mouse			
B16BL6	murine monoclonal anti-p21^{ras} antibody	e.p.; FACS; labelling of *ras* oncogene product, p21	104, 110
Muscle			
rat			
L6 skeletal muscle myoblasts	guanine nucleotides, orthovanadate	e.p.; insulin binding; tyrosine phosphorylation	357
Myelomas & Plasmacytomas			
mouse			
P3X63Ag8 myeloma	FITC-dextran 40 kDa, [^{125}I]-labeled-C-protein, SFV core protein	e.p.; protein synthesis; uptake and resealing kinetics	205
SP2/0-Ag14 myeloma	bisbenzimide	e.p.; asymmetric uptake	71, 148
Neuroblastoma			
human			
SH-SY5Y	phosphatase analogous of [Ins(1,4,5)P$_3$], GTP-γ-S	e.p.; inositol phosphate metabolism	358
Neutrophils			
guinea pig			
neutrophils	F$^-$, ATP, NADPH, protein kinase inhibitors	e.p.; respiratory burst	242

human neutrophils	$[\gamma\text{-}^{32}P]$-ATP	e.p.; lipocortins	359
	$[^{32}P]$-ATP, protein kinase inhibitor peptide	e.p.; inhibitory pathways; therapeutic restoration	360
	$[^{45}Ca^{2+}]$, auranofin	e.p.; chemotactic factors; inflammatory response	361
	okadaic acid, staurosporine	e.p.; protein phosphatases in cell activation	362
	$[Ca^{2+}]_e$, guanine nucleotides	e.p.; degranulation	363
	$[Ca^{2+}]_e$	e.p.; FACS; compared to ionophore loading; actin	364
	$[Ca^{2+}]$, $[Mg^{2+}]$, nucleotides, GTP-γ-S	e.p.; secretion regulation; actin; degranulation	365
	$[Ca^{2+}]_e$, $[Mg^{2+}]_e$	e.p.; chemiluminescence	356
	$[Ca^{2+}]_e$, $[Mg^{2+}]_e$, ATP, guanine nucleotides	e.p.; oxidative response elicited by arachidonic acid	366
	ATP, cyclic nucleotides	e.p.; secretion	367
	guanine nucleotides, AlF_4^-	e.p.; FACS; microfilaments	250
	propidium iodide, NADPH, chymotrypsin inhibitor zLYCK	e.p.; chemoattractant fMLP; phospholipase D activation	368
	NADPH, $[Ca^{2+}]_e$, protein kinase C inhibitor H7	e.p.; chemoattractants; respiratory burst; FACS	369
	protein-kinase C activators, fMLP, PMA, okadaic acid	e.p.; phorbol ester-induced actin assembly	370
	protein kinase C pseudo-substrates, release of marker enzymes	e.p.; complement receptor-mediated accumulation of diglyceride	249
	synthetic peptides, PMA, fMLP	e.p.; respiratory burst	260
	vanadate, AlF_4^-, nucleotides	e.p.; respiratory burst	371

TABLE 1 (continued)
Electroinjection of Exogeneous Material

Tissue/Species Tumor/Species/Site	Exogeneous material	Remarks/Objective	Ref.
rabbit			
neutrophils	$[Ca^{2+}]_e$, nucleotides	e.p.; exocytosis	372
	$[Ca^{2+}]_e$, Ca^{2+}-antagonists	e.p.; chemotaxis	251
	$[Ca^{2+}]_e$, $[Mg^{2+}]_e$, $[Sr^{2+}]_e$, $[Ba^{2+}]_e$, GTP-γ-S, fMLP	e.p.; exocytosis	373
rat			
neutrophils	propidium iodide	r.p.; drug carrier	227
Oocytes			
mouse			
oocytes	$[Ca^{2+}]_e$	r.p.; oocyte activation	374
	$[Ca^{2+}]_e$, Ins(1,4,5)P_3	r.p.; $[Ca^{2+}]_i$-oscillations	375
Ovary			
hamster, Chinese			
CHO	pancreatic DNase I	e.p.; cell cycle; chromosomal aberrations	268
	anti-α-tubulin, propidium iodide, ethidium bromide, [^{35}S]-CHO cell protein	r.f.p. and centrifugation; microtubuli; nuclei	92
	trypan blue	r.p.; permeability; ^{31}P-NMR spectroscopy; phospholipids	213
	[α-^{32}P]-deoxyribonucleoside triphosphate (dATP)	r.p.; DNA synthesis	278
	catalase, ovalbumin, histone HI, IgM-coated colloidal particles	e.p.; vesicle formation; electrointernalization	218
	restriction enzyme AluI	e.p.; chromosomal aberrations	376

CHO-6	restriction enzymes EcoRI, DraI, HaeIII, ScaI	e.p.; chromosome damage	270
CHO-K1	restriction enzyme PvuII	e.p.; chromosome damage	377
	restriction enzymes AluI, Sau3AI	e.p.; chromosome damage	269, 378
	restriction enzymes PvuII, BamHI, EcoRI, adenine analogous	e.p.; chromosome damage; mutagenic effects	379
	restriction enzymes BamHI, AluI, HaeIII, PvuII, HinfI, Sau3AI,	e.p.; chromosome damage; mutagenic effects	271
CHO-WTT	Ca^{2+}-influx, trypan blue	e.p.; mechanism; membrane organization	65
	ethidium bromide, propidium iodide, FITC-dextran (3.8 kDa)	e.p.; permeability; uptake and release kinetics	380

Pancreas

rat

pancreatic acini	$[Ca^{2+}]_e$, heparin, phorbol ester, cyclic nuleotides	e.p.; IP_3-induced Ca^{2+}-release	381
pancreatic islet cells	$[Ca^{2+}]_e$, dyes, glycerol phosphate, ATP	e.p.; mitochondrial FAD-linked GPDH; regulation	341
pancreatic islets	$[^{14}C]$-sucrose, cytochalasin B, vinblastine sulphate, $[Ca^{2+}]_e$	e.p.; insulin secretion; microfilaments	85
	$[^{32}P]$-arachidonic acid, $[Ca^{2+}]_e$	e.p.; β-cell function; protein phosphorylation	247
	$[^{32}P]$-ATP, $[Ca^{2+}]_e$, cAMP, phorbol ester	e.p.; β-cell function; protein phosphorylation	244
	glucose metabolites, $[^{14}C]$-urea, $[^3H]$-sucrose	e.p.; insulin release; exocytosis	59
	$[Ca^{2+}]_e$, phorbol ester	e.p.; insulin secretion; protein phosphorylation	382
	$[Ca^{2+}]_e$, phorbol ester, ATP, cAMP, IBMX, $[Mg^{2+}]_e$	e.p.; insulin release; protein phosphorylation	245
	$[Ca^{2+}]_e$, Russel's viper venom	e.p.; insulin release	246

TABLE 1 (continued)
Electroinjection of Exogenous Material

Tissue/Species Tumor/Species/Site	Exogenous material	Remarks/Objective	Ref.
pancreatic islets	$[Ca^{2+}]_e$, cAMP, methylamine, cystamine, glycine methyl-ester, etc.	e.p.; insulin release; involvement of transglutaminase	383
	$[Ca^{2+}]_e$, cGMP, cAMP	e.p.; nitric oxide; insulin	384
	okadaic acid, I-nor-okadone, cAMP, $[^{32}P]$-ATP	e.p.; insulin secretion; protein phosphorylation	385
Phaeochromocytoma			
PC-12	FITC-, lissamin-rhodamin-BSA, lact-/egg albumin, antibodies	e.p.; FACS; electrointernalization	103
Platelets			
bovine			
platelets	$[Ca^{2+}]_e$, protease inhibitors	e.p.; serotonin secretion	386
human			
platelets	$[Ca^{2+}]_e$, guanine nucleotides, nucleosides, thrombin	e.p.; secretion from electro-permeabilized cells	387–389
	$[Ca^{2+}]_e$, phorbol ester, guanine nucleotides, thrombin	e.p.; dense and α-granule secretion	390
	$[Ca^{2+}]_e$, guanine nucleotides, staurosporine	e.p.; adhesion to collagen	391
	Lucifer Yellow, adenine nucleotides, ^{111}Indium oxide	e.p.; uptake; release; integrity; drug carrier	53

mouse			
platelets	Lucifer Yellow, adenine nucleotides, ^{111}Indium oxide	e.p.; uptake; release; integrity; drug carrier	392
rat			
platelets	Lucifer Yellow, adenine nucleotides, ^{111}Indium oxide	e.p.; uptake; release; integrity; drug carrier	53, 392
Sperm			
bovine			
sperm	[^{32}P]-plasmid DNA	e.p.; used for fertilization of oocytes	393
human			
sperm	$[Ca^{2+}]_e$	e.p.; induction of acrosome reaction	394

[a] listed alphabetically.

[b] listed according to the Tissue Index and Tumor Index of the American Type Culture Collection catalog 1988.

Note: Abbreviations: e.p. = exponentially decaying pulse, MAb = monoclonal antibody, r.f.p.= radio frequency (oscillating) pulse, r.p. = rectangular pulse.

TABLE 2
Electrotransfection

Plant Kingdom[a]

Tissue/Species Tumor/Species/Site	Construct	Remarks/Objective	Ref.
Algae			
Chlamydomonas rheinhardtii (137 c⁺)	pCAMVCAT, plasmids of 3–12 kb size	e.p.; t.e.; walled cells; electrointernalization?	395
Chlamydomonas rheinhardtii (cw-92)	pCAMVCAT	e.p.; t.e.; wall-less cells	395
Chlamydomonas rheinhardtii (mutants nit1-305; cw nit1-305)	pMN24	e.p.; s.t.; walled cells; expression of nitrate reductase gene	395
Higher plants			
Alnus incana (PPL)	pBI221	e.p.; sc; t.e.; $[PEG]_e$; GUS gene expression	275
Avena sativa (PPL)	pMON8678, GUS mRNA	e.p.; t.e.; GUS gene; expression; $[spermidine]_e$	396
Beta vulgaris (PPL) (var. R45r4 and M1 cells)	pDW2	e.p.; r.p.; r.f.p.; t.e.; CAT activity	231
Beta vulgaris (PPL)	p35SCN, pABDI	e.p.; t.e.; enzyme treatment	193
	p35SCN	r.p.; t.e.; CAT activity	397
	NPTII gene	e.p.; s.t.; gene obtained from digested tobacco genomic DNA	398
	pABDI	r.p.; uptake studies	93
Brassica campestris (PPL)	pDW2	r.p.; t.e.; field conditions	399
Brassica napus (PPL)	pDW2	r.p.; t.e.; field conditions	399

Organism	Plasmid/Construct	Conditions	Ref.
Brassica napus (PPL) (cv. Brutor)	pKSIV.35, pIB16.1, pHP23b, pGL2	e.p.; s.t.; cotransformation; transgenic plants	400
Brassica napus (PPL) ssp. oleifera (cv. Hanna)	pABD1	e.p.; s.t.; l; transgenic plants	401
Brassica oleracea (PPL) (var. *botrytis*)	pRT99GUS	e.p.; sc/c; t.e.; s.t.; transgenic plants	402
Chenopodium quinoa (PPL)	NPTII gene	r.p.; s.t.; EP vs. PEG and *Agrobacterium* transfer	403
	biological active transcripts from cDNA of GFLV-RNAs	e.p.; t.e.; role of viral proteins in virus multiplication	404
Chrysanthemum morifolium (PPL) (cv. Shuuhou no Chikara)	TMV RNA	r.p.; t.e.; host range of TMV; maximum virus yield	405
Dactylis glomerata (PPL)	pCIB709	e.p.; s.t.; l; heat shock and field; transgenic plants	406
Daucus carota (PPL) (TC)	pMON145'	e.p.; r.p; t.e.; CAT activity	94
Daucus carota (PPL) (W001C)	pIC35/A, p35SOcs, ICP plasmids	e.p.; t.e.; cotransfection ICP RNA stability in plants	407
	p35SOcs	e.p.; t.e.; PEG addition; RNA transcripts	408
Daucus carota (PPL) (W001C + cv Juwarot)	pUC, DC8:GUS constructs	e.p.; t.e.; transcriptional regulation of DC8-gene	409
Daucus carota (PPL) (W001C + RCWC-1)	sense/antisense CAT gene plasmids	e.p.; sc/l; t.e.; effects on gene expression	410–415
Glycine canescens (PPL)	pCMC1021	e.p.; s.t.; calli recovery	416
Glycine max (PPL)	CAT/GUS gene plasmids	e.p.; t.e.; factors affecting gene expression	94, 417
	GUS gene constructs	e.p.; t.e.; β-conglycinin expression	418
	pZA300	e.p.; sc; heat shock and field; shoot regeneration	419

TABLE 2 (continued)
Electrotransfection

Tissue/Species Tumor/Species/Site	Construct	Remarks/Objective	Ref.
Hordeum vulgare (PPL) (cv. Himalaya)	pGSJ290, pHTT246	e.p.; t.e.; NPTII activity	420
Hordeum vulgare (PPL) (cv. New Golden)	TMV RNA	r.p.; t.e.; host range of TMV; maximum virus yield	405
Lycopersicon esculentum (PPL) (cv. Kurihara)	TMV RNA	r.p.; t.e.; virus host range; maximum virus yield	405
Lycopersicon esculentum (PPL) (cv. Petit Tomato)	CAT gene plasmids, CMV RNA	e.p.; t.e.; conditions for optimum yield	421
Lycopersicon peruvianum (PPL)	TMV RNA	r.p.; t.e.; host range	405
Medicago sativa (PPL)	p35SGUS, p35Snpt	e.p.; s.t.; EP vs. PEG; EP vs. heat shock	422
Medicago sativa (PPL) (cv. Calwest 475)	DNA constructs pCHC1, *pMaeI*-1, pCHP1	e.p.; t.e.; cotransfection r.p.; t.e.; cotransfection response to fungal elicitor	423 424
Nicotiana gossei (pollen)	pBI221	r.p.; l; t.e.; GUS activation	425, 426
Nicotiana plumbaginifolia (PPL) (NpT5)	pCAMVCAT	e.p.; sc; t.e.; yield	290
Nicotiana plumbaginifolia (PPL)	BMV, CCMV, CCMV RNA	e.p.; EP vs. PEG	427
	p35SGUS	e.p.; t.e.; heat shock; not beneficial	422
Nicotiana tabacum (PPL) (BY-2)	*in vitro* transcripts of TMV RNA	e.p.; low concentrations of viral RNA required	428
	TMV RNA, CMV RNA, virus, pCAMVCAT	e.p.; l; c; t.e.; s.t.; cotransfection	429, 430
	CHN-GUS constructs	e.p.; t.e.; chitinase gene	431

Organism	DNA/RNA	Conditions	Ref.
	GUS gene plasmids	e.p.; t.e.; β-conglycinin expression	418
	CAT gene plasmids	e.p.; t.e.; s.t.; field conditions	432
Nicotiana tabacum (PPL) (NT1)	CAT gene plasmids	e.p.; l; t.e.; inhibition of transgene expression	433
	pMON213, pCPI	e.p.; l; t.e.; homo-/non-homologous transformation	434
Nicotiana tabacum (PPL) (var. T x D)	pDW2	e.p.; r.p.; r.f.p.; t.e.; CAT activity	231
Nicotiana tabacum (PPL) (cv. Xanthi; XHFD8)	pUBI-CAT, p35S-CAT, pGP229	e.p.; t.e.; expression of ubiquitin-CAT gene	435
	suppressor tRNA genes, GUS genes	e.p.; t.e.; cotransfection; suppression of stop codons	436
Nicotiana tabacum (PPL) (cv. W38 + SR1)	pLGneo213, pABD1	r.p.; s.t.; EP vs. PEG	437
Nicotiana tabacum (PPL) (cv. Burley 21)	potyviral RNA (TVMV, PVY)	e.p.; EP vs. liposome fusion	276
Nicotiana tabacum (PPL) (cv. Petit Havana SR1)	pABDI	e.p.; l/c; s.t.; heat shock; PEG	438
	pHP23, HMW GH50 genomic DNA pRAL6130, pRAL6131 pRAL6130ss, pRAL6131ss TMV, TMV-Lsl RNA, TMV-L RNA	e.p.; s.t.; transgenic plants	398
		r.p.; ds, ss; t.e.; s.t.; calli stage	439
		r.p.; field and media conditions; yield	440
Nicotiana tabacum (PPL) (var. Samsun)	TMV-RNA	e.p.; injection into maceroenzyme-treated cells	441
Nicotiana tabacum (PPL) (cv. Wisconsin 38)	pNOSCAT, pCaMVCAT	e.p.; t.e.; transfection yield	442

TABLE 2 (continued)
Electrotransfection

Tissue/Species Tumor/Species/Site	Construct	Remarks/Objective	Ref.
Nicotiana tabacum (PPL) (cv. Xanthi)	CCMV	e.p.; EP vs. PEG	427
	pCGN7150, pCGN7394	e.p.; t.e.; cotransfection; efficacy of exotoxin A	442
	pMON200	e.p.; l; transgenic plants	414, 443
	TMV RNA, CMV RNA	r.p.; field and media conditions; yield	444
	T7 RNA polymerase genes (T7P1, T7P2)	e.p.; polymerase-mediated gene expression system	445
Nicotiana tabacum (PPL) (cv. Xanthi nc)	CMV RNA	e.p.; r.p.; effect of pulse shape on RNA incorporation	230
Nicotiana tabacum (PPL) (cv. Xanthi NN)	TMV RNA	r.p.; field conditions; continuous injection	88, 89
Oryza sativa (PPL)	TMV, CMV	e.p.; cotransfection	429
Oryza sativa (PPL) (C5924)	TMV RNA, CMV RNA	e.p.; cotransfection	446
Oryza sativa (PPL) (cv. Tai-nung 67)	p35S-CAT, pCopia-CAT	e.p.; r.p.; t.e.; inheritance + expression of genes	447
Oryza sativa (PPL) (cv. Taipei 309)	pCaMVNEO	e.p.; plant regeneration	448
	pNEOGUSS15	e.p.; transgenic plants	449
	p2′-nptII, p35S-nptII, pext-nptII, etc.	e.p.; sc/l; t.e.; s.t.; transgenic plants; EP of leaf base	450
	pMON806	e.p.; t.e.; EP vs. PEG; transgenic plants	451

Species	Construct	Description	Ref.
Panicum maximum (PPL)	pMON145'	e.p.; t.e.; CAT activity	94
Pennisetum purpureum (PPL)	pMON806	e.p.; c; callus lines	452
Persea americana (PPL)	pMON145'	e.p.; t.e.; CAT activity	94
(cv. Hass)	pCaMVCAT	e.p.; sc; t.e.	453
Petunia hybrida (PPL) (cv. Mitchell)	pMON243, pMON145'	e.p.; r.p.; CAT activity	94
Petunia hybrida (PPL) (line 3704)	biological active DNA constructs	e.p.; sc; l; ss; yield increased by heat shock	454
Phaseolus vulgaris (PPL) (cv. Canadian Wonder)	pCHC1, pMael-1	e.p.; t.e.; cotransfection; silencer region	455
Picea glauca (PPL)	pBI221, pCaMVCN	e.p.; l/c; t.e.; comparison of linearized/circular DNA	456
Solanum brevidens (PPL)	CAT/GUS gene plasmids	e.p.; t.e.; s.t.; CAT/GUS activity	457, 458
Solanum tuberosum (PPL) (AEC-1)	pBI221	e.p.; r.p.; t.e.; yield	228
Solanum tuberosum (PPL) (SVP11 cv. Astarte)	pBI221, pGBSSO.8, pGBSSO.4	e.p.; r.p.; t.e.; s.t.; gene regulation	458
Solanum tuberosum (PPL) (cv. Désirée)	CAT/GUS gene plasmids	e.p.; t.e.; CAT/GUS activity	457
Sorghum bicolor (PPL)	p35S-CAT, pCopia-CAT	e.p.; r.p.; t.e.	447
Triticum aestivum (PPL) (cv. Anza)	CAT gene plasmids	e.p.; t.e.; gene expression using maize intron	459
Triticum aestivum (cv. Fidel)	p2'-nptII, p35S-nptII	e.p.; sc/l; t.e.; intact cells	450
Triticum aestivum (cv. Sonora)	pBC17, pAct1-D	e.p.; t.e.; electroinjection into zygotic embryos	460
Triticum monococcum (PPL)	p35SGUS, p35Snpt	e.p.; s.t.; expression/ cell cycle	422
	pMON145'	e.p.; t.e.; CAT activity	94
	pMON410, pMON273, pMON145'	e.p.; c; callus lines	452

TABLE 2 (continued)
Electrotransfection

Tissue/Species Tumor/Species/Site	Construct	Remarks/Objective	Ref.
Triticum monococcum (PPL) (R-TM 1066)	p35S-CAT, pCopia-CAT	e.p.; r.p.; t.e.	447
Vigna aconitifolia (PPL) (cv. IPMCO-560)	pLGVneo2103, pCAP212	r.p.; t.e.; s.t.; injection through wall not successful	461
Vigna aconitifolia (PPL) (cvs. IPMCO-560/-909)	pLGVneo2103, pABD1	r.p.; s.t.; EP vs. PEG	437
Vigna sesquipedalis (PPL) (cv. Kurodane-Sanjaku)	TMV RNA	r.p.; field conditions; continuous injection	89
Vigna unguiculata (Walp)	p35SGUSIND/pUC8, pPAL2-GUS	e.p.; injection into mature embryos	462, 463
Vinca rosea (PPL)	GUS gene constructs	e.p.; t.e.; β-conglycinin expression	418
Zea mays (PPL)	TMV RNA	e.p.; viral RNA infection	430
Zea mays (2418)	p35S-CAT	e.p.; t.e.; *ocs*-elements	464
	p2′-nptII, p35S-nptII	e.p.; sc; t.e.; injection into intact cells	450
Zea mays (PPL) (cv. Black Mexican Sweet)	p35SNEO, p35SHPT	e.p.; sc; s.t.; field/media conditions; calli	285
	CAT gene plasmids	e.p.; t.e.; CAT gene expression: intron-enhanced	465
	pUBI-CAT, p35S-CAT, pGP229	e.p.; t.e.; maize polyubiquitin genes	435
	pNOSCAT, pCaMVCAT	e.p.; t.e.; yield	412
Zea mays (PPL) (BMSI)	pCAMVNEO	e.p.; sc; s.t.; calli	466

Zea mays (PPL) (BMS XII-II)	CAT/GUS gene plasmids	e.p.; t.e.; *Adh1* gene	467
Zea mays (H99, pa91)	pDE108	e.p.; l; use of pre-treated zygotic embryos and calli	468
Zea mays (Pioneer hybrid P3377)	pNIU22, pNIU24, pAIICN, pUBI-CAT	e.p.; t.e.; microspores; EP vs. PEG	469
Fungi			
Aspergillus awamori (PPL) (GC12)	pUC4ΔAB-*argB*	e.p.; l; isolation of PEPA-deficient mutants	470
Aspergillus awamori (PPL) (GC12, GCΔGAM23)	pGR1, pGR3, pGAMpR	e.p.; yield; EP vs. PEG; chymosin production	471
Aspergillus awamori (PPL) (GC12, GC5)	pANG1, PBH2	e.p.; s.t.; yield; EP vs. PEG	472
Aspergillus niger (PPL) (GC44)	pMW30	e.p.; s.t., yield; EP vs. PEG	472, 473
Candida maltosa (G374, G587)	pCrLPII, pCCD5, pCipA15, pCipLI, pCipLL1, pCrAp2	e.p.; c/l; cotransfection; EP vs. LiCl	474
Dictyostelium discoideum (Ax-2)	VII-Luc, cath-Luc8, VII-Luc antisense	e.p.; sc; t.e.; s.t.; field conditions	475
Dictyostelium discoideum (Ax-4)	pDE109	e.p.; hygromycin resistance as a marker	476
Dictyostelium discoideum (KAx-3)	VII-Luc, cath-Luc8, VII-Luc antisense	e.p.; sc; t.e.; s.t.; field conditions	475
Dictyostelium discoideum (KAx-3, HPS400)	pATANB43, Thy1P1, Thy1P42	e.p.; complementation of thymidine deficiency	477
Dictyostelium discoideum (Synag7)	VII-Luc	e.p.; sc; t.e.; cellular differentiation	475
Neurospora crassa (74A)	pCSN44, qA-2$^+$	e.p.; s.t.; pre-treated with β-glucuronidase	478
Physarum polycephalum (LU352, LU381)	pPardC-CAT, pPardC-CATΔT	e.p.; t.e.; heat shock not beneficial; actin promoter	479

TABLE 2 (continued)
Electrotransfection

Tissue/Species Tumor/Species/Site	Construct	Remarks/Objective	Ref.
Saccharomyces carlsbergensis (PPL) (34)	killer dsRNA	e.p.; l; s.t.; use of *rho⁻* mutants beneficial	480
Saccharomyces cerevisiae (AB1380)	pYAC4	e.p.; c/l; yield of transformants	481
Saccharomyces cerevisiae (PPL) (AH215, AS-4/H2, T158)	killer dsRNA	e.p.; l; s.t.; transmission of killer activity	482
Saccharomyces cerevisiae (AH22, ALDI)	pJDB248, pJDB219	e.p.; heritable damage; slow growth phenotype	181
Saccharomyces cerevisiae (PPL) (AS.4/H₂-1, AS.4/H₂/K⁺₁, AS.4/H₂, AS.4/H₂/21, T158C)	killer dsRNA	e.p.; l; s.t.; transmission of killer acitivity into industrial strains	480
Saccharomyces cerevisiae (BWG1-7A)	episomal plasmids of various sizes	e.p.; s.t.; protocol for transformation	483
Saccharomyces cerevisiae (CM1)	pBR-u3	e.p.; s.t.; yield; EP vs. spheroblast method	484
Saccharomyces cerevisiae (DBY745, KY114, TDY718)	URA3 and His3 genes	e.p.; l; PEG added: EP vs. alkaline transfection	485
Saccharomyces cerevisiae (PPL) (D13-1A)	YRP7	e.p.; c; field conditions; breakdown voltage	486
Saccharomyces cerevisiae (EH3714-2B, MG5123-6B)	episomal plasmids (YEp352, YEp24)	e.p.; transformation protocol	487
Saccharomyces cerevisiae (FF18237; aSSIII)	pTSUK	r.p.; pre-treatment with DTT increases yield	488
Saccharomyces cerevisiae (KK4)	YEP13	e.p.; transformation procedure	489

Saccharomyces cerevisiae (JI26A, M12B)	YRP7, YEP24	e.p.; transformation procedure	490
Saccharomyces cerevisiae (SE7-6)	YRP7, pYE(CEN3)30	e.p.; s.t.; yield; EP vs. LiCl method	484
Saccharomyces cerevisiae (S288C)	tRNA^{Lys}(CUU)	r.p.; PEG added; tRNA mitochondrially-imported	491
Schizosaccharomyces pombe (h-ura4-294)	pFL20	e.p.; EP vs. heat shock; LiCl and PEG methods	492
Yarrowia lipolytica (INAG 33234)	pINA176	r.p.; field conditions; transformant yield	488
Animal Kingdom			
Protozoa			
Giardia lamblia WB	ssRNA	e.p.; ss; infection	493
Leishmania brasiliensis	pALT1-1	e.p.; t.e.; CAT activity	494
Leishmania enriettii	pALT1-1 + derivatives	e.p.; t.e.; α-tubulin gene	494
Leishmania major	pALT1-1	e.p.; t.e.; CAT activity	494
Tetrahymena thermophila (CU427)	rDNA	e.p.; cellular nuclease problem adressed	495
Insects			
Drosophila (embryo)	*ma-l*-phage DNA	e.p.; t.e.; AO activity	496
Spodoptera frugiperda (insect cells)	AcNPV DNA, pAcRP23-LacZ	e.p.; cotransfection; EP vs. Ca^{2+}-phosphate	497
Vertebrate			
Fish			
Oryzias latipes (eggs)	pMV-GH	r.p.; l; transgenic fish; EP vs. microinjection	498
Carrasius auratus RBCF1 cells	pSV2-*neo*	e.p.; s.t.; malignant transformation	499

TABLE 2 (continued)
Electrotransfection

Tissue/Species Tumor/Species/Site	Construct	Remarks/Objective	Ref.
Mammalian Cells[b]			
Bone marrow			
chicken			
HD3 (virus transformed erythroid cell line)	pRSV-luc	e.p.; c; carrier DNA important	500
human			
stem cells	pRSV2CAT, pSV7-*neo*gpt	e.p.; l; s.t.; conditions	501
mouse			
stem cells	pRSVCAT, RSVCAT-AaT2, pSV2-*neo*	e.p.; l; cotransfection; *in vivo* expression	502
Carcinoma			
human			
breast			
MCF-7	pSVX, pSVX021	e.p.; s.t.; p120 expression	503
colon			
GEO	TGFα-cDNA	e.p.; s.t.; tumor formation	504
cervix			
HeLa	pMT-hGH-SV2	e.p.; transfectant yield; EP vs. Ca²⁺-phosphate	505
	pRSV-gpt	e.p.; field conditions	229
	pSV2-*neo*	e.p.; l; field conditions	506
larynx			
HEp-2	SV40 DNA	e.p.; field conditions	277
mouse			
mammary			
FM3A	pSV2-*neo*, pTK1	e.p.; l; field conditions	506

Connective tissue
human

GM0847 fibroblasts (HPRT⁻, SV40 transformed)	pLAMP, pSV2-*neo*	e.p.; l; transfectant yield	507

mouse

L-cells	pSV2-*neo*	e.p.; l/c; s.t.; field conditions	187, 287
L-cells	pSV2-*neo*	e.p.; l; s.t.; hypoosmolar media: high yield	286
	pSV7-*neogpt*,* mouse κ-Ecogpt*	e.p.; l; transcription of the Ecogpt* gene	508
Ltk⁻	pRSV-L	e.p.; s.t.; field conditions	509
	pHSV106, ssMTK1 DNA, HindIII, XbaI, XhoI	e.p.; l/c; cotransfer with HindIII etc. enhances s.t.	510
	pAGO	s.c.; l; Mg^{2+} reduces or prevents gene transfer	313
	pHSV106	e.p.; s.t.; synchronized cells; G_2/M: high yield	511
L-MTK⁻	pX1TK	e.p.; t.e.; s.t.	512

rat

F111 fibroblasts (polyoma virus transformed)	pY3	e.p.; s.t.; pulsing of adherent cells	332

Embryo / fetus, whole
mouse

NIH3T3	CAT gene	e.p.; t.e.; medium osmolality	284
	pMT-hGH-SV2	e.p.; transfectant yield; EP vs. Ca^{2+}-phosphate	505
	pSV2-*neo-ras*	e.p.; s.t.; malignant transformation	499
	pSV2-*neo*, pUCEJ	e.p.; t.e.; s.t.; cotransfection	512
	pSV2-*neo*	r.f.p.; c; asymmetric uptake	149
	pSV2-*neo*	e.p.; l; field conditions	506

TABLE 2 (continued)
Electrotransfection

Tissue/Species Tumor/Species/Site	Construct	Remarks/Objective	Ref.
rat			
Rat-1 (infected with RLV)	pLZRNL, pRc/RSV/β-gal	e.p.; l; yield; EP vs. other transfection methods	513
Enamel Organ			
mouse			
epithelial cells	pRSV-T	e.p.; l; large-T-antigen expression	514
Erythrocytes			
human			
erythrocytes	SV40 DNA, RNA	e.p.; sc/l; field conditions	56
Hepatoma			
human			
5123, FT02B	CAT gene	e.p.; t.e.; medium osmolality	284
rat			
H4AzC2	pA10CAT2, pSV2CAT	e.p.; l; t.e. EP vs. other transfection methods	515
Kidney			
monkey, African green			
COS	CAT gene plasmids	e.p.; t.e.; CAT activity	415
COS-1	pKTH539, pKTH539(GP8)	r.p.; s.t.; β-galactosidase expression	516
	pCH110	e.p.; EP vs. other transfection methods	517

Cell line	DNA	Method/conditions	Ref.
COS-M6	GST cDNA	e.p.; sc; t.e.; s.t.; FACS	101
	pGTA12cDV	e.p.; sc; t.e.; FACS; EP vs. other transfection methods	107
	pSV2CAT	r.f.p.; t.e.; CAT activity	236
CV-1	GST cDNA	e.p.; sc; t.e.; FACS	105
	pMT-hGH-SV2	e.p.; transfectant yield; EP vs. Ca^{2+}-phosphate	505
	pSV3-*neo*	r.p.; s.t.; EP vs. Ca^{2+}-phosphate method	518
	SV40 DNA	e.p.; field conditions	277
pig			
PK-15 (nucleoside transport deficient mutant)	pSV2-*neo*	e.p.; l; EP vs. lipofectin	519
Leukemias & Lymphomas			
human			
10F2 (B-EBV$^+$ lymphoblastoid cell line)	LP/pUC13	r.p.; c; field conditions	226
8166 (transformed T-cell line)	LP/pUC13	r.p.; c; field conditions	226
BL18, BL65(Burkitt lymphomas)	pSV7-*neogpt**	e.p.; l; transfection conditions	508
GM4672 (lymphoblastoid cell line)	pRSVCAT, pSV2CAT	e.p.; sc/l; t.e.; s.t.; CAT activity vs. yield of s.t.	520
HEL (erythroleukemic cell line)	pEMSV-cat	e.p.; l/sc; t.e.; s.t.; supercoiled vs. linear DNA	521
HUT-78 cells (cutaneous T-cell lymphoma)	pRSV-gpt	e.p.; field conditions	229
Jurkat (clone E6-1)	SPR cDNA	e.p.; neurotransmitter substance P	522
K562 (myeloid leukemic cell line)	pRSV-luc	e.p.; l/sc; t.e.; s.t.; field conditions	500

TABLE 2 (continued)
Electrotransfection

Tissue/Species Tumor/Species/Site	Construct	Remarks/Objective	Ref.
K562 (myeloid leukemic cell line)	pEMSVCAT, pEMSV-*neo*	e.p.; sc/l; t.e.; s.t.; EP vs. Ca²⁺-phosphate	521
	pRSV-*neo*, pN-*ras*(Fe) pN-*ras*, pN-*ras*(HT-JT)	e.p.; l; cotransfection; differentiation	523
	pᴬ-γ-*neo*, pRSV-*neo*	e.p.; r.p.; t.e.; γ-globin gene promoter	524
	pMoZtk	r.p.; c; t.e.; FACS; EP of synchronized cells	525
	pREP2	e.p.; FACS; CD8 transfectants	526
	LP/pUC13	r.p.; c; uptake studies	226
	pRSV-luc	e.p.; c; t.e.; carrier DNA important	500
K562 (subclone SA1)	pLW4, Homer 6 plasmid	e.p.; s.t.; conditions	527
KG-1 (acute myelogeneous leukemic cell line)	pEMSVCAT	e.p.; l/sc; t.e.; s.t.; supercoiled vs. linear DNA	521
LAZ509 (EBV-transformed lymphoblastoid cell line)	pSV7-*neogpt**	e.p.; l; transfection conditions	508
Molt13	pRSV-luc, TCR-α enhancer-luc (Eα)	e.p.; c; t.e.; carrier DNA important	500
NC-37 (virus transformed lymphoblastoid cell line)	pEMSVCAT	e.p.; l/sc; t.e.; s.t.; EP vs. Ca²⁺-phosphate	521
Peer (T-cell line)	pRSV-luc	e.p.; c; t.e.; DNA carrier important	500

Raji (Burkitt's lymphoma cell line)	EBV-based vector	e.p.; transfection in G2/M-phase	528
U937	pSV2-neo	e.p.; l; conditions	506
	pREP2	e.p.; FACS; CD8•DF2/REP2 transfectants	526
mouse			
1881 (transformed pre-B cell lymphoma)	pSV7-neogpt, mouse κ-Ecogpt*	e.p.; l/sc; s.t.; Ig gene expression	508
BW (T-cell lymphoma)	DF13:2/MSG	e.p.; l; DAF expression; glycolipid anchoring; CD8	526
BW5147 lymphoma	pRSV-luc	e.p.; c; t.e.; carrier DNA important	500
	pRSVCAT, pSV2CAT pSV7-neogpt*	e.p.; sc/l; t.e.; s.t.; CAT activity	520
DPL-16-1 (erythroleukemic cell line)	p53^{ts}-neo, pSV2-neo	e.p.; l; FACS; p53 induced cell death	529
Hu 11 (erythroleukemic mouse cell line with the human chromosome 11)	pΔβ117	e.p.; l; field conditions; transfectant yield	507
L1210	pSV2-neo	e.p.; l; field conditions	506
L51878Y lymphoma (var. XUMI)	pSV2-neo	e.p.; l; conditions	506
M12 (spontaneous B-cell lymphoma)	pRSVCAT, pSV2CAT, pSV2-neo; pSV2-gpt, pSV7-neogpt*	e.p.; l/sc; t.e.; s.t.; cotransfection; no chloroquine effect	520
		e.p.; l; transfection conditions	508
SKDHL2A (Burkitt's lymphoma cell line)	pRSVCAT, pSV2CAT, pSV7-neogpt*	e.p.; sc/l; t.e.; s.t.; CAT activity vs. s.t.	520
Liver			
rat			
hepatocytes	pA10CAT2, pSV2CAT	e.p.; l; t.e.	515

TABLE 2 (continued)
Electrotransfection

Tissue/Species Tumor/Species/Site	Construct	Remarks/Objective	Ref.
Lung			
human			
HSC172-fibroblasts	pRSV-IL2R	e.p.; s.t.; EP of G2/M phase	223
IMR-90	pMT-hGH-SV2 DNA	e.p.; transfection yield; EP vs. Ca²⁺-phosphate	505
WI-38-VA13 (fibroblasts, fetal lung)	pMT-hGH-SV2, pRSV-*neo*	e.p.; sc/l/m; yield; EP vs. Ca²⁺-phosphate	505
	pSV2-*neo*	e.p.; l; conditions	506
Lymphocytes			
mouse			
CTLL (cytotoxic T-cell line)	pSV7-*neogpt*˙	e.p.; l; transfection conditions	508
Myelomas & Plasmacytomas			
human			
BD7-S17 (T cell hybridoma)	CD8 gene construct (pLyt2L HMR)	e.p.; l; s.t.; increased T-cell recognition	530
mouse			
1739, 2008 (hybridoma cell lines)	pAG60, pHEBO	e.p.; s.t.; improvement of MAb production	531
FDC-P2 (myeloid cell line)	c-H-*ras* gene/BMG vector, pRSV-*neo*, pHs-49, Mock	e.p.; cotransfection; Il-3 independence	532
J558L (var. of mouse myeloma 1558)	pZEM214	e.p.; t.e.; s.t.; pulse equipment	432
	pLW4	e.p.; t.e.; conditions	533
	pSV2CAT, pSV7-*neogpt*˙	e.p.; sc/l; t.e.; s.t.; CAT activity vs. s.t.	520

F558L (var. of mouse myeloma 1558)	pRSV-L	e.p.; s.t.; conditions	509
MPC11 (myeloma)	pLW4, pTKC-Mo9	e.p.; t.e.; field conditions	533
P3X63-Ag8 sublines	pSV2CAT, pSV7-*neogpt**, pLW4	e.p.; sc/l; t.e.; s.t.; field conditions	520, 533
S194 (myeloma)	pRSV-L	e.p.; t.e.; s.t.	509
SP2/0-Ag14 (myeloma)	pRSSCAT, pSV2CAT pSV7-*neogpt**	e.p.; sc/l; t.e.; s.t.; CAT activity vs. s.t.	520
	pSV2-*neo*	e.p.; l; s.t.; field conditions for EP	288
tPA7-8-4, tPA12-5-3, Y22/HAT^s (hybridomas)	pSV2-*neo*	e.p.; c/l; selection for quadroma cell lines	534
Macrophages *mouse*			
J774.A1	pSV2-*neo*	e.p.; l; s.t.; hypoosmolar media: high yield	286
Nervous System *rat*			
B50	pRSV-L	e.p.; s.t.; uptake	509
	pSV2-T8	r.p.; FACS; Thy-1 glyco-protein expression	535
Nose *human*			
epithelial cells	pRSVCAT	e.p.; t.e.; yield; EP vs. Ca^{2+}-phosphate	536
Ovary *hamster, Chinese*			
CHO	DF13:2/MSG	e.p.; l; DAF expression; glycolipid anchoring of CD8	526
	pAGO	e.p.; r.p.; pulsing of cell suspensions or monolayers	537
CHO (DHFR⁻)	pSV2-DHFR, pJODTPA	e.p.; l/sc; introduction of high-copy number DNA	538
CHO-K1	pSV2-*neo*	e.p.; sc/l; yield	539

TABLE 2 (continued)
Electrotransfection

Tissue/Species Tumor/Species/Site	Construct	Remarks/Objective	Ref.
CHO-K1c	pSV2-gpt△	e.p.; l; carrier DNA effect	540
CHO-WTT	pSV2CAT	r.p.; field conditions	541
Phaeochromocytoma			
rat			
PC-12	pRSV-L	e.p.; t.e.; s.t.; field conditions	509
Skin			
human			
A8 fibroblasts	pRSV-IL2R, pRSVCAT	e.p.; t.e.; FACS; synchronized cells	223
XP12RO fibroblasts (xeroderma pigmentosum)	pMT-hGH-SV2, pRSVCAT, pRSV-*neo*, pRSV-gpt pSV2-*neo*	e.p.; sc/l/m; EP vs. Ca²⁺-phosphate e.p.; l; yield	505 506
mouse			
MK-1 (epidermal cells)	pSV2-CAT, pSV2-*neo*	r.p.; l; t.e.; s.t.; transfection yield	542
skin cells (newborn mice)	pSV3-*neo*, pHEB4	e.p.; sc; EP of a pleat of skin	543
rat			
fibroblasts	pLZRNL, pRC/RSV/β-gal	e.p.; l; yield; EP vs. other transfection methods	513

ᵃ listed alphabetically.

ᵇ listed according to the Tissue Index and Tumor Index of the American Type Culture Collection catalog 1988.

Note: Abbreviations: c = circular, ccc = circular covalently closed, ds = double stranded, e.p. = exponentially decaying pulse, EP = electropermeabilization, l = linear, m = mock, MAb = monoclonal antibody, PPL = protoplasts, r.f.p. = radio frequency (oscillating) pulse, r.p. = rectangular pulse, s.t. = stable transformants, sc = supercoiled, ss = single stranded, t.e. = transient expression.

VIII. POTENTIALS OF ELECTROPERMEABILIZATION IN DIAGNOSIS AND THERAPY

Conventional forms of drug dosing for human diagnostic and therapeutic applications (particularly those involving newer biotechnological products) are limited by several factors including:

1. short circulation time due to rapid clearance from the blood circulation,
2. rapid biodegradation,
3. the inability to direct drugs to specific tissue and cellular sites.

High concentrations of drugs together with high frequencies of drug administration are often required to maintain levels in the therapeutic range. This may result in vastly increased costs, detrimental side effects, and the necessity for invasive drug delivery systems, especially for pharmacological agents with short biological half-lives.

These problems have motivated research and development of drug delivery systems that meet the following criteria:

1. continuous control of spatial and temporal delivery patterns,
2. extended operational delivery times.

The superiority of controlled drug release systems over conventional dosage systems lies in safety, i.e., less severe side effects and obviation of continuous intravenous therapy, therapeutic effectiveness, and reduced costs. In addition, the controlled spatial and temporal administration of a specific drug achieved by drug delivery systems is expected to bring about a pronounced alteration in the pharmaceutical spectrum. The profusion of newer biotechnological reagents including interferons, prostaglandins, interleukins, and monoclonal antibodies provides additional impetus for this research.

Much of the research and development in the field of therapeutic drug delivery systems has been directed toward the encapsulation of drugs and exogeneous enzymes in synthetic membrane capsules and liposomes.[545] The permeability of the membrane envelope can both protect the drug from the external environment and modulate the spatial and temporal release of the drug into the body's fluid and tissue. Immunological reactions, lack of selectivity, and the life span of these carrier systems are only some of the problems which still have to be solved. Electromanipulation offers several advantages for the solution of these problems.

The problem of biocompatibility can be easily circumvented if autologous erythrocytes, lymphocytes, neutrophils, or platelets are used as drug delivery systems.[14,41,42,45,46,53,222,227,253,339,340,392] Clinically relevant amounts of (almost membrane-impermeable) drugs and exogeneous enzymes can be entrapped inside these cells by the electropermeabilization technique. Animal studies have shown that electrically loaded red blood cells and platelets showed essentially normal life-spans when re-infused into the blood circulation.[186,340] The life-span of the loaded red cells depended on the field strength and duration time of the applied external field pulse, i.e., on the reversibility of the electric field-induced perturbations within the membrane structure.

Target-specificity of the erythrocyte and the lymphocyte carrier system can be achieved by entrapment of small magnetic particles simultaneously with the drug within the cells to allow direction of the cells to their target with an external magnetic field.[339,340] The recent progress in electric field-mediated immortalization of a small input number of human lymphocytes (see Chapter 4) opens further possibilities for the use of highly specific membrane-bound monoclonal antibodies as "specific markers" to guide the electro-cell carrier systems to the target. Due to their spontaneous accumulation in inflamed or infected areas electroloaded blood phagocytes are *a priori* potential drug carriers with specific targeting of inflammation areas.[227]

Most lipid-soluble drugs can permeate intact cell membranes, albeit slowly. Continuous drug release from electromanipulated cell carrier systems after accumulation at the target site should not, therefore, present a severe problem. Furthermore, the release kinetics can be altered, by coupling the drug to other (impermeable) chemical groups (e.g., to sugar moieties[186]) as required.

Electroimplantation or electroinsertion of xenoproteins (antigens, receptors, etc.) into the red blood cell membrane is another promising way[188] to use living cells as long life-span carriers in the blood circulation for various applications (e.g., study of functional activity of the brain in combination with nuclear magnetic resonance imaging). Electroimplantation of the CD4 receptor of the helper lymphocytes and of xeno-glycophorin in red blood cell membranes has been successfully performed by application of a train of 4 pulses of 2.1 kV cm^{-1} strength and 1 μs duration at 37°C.[188,197,198] This field strength is high enough to exceed the breakdown voltage of about 0.5 V at this temperature (see above). The pulse duration time is also not lethal under these conditions because of the very rapid resealing process. The fast resealing of the membrane seems to be responsible for the localization and incorporation of the xeno-proteins[21] within the membrane.

Electrochemotherapy (i.e. electropermeabilization of tissue and/or cells) is an alternative approach to the use of conventional electropermeabilization technique for medical applications, specifically for tumor treatment.[546-549] With this technique, a train of electric field pulses (which are sufficiently strong enough to permeabilize the membranes) are applied to the tumor tissue via inserted electrodes. Cytotoxic, membrane-impermeable drugs (such as bleomycin) are administered before the field treatment. Electropermeabilization of the tumor cells facilitates the uptake of the cytotoxic substances, whereas the non-malignant cells in the surrounding area are not subjected to the field pulses, and are thus less affected by the drug. Bleomycin treatment of transplanted and spontaneous mammary tumors in mice and of patients with cutaneous metastases of head and neck squamous cell carcinomas has resulted in successful regression of primary and metastatic tumor growth, respectively. The potentiation of the anti-tumor efficiency of bleomycin due to the pretreatment with electric breakdown pulses simultaneously resulted in a reduction of the total administered dosage.

Electrically mediated delivery systems and electrochemotherapy have great promise for the therapy of tumors. These techniques also provide new possibilities for the study of the pathways, the reactions, and the biodegradation of pharmaceutical drugs in the body.

IX. CONCLUSIONS

Discovery of electric reversible membrane breakdown led to the development of the powerful cell electromanipulation techniques. The time interval from the early pioneering work in this field to the present is more than two decades. During the last decade the field has grown explosively as electrophysiologists, biophysicists, biologists, biotechnologists, and physicians have adapted the techniques for a bewildering array of applications. Many of these techniques have taken their place as the "methods of choice." Such standard applications now include electrofusion for hybridoma production particularly in situations where cell numbers are limiting (see Chapter 4), as well as electrotransfection of both prokaryotic (see Chapter 2) and eukaryotic cells (as described in this chapter).

Despite intensive research in the field, however, the potential of these techniques remains enormous. The "surface has barely been skimmed" — with a considerable number of exciting new technologies now on the horizon. Therapy of neoplasms, drug delivery by autologous "electroloaded" cells, and single cell selection for transformation are only a few of the more promising applications now moving to the fore. One of these worthy of mention may be the use of electrically transformed autologous cells for directed magnetic resonance imaging of blood flow in the brain. Many others are currently under development as well.

We have seen that a thorough understanding of the underlying biophysical principles is a prerequisite to the best use of this technology. It is much more difficult to troubleshoot problems that arise in such applications if one cannot fall back upon a solid foundation of first principles. Deviation from basic principles in favor of an empirical approach has led to the introduction of several faulty concepts and to a blockage of scientific progress. An example of this is the term "electrocompetence." The field effects underlying electric breakdown are universal and therefore all cells are "electrocompetent." However, the appropriate conditions required to perform specific electromanipulations may be substantially different in various cell types as outlined above. This is particularly true for walled cells such as bacteria.

As noted above, one might "get away" with such an approach if the number of input cells is not limiting. This may not be possible with the critical new applications described above. Additional progress will require continued rigorous investigation and elucidation of the mechanisms underlying the observed phenomena using information like that presented in Chapters 3–6. The reader who wishes to adapt these techniques to his own research is advised to bear in mind the principles outlined here and elsewhere in this book. They will prove most helpful in achieving the desired end as well as in obtaining optimal results.

REFERENCES

1. **Kraus, K.,** Bakterien im elektrischen Kraftfeld, *Zentralbl. Bakteriol. Parasit. Infekt. Hyg.,* 124, 64, 1932.
2. **Muth, E.,** Über die Erscheinung der Perlschnurkettenbildung von Emulsionspartikelchen unter Einwirkung eines Wechselfeldes, *Kolloid Z.,* 41, 97, 1927.
3. **Liebesny, P.,** Athermic short wave therapy, *Arch. Phys. Ther.,* 19, 736, 1938.
4. **Krasny-Ergen, W.,** Nicht-thermische Wirkungen elektrischer Schwingungen auf Kolloide, *Hochfrequenztech. Elektroak.,* 40, 126, 1936.
5. **Zimmermann, U., Schulz, J. and Pilwat, G.,** Transcellular ion flow in *E. coli* B and electrical sizing of bacteria, *Biophys. J.,* 13, 1005, 1973.
6. **Zimmermann, U., Pilwat, G. and Riemann, F.,** Dielectric breakdown of cell membranes, *Biophys. J.,* 14, 881, 1974.
7. **Zimmermann, U., Pilwat, G. and Riemann, F.,** Reversible dielectric breakdown of cell membranes by electrostatic fields, *Z. Naturforsch.,* 29c, 304, 1974.
8. **Zimmermann, U., Büchner, K.-H. and Arnold, W. M.,** Electrofusion of cells: recent developments and relevance for evolution, in *Charge and Field Effects in Biosystems,* Allen, M. J. and Usherwood, P. N. R., Eds., Abacus Press, Turnbridge Wells, 1984, 293.
9. **Steponkus, P. L., Stout, D. G., Wolfe, J. and Lovelace, R. V. E.,** Possible role of transient electric fields in freezing-induced membrane destabilization, *J. Membrane Biol.,* 85, 191, 1985.
10. **Zimmermann, U. and Küppers, G.,** Cell fusion by electromagnetic waves and its possible relevance for evolution, *Naturwissenschaften,* 70, 568, 1983.
11. **Küppers, G. and Zimmermann, U.,** Cell fusion by spark discharge and its relevance for evolutionary processes, *FEBS Lett.,* 164, 323, 1983.
12. **Schwartz, R. M. and Dayhoff, M. O.,** Origins of procaryotes, eukaryotes, mitochondria and chloroplasts, *Science,* 199, 395, 1978.
13. **Czihak, G., Langer, H. and Ziegler, H.,** *Biologie,* 3rd edition, Springer-Verlag, Berlin, 1984, 135.
14. **Vienken, J., Jeltsch, E. and Zimmermann, U.,** Penetration and entrapment of large particles in erythrocytes by electrical breakdown techniques, *Cytobiologie,* 17, 182, 1978.
15. **Zimmermann, U., Schnettler, R., Klöck, G., Watzka, H., Donath, E. and Glaser, R. W.,** Mechanisms of electrostimulated uptake of macromolecules into living cells, *Naturwissenschaften,* 77, 543, 1990.
16. **Zimmermann, U., Küppers, G. and Salhani, N.,** Electric field-induced release of chloroplasts from plant protoplasts, *Naturwissenschaften,* 69, 451, 1982.
17. **Zimmermann, U.,** Electrofusion of cells: principles and industrial potential, *Trends Biotechnol.,* 1, 149, 1983.
18. **Zimmermann, U. and Stopper, H.,** Elektrofusion und Elektropermeabilisierung von Zellen, *Physik in unserer Zeit,* 18, 169, 1987.

19. **Zimmermann, U., Pilwat, G., Beckers, F. and Riemann, F.,** Effects of external electrical fields on cell membranes, *Bioelectrochem. Bioenerg.*, 3, 58, 1976.
20. **Zimmermann, U., Scheurich, P., Pilwat, G. and Benz, R.,** Cells with manipulated functions: new perspectives for cell biology, medicine and technology, *Angewandte Chemie*, Int. Ed. Engl., 20, 325, 1981.
21. **Zimmermann, U.,** Electric field-mediated fusion and related electrical phenomena, *Biochim. Biophys. Acta*, 694, 227, 1982.
22. **Zimmermann, U. and Vienken, J.,** Electric field-induced cell-to-cell fusion, *J. Membrane Biol.*, 67, 165, 1982.
23. **Zimmermann, U., Vienken, J. and Pilwat, G.,** Electrofusion of cells, in *Investigative Microtechniques in Medicine and Biology*, Vol. 1, Chayen, J. and Bitensky, L., Eds., Marcel Dekker, New York, 1984, 89.
24. **Zimmermann, U., Vienken, J., Halfmann, J. and Emeis, C. C.,** Electrofusion: a novel hybridization technique, in *Advances in Biotechnological Processes*, Vol. 4, Mizrahi, A. and van Wenzel, A., Eds., Alan R. Liss, New York, 1985, 79.
25. **Zimmermann, U.,** Electrical breakdown, electropermeabilization and electrofusion, *Rev. Physiol. Biochem. Pharmacol.*, 105, 175, 1986.
26. **Zimmermann, U. and Urnovitz, H. B.,** Principles of electrofusion and electropermeabilization, *Meth. Enzymol.*, 151, 194, 1987.
27. **Zimmermann, U., Gessner, P., Wander, M. and Foung, S. K. H.,** Electroinjection and electrofusion in hypo-osmolar solution, in *Electromanipulation in Hybridoma Technology*, Borrebaeck, C. A. K. and Hagen, I., Eds., Stockton Press, New York, 1990, 1.
28. **Zimmermann, U.,** Electrofusion and electropermeabilization in genetic engineering, in *Membrane Fusion*, Wilschut, J. and Hoekstra, D., Eds., Marcel Dekker, New York, 1991, 665.
29. **Neil, G. A. and Zimmermann, U.,** Electrofusion, *Meth. Enzymol.*, 220, 174, 1993.
30. **Neil, G. A. and Zimmermann, U.,** Electroinjection, *Meth. Enzymol.*, 221, 339, 1993.
31. **Coster, H. G. L. and Zimmermann, U.,** Direct demonstration of dielectric breakdown in the membranes of *Valonia utricularis*, *Z. Naturforsch.*, 30c, 77, 1975.
32. **Coster, H. G. L. and Zimmermann, U.,** The mechanism of electrical breakdown in the membranes of *Valonia utricularis*, *J. Membrane Biol.*, 22, 73, 1975.
33. **Coster, H. G. L. and Zimmermann, U.,** Dielectric breakdown in the membranes of *Valonia utricularis*. The role of energy dissipation, *Biochim. Biophys. Acta*, 382, 410, 1975.
34. **Zimmermann, U. and Benz, R.,** Dependence of the electrical breakdown voltage on the charging time in *Valonia utricularis*, *J. Membrane Biol.*, 53, 33, 1980.
35. **Benz, R. and Zimmermann, U.,** High electric field effects on the cell membranes of *Halicystis parvula*, *Planta*, 152, 314, 1981.
36. **Benz, R. and Conti, F.,** Reversible electrical breakdown of squid giant axon membrane, *Biochim. Biophys. Acta*, 645, 115, 1981.
37. **Wehner, G., Friedmann, B. and Zimmermann, U.,** Biphasic voltage relaxation pattern observed in cells of *Eremosphaera viridis* after injection of charge-pulses of short duration: detection of tip clogging of intracellular microelectrodes by charge pulse technique, *Biochim. Biophys. Acta*, 1027, 105, 1990.
38. **Pilwat, G. and Zimmermann, U.,** Comments on "Erythrocyte and ghost cytoplasmic resistivity and voltage-dependent apparent size", *Biophys. J.*, 48, 671, 1985.
39. **Grover, N. B., Ben-Sasson, S. A. and Naaman, J.,** Electrical sizing of particles in suspensions. V. High electric fields, *Anal. Quant. Cytol.*, 4, 302, 1982.
40. **Zimmermann, U., Groves, M., Schnabl, H. and Pilwat, G.,** Development of a new Coulter Counter system: measurement of the volume, internal conductivity, and dielectric breakdown voltage of a single guard cell protoplast of *Vicia faba*, *J. Membrane Biol.*, 52, 37, 1980.
41. **Zimmermann, U., Pilwat, G. and Esser, B.,** The effect of encapsulation in red blood cells on the distribution of methotrexate in mice, *J. Clin. Chem. Clin. Biochem.*, 16, 135, 1978.

42. **Zimmermann, U., Riemann, F. and Pilwat, G.,** Enzyme loading of electrically homogeneous human red blood cell ghosts prepared by dielectric breakdown, *Biochim. Biophys. Acta,* 436, 460, 1976.

43. **Zimmermann, U., Pilwat, G. and Riemann, F.,** Preparation of erythrocyte ghosts by dielectric breakdown of the cell membrane, *Biochim. Biophys. Acta,* 375, 209, 1975.

44. **Saleemuddin, M., Zimmermann, U. and Schneeweiß, F.,** Preparation of human erythrocyte ghosts in isotonic solution: haemoglobin content and polypeptide composition, *Z. Naturforsch.,* 32c, 627, 1977.

45. **Zimmermann, U. and Pilwat, G.,** Dielectric breakdown of the cell membranes: a new tool in clinical diagnosis and therapy, *Biomed. Technik.,* 20, 193, 1975.

46. **Zimmermann, U. and Pilwat, G.,** Organ specific application of drugs by means of cellular capsule systems, *Z. Naturforsch.,* 31c, 732, 1976.

47. **Steinbiss, H.-H.,** The dielectric breakdown in the plasmalemma of *Valerianella locusta* protoplasts, *Z. Pflanzenphysiol.,* 88, 95, 1978.

48. **Gauger, B. and Bentrup, F. W.,** A study of dielectric membrane breakdown in the *Fucus* egg, *J. Membrane Biol.,* 48, 249, 1979.

49. **Baker, P. F., Knight, D. E. and Whitaker, M. J.,** The relation between ionized calcium and cortical granule exocytosis in eggs of the sea urchin *Echinus esculentus, Proc. R. Soc. London,* B 207, 149, 1980.

50. **Baker, P. F. and Knight, D. E.,** Calcium control of exocytosis and endocytosis in bovine adrenal medullary cells, *Phil. Trans. R. Soc. London,* B 296, 83, 1981.

51. **Gordon, P. B. and Seglen, P. O.,** Autophagic sequestration of [^{14}C]sucrose introduced into rat hepatocytes by reversible electro-permeabilization, *Exp. Cell Res.,* 142, 1, 1982.

52. **Serpersu, E. H., Kinosita, K. Jr. and Tsong, T. Y.,** Reversible and irreversible modification of erythrocyte membrane permeability by electric field, *Biochim. Biophys. Acta,* 812, 779, 1985.

53. **Hughes, K. and Crawford, N.,** Reversible electropermeabilization of human and rat blood platelets: evaluation of morphological and functional integrity *"in vitro"* and *"in vivo"*, *Biochim. Biophys. Acta,* 981, 277, 1989.

54. **Riemann, F., Zimmermann, U. and Pilwat, G.,** Release and uptake of haemoglobin and ions in red blood cells induced by dielectric breakdown, *Biochim. Biophys. Acta,* 394, 449, 1975.

55. **Zimmermann, U., Pilwat, G. Holzapfel, C. and Rosenheck, K.,** Electrical hemolysis of human and bovine red blood cells, *J. Membrane Biol.,* 30, 135, 1976.

56. **Auer, D., Brandner, G. and Bodemer, W.,** Dielectric breakdown of the red blood cell membrane and uptake of SV 40 DNA and mammalian cell RNA, *Naturwissenschaften,* 63, 391, 1976.

57. **Kinosita, K. Jr. and Tsong, T. Y.,** Formation and resealing of pores of controlled sizes in human erythrocyte membranes, *Nature,* 268, 438, 1977.

58. **Kinosita, K. Jr. and Tsong, T.Y.,** Voltage-induced pore formation and hemolysis of human erythrocytes, *Biochim. Biophys. Acta,* 471, 227, 1977.

59. **Pace, C. S., Tarvin, J. T., Neighbors, A. S., Pirkle, J. A. and Greider, M. H.,** Use of a high voltage technique to determine the molecular requirements for exocytosis in islet cells, *Diabetes,* 29, 911, 1980.

60. **Mehrle, W., Zimmermann, U. and Hampp, R.,** Evidence for asymmetrical uptake of fluorescent dyes through electro-permeabilized membranes of *Avena* mesophyll protoplasts, *FEBS Lett.,* 185, 89, 1985.

61. **Mishra, K. P., Joshua, D. C. and Bhatia, C. R.,** *In vitro* electroporation of tobacco pollen, *Plant Sci.,* 52, 135, 1987.

62. **Liang, H., Purucker, W. J., Stenger, D. A., Kubiniec, R. T. and Hui, S. W.,** Uptake of fluorescence-labeled dextrans by 10T 1/2 fibroblasts following permeation by rectangular and exponential-decay electric field pulses, *BioTechniques,* 6, 550, 1988.

63. **Tekle, E., Astumian, R. D. and Chock, P. B.,** Electro-permeabilization of cell membranes: effect of the resting membrane potential, *Biochem. Biophys. Res. Commun.*, 172, 282, 1990.
64. **Rols, M.-P. and Teissié, J.,** Electropermeabilization of mammalian cells. Quantitative analysis of the phenomenon, *Biophys. J.*, 58, 1089, 1990.
65. **Rols, M.-P., Dahhou, F., Mishra, K. P. and Teissié, J.,** Control of electric field induced cell membrane permeabilization by membrane order, *Biochemistry*, 29, 2960, 1990.
66. **Muraji, M., Tatebe, W., Konishi, T. and Fujii, T.,** Effect of electrical energy on the electropermeabilization of yeast cells, *Bioelectrochem. Bioenerg.*, 31, 77, 1993.
67. **Sukhorukov, V. L., Djuzenova, C. S., Arnold, W. M. and Zimmermann, U.,** DNA, protein, and plasma-membrane incorporation by arrested mammalian cells, *J. Membrane Biol.*, 142, 77, 1994.
68. **Sukhorukov, V. L., Djuzenova, C. S., Frank, H., Arnold, W. M. and Zimmermann, U.,** Electropermeabilization and fluorescent tracer exchange: the role of whole cell capacitance, *Cytometry*, 21, 230, 1995.
69. **Gratzl, M.,** Permeabilized cells. An approach to the study of exocytosis, in *Membrane Fusion*, Wilschut, J. and Hoekstra, D., Eds., Marcel Dekker, New York, 1991, 553.
70. **Mehrle, W., Hampp, R. and Zimmermann, U.,** Electric pulse induced membrane permeabilization. Spatial orientation and kinetics of solute efflux in freely suspended and dielectrophoretically aligned plant mesophyll protoplasts, *Biochim. Biophys. Acta*, 978, 267, 1989.
71. **Zimmermann, U., Arnold, W. M. and Mehrle, W.,** Biophysics of electroinjection and electrofusion, *J. Electrostatics*, 21, 309, 1988.
72. **Gross, D., Loew, L. M. and Webb, W. W.,** Optical imaging of cell membrane potential changes induced by applied electric fields, *Biophys. J.*, 50, 339, 1986.
73. **Ehrenberg, B., Farkas, D. L., Fluhler, E. N., Lojewska, Z. and Loew, L. M.,** Membrane potential induced by external electric field pulses can be followed with a potentiometric dye, *Biophys. J.*, 51, 833, 1987.
74. **Kinosita, K. Jr., Ashikawa, I., Saita, N., Yoshimura, H., Itoh, H., Nagayama, K. and Ikegami, A.,** Electroporation of cell membrane visualized under a pulsed-laser fluorescence microscope, *Biophys. J.*, 53, 1015, 1988.
75. **Kinosita, K. Jr., Ashikawa, J., Hibino, M., Shigemori, M., Yoshimura, H., Itoh, H., Nagayama, K. and Ikegami, A.,** Submicrosecond imaging under a pulsed-laser fluorescence microscope, *Proc. SPIE, Time-Resolved Laser Spectroscopy in Biochemistry*, 909, 271, 1988.
76. **Itoh, H., Hibino, M., Shigemori, M., Koishi, M., Takahashi, A., Hayakawa, T. and Kinosita, K. Jr.,** Multi-shot pulsed laser fluorescence microscope system, *Proc. SPIE, Time-Resolved Laser Spectroscopy in Biochemistry II*, 1204, 49, 1990.
77. **Kinosita, K. Jr., Itoh, H., Ishiwata, S., Hirano, K., Nishizaka, T. and Hayakawa, T.,** Dual-view microscopy with a single camera: real-time imaging of molecular orientations and calcium, *J. Cell Biol.*, 115, 67, 1991.
78. **Hibino, M., Shigemori, M., Itoh, H., Nagayama, K. and Kinosita, K. Jr.,** Membrane conductance of an electroporated cell analyzed by submicrosecond imaging of transmembrane potential, *Biophys. J.*, 59, 209, 1991.
79. **Kinosita, K., Hibino, M., Itoh, H., Shigemori, M., Hirano, K., Kirino, Y. and Hayakawa, T.,** Events of membrane electroporation visualized on a time scale from microsecond to seconds, in *Guide to Electroporation and Electrofusion*, Chang, D. C., Chassy, B. M., Saunders, J. A. and Sowers, A. E., Eds., Academic Press, San Diego, 1992, 29.
80. **Hibino, M., Itoh, H. and Kinosita, K. Jr.,** Time courses of cell electroporation as revealed by submicrosecond imaging of transmembrane potential, *Biophys. J.*, 64, 1789, 1993.
81. **O'Neill, R. J. and Tung, L.,** Cell-attached patch clamp study of the electropermeabilization of amphibian cardiac-cells, *Biophys. J.*, 59, 1028, 1991.
82. **Chen, W. and Lee, R. C.,** An improved double vaseline gap voltage clamp to study electroporated skeletal muscle fibers, *Biophys. J.*, 66, 700, 1994.

83. **Ghosh, P. M., Keese, C. R. and Giaever, I.,** Monitoring electropermeabilization in the plasma membrane of adherent mammalian cells, *Biophys. J.,* 64, 1602, 1993.

84. **Pilwat, G. and Zimmermann, U.,** Determination of intracellular conductivity from electrical breakdown measurements, *Biochim. Biophys. Acta,* 820, 305, 1985.

85. **Yaseen, M. A., Pedley, K. C. and Howell, S. L.,** Regulation of insulin secretion from islets of Langerhans rendered permeable by electric discharge, *Biochem. J.,* 206, 81, 1982.

86. **Knight, D. E. and Baker, P. F.,** Calcium-dependence of catecholamine release from bovine adrenal medullary cells after exposure to intense electric fields, *J. Membrane Biol.,* 68, 107, 1982.

87. **Baker, P. F. and Knight, D. E.,** High-voltage techniques for gaining access to the interior of cells: application to the study of exocytosis and membrane turnover, *Meth. Enzymol.,* 98, 28, 1983.

88. **Hibi, T., Kano, H., Sugiura, M., Kazami, T. and Kimura, S.,** High efficiency electro-transfection of tobacco mesophyll protoplasts with tobacco mosaic virus RNA, *J. Gen. Virol.,* 67, 2037, 1986.

89. **Hibi, T., Kano, H., Sugiura, M., Kazami, T. and Kimura, S.,** High-speed electro-fusion and electro-transfection of plant protoplasts by a continuous flow electro-manipulator, *Plant Cell Rep.,* 7, 153, 1988.

90. **Melvik, J. E., Dornish, J. M. and Pettersen, E. O.,** The binding of *cis*-dichlorodiammineplatinum (II) to extracellular and intracellular compounds in relation to drug uptake and cytotoxicity *in vitro, Br. J. Cancer,* 66, 260, 1992.

91. **Schwister, K. and Deuticke, B.,** Formation and properties of aqueous leaks induced in human erythrocytes by electrical breakdown, *Biochim. Biophys. Acta,* 816, 332, 1985.

92. **Kim, D., Lee, Y. J., Rausch, C. M. and Borrelli, M. J.,** Electroporation of extraneous proteins into CHO cells: increased efficacy by utilizing centrifugal force and microsecond electrical pulses, *Exp. Cell Res.,* 197, 207, 1991.

93. **Lindsey, K. and Jones, M. G. K.,** The permeability of electroporated cells and protoplasts of sugar beet, *Planta,* 172, 346, 1987.

94. **Hauptmann, R. M., Ozias-Akins, P., Vasil, V., Tabaeizadeh, Z., Rogers, S. G., Horsch, R. B., Vasil, I. K. and Fraley, R. T.,** Transient expression of electroporated DNA in monocotyledonous and dicotyledonous species, *Plant Cell Rep.,* 6, 265, 1987.

95. **Hollon, T. and Yoshimura, F. K.,** A nonuniform electrical field electroporation chamber design, *Anal. Biochem.,* 182, 253, 1989.

96. **Kwee, S., Nielsen, H. V. and Celis, J. E.,** Electropermeabilization of human cultured cells grown in monolayers. Incorporation of monoclonal antibodies, *Bioelectrochem. Bioenerg.,* 23, 65, 1990.

97. **Bartoletti, D. C., Harrison, G. I. and Weaver, J. C.,** The number of molecules taken up by electroporated cells: quantitative determination, *FEBS Lett.,* 256, 4, 1989.

98. **Weaver, J. C., Harrison, G. I., Bliss, J. G., Mourant, J. R. and Powell, K. T.,** Electroporation: high frequency of occurrence of a transient high-permeability state in erythrocytes and intact yeast, *FEBS Lett.,* 229, 30, 1988.

99. **Bliss, J. G., Harrison, G. I., Mourant, J. R., Powell, K. T. and Weaver, J. C.,** Electroporation: the population distribution of macromolecular uptake and shape changes in red blood cells following a single 50 µs square wave pulse, *Bioelectrochem. Bioenerg.,* 20, 57, 1988.

100. **Chakrabarti, R., Wylie, D. E. and Schuster, S. M.,** Transfer of monoclonal antibodies into mammalian cells by electroporation, *J. Biol. Chem.,* 264, 15494, 1989.

101. **Puchalski, R. B. and Fahl, W. E.,** Expression of recombinant glutathione S-transferase πi, Ya, or Yb_1 confers resistance to alkylating agents, *Proc. Natl. Acad. Sci. U.S.A.,* 87, 2443, 1990.

102. **Graziadei, L., Burfeind, P. and Bar-Sagi, D.,** Introduction of unlabeled proteins into living cells by electroporation and isolation of viable protein-loaded cells using dextran-fluorescein isothiocyanate as a marker for protein uptake, *Anal. Biochem.,* 194, 198, 1991.

103. **Wilson, A. K., Horwitz, J. and de Lanerolle, P.,** Evaluation of the electroinjection method for introducing proteins into living cells, *Am. J. Physiol.,* 260, C355, 1991.

104. **Berglund, D. L. and Starkey, J. R.**, Introduction of antibody into viable cells using electroporation, *Cytometry*, 12, 64, 1991.
105. **Puchalski, R. B., Manoharan, T. H., Lathrop, A. L. and Fahl, W. E.**, Recombinant glutathione S-transferase (GST) expressing cells purified by flow cytometry on the basis of a GST-catalyzed intracellular conjugation of glutathione to monochlorobimane, *Cytometry*, 12, 651, 1991.
106. **Glogauer, M. and McCulloch, C. A. G.**, Introduction of large molecules into viable fibroblasts by electroporation: optimization of loading and identification of labeled cellular compartments, *Exp. Cell Res.*, 200, 227, 1992.
107. **Puchalski, R. B. and Fahl, W. E.**, Gene transfer by electroporation, lipofection, and DEAE-dextran transfection: compatibility with cell-sorting by flow cytometry, *Cytometry*, 13, 23, 1992.
108. **Dinchuk, J. E., Kelley, K. A. and Callahan, G. N.**, Flow cytometric analysis of transport activity in lymphocytes electroporated with a fluorescent organic anion dye, *J. Immunol. Meth.*, 155, 257, 1992.
109. **Djuzenova, C. S., Sukhorukov, V. L., Klöck, G., Arnold, W. M. and Zimmermann, U.**, Effect of electric field pulses on the viability and on the membrane-bound immunoglobulins of LPS-activated murine B-lymphocytes: correlation with the cell cycle, *Cytometry*, 15, 35, 1994.
110. **Berglund, D. L. and Starkey, J. R.**, Isolation of viable tumor cells following introduction of labelled antibody to an intracellular oncogene product using electroporation, *J. Immunol. Meth.*, 125, 79, 1989.
111. **Bakker Schut, T. C., de Grooth, B. G. and Greve, J.**, A new principle of cell sorting by using selective electroporation in a modified flow cytometer, *Cytometry*, 11, 659, 1990.
112. **Kinosita, K. Jr. and Tsong, T.**, Survival of sucrose-loaded erythrocytes in the circulation, *Nature*, 272, 258, 1978.
113. **Pilwat, G., Hampp, R. and Zimmermann, U.**, Electrical field effects induced in membranes of developing chloroplasts, *Planta*, 147, 396, 1980.
114. **Güveli, D. E.**, Fusion of mitochondria, *J. Biochem.*, 102, 1329, 1987.
115. **Zimmermann, U., Pilwat, G., and Richter, H.-P.**, Electric-field-stimulated fusion: Increased field stability of cells induced by pronase, *Naturwissenschaften*, 68, 577, 1981.
116. **Song, L., Ahkong, Q. F., Georgescauld, D. and Lucy, J. A.**, Membrane fusion without cytoplasmic fusion (hemi-fusion) in erythrocytes that are subjected to electrical breakdown, *Biochim. Biophys. Acta*, 1065, 54, 1991.
117. **Zimmermann, U., Pilwat, G. and Riemann, F.**, Dielectric breakdown of cell membranes, in *Membrane Transport in Plants*, Zimmermann, U. and Dainty, J., Eds., Springer-Verlag, Berlin, 1974, 146.
118. **Pilwat, G., Zimmermann, U. and Riemann, F.**, Dielectric breakdown measurements of human and bovine erythrocyte membranes using benzyl alcohol as a probe molecule, *Biochim. Biophys. Acta*, 406, 424, 1975.
119. **Zhelev, D. V., Dimitrov, D. S. and Doinov, P.**, Correlation between physical parameters in electrofusion and electroporation of protoplasts, *Bioelectrochem. Bioenerg.*, 20, 155, 1988.
120. **Péqueux, A., Gilles, R., Pilwat, G. and Zimmermann, U.**, Pressure-induced variations of K$^+$-permeability as related to a possible reversible electrical breakdown in human erythrocytes, *Experientia*, 36, 565, 1980.
121. **Zimmermann, U., Pilwat, G., Péqueux, A. and Gilles, R.**, Electro-mechanical properties of human erythrocyte membranes: the pressure-dependence of potassium permeability, *J. Membrane Biol.*, 54, 103, 1980.
122. **Zimmermann, U.**, Pressure mediated osmoregulatory processes and pressure sensing mechanism, in *Animals and Environmental Fitness*, Gilles, R., Ed., Pergamon Press, Oxford, 1980, 441.
123. **Coster, H. G. L. and Zimmermann, U.**, Transduction of turgor pressure by cell membrane compression, *Z. Naturforsch.*, 31c, 461, 1976.

124. **Zimmermann, U., Beckers, F. and Coster, H. G. L.,** The effect of pressure on the electrical breakdown in the membranes of *Valonia utricularis, Biochim. Biophys. Acta,* 464, 399, 1977.

125. **Coster, H. G. L., Steudle, E. and Zimmermann, U.,** Turgor pressure sensing in plant cell membranes, *Plant Physiol.,* 58, 636, 1977.

126. **Zhelev, D. V., Dimitrov, D. S. and Tsoneva, I.,** Electrical breakdown of protoplast membranes under different osmotic pressures, *Bioelectrochem. Bioenerg.,* 19, 217, 1988.

127. **Rols, M.-P. and Teissié, J.,** Modulation of electrically induced permeabilization and fusion of Chinese hamster ovary cells by osmotic pressure, *Biochemistry,* 29, 4561, 1990.

128. **Zimmermann, U.,** Physics of turgor- and osmoregulation, *Ann. Rev. Plant Physiol.,* 29, 121, 1978.

129. **Sale, A. J. H. and Hamilton, W. A.,** Effects of high electric fields on micro-organisms. III. Lysis of erythrocytes and protoplasts, *Biochim. Biophys. Acta,* 163, 37, 1968.

130. **Turnbull, R. J.,** An alternate explanation for the permeability changes induced by electrical impulses in vesicular membranes, *J. Membrane Biol.,* 14, 193, 1973.

131. **Neumann, E. and Rosenheck, K.,** Potential difference across vesicular membranes, *J. Membrane Biol.,* 14, 194, 1973.

132. **Jeltsch, E. and Zimmermann, U.,** Particles in a homogeneous electrical field: a model for the electrical breakdown of living cells in a coulter counter, *Bioelectrochem. Bioenerg.,* 6, 349, 1979.

133. **Grover, N. B., Naaman, J., Ben-Sasson, S. and Doljanski, F.,** Electrical sizing of particles in suspensions, *Biophys. J.,* 9, 1398, 1969.

134. **Wang, J., Zimmermann, U. and Benz, R.,** Contribution of electrogenic ion transport to impedance of the algae *Valonia utricularis* and artificial membranes, *Biophys. J.,* 67, 1582, 1994.

135. **Zimmermann, U.,** Electrofusion of cells, in *Methods of Hybridoma Formation,* Bartal, A. H. and Hirshaut, Y., Eds., Humana Press, Clifton, 1987, 97.

136. **Zimmermann, U., Klöck, G., Gessner, P., Sammons, D. W. and Neil, G. A.,** Microscale production of hybridomas by hypo-osmolar electrofusion, *Hum. Antibod. Hybridomas,* 3, 14, 1992.

137. **Grosse, C. and Schwan, H. P.,** Cellular membrane potentials induced by alternating fields, *Biophys. J.,* 63, 1632, 1992.

138. **Teissié, J. and Rols, M.-P.,** An experimental evaluation of the critical potential difference inducing cell membrane electropermeabilization, *Biophys. J.,* 65, 409, 1993.

139. **Sukhorukov, V. L., Arnold, W. M. and Zimmermann, U.,** Hypotonically induced changes in the plasma membrane of cultured mammalian cells, *J. Membrane Biol.,* 132, 27, 1993.

140. **Neumann, E. and Rosenheck, K.,** Permeability changes induced by electric impulses in vesicular membranes, *J. Membrane Biol.,* 10, 279, 1972.

141. **Rosenheck, K., Lindner, P. and Pecht, I.,** Effect of electric fields on light-scattering and fluorescence of chromaffin granules, *J. Membrane Biol.,* 20, 1, 1975.

142. **Lindner, P., Neumann, E. and Rosenheck, K.,** Kinetics of permeability changes induced by electric impulses in chromaffin granules, *J. Membrane Biol.,* 32, 231, 1977.

143. **Mehrle, W., Naton, B. and Hampp, R.,** Determination of physical membrane properties of plant cell protoplasts via the electrofusion technique: prediction of optimal fusion yields and protoplast viability, *Plant Cell Rep.,* 8, 687, 1990.

144. **Broda, H. G., Schnettler, R. and Zimmermann, U.,** Parameters controlling yeast hybrid yield in electrofusion: the relevance of pre-incubation and the skewness of the size distributions of both fusion partners, *Biochim. Biophys. Acta,* 899, 25, 1987.

145. **Kowalski, M., Hannig, K., Klöck, G., Gessner, P., Zimmermann, U., Neil, G. A. and Sammons, D. W.,** Electrofused mammalian cells analyzed by free-flow electrophoresis, *BioTechniques,* 9, 332, 1990.

146. **Farkas, D. L., Korenstein, R. and Malkin, S.,** Electroselection in the photosynthetic membrane: polarized luminescence induced by an external electric field, *FEBS Lett.*, 120, 236, 1980.

147. **Rossignol, D. P., Decker, G. L., Lennarz, W. J., Tsong, T. Y. and Teissié, J.,** Induction of calcium-dependent, localized cortical granule breakdown in sea-urchin eggs by voltage pulsation, *Biochim. Biophys. Acta*, 763, 346, 1983.

148. **Zimmermann, U. and Stopper, H.,** Electrofusion and electropermeabilization of cells, in *Biomembrane and Receptor Mechanisms*, Vol. 7, Bertoli, E., Chapman, D., Cambria, A. and Scapagnini, U., Eds., Fidia Research Series, Liviana Press, Padova, and Springer-Verlag, Berlin, 1987, 371.

149. **Tekle, E., Astumian, R. D. and Chock, P. B.,** Electroporation by using bipolar oscillating electric field: An improved method for DNA transfection of NIH 3T3 cells, *Proc. Natl. Acad. Sci. U.S.A.*, 88, 4230, 1991.

150. **Sowers, A. E. and Lieber, M. R.,** Electropore diameters, lifetime, numbers and locations in individual erythrocyte ghosts, *FEBS Lett.*, 205, 179, 1986.

151. **Sukhorukov, V. L., Djuzenova, C. S., Frank, H. and Zimmermann, U.,** unpublished results.

152. **Benz, R., Beckers, F. and Zimmermann, U.,** Reversible electrical breakdown of lipid bilayer membranes: A charge-pulse relaxation study, *J. Membrane Biol.*, 48, 181, 1979.

153. **Glaser, A. and Donath, E.,** Osmotically-induced vesicle-membrane fusion in plant protoplasts, *Studia biophysica*, 127, 129, 1988.

154. **Saulis, G.,** Cell electroporation. Part 3. Theoretical investigation of the appearance of asymmetric distribution of pores on the cell and their further evolution, *Bioelectrochem. Bioenerg.*, 32, 249, 1993.

155. **Saulis, G. and Venslauskas, M. S.,** Cell electroporation. Part 1. Theoretical simulation of the process of pore formation in a cell, *Bioelectrochem. Bioenerg.*, 32, 221, 1993.

156. **Saulis, G. and Venslauskas, M. S.,** Cell electroporation. Part 2. Experimental measurements of the kinetics of pore formation in human erythrocytes, *Bioelectrochem. Bioenerg.*, 32, 237, 1993.

157. **Gross, D.,** Electromobile surface charge alters membrane potential induced by applied electric fields, *Biophys. J.*, 54, 879, 1988.

158. **Sowers, A. E.,** The mechanism of electroporation and electrofusion in erythrocyte membranes, in *Electroporation and Electrofusion in Cell Biology*, Neumann, E., Sowers, A. E. and Jordan, C. A., Eds., Plenum Publishing, New York, 1989, 229.

159. **Dimitrov, D. S. and Sowers, A. E.,** Membrane electroporation — fast molecular exchange by electroosmosis, *Biochim. Biophys. Acta*, 1022, 381, 1990.

160. **Sowers, A. E.,** Mechanisms of electroporation and electrofusion, in *Guide to Electroporation and Electrofusion*, Chang, D. C., Chassy, B. M., Saunders, J. A. and Sowers, A. E., Eds., Academic Press, San Diego, 1992, 119.

161. **Coster, H. G. L.,** A quantitative analysis of the voltage current relationship of fixed charge membranes and the associated property of "punch through", *Biophys. J.*, 5, 669, 1965.

162. **Coster, H. G. L.,** The role of pH in the punch through effect in the electrical characteristics of *Chara australis*, *Aust. J. Biol. Sci.*, 22, 365, 1976.

163. **Findlay, G. P. and Hope, A. B.,** Electrical properties of plant cells: methods and findings, in *Encyclopedia of Plant Physiology, New series. Transport in Plants II, Part A, Cells*, Vol. 2, Lüttge, U. and Pitman, M. G., Eds., Springer-Verlag, Berlin, 1976, 53.

164. **Barry, P. H. and Hope, A. B.,** Electroosmosis in membranes: effects of unstirred layers and transport numbers. I. Theory, *Biophys. J.*, 9, 700, 1969.

165. **Barry, P. H. and Hope, A. B.,** Electroosmosis in membranes: effects of unstirred layers and transport numbers. II. Experimental, *Biophys. J.*, 9, 729, 1969.

166. **Gyenes, M.,** Reversible electrical breakdown in *Valonia ventricosa* regarded as an action potential, *Biochem. Physiol. Pflanzen*, 180, 689, 1985.

167. **Shimmen, T. and Tazawa, M.,** Permeabilization of *Nitella* internodal cell with electrical pulses, *Protoplasma*, 117, 93, 1983.

168. **Joersbo, M. and Brunstedt, J.,** Electroporation: mechanism and transient expression, stable transformation and biological effects in plant protoplasts, *Physiol. Plant.*, 81, 256, 1991.

169. **Marszalek, P., Liu, D.-S. and Tsong, T. Y.,** Schwan equation and transmembrane potential induced by alternating electric field, *Biophys. J.*, 58, 1053, 1990.

170. **Sale, A. J. H. and Hamilton, W. A.,** Effects of high electric fields on microorganisms. I. Killing of bacteria and yeasts, *Biochim. Biophys. Acta*, 148, 781, 1967.

171. **Hamilton, W. A., and Sale, A. J. H.,** Effects of high electric fields on microorganisms. II. Mechanism of action of the lethal effect, *Biochim. Biophys. Acta*, 148, 789, 1967.

172. **Hülsheger, H., Potel, J. and Niemann, E.-G.,** Killing of bacteria with electric pulses of high field strength, *Radiat. Environ. Biophys.*, 20, 53, 1981.

173. **Hülsheger, H. and Niemann, E.-G.,** Lethal effects of high-voltage pulses on *E. coli* K12, *Radiat. Environ. Biophys.*, 18, 281, 1980.

174. **Zhuk, Y. G.,** Disinfection of water with pulsed electric discharges (Russian), *Zhurnal Microbiologii, Epidemiologii i immunobiologii*, 48, 99, 1971.

175. **Zhuk, Y. G.,** Disinfection of water in northern regions by pulsed electric discharges (Russian), *Gigiena i Sanitariya*, 38, 8, 1973.

176. **Pavlovich, S. A., Osipov, G. P., Tofilo, P. P. and Voronkina, M. I.,** Influence of pulsed electric discharges on microorganisms in water (Russian), *Gigiena i Sanitariya*, 40, 110, 1975.

177. **Zhuk, Y. G.,** Disinfecting properties of pulsed electric discharges acting on a two-component bacterial mixture (Russian), *Zhurnal Microbiologii, Epidemiologii i immunobiologii*, 7, 150, 1977.

178. **Sakurauchi, Y. and Kondo, E.,** Bactericidal effect of high electric fields on microorganisms (Japanese), *Nippon Nogeikagaku Kaishi*, 54, 837, 1980.

179. **Doevenspeck, M.,** Influencing cells and cell walls by electrostatic impulses, *Fleischwirtschaft*, 12, 986, 1961.

180. **Nebola, M.,** Cell proliferation and chromosomal characteristics following electromanipulation, *Studia biophysica*, 137, 183, 1990.

181. **Danhash, N., Gardner, D. C. J. and Oliver, S. G.,** Heritable damage to yeast caused by transformation, *Bio/Technology*, 9, 179, 1991.

182. **Sabelnikov, A. G. and Cymbalyuk, E. S.,** Clotting of donor DNA during electroporation of prokaryotes, *Bioelectrochem. Bioenerg.*, 24, 313, 1990.

183. **Tamiya, E., Nakajima, Y., Kamioka, H., Suzuki, M. and Karube, I.,** DNA cleavage based on high voltage electric pulse, *FEBS Lett.*, 234, 357, 1988.

184. **Prausnitz, M. R., Lau, B. S., Milano, C. D., Conner, S., Langer, R. and Weaver, J. C.,** A quantitative study of electroporation showing a plateau in net molecular transport, *Biophys. J.*, 65, 414, 1993.

185. **Tsong, T. Y. and Kingsley, E.,** Hemolysis of human erythrocyte induced by a rapid temperature jump, *J. Biol. Chem.*, 250, 786, 1975.

186. **Tsong, T. Y.,** Voltage modulation of membrane permeability and energy utilization in cells, *Biosci. Rep.*, 3, 487, 1983.

187. **Stopper, H., Jones, H. and Zimmermann, U.,** Large scale transfection of mouse L-cells by electropermeabilization, *Biochim. Biophys. Acta*, 900, 38, 1987.

188. **Mouneimne, Y., Tosi, P.-F., Barhoumi, R. and Nicolau, C.,** Electroinsertion of full length recombinant CD4 into red blood cell membrane, *Biochim. Biophys. Acta*, 1027, 53, 1990.

189. **Chernomordik, L. V.,** Electropores in lipid bilayers and cell membranes, in *Guide to Electroporation and Electrofusion*, Chang, D. C., Chassy, B. M., Saunders, J. A. and Sowers, A. E., Eds., Academic Press, San Diego, 1992, 63.

190. **Poo, M.,** *In situ* electrophoresis of membrane components, *Ann. Rev. Biophys. Bioeng.*, 10, 245, 1981.

191. **Trevors, J. T., Chassy, B. M., Dower, W. J. and Blaschek, H. P.,** Electrotransformation of bacteria by plasmid DNA, in *Guide to Electroporation and Electrofusion*, Chang, D. C., Chassy, B. M., Saunders, J. A. and Sowers, A. E., Eds., Academic Press, San Diego, 1992, 265.

192. **Wirth, R.,** Elektroporation: eine alternative Methode zur Transformation von Bakterien mit Plasmid DNA, *Forum Mikrobiologie*, 11, 507, 1989.

193. **Lindsey, K. and Jones, M. G. K.,** Electroporation of cells, *Physiol. Plant.*, 79, 168, 1990.

194. **Büchner, K.-H. and Zimmermann, U.,** Water relations of immobilized giant algal cells, *Planta*, 154, 318, 1982.

195. **Guggino, S. and Gutknecht, J.,** Turgor regulation in *Valonia macrophysa* after acute hypoosmotic shock, in *Plant Membrane Transport: Current Conceptual Issues*, Spanswick, R. M., Lucas, W. J. and Dainty, J., Eds., Elsevier/North-Holland, Amsterdam, 1980, 495.

196. **Xie, T.-D., Sun, L. and Tsong, T. Y.,** Study of mechanisms of electric field induced DNA transfection I. DNA entry by surface binding and diffusion through membrane pores, *Biophys. J.*, 58, 13, 1990.

197. **Mouneimne, Y., Tosi, P.-F., Gazitt, Y. and Nicolau, C.,** Electroinsertion of xeno-glycophorin into the red blood cell membrane, *Biochem. Biophys. Res. Commun.*, 159, 34, 1989.

198. **Mouneimne, Y., Tosi, P.-F., Barhoumi, R. and Nicolau, C.,** Electroinsertion of xeno-proteins in red blood cell membranes yields a long lived protein carrier in circulation, *Biochim. Biophys. Acta*, 1066, 83, 1991.

199. **Schwan, H. P.,** Electrical properties of tissue and cell suspensions, in *Advances in Biological and Medical Physics*, Vol. 5, Lawrence, J. H. and Tobias, C. E., Eds., Academic Press, New York, 1957, 147.

200. **Bernhardt, J. and Pauly, H.,** On the generation of a potential difference across the membranes of ellipsoidal cells in an alternating electrical field, *Biophysik*, 10, 89, 1973.

201. **Holzapfel, C., Vienken, J. and Zimmermann, U.,** Rotation of cells in an alternating electric field: theory and experimental proof, *J. Membrane Biol.*, 67, 13, 1982.

202. **Rols, M.-P. and Teissié, J.,** Ionic-strength modulation of electrically induced permeabilization and associated fusion of mammalian cells, *Eur. J. Biochem.*, 179, 109, 1989.

203. **Gneno, R., Azzar, G., Got, R. and Roux, B.,** Temperature and external electric field influence in membrane permeability of *Babesia* infected erythrocytes, *Int. J. Biochem.*, 19, 1069, 1987.

204. **Tsong, T. Y.,** On electroporation of cell membranes and some related phenomena, *Bioelectrochem. Bioenerg.*, 24, 271, 1990.

205. **Michel, M. R., Elgizoli, M., Koblet, H. and Kempf, Ch.,** Diffusion loading conditions determine recovery of protein synthesis in electroporated P3X63Ag8 cells, *Experientia*, 44, 199, 1988.

206. **Gordon, P. B. and Seglen, P.O.,** Electropermeabilization: getting inside the cell to study autophagy, *Hepatology*, 8, 418, 1988.

207. **Teissié, J. and Tsong, T. Y.,** Electric field induced transient pores in phospholipid bilayer vesicles, *Biochemistry*, 20, 1548, 1981.

208. **Adam, G., Läuger, P. and Stark, G.,** *Physikalische Chemie und Biophysik*, 2nd edition, Springer-Verlag, Berlin, 1988.

209. **Devaux, P. F.,** Static and dynamic lipid asymmetry in cell membranes, *Biochemistry*, 30, 1163, 1991.

210. **Dressler, V., Schwister, K., Haest, C. W. M. and Deuticke, B.,** Dielectric breakdown of the erythrocyte membrane enhances transbilayer mobility of phospholipids, *Biochim. Biophys. Acta*, 732, 304, 1983.

211. **Song, L. Y., Baldwin, J. M., O'Reilly, R. and Lucy, J. A.,** Relationships between the surface exposure of acidic phospholipids and cell fusion in erythrocytes subjected to electrical breakdown, *Biochim. Biophys. Acta*, 1104, 1, 1992.

212. **Hahn-Hägerdal, B., Hosono, K., Zachrisson, A. and Bornman, C. H.,** Polyethylene glycol and electric field treatment of plant protoplasts: characterization of some membrane-properties, *Physiol. Plant.*, 67, 359, 1986.

213. **Lopez, A., Rols, M.-P. and Teissié, J.,** ^{31}P NMR Analysis of membrane phospholipid organization in viable, reversibly electropermeabilized Chinese hamster ovary cells, *Biochemistry*, 27, 1222, 1988.

214. **Biedinger, U., Youngman, R. J., and Schnabl, H.,** Differential effects of electrofusion and electropermeabilization parameters on the membrane integrity of plant protoplasts, *Planta*, 180, 598, 1990.

215. **Sowers, A. E.,** A long lived fusogenic state is induced in erythrocyte ghosts by electric pulses, *J. Cell Biol.*, 102, 1358, 1986.

216. **Chauvin, F., Astumian, R. D. and Tsong, T. Y.,** Voltage sensing of mitochondrial ATPase in pulsed electric field induced ATP synthesis, *Biophys. J.*, 51, 243a, 1987.

217. **Tsong, T. Y. and Astumian, R. D.,** Electroconformational coupling and membrane protein function, *Prog. Biophys. Molec. Biol.*, 50, 1, 1987.

218. **Lambert, H., Pankov, R., Gauthier, J. and Hancock, R.,** Electroporation-mediated uptake of proteins into mammalian cells, *Biochem. Cell Biol.*, 68, 729, 1990.

219. **Glogauer, M., Lee, W. and McCulloch, C. A. G.,** Induced endocytosis in human fibroblasts by electrical fields, *Exp. Cell Res.*, 208, 232, 1993.

220. **Chernomordik, L. V., Sokolov, A. V. and Budker, V. G.,** Electrostimulated uptake of DNA by liposomes, *Biochim. Biophys. Acta*, 1024, 179, 1990.

221. **Schüssler, W. and Ruhenstroth-Bauer, G.,** Stomatocytosis of latex particles (0.26 μm) by rat erythrocytes by the electrical breakdown technique, *Blut*, 49, 213, 1984.

222. **Zimmermann, U., Pilwat, G. and Vienken, J.,** Erythrocytes and lymphocytes as drug carrier systems: techniques for entrapment of drugs in living cells, in *Recent Results in Cancer Research*, 75, Mathé, G. and Muggia, F. M., Eds., Springer-Verlag, Berlin, 1980, 252.

223. **Goldstein, S., Fordis, C. M. and Howard, B. H.,** Enhanced transfection efficiency and improved cell survival after electroporation of G2/M synchronized cells and treatment with sodium butyrate, *Nucleic Acids Res.*, 17, 3959, 1989.

224. **Pliquett, F. and Wunderlich, S.,** Relationship between cell parameters and pulse deformation due to these cells as well as its change after electrically induced membrane breakdown, *Bioelectrochem. Bioenerg.*, 10, 467, 1980.

225. **Joersbo, M., Brunstedt, J. and Floto, F.,** Quantitative relationship between parameters of electroporation, *J. Plant Physiol.*, 137, 169, 1990.

226. **Presse, F., Quillet, A., Mir, L., Marchiol-Fournigault, C., Feunteun, J. and Fradelizi, D.,** An improved electrotransfection method using square shaped electric impulsions, *Biochem. Biophys. Res. Commun.*, 151, 982, 1988.

227. **Sixou, S. and Teissié, J.,** *In vivo* targeting of inflamed areas by electroloaded neutrophils, *Biochem. Biophys. Res. Commun.*, 186, 860, 1992.

228. **Van der Steege, G. and Tempelaar, M. J.,** Comparison of electric field mediated DNA-uptake- and fusion properties of protoplasts of *Solanum tuberosum, Plant Sci.,* 69, 103, 1990.

229. **Kubiniec, R. T., Liang, H. and Hui, S. W.,** Effects of pulse length and pulse strength on transfection by electroporation, *BioTechniques*, 8, 18, 1990.

230. **Saunders, J. A., Rhodes Smith, C. and Kaper, J. M.,** Effects of electroporation pulse wave on the incorporation of viral RNA into tobacco protoplasts, *BioTechniques*, 7, 1124, 1989.

231. **Joersbo, M. and Brunstedt, J.,** Direct gene transfer to plant protoplasts by electroporation by alternating, rectangular and exponentially decaying pulses, *Plant Cell Rep.*, 8, 701, 1990.

232. **Fuhr, G., Müller, T., Wagner, A. and Donath, E.,** Electrorotation of oat protoplasts before and after fusion, *Plant Cell Physiol.*, 28, 549, 1987.

233. **Gimsa, J., Marszalek, P., Loewe, U. and Tsong, T. Y.,** Electroporation in rotating electric fields, *Bioelectrochem. Bioenerg.*, 29, 81, 1992.
234. **Scheurich, P. and Zimmermann, U.,** Membrane fusion and deformation of red blood cells by electric fields, *Z. Naturforsch.*, 35c, 1081, 1980.
235. **Bertsche, U., Mader, A. and Zimmermann, U.,** Nuclear membrane fusion in electrofused mammalian cells, *Biochim. Biophys. Acta*, 939, 509, 1988.
236. **Chang, D. C.,** Cell poration and cell fusion using an oscillating electric field, *Biophys. J.*, 56, 641, 1989.
237. **Chang, D. C., Hunt, J. R., Zheng, Q. and Gao, P.-Q.,** Electroporation and electrofusion using a pulsed radio-frequency electric field, in *Guide to Electroporation and Electrofusion*, Chang, D. C., Chassy, B. M, Saunders, J. A. and Sowers, A. E., Eds., Academic Press, San Diego, 1992, 303.
238. **Boitano, S., Dirksen, E. R. and Sanderson, M. J.,** Intercellular propagation of calcium waves mediated by inositol trisphosphate, *Science*, 258, 292, 1992.
239. **Swezey, R. R. and Epel, D.,** Stable, resealable pores formed in sea urchin eggs by electric discharge (electroporation) permit substrate loading for assay of enzymes *in vivo*, *Cell Regulation*, 1, 65, 1989.
240. **Sixou, S. and Teissié, J.,** Specific electropermeabilization of leucocytes in a blood sample and application to large volumes of cells, *Biochim. Biophys. Acta*, 1028, 154, 1990.
241. **Fritz, G. and Kaina, B.,** Stress factors affecting expression of 0^6-methylguanine-DNA methyltransferase mRNA in rat hepatoma cells, *Biochim. Biophys. Acta*, 1171, 35, 1992.
242. **Hartfield, P. J. and Robinson, J. M.,** Fluoride-mediated activation of the respiratory burst in electropermeabilized neutrophils, *Biochim. Biophys. Acta*, 1054, 176, 1990.
243. **Regateiro, F. J., Carvalho, C. M., Ferreira, I. L., Bairos, V. A. and Carvalho, A. P.,** Calcium stores in electropermeabilized HL-60 cells before and after differentiation, *Cell. Signall.*, 3, 41, 1991.
244. **Jones, P. M., Salmon, D. M. W. and Howell, S. L.,** Protein phosphorylation in electrically permeabilized islets of Langerhans, *Biochem. J.*, 254, 397, 1988.
245. **Jones, P. M., Persaud, S. J. and Howell, S. L.,** Ca^{2+}-induced insulin secretion from electrically permeabilized islets. Loss of the Ca^{2+}-induced secretory response is accompanied by loss of Ca^{2+}-induced protein phosphorylation, *Biochem. J.*, 285, 973, 1992.
246. **Jones, P. M. and Mann, F. M.,** Russel's viper venom stimulates insulin secretion from rat islets of Langerhans, *J. Endocrinol.*, 136, 27, 1993.
247. **Basudev, H., Jones, P. M., Persaud, S. J. and Howell, S. L.,** Arachidonic acid induces phosphorylation of an 18 kDa protein in electrically permeabilized rat islets of Langerhans, *FEBS Lett.*, 296, 69, 1992.
248. **Loomis-Husselbee, J. W. and Dawson, A. P.,** A steady-state mechanism can account for the properties of inositol 2,4,5-trisphosphate-stimulated Ca^{2+} release from permeabilized L1210 cells, *Biochem. J.*, 289, 861, 1993.
249. **Fällman, M., Gullberg, M., Hellberg, C. and Andersson, T.,** Complement receptor-mediated phagocytosis is associated with accumulation of phosphatidylcholine-derived diglyceride in human neutrophils. Involvement of phospholipase D and direct evidence for a positive feedback signal of protein kinase C, *J. Biol. Chem.*, 267, 2656, 1992.
250. **Downey, G. P., Chan, C. K. and Grinstein, S.,** Actin assembly in electropermeabilized neutrophils: Role of G-proteins, *Biochem. Biophys. Res. Commun.*, 164, 700, 1989.
251. **Elferink, J. G. R., Boonen, G. J. J. C. and de Koster, B. M.,** The role of calcium in neutrophil migration: the effect of calcium and calcium-antagonists in electroporated neutrophils, *Biochem. Biophys. Res. Commun.*, 182, 864, 1992.
252. **van den Hoff, M. J. B., Moorman, A. F. M. and Lamers, W. H.,** Electroporation in 'intracellular' buffer increases cell survival, *Nucleic Acids Res.*, 20, 2902, 1992.
253. **Zimmermann, U., Vienken, J. and Pilwat, G.,** Development of drug carrier systems: electrical field induced effects in cell membranes, *Bioelectrochem. Bioenerg.*, 7, 553, 1980.

254. **Schneeweiss, F., Zimmermann, U. and Saleemuddin, M.,** Preparation of uniform haemoglobin free human erythrocyte ghosts in isotonic solution, *Biochim. Biophys. Acta,* 466, 373, 1977.

255. **Saleemuddin, M. and Zimmermann, U.,** Use of glyceraldehyde-3-phosphate dehydrogenase-depleted human erythrocyte ghosts as specific high affinity adsorbents for the purification of glyceraldehyde-3-phosphate dehydrogenase from various tissues, *Biochim. Biophys. Acta,* 527, 182, 1978.

256. **Gupta, H. S., Rech, E. L., Cocking, E. C. and Davey, M. R.,** Electroporation and heat shock stimulate division of protoplasts of *Pennisetum squamulatum, J. Plant Physiol.,* 133, 457, 1988.

257. **Rech, E. L., Ochatt, S. J., Chand, P. K., Davey, M. R., Mulligan, B. J. and Power, J. B.,** Electroporation increases DNA synthesis in cultured plant protoplasts, *Biotechnology,* 6, 1091, 1988.

258. **Chand, P. K., Ochatt, S. J., Rech, E. L., Power, J. B. and Davey, M. R.,** Electroporation stimulates plant regeneration from protoplasts of the woody medicinal species *Solanum dulcamara L., J. Exp. Bot.,* 206, 1267, 1988.

259. **Tomov, T. Ch., Tsoneva, I. Ch. and Doncheva, J. Ch.,** Electrical stability of erythrocytes in the presence of divalent cations, *Biosci. Rep.,* 8, 421, 1988.

260. **Rotrosen, D., Kleinberg, M. E., Nunoi, H., Leto, T., Gallin, J. I. and Malech, H. L.,** Evidence for a functional cytoplasmic domain of phagocyte oxidase cytochrome b_{558}, *J. Biol. Chem.,* 265, 8745, 1990.

261. **Bazile, D., Mir, L. M. and Paoletti, C.,** Voltage-dependent introduction of a d[α]octothymidylate into electropermeabilized cells, *Biochem. Biophys. Res. Commun.,* 159, 633, 1989.

262. **Mosior, M., Mikolajczak, A. and Gomulkiewicz, J.,** The effect of ATP on the order and the mobility of lipids in the bovine erythrocyte membrane, *Biochim. Biophys. Acta,* 1022, 361, 1990.

263. **Verhoek-Köhler, B., Hampp, R., Ziegler, H. and Zimmermann, U.,** Electro-fusion of mesophyll protoplasts of *Avena sativa, Planta,* 158, 199, 1983.

264. **Schoen, C. D., Bruin, T., Arents, J. C. and van Driel, R.,** Regulation of adenylate cyclase in electropermeabilized *Dictyostelium discoideum* cells, *Exp. Cell Res.,* 199, 162, 1992.

265. **Larochelle, D. A. and Epel, D.,** *In vivo* protein phosphorylation and labeling of ATP in sea urchin eggs loaded with $^{32}PO_4$ via electroporation, *Develop. Biol.,* 148, 156, 1991.

266. **Teissié, J.,** Effects of electric fields and currents on living cells and their potential use in biotechnology: a survey, *Bioelectrochem. Bioenerg.,* 20, 133, 1988.

267. **Maccarrone, M., Veldink, G. A., and Vliegenthart, F. G.,** Inhibition of lipoxygenase activity in lentil protoplasts by monoclonal antibodies introduced into the cells via electroporation, *Eur. J. Biochem.,* 205, 995, 1992.

268. **Folle, G. A., Johannes, C. and Obe, G.,** Induction of chromosomal aberrations by DNase I., *Int. J. Radiat. Biol.,* 59, 1371, 1991.

269. **Winegar, R. A., Phillips, J. W., Youngblom, J. H. and Morgan, W. F.,** Cell electroporation is a highly efficient method for introducing restriction endonucleases into cells, *Mutation Res.,* 225, 49, 1989.

270. **Cortés, F. and Ortiz, T.,** Chromosome damage induced by restriction endonucleases recognizing thymine-rich DNA sequences in electroporated CHO cells, *Int. J. Radiat. Biol.,* 61, 323, 1992.

271. **Kinashi, Y., Nagasawa, H. and Little, J. B.,** Mutagenic effects of restriction enzymes in Chinese hamster cells: evidence for high mutagenicity of Sau3AI at the *hprt* locus, *Mutat. Res.,* 285, 251, 1993.

272. **Langlet, C. and Schmitt-Verhulst, A. M.,** Electroporation of CTL clones: a useful method to investigate signalling pathways leading to the expression of effector functions, *J. Immunol. Meth.,* 151, 107, 1992.

273. **Chen, W., Carbone, F. R. and McCluskey, J.**, Electroporation and commercial liposomes efficiently deliver soluble protein into the MHC class I presentation pathway. Priming *in vitro* and *in vivo* for class I-restricted recognition of soluble antigen, *J. Immunol. Meth.*, 160, 49, 1993.

274. **Rutter, G. A. and Denton, R. M.**, Effects of insulin and guanosine 5'[γ-thio]triphosphate on fatty acid sythesis and lipolysis within electropermeabilized fat-cells, *Biochem. J.*, 281, 431, 1992.

275. **Séguin, A. and Lalonde, M.**, Gene transfer by electroporation in betulaceae protoplasts: *Alnus incana, Plant Cell Reports*, 7, 367, 1988.

276. **Luciano, C. S., Rhoads, R. E. and Shaw, J. G.**, Synthesis of potyviral RNA and proteins in tobacco mesophyll protoplasts inoculated by electroporation, *Plant Sci.*, 51, 295, 1987.

277. **Knutson, J. C. and Yee, D.**, Electroporation: parameters affecting transfer of DNA into mammalian cells, *Anal. Biochem.*, 164, 44, 1987.

278. **Sokoloski, J. A., Jastreboff, M. M., Bertino, J. R., Sartorelli, A. C. and Narayanan, R.**, Introduction of deoxyribonucleoside triphosphates into intact cells by electroporation, *Anal. Biochem.*, 158, 272, 1986.

279. **Potter, H.**, Electroporation in biology: methods, applications, and instrumentation, *Anal. Biochem.*, 174, 361, 1988.

280. **Rutter, G. A., Diggle, T. A. and Denton, R. M.**, Regulation of pyruvate dehydrogenase by insulin and polyamines within electropermeabilized fat-cells and isolated mitochondria, *Biochem. J.*, 285, 435, 1992.

281. **Gimmler, H., Treffny, B., Kowalski, M. and Zimmermann, U.**, The resistance of *Dunaliella acidophila* against heavy metals: the importance of the Zeta potential, *J. Plant Physiol.*, 138, 708, 1991.

282. **Loomis-Husselbee, J., Cullen, P. J., Irvine, R. F. and Dawson, A. P.**, Electroporation can cause artefacts due to solubilization of cations from the electrode plates, *Biochem. J.*, 277, 883, 1991.

283. **Stenger, D. A., Kubiniec, R. T., Purucker, W. J., Liang, H. and Hui, S. W.**, Optimization of electrofusion parameters for efficient production of murine hybridomas, *Hybridoma*, 7, 505, 1988.

284. **van den Hoff, M. J. B., Labruyère, W. T., Moorman, A. F. M. and Lamers, W. H.**, The osmolarity of the electroporation medium affects the transient expression of genes, *Nucleic Acids Res.*, 18, 6464, 1990.

285. **Huang, Y. W. and Dennis, E. S.**, Factors influencing stable transformation of maize protoplasts by electroporation, *Plant Cell Tissue Organ Culture*, 18, 281, 1989.

286. **Däumler, R. and Zimmermann, U.**, High yields of stable transformants by hypo-osmolar plasmid electroinjection, *J. Immunol. Meth.*, 122, 203, 1989.

287. **Stopper, H., Zimmermann, U. and Wecker, E.**, High yields of DNA-transfer into mouse-L-cells by electropermeabilization, *Z. Naturforsch.*, 40c, 929, 1985.

288. **Stopper, H., Zimmermann, U. and Neil, G. A.**, Increased efficiency of transfection of murine hybridoma cells with DNA by electro-permeabilization, *J. Immunol. Meth.*, 109, 145, 1988.

289. **Orlowski, S., Belehradek, J. Jr., Paoletti, C. and Mir, L. M.**, Transient electropermeabilization of cells in culture. Increase of the cytotoxicity of anticancer drugs, *Biochem. Pharmacol.*, 37, 4727, 1988.

290. **Taylor, B. H. and Larkin, P. J.**, Analysis of electroporation efficiency in plant protoplasts, *Aust. J. Biotech.*, 1, 52, 1988.

291. **Lipowsky, R.**, The physics of flexible membranes, in *Festkörperprobleme. Advances in Solid State Physics*, Vol. 32. Rössler, U., Ed., Vieweg, Braunschweig, Wiesbaden, 1992, 19.

292. **Lipowsky, R.**, Budding of membranes induced by intramembrane domains, *J. Phys. II France*, 2, 1825, 1992.

293. **Lipowsky, R.**, Domain-induced budding of fluid membranes, *Biophys. J.*, 64, 1133, 1993.

294. **Seifert, U., Miao, L., Döbereiner, H.-G. and Wortis, M.,** Budding transition for bilayer fluid vesicles with area-difference elasticity, in *Springer Proceedings in Physics* Vol. 66: The structure and conformation of amphiphilic membranes, Lipowsky, R., Richter, D. and Kremer, K., Eds., Springer-Verlag, Berlin, 1992, 93.

295. **Deuticke, B. and Schwister, K.,** Leaks induced by electrical breakdown in the erythrocyte membrane, in *Electroporation and Electrofusion in Cell Biology*, Neumann, E., Sowers, A. E. and Jordan, C. A., Eds., Plenum Press, New York, 1989, 127.

296. **Needham, D. and Hochmuth, R. M.,** Electro-mechanical permeabilization of lipid vesicles: role of membrane tension and compressibility, *Biophys. J.*, 55, 1001, 1989.

297. **Dimitrov, D. S.,** Electric field-induced breakdown of lipid bilayers and cell membranes: a thin viscoelastic film model, *J. Membrane Biol.*, 78, 53, 1984.

298. **Dimitrov, D. S. and Jain, R. K.,** Membrane stability, *Biochim. Biophys. Acta*, 779, 437, 1984.

299. **Weaver, J. C. and Powell, K. T.,** Theory of electroporation, in *Electroporation and Electrofusion in Cell Biology*, Neumann, E., Sowers, A. E. and Jordan, C. A., Eds., Plenum Press, New York, 1989, 111.

300. **Chang, D. C. and Reese, T. S.,** Changes in membrane structure induced by electroporation as revealed by rapid-freezing electron microscopy, *Biophys. J.*, 58, 1, 1990.

301. **Chang, D. C.,** Structure and dynamics of electric field-induced membrane pores as revealed by rapid-freezing electron microscopy, in *Guide to Electroporation and Electrofusion*, Chang, D. C., Chassy, B. M., Saunders, J. A. and Sowers, A. E., Eds., Academic Press, San Diego, 1992, 9.

302. **Stenger, D. A. and Hui, S. W.,** The kinetics of ultrastructural changes during electrically induced fusion of human erythrocytes, *J. Membrane Biol.*, 93, 43, 1986.

303. **Adelman, W. J. Jr.,** *Biophysics and Physiology of Excitable Membranes*, Van Nostrand Reinhold Company, New York, 1971.

304. **Carraway, K. L. and Carothers Carraway, C. A.,** Membrane-cytoskeleton interactions in animal cells, *Biochim. Biophys. Acta*, 988, 147, 1989.

305. **Tsong, T. Y.,** Time sequence of molecular events in electroporation, in *Guide to Electroporation and Electrofusion*, Chang, D. C., Chassy, B. M., Saunders, J. A., and Sowers, A. E., Eds., Academic Press, San Diego, 1992, 47.

306. **Tsong, T. Y.,** Electroporation of cell membranes, *Biophys J.*, 60, 297, 1991.

307. **Escande-Géraud, M. L., Rols, M.-P., Dupont, M. A., Gas, N. and Teissié, J.,** Reversible plasma membrane ultrastructural changes correlated with electropermeabilization in Chinese hamster ovary cells, *Biochim. Biophys. Acta*, 939, 247, 1988.

308. **Gass, G. V. and Chernomordik, L. V.,** Reversible large-scale deformations in the membranes of electrically-treated cells: electroinduced bleb formation, *Biochim. Biophys. Acta*, 1023, 1, 1990.

309. **Ginsburg, H. and Stein, W. D.,** Biophysical analysis of novel transport pathways induced in red blood cell membranes, *J. Membrane Biol.*, 96, 1, 1987.

310. **Renkin, E. M.,** Filtration, diffusion, and molecular sieving through porous cellullose membranes, *J. Gen. Physiol.*, 38, 225, 1955.

311. **Stein, W. D.,** *Transport and Diffusion across Cell Membranes*, Academic Press, Orlando, 1986.

312. **Chernomordik, L. V., Sukharev, S. I., Popov, S. V., Pastushenko, V. F., Sokirko, A. V., Abidor, I. G. and Chizmadzhev, Y. A.,** The electrical breakdown of cell and lipid membranes: the similarity of phenomenologies, *Biochim. Biophys. Acta*, 902, 360, 1987.

313. **Neumann, E., Schaefer-Ridder, M., Wang, Y. and Hofschneider, P. H.,** Gene transfer into mouse lyoma cells by electroporation in high electric fields, *EMBO J.*, 1, 841, 1982.

314. **Neumann, E. and Boldt, E.,** Membrane electroporation: the dye method to determine the cell membrane conductivity, in *Horizons in Membrane Biotechnology*, Nicolau, C. and Chapman, D., Eds., Wiley-Liss, New York, 1990, 69.

315. **Traas, J. A., Doonan, J. H., Rawlins, D. J., Shaw, P. J., Watts, J. and Lloyd, C. W.,** An actin network is present in the cytoplasm throughout the cell cycle of carrot cells and associates with the dividing nucleus, *J. Cell Biol.*, 105, 387, 1987.

316. **Scheuerlein, R., Schmidt, K., Poenie, M. and Roux, S. J.,** Determination of cytoplasmic calcium concentration in *Dryopteris* spores. A developmentally non-disruptive technique for loading of the calcium indicator fura-2, *Planta*, 184, 166, 1991.

317. **Obermeyer, G. and Weisenseel, M. H.,** Introduction of impermeable molecules into pollen grains by electroporation, *Protoplasma*, 187, 132, 1995.

318. **Gilroy, S., Hughes, W. A. and Trewavas, A. J.,** The measurement of intracellular calcium levels in protoplasts from higher plant cells, *FEBS Lett.*, 199, 217, 1986.

319. **van Haastert, P. J. M., De Vries, M. J., Penning, L. C., Roovers, E., Van Der Kaay, J., Erneux, C. and Van Lookeren Campagne, M. M.,** Chemoattractant and guanosine 5′[γ-thio]triphosphate induce the accumulation of inositol 1,4,5-trisphosphate in *Dictyostelium* cells that are labelled with [^{3}H]inositol by electroporation, *Biochem. J.*, 258, 577, 1989.

320. **Valkema, R. and Van Haastert, P. J. M.,** Inhibition of receptor-stimulated guanylyl cyclase by intracellular calcium ions in *Dictyostelium* cells, *Biochem. Biophys. Res. Commun.*, 186, 263, 1992.

321. **Uno, I., Fukami, K., Kato, H., Takenawa, T. and Ishikawa, T.,** Essential role for phosphatidylinositol 4,5-biphosphate in yeast cell proliferation, *Nature,* 333, 188, 1988.

322. **Shibata, H., Robinson, F. W., Benzing, C. F. and Kono, T.,** Evidence that protein kinase C may not be involved in the insulin action on cAMP phosphodiesterase: studies with electroporated rat adipocytes that were highly responsive to insulin, *Arch. Biochem. Biophys.*, 285, 97, 1991.

323. **Standaert, M. L., Sasse, J., Cooper, D. R. and Farese, R. V.,** Protein kinase C(19–31) pseudosubstrate inhibition of insulin action in rat adipocytes, *FEBS Lett.*, 282, 139, 1991.

324. **Cooper, D. R., Watson, J. E., Hernandez, H., Yu, B., Standaert, M. L., Ways, D. K., Arnold, T. T., Ishizuka, T. and Farese, R. V.,** Direct evidence for protein kinase C involvement in insulin-stimulated hexose uptake, *Biochem. Biophys. Res. Commun.*, 188, 142, 1992.

325. **von Grafenstein, H. and Knight, D. E.,** Membrane recapture and early triggered secretion from the newly formed endocytotic compartment in bovine chromaffin cells, *J. Physiol.*, 453, 15, 1992.

326. **Bartels, F. and Bigalke, H.,** Restoration of exocytosis occurs after inactivation of intra-cellular tetanus toxin, *Infec. Immun.,* 60, 302, 1992.

327. **Knight, D. E. and Baker, P. F.,** Guanine nucleotides and Ca-dependent exocytosis. Studies on two adrenal cell preparations, *FEBS Lett.*, 189, 345, 1985.

328. **Knight, D. E. and Baker, P. F.,** The phorbol ester TPA increases the affinity of exocytosis for calcium in 'leaky' adrenal medullary cells, *FEBS Lett.*, 160, 98, 1983.

329. **Kwee, S. and Celis, J. E.,** Electroporation as a tool for studying cell proliferation and DNA synthesis in human cultured cells grown in monolayers, *Bioelectrochem. Bioenerg.*, 25, 325, 1991.

330. **Watson, P. H., Pon, R. T. and Shiu, R. P. C.,** Inhibition of cell adhesion to plastic substratum by phosphorothioate oligonucleotide, *Exp. Cell Res.*, 202, 391, 1992.

331. **Bidawid-Woodroffe, S., Sullivan-Tailyour, G., Roig-Farran, P. and Garnett, H. M.,** Increased cytomegalovirus infection of human fibroblast and endothelial cells by electroporation, *J. Virol. Meth.*, 38, 167, 1992.

332. **Raptis, L. and Firth, K. L.,** Laboratory methods. Electroporation of adherent cells *in situ*, *DNA Cell Biol.*, 9, 615, 1990.

333. **Chang, C., Biedermann, K. A., Mezzina, M. and Brown, J. M.,** Characterization of the DNA double strand break repair defect in SCID mice, *Cancer Res.*, 53, 1244, 1993.

334. **Borek, C., Ong, A., Morgan, W. F. and Cleaver, J. E.,** Morphological transformation of 10T1/2 mouse embryo cells can be initiated by DNA double-strand breaks alone, *Mol. Carcinog.*, 4, 243, 1991.

335. **Tanti, J.-F., Gautier, N., Cormont, M., Baron, V., van Obberghen, E. and Le Marchand-Brustel, Y.,** Potential involvement of the carboxy-terminus of the Glut 1 transporter in glucose transport, *Endocrinology*, 131, 2319, 1992.

336. **Plevin, R., Palmer, S., Gardner, S. D. and Wakelam, M. J. O.,** Regulation of bombesin-stimulated inositol 1,4,5-trisphosphate generation in Swiss 3T3 fibroblasts by a guanine-nucleotide-binding protein, *Biochem. J.*, 268, 605, 1990.

337. **Dagher, S. F., Conrad, S. E., Werner, E. A. and Patterson, R. J.,** Phenotypic conversion of TK-deficient cells following electroporation of functional TK enzyme, *Exp. Cell Res.*, 198, 36, 1992.

338. **Kruse, C. A., Mierau, G. W. and James, G. T.,** Methotrexate loading of red cell carriers by osmotic stress and electric-pulse methods: ultrastructural observations, *Biotechnol. Appl. Biochem.*, 11, 571, 1989.

339. **Zimmermann, U.,** Targeted drugs, *Labo-Pharma*, 31, 69, 1983.

340. **Zimmermann, U.,** Cellular drug-carrier systems and their possible targeting, in *Targeted Drugs*, Goldberg, E., Ed., John Wiley & Sons, New York, 1983, 153.

341. **Rutter, G. A., Pralong, W.-F., and Wollheim, C. B.,** Regulation of mitochondrial glycerol-phosphate dehydrogenase by Ca^{2+} within electropermeabilized insulin-secreting cells (INS-1), *Biochim. Biophys. Acta*, 1175, 107, 1992.

342. **Bases, R. and Mendez, F.,** Repair of ionizing radiation damage in primate αDNA transfected into rat cells, *Int. J. Radiat. Biol.*, 62, 21, 1992.

343. **Hashimoto, K., Tatsumi, N. and Okuda, K.,** Introduction of phalloidin labelled with fluorescein isothiocyanate into living polymorphonuclear leukocytes by electroporation, *J. Biochem. Biophys. Meth.*, 19, 143, 1989.

344. **Kubota, M., Yorifuji, T., Hashimoto, H., Shimizu, T. and Mikawa, H.,** Metabolism and toxicity of electroporated 1-beta-D-arabinofuranosylcytosine triphosphate in a human leukemia cell line, *Jpn. J. Cancer Res.*, 81, 1314, 1990.

345. **Xing, M. and Mattera, R.,** Phosphorylation-dependent regulation of phospholipase A_2 by G-proteins and Ca^{2+} in HL60 granulocytes, *J. Biol. Chem.*, 267, 25966, 1992.

346. **Bourgoin, S. and Grinstein, S.,** Peroxides of vanadate induce activation of phospholipase D in HL-60 Cells, *J. Biol. Chem.*, 267, 11908, 1992.

347. **Kashanchi, F., Duvall, J. F. and Brady, J. N.,** Electroporation of viral transactivator proteins into lymphocyte suspension cells, *Nucleic Acids Res.*, 20, 4673, 1992.

348. **Winegar, R. A., Lutze, L. H., Rufer, J. T. and Morgan, W. F.,** Spectrum of mutations produced by specific types of restriction enzyme-induced double-strand breaks, *Mutagenesis*, 7, 439, 1992.

349. **Parrish, D. D., Lambert, W. C. and Lambert, M. W.,** Xeroderma pigmentosum endonuclease complexes show reduced activity on and affinity for psoralen cross-linked nucleosomal DNA, *Mutat. Res.*, 273, 157, 1992.

350. **Lang, P., Guizani, L., Vitté-Mony, I., Stancou, R., Dorseuil, O., Gacon, G. and Bertoglio, J.,** ADP-ribosylation of the *ras*-related, GTP-binding protein RhoA inhibits lymphocyte-mediated cytotoxicity, *J. Biol. Chem.*, 267, 11677, 1992.

351. **Harding, C. V. III.,** Electroporation of exogenous antigen into the cytosol for antigen processing and class I major histocompatibility complex (MHC) presentation: weak base amines and hypothermia (18°C) inhibit the class I MHC processing pathway, *Eur. J. Immunol.*, 22, 1865, 1992.

352. **Holen, I., Gordon, P. B. and Seglen, P. O.,** Protein kinase-dependent effects of okadaic acid on hepatocytic autophagy and cytoskeletal integrity, *Biochem. J.*, 284, 633, 1992.

353. **Poddevin, B., Orlowski, S., Belehradek, J. Jr. and Mir, L. M.,** Very high cytotoxicity of bleomycin introduced into the cytosol of cells in culture, *Biochem. Pharmacol.*, 42, S67, 1991.

354. **Wang, P., Nishihata, J., Makishima, F., Moriishi, K., Syuto, B., Toyoshima, S. and Osawa, T.,** Low-molecular-weight GTP-binding proteins serving as ADP-ribosylation substrate for ADP-ribosyltransferase from *Clostridium botulinum* and their relation to phosphoinositides metabolism in thymocytes, *J. Biochem.*, 108, 879, 1990.

355. **Lukacs, G. L., Rotstein, O. D. and Grinstein, S.,** Phagosomal acidification is mediated by a vacuolar-type H⁺-ATPase in murine macrophages, *J. Biol. Chem.*, 265, 21099, 1990.

356. **Malinin, V. S., Sharov, V. S., Putvinsky, A. V., Osipov, A. N. and Vladimirov, Y. A.,** Chemiluminescent reactions of phagocytes induced by electroporation. The role of Ca^{2+} and Mg^{2+} ions, *Bioelectrochem. Bioenerg.*, 22, 37, 1989.

357. **Burdett, E., Mills, G. B. and Klip, A.,** Effect of GTP γ S on insulin binding and tyrosine phosphorylation in liver membranes and L6 muscle cells, *Am. J. Physiol.*, 258, C99, 1990.

358. **Wojcikiewicz, R. J. H., Cooke, A. M., Potter, B. V. L. and Nahorski, S. R.,** Inhibition of inositol 1,4,5-triphosphate metabolism in permeabilized SH-SY5Y human neuroblastoma cells by a phosphorothioate-containing analogue of inositol 1,4,5-trisphosphate, *Eur. J. Biochem.*, 192, 459, 1990.

359. **Stoehr, S. J., Smolen, J. E. and Suchard, S. J.,** Lipocortins are major substrates for protein kinase C in extracts of human neutrophils, *J. Immunol.*, 144, 3936, 1990.

360. **Mueller, H., Montoya, B. and Sklar, L. A.,** Reversal of inhibitory pathways in neutrophils by protein kinase antagonists: a rational approach to the restoration of depressed cell function?, *J. Leukocyte Biol.*, 52, 400, 1992.

361. **Fällman, M., Bergstrand, H. and Andersson, T.,** Auranofin dissociates chemotactic peptide-induced generation of inositol 1,4,5-trisphosphate from the subsequent mobilization of intracellular calcium in intact human neutrophils, *Biochim. Biophys. Acta*, 1055, 173, 1990.

362. **Lu, D. J., Takai, A., Leto, T. L. and Grinstein, S.,** Modulation of neutrophil activation by okadaic acid, a protein phosphatase inhibitor, *Am. J. Physiol.*, 262, C39, 1992.

363. **Niessen, H. W. M. and Verhoeven, A. J.,** Differential up-regulation of specific and azurophilic granule membrane markers in electropermeabilized neutrophils, *Cell. Signall.*, 4, 501, 1992.

364. **Downey, G. P., Chan, C. K., Trudel, S. and Grinstein, S.,** Actin assembly in electropermeabilized neutrophils: role of intracellular calcium, *J. Cell. Biol.*, 110, 1975, 1990.

365. **Smolen, J. E., Stoehr, S. J., Kuczynski, B., Koh, E. K. and Omann, G. M.,** Dual effects of guanosine 5′-[γ-thio]triphosphate on secretion by electroporated human neutrophils, *Biochem. J.*, 279, 657, 1991.

366. **Lu, D. J. and Grinstein, S.,** ATP and guanine nucleotide dependence of neutrophil activation. Evidence for the involvement of two distinct GTP-binding proteins, *J. Biol. Chem.*, 265, 13721, 1990.

367. **Smolen, J. E., Stoehr, S. J. and Kuczynski, B.,** Cyclic AMP inhibits secretion from electroporated human neutrophils, *J. Leukoc. Biol.*, 49, 172, 1991.

368. **Kessels, G. C. R., Gervaix, A., Lew, P. D. and Verhoeven, A. J.,** The chymotrypsin inhibitor carbobenzyloxy-leucine-tyrosine-chloromethylketone interferes with phospholipase D activation induced by formyl-methionyl-leucyl-phenylalanine in human neutrophils, *J. Biol. Chem.*, 266, 15870, 1991.

369. **Grinstein, S. and Furuya, W.,** Receptor-mediated activation of electropermeabilized neutrophils, *J. Biol. Chem.*, 263, 1779, 1988.

370. **Downey, G. P., Chan, C. K., Lea, P., Takai, A. and Grinstein, S.,** Phorbol ester-induced actin assembly in neutrophils: role of protein kinase C., *J. Cell Biol.*, 116, 695, 1992.

371. **Grinstein, S., Furuya, W., Lu, D. J. and Mills, G. B.,** Vanadate stimulates oxygen consumption and tyrosine phosphorylation in electropermeabilized human neutrophils, *J. Biol. Chem.*, 265, 318, 1990.

372. **Boonen, G. J. J. C., van Steveninck, J., Dubbelman, T. M. A. R., van den Broek, P. J. A. and Elferink, J. G. R.,** Exocytosis in electropermeabilized neutrophils. Responsiveness to calcium and guanosine 5′-[γ-thio]triphosphate, *Biochem. J.*, 287, 695, 1992.

373. **Boonen, G. J. J. C., van Steveninck, J. and Elferink, J. G. R.,** Strontium and barium induce exocytosis in electropermeabilized neutrophils, *Biochim. Biophys. Acta*, 1175, 155, 1993.

374. **Rickords, L. F. and White, K. L.,** Electrofusion-induced intracellular Ca^{2+} flux and its effect on murine oocyte activation, *Molec. Reprod. Develop.*, 31, 152, 1992.

375. **Rickords, L. F. and White, K. L.,** Electroporation of inositol 1,4,5-triphosphate induces repetitive calcium oscillations in murine oocytes, *J. Exp. Zool.*, 265, 178, 1993.

376. **Johannes, C. and Obe, G.,** Induction of chromosomal aberrations with the restriction endonuclease AluI in Chinese hamster ovary cells: comparison of different treatment methods, *Int. J. Radiat. Biol.*, 59, 1379, 1991.

377. **Costa, N. D. and Bryant, P. E.,** The induction of DNA double-strand breaks in CHO cells by *Pvu* II: kinetics using neutral filter elution (pH 9.6), *Int. J. Radiat. Biol.*, 57, 933, 1990.

378. **Morgan, W. F., Valcarel, E. R., Columna, E. A., Winegar, R. A. and Yates, B. L.,** Induction of chromosome damage by restriction enzymes during mitosis, *Radiat. Res.*, 127, 101, 1991.

379. **Costa, N. D. and Bryant, P. E.,** Differences in accumulation of blunt- and cohesive-ended double-strand breaks generated by restriction endonucleases in electroporated CHO cells, *Mutation Res., DNA Repair*, 254, 239, 1991.

380. **Sixou, S. and Teissié, J.,** Exogenous uptake and release of molecules by electroloaded cells: a digitized videomicroscopy study, *Bioelectrochem. Bioenerg.*, 31, 237, 1993.

381. **Arita, Y., Kimura, T., Ogami, Y. and Nawata, H.,** Phorbol ester attenuates inositol 1,4,5-triphosphate-induced Ca^{2+} release in electropermeabilized rat pancreatic acini, *Res. Exp. Med.*, 192, 295, 1992.

382. **Jones, P. M., Persaud, S. J. and Howell, S. L.,** Insulin secretion and protein phosphorylation in PKC-depleted islets of Langerhans, *Life Sci.*, 50, 761, 1992.

383. **Lindsay, M. A., Bungay, P. J. and Griffin, M.,** Transglutaminase involvement in the secretion of insulin from electropermeabilized rat islets of Langerhans, *Biosci. Rep.*, 10, 557, 1990.

384. **Jones, P. M., Persaud, S. J., Bjaaland, T., Pearson, J. D. and Howell, S. L.,** Nitric oxide is not involved in the initiation of insulin secretion from rat islets of Langerhans, *Diabetologia*, 35, 1020, 1992.

385. **Ratcliff, H. and Jones, P. M.,** Effects of okadaic acid on insulin secretion from rat islets of Langerhans, *Biochim. Biophys. Acta*, 1175, 188, 1993.

386. **Morimoto, T., Oho, C., Ueda, M., Ogihara, S. and Takisawa, H.,** Repression of serotonin secretion by an endogenous Ca^{2+}-activated protease in electropermeabilized bovine platelets, *J. Biochem.*, 108, 311, 1990.

387. **Knight, D. E. and Scrutton, M. C.,** Effects of guanine nucleotides on the properties of 5-hydroxytryptamine secretion from electropermeabilized human platelets, *Eur. J. Biochem.*, 160, 183, 1986.

388. **Haslam, R. J. and Davidson, M. M. L.,** Guanine nucleotides decrease the free $[Ca^{2+}]$ required for secretion of serotonin from permeabilized blood platelets. Evidence of a role for a GTP-binding protein in platelet activation, *FEBS Lett.*, 174, 90, 1984.

389. **Athayde, C. M. and Scrutton, M. C.,** Guanine nucleotides and Ca^{2+}-dependent lysosomal secretion in electropermeabilized human platelets, *Eur. J. Biochem.*, 189, 647, 1990.

390. **Coorssen, J. R., Davidson, M. M. L. and Haslam, R. J.,** Factors affecting dense and α-granule secretion from electropermeabilized human platelets: Ca^{2+}-independent actions of phorbol ester and GTP-γ-S., *Cell Regul.*, 1, 1027, 1990.

391. **Daniel, J. L., Dangelmaier, C. and Smith, J. B.,** Evidence that adhesion of electrically permeabilized platelets to collagen is mediated by guanine nucleotide regulatory proteins, *Biochem. J.*, 286, 701, 1992.

392. **Hughes, K. and Crawford, N.,** Reversibly electropermeabilized platelets: potential use as vehicles for drug delivery, *Biochem. Soc. Trans.*, 18, 871, 1990.

393. **Gagné, M. B., Pothier, F. and Sirard, M.-A.,** Electroporation of bovine spermatozoa to carry foreign DNA in oocytes, *Mol. Reprod. Dev.*, 29, 6, 1991.

394. **Tomkins, P. T. and Houghton, J. A.,** The rapid induction of the acrosome reaction of human spermatozoa by electropermeabilization, *Fertil. Steril.*, 50, 329, 1988.

395. **Brown, L. E., Sprecher, S. L. and Keller, L. R.,** Introduction of exogenous DNA into *Chlamydomonas reinhardtii* by electroporation, *Mol. Cell. Biol.*, 11, 2328, 1991.

396. **Higgs, D. C. and Colbert, J. T.,** β-glucuronidase gene expression and mRNA stability in oat protoplasts, *Plant Cell Rep.,* 12, 445, 1993.
397. **Lindsey, K. and Jones, M. G. K.,** Transient gene expression in electroporated protoplasts and intact cells of sugar beet, *Plant Mol. Biol.,* 10, 43, 1987.
398. **Gallois, P., Lindsey, K., Malone, R., Kreis, M. and Jones, M. G. K.,** Gene rescue in plants by direct gene transfer of total genomic DNA into protoplasts, *Nucleic Acids Res.,* 20, 3977, 1992.
399. **Rouan, D., Montané, M.-H., Alibert, G. and Teissié, J.,** Relationship between protoplast size and critical field strength in protoplast electropulsing and application of reliable DNA uptake in *Brassica, Plant Cell Rep.,* 10, 139, 1991.
400. **Hervé, C., Rouan, D., Guerche, P., Montané, M.-H. and Yot, P.,** Molecular analysis of transgenic rapeseed plants obtained by direct transfer of two separate plasmids containing, respectively, the cauliflower mosaic virus coat protein gene and a selectable marker gene, *Plant Sci.,* 91, 181, 1993.
401. **Guerche, P., Charbonnier, M., Jouanin, L., Tourneur, C., Paszkowski, J. and Pelletier, G.,** Direct gene transfer by electroporation in *Brassica napus, Plant Sci.,* 52, 111, 1987.
402. **Bergman, P. and Glimelius, K.,** Electroporation of rapeseed protoplasts — transient and stable transformation, *Physiologia Plantarum,* 88, 604, 1993.
403. **Eimert, K. and Siegemund, F.,** Transformation of cauliflower (*Brassica oleracea* L. var. *botrytis*) — an experimental survey, *Plant Mol. Biol.,* 19, 485, 1992.
404. **Viry, M., Serghini, M. A., Hans, F., Ritzenthaler, C., Pinck, M. and Pinck, L.,** Biologically active transcripts from cloned cDNA of genomic grapevine fanleaf nepovirus RNAs, *J. Gen. Virol.,* 74, 169, 1993.
405. **Matsunaga, R., Sawamura, K., De Kok, M., Makino, T., Miki, K., Kojima, M., Tsuchizaki, T. and Hibi, T.,** Electrotransfection of protoplasts from tomato, wild tomato, barley and chrysanthemum with tobacco mosaic virus RNA, *J. Gen. Virol.,* 73, 763, 1992.
406. **Horn, M. E., Shillito, R. D., Conger, B. V. and Harms, C. T.,** Transgenic plants of Orchardgrass (*Dactylis glomerata* L.) from protoplasts, *Plant Cell Rep.,* 7, 469, 1988.
407. **Murray, E. E., Rocheleau, T., Eberle, M., Stock, C., Sekar, V. and Adang, M.,** Analysis of unstable RNA transcripts of insecticidal crystal protein genes of *Bacillus thuringiensis* in transgenic plants and electroporated protoplasts, *Plant Mol. Biol.,* 16, 1035, 1991.
408. **Murray, E. E., Buchholz, W. G. and Bowen, B.,** Direct analysis of RNA transcripts in electroporated carrot protoplasts, *Plant Cell Rep.,* 9, 129, 1990.
409. **Goupil, P., Hatzopoulos, P., Franz, G., Hempel, F. D., You, R. and Sung, Z. R.,** Transcriptional regulation of a seed specific carrot gene, DC8, *Plant Mol. Biol.,* 18, 1049, 1992.
410. **Ecker, J. R. and Davis, R. W.,** Inhibition of gene expression in plant cells by expression of antisense RNA, *Proc. Natl. Acad. Sci. U.S.A.,* 83, 5372, 1986.
411. **Boston, R. S., Becwar, M. R., Ryan, R. D., Goldsbrough, P. B., Larkins, B. A. and Hodges, T. K.,** Expression from heterologous promoters in electroporated carrot protolasts, *Plant Physiol.,* 83, 742, 1987.
412. **Fromm, M., Taylor, L. P. and Walbot, V.,** Expression of genes transferred into monocot and dicot plant cells by electroporation, *Proc. Natl. Acad. Sci. U.S.A.,* 82, 5824, 1985.
413. **Bourque, J. E. and Folk, W. R.,** Suppression of gene expression in plant cells utilizing antisense sequences transcribed by RNA polymerase III, *Plant Mol. Biol.,* 19, 641, 1992.
414. **Bates, G. W., Piastuch, W., Riggs, C. D. and Rabussay, D.,** Electroporation for DNA delivery to plant protoplasts, *Plant Cell Tissue Organ Culture,* 12, 213, 1988.
415. **Callis, J., Fromm, M. and Walbot, V.,** Expression of mRNA electroporated into plant and animal cells, *Nucleic Acids Res.,* 15, 5823, 1987.
416. **Christou, P., Murphy, J. E. and Swain, W. F.,** Stable transformation of soybean by electroporation and root formation from transformed callus, *Proc. Natl. Acad. Sci. U.S.A.,* 84, 3962, 1987.
417. **Dhir, S. H., Dhir, S., Hepburn, A. and Widholm, J. M.,** Factors affecting transient gene expression in electroporated *Glycine max* protoplasts, *Plant Cell Rep.,* 10, 106, 1991.

418. **Fujiwara, T., Naito, S., Chino, M. and Nagata, T.,** Electroporated protoplasts express seed specific gene promoters, *Plant Cell Rep.,* 9, 602, 1991.

419. **Dhir, S. H., Dhir, S., Sturtevant, A. P. and Widholm, J. M.,** Regeneration of transformed shoots from electroporated soybean (*Glycine max* (L.) Merr.) protoplasts, *Plant Cell Rep.,* 10, 97, 1991.

420. **Salmenkallio, M., Hannus, R., Teeri, T. H. and Kauppinen, V.,** Regulation of α-amylase promoter by gibberellic acid and abscisic acid in barley protoplasts transformed by electroporation, *Plant Cell Rep.,* 9, 352, 1990.

421. **Tsukada, M., Kusano, T. and Kitagawa, Y.,** Introduction of foreign genes into tomato protoplasts by electroporation, *Plant Cell Physiol.,* 30, 599, 1989.

422. **Larkin, P. J., Taylor, B. H., Gersmann, M. and Brettell, R. I. S.,** Direct gene transfer to protoplasts, *Aust. J. Plant Physiol.,* 17, 291, 1990.

423. **Loake, G. J., Faktor, O., Lamb, C. J. and Dixon, R. A.,** Combination of H-box [CCTACC(N)$_7$CT] and G-box [CACGTG] *cis* elements is necessary for feed-forward stimulation of a chalcone synthase promoter by the phenylpropanoid-pathway intermediate *p*-coumaric acid, *Proc. Natl. Acad. Sci. U.S.A.,* 89, 9230, 1992.

424. **Choudhary, A. D., Lamb, C. J. and Dixon, R. A.,** Stress responses in alfalfa (*Medicago sativa* L.) VI. Differential responsiveness of chalcone synthase induction to fungal elicitor of glutathione in electroporated protoplasts, *Plant Physiol.,* 94, 1802, 1990.

425. **Abdul-Baki, A. A., Saunders, J. A., Matthews, B. F. and Pittarelli, G. W.,** DNA uptake during electroporation of germinating pollen grains, *Plant Sci.,* 70, 181, 1990.

426. **Matthews, B. F., Abdul-Baki, A. A. and Saunders, J. A.,** Expression of a foreign gene in electroporated pollen grains of tobacco, *Sex Plant Reprod.,* 3, 147, 1990.

427. **Watts, J. W., King, J. M. and Stacey, N. J.,** Inoculation of protoplasts with viruses by electroporation, *Virology,* 157, 40, 1987.

428. **Watanabe, Y., Meshi, T. and Okada, Y.,** Infection of tobacco protoplasts with *in vitro* transcribed tobacco mosaic virus RNA using an improved electroporation method, *FEBS Lett.,* 219, 65, 1987.

429. **Nagata, T.,** Cell biological aspects of gene delivery into plant protoplasts by electroporation, *Int. Rev. Cytol.,* 116, 229, 1989.

430. **Okada, K., Nagata, T. and Takebe, I.,** Introduction of functional RNA into plant protoplasts by electroporation, *Plant Cell Physiol.,* 27, 619, 1986.

431. **Fukuda, Y., Ohme, M. and Shinshi, H.,** Gene structure and expression of a tobacco endochitinase gene in suspension-cultured tobacco cells, *Plant Mol. Biol.,* 16, 1, 1991.

432. **Bradshaw, H. D. Jr., Parson, W. W., Sheffer, M., Lioubin, P. J., Mulvihill, E. R. and Gordon, M. P.,** Design, construction, and use of an electroporator for plant protoplasts and animal cells, *Anal. Biochem.,* 166, 342, 1987.

433. **Paszty, C. J. R. and Lurquin, P. F.,** Inhibition of transgene expression in plant protoplasts by the presence in *cis* of an opposing 3'-promoter, *Plant Sci.,* 72, 69, 1990.

434. **Lurquin, P. F. and Paszty, C.,** Electroporation of tobacco protoplasts with homologous and nonhomologous transformation vectors, *J. Plant Physiol.,* 133, 332, 1988.

435. **Christensen, A. H., Sharrock, R. A. and Quail, P. H.,** Maize polyubiquitin genes: structure, thermal perturbation of expression and transcript splicing, and promoter activity following transfer to protoplasts by electroporation, *Plant Mol. Biol.,* 18, 675, 1992.

436. **Carneiro, V. T. C., Pelletier, G. and Small I.,** Transfer RNA-mediated suppression of stop codons in protoplasts and transgenic plants, *Plant Mol. Biol.,* 22, 681, 1993.

437. **Köhler, F., Golz, C., Eapen, S. and Schieder, O.,** Influence of plant cultivar and plasmid-DNA on transformation rates in tobacco and moth bean, *Plant Sci.,* 53, 87, 1987.

438. **Shillito, R. D., Saul, M. W., Paszkowski, J., Müller, M. and Potrykus, I.,** High efficiency direct gene transfer to plants, *Bio/Technology,* 3, 10, 1985.

439. **Rodenburg, K. W., de Groot, M. J. A., Schilperoort, R. A. and Hooykaas, P. J. J.,** Single-stranded DNA used as an efficient new vehicle for transformation of plant protoplasts, *Plant Mol. Biol.,* 13, 711, 1989.

440. **Nishiguchi, M., Sato, T. and Motoyoshi, F.,** An improved method for electroporation in plant protoplasts: infection of tobacco protoplasts by tobacco mosaic virus particles, *Plant Cell Rep.,* 6, 90, 1987.

441. **Morikawa, H., Iida, A., Matsui, C., Ikegami, M. and Yamada, Y.,** Gene transfer into intact plant cells by electroinjection through cell walls and membranes, *Gene,* 41, 121, 1986.

442. **Koning, A., Jones, A., Fillatti, J. J., and Lassner, M. W.,** Arrest of embryo development in *Brassica napus* mediated by modified *Pseudomonas aeruginosa* exotoxin A, *Plant Mol. Biol.,* 18, 247, 1992.

443. **Riggs, C. D. and Bates, G. W.,** Stable transformation of tobacco by electroporation: evidence for plasmid concatenation, *Proc. Natl. Acad. Sci. U.S.A.,* 83, 5602, 1986.

444. **Nishiguchi, M., Langridge, W. H. R., Szalay, A. A. and Zaitlin, M.,** Electroporation-mediated infection of tobacco leaf protoplasts with tobacco mosaic virus RNA and cucumber mosaic virus RNA, *Plant Cell Rep.,* 5, 57, 1986.

445. **Lassner, M. W., Jones, A., Daubert, S. and Comai, L.,** Targeting of T7 RNA polymerase to tobacco nuclei mediated by an SV40 nuclear location signal, *Plant Mol. Biol.,* 17, 229, 1991.

446. **Okada, K., Nagata, T. and Takebe, I.,** Co-electroporation of rice protoplasts with RNAs of cucumber mosaic and tobacco mosaic viruses, *Plant Cell Rep.,* 7, 333, 1988.

447. **Ou-Lee, T. M., Turgeon, R. and Wu, R.,** Expression of a foreign gene linked to either a plant-virus or a *Drosophila* promoter, after electroporation of protoplasts of rice, wheat, and sorghum, *Proc. Natl. Acad. Sci. U.S.A.,* 83, 6815, 1986.

448. **Zhang, H. M., Yang, H., Rech, E. L., Golds, T. J., Davis, A. S., Mulligan, B. J., Cocking, E. C. and Davey, M. R.,** Transgenic rice plants produced by electroporation-mediated plasmid uptake into protoplasts, *Plant Cell Rep.,* 7, 379, 1988.

449. **Battraw, M. J. and Hall, T. C.,** Histochemical analysis of CaMV 35S promoter-β-glucuronidase gene expression in transgenic rice plant, *Plant Mol. Biol.,* 15, 527, 1990.

450. **Dekeyser, R. A., Claes, B., De Rycke, R. M. U., Habets, M. E., Van Montagu, M. C. and Caplan, A. B.,** Transient gene expression in intact and organized rice tissues, *Plant Cell,* 2, 591, 1990.

451. **Meijer, E. G. M., Shilperoort, R. A., Rueb, S., van Os-Ruygrok, P. E. and Hensgens, L. A. M.,** Transgenic rice cell lines and plants: expression of transferred chimeric genes, *Plant Mol. Biol.,* 16, 807, 1991.

452. **Hauptmann, R. M., Vasil, V., Ozias-Akins, P., Tabaeizadeh, Z., Rogers, S. G., Fraley, R. T., Horsch, R. B. and Vasil, I. K.,** Evaluation of selectable markers for obtaining stable transformants in the gramineae, *Plant Physiol.,* 86, 602, 1988.

453. **Percival, F. W., Cass, L. G., Bozak, K. R. and Christoffersen, R. E.,** Avocado fruit protoplasts: a cellular model system for ripening studies, *Plant Cell Rep.,* 10, 512, 1991.

454. **Zakai, N., Ballas, N., Hershkovitz, M., Broido, S., Ram, R. and Loyter, A.,** Transient gene expression of foreign genes in preheated protoplasts: stimulation of expression of transfected genes lacking heat shock elements, *Plant Mol. Biol.,* 21, 823, 1993.

455. **Lawton, M. A., Dean, S. M., Dron, M., Kooter, J. M., Kragh, K. M., Harrison, M. J., Yu, L., Tanguay, L., Dixon, R. A. and Lamb, C. J.,** Silencer region of a chalcone synthase promoter contains multiple binding sites for a factor, SBF-1, closely related to GT-1, *Plant Mol. Biol.,* 16, 235, 1991.

456. **Bekkaoui, F., Pilon, M., Laine, E., Raju, D. S. S., Crosby, W. L. and Dunstan, D. I.,** Transient gene expression in electroporated *Picea glauca* protoplasts, *Plant Cell Rep.,* 7, 481, 1988.

457. **Jones, H., Ooms, G. and Jones, M. G. K.,** Transient gene expression in electroporated *Solanum* protoplasts, *Plant Mol. Biol.,* 13, 503, 1989.

458. **van der Steege, G., Nieboer, M., Swaving, J. and Tempelaar, M. J.,** Potato granule-bound starch synthase promoter-controlled GUS expression: regulation of expression after transient and stable transformation, *Plant Mol. Biol.*, 20, 19, 1992.

459. **Oard, J. H., Paige, D. and Dvorak, J.,** Chimeric gene expression using maize intron in cultured cells of breadwheat, *Plant Cell Rep.*, 8, 156, 1989.

460. **Klöti, A., Iglesias, V. A., Wünn, J., Burkhardt, P. K., Datta, S. K. and Potrykus, I.,** Gene transfer by electroporation into intact scutellum cells of wheat embryos, *Plant Cell Rep.*, 12, 671, 1993.

461. **Köhler, F., Golz, C., Eapen, S., Kohn, H. and Schieder, O.,** Stable transformation of moth bean *Vigna aconitifolia* via direct gene transfer, *Plant Cell Rep.*, 6, 313, 1987.

462. **Penza, R., Akella, V. and Lurquin, P. F.,** Transient expression and histological localization of a *gus* chimeric gene after direct transfer to mature cowpea embryos, *BioTechniques*, 13, 576, 1992.

463. **Akella, V. and Lurquin, P. F.,** Expression in cowpea seedlings of chimeric transgenes after electroporation into seed-derived embryos, *Plant Cell Rep.*, 12, 110, 1993.

464. **Fox, P. C., Vasil, V., Vasil, I. K. and Gurley, W. B.,** Multiple *ocs*-like elements required for efficient transcription of the mannopine synthase gene of T-DNA in maize protoplasts, *Plant Mol. Biol.*, 20, 219, 1992.

465. **Mascarenhas, D., Mettler, I. J., Pierce, D. A. and Lowe, H. W.,** Intron-mediated enhancement of heterologous gene expression in maize, *Plant Mol. Biol.*, 15, 913, 1990.

466. **Fromm, M. E., Taylor, L. P. and Walbot, V.,** Stable transformation of maize after gene transfer by electroporation, *Nature*, 319, 791, 1986.

467. **Olive, M. R., Walker, J. C., Singh, K., Dennis, E. S. and Peacock, W. J.,** Functional properties of the anaerobic responsive element of the maize *Adh 1* gene, *Plant Mol. Biol.*, 15, 593, 1990.

468. **D'Halluin, K., Bonne, E., Bossut, M., Beuckeleer, M. and Leemans, J.,** Transgenic maize plants by tissue electroporation, *Plant Cell*, 4, 1495, 1992.

469. **Fennell, A. and Hauptmann, R.,** Electroporation and PEG delivery of DNA into maize microspores, *Plant Cell Rep.,* 11, 567, 1992.

470. **Berka, R. M., Ward, M., Wilson, L. J., Hayenga, K. J., Kodama, K. H., Carlomagno, L. P. and Thompson, S. A.,** Molecular cloning and deletion of the gene encoding aspergillopepsin A from *Aspergillus awamori*, *Gene*, 86, 153, 1990.

471. **Ward, M., Wilson, L. J., Kodama, K. H., Rey, M. W. and Berka, R. M.,** Improved production of chymosin in *Aspergillus* by expression as a glucoamylase-chymosin fusion, *Bio/Technology*, 8, 435, 1990.

472. **Ward, M., Kodama, K. H. and Wilson, L. J.,** Transformation of *Aspergillus awamori* and *A. niger* by electroporation, *Exptl. Mycol.*, 13, 289, 1989.

473. **Ward, M., Wilson, L. J., Carmona, C. L. and Turner, G.,** The oli C3 gene of *Aspergillus niger:* isolation, sequence and use as a selectable marker for transformation, *Curr. Genet.*, 14, 37, 1988.

474. **Kasüske, A., Wedler, H., Schulze, S. and Becher, D.,** Efficient electropulse transformation of intact *Candida maltosa* cells by different homologous vector plasmids, *Yeast,* 8, 691, 1992.

475. **Howard, P. K., Ahern, K. G. and Firtel, R. A.,** Establishment of a transient expression system for *Dictyostelium discoideum*, *Nucleic Acids Res.*, 16, 2613, 1988.

476. **Egelhoff, T. T., Brown, S. S., Manstein, D. J. and Spudich, J. A.,** Hygromycin resistance as a selectable marker in *Dictyostelium discoideum*, *Mol. Cell. Biol.*, 9, 1965, 1989.

477. **Dynes, J. L. and Firtel, R. A.,** Molecular complementation of a genetic marker in *Dictyostelium* using a genomic DNA library, *Proc. Natl. Acad. Sci. U.S.A.*, 86, 7966, 1989.

478. **Chakraborty, B. N. and Kapoor, M.,** Transformation of filamentous fungi by electroporation, *Nucleic Acids Res.*, 18, 6737, 1990.

479. **Burland, T. G., Bailey, J., Adam, L., Mukhopadhyay, M. J., Dove, W. F. and Pallotta, D.,** Transient expression in *Physarum* of a chloramphenicol-acetyltransferase gene under the control of actin gene promoters, *Curr. Genet.*, 21, 393, 1992.

480. **Salek, A., Schnettler, R. and Zimmermann, U.,** Stably inherited killer activity in industrial yeast strains obtained by electrotransformation, *FEMS Microbiol. Lett.*, 96, 103, 1992.

481. **Rech, E. L., Dobson, M. J., Davey, M. R. and Mulligan, B. J.,** Introduction of a yeast artificial chromosome vector into *Saccharomyces cerevisiae* cells by electroporation, *Nucleic Acids Res.*, 18, 1313, 1990.

482. **Salek, A., Schnettler, R. and Zimmermann, U.,** Transmission of killer activity into laboratory and industrial strains of *Saccharomyces cerevisiae* by electroinjection, *FEMS Microbiol. Lett.*, 70, 67, 1990.

483. **Becker, D. M. and Guarente, L.,** High-efficiency transformation of yeast by electroporation, *Meth. Enzymol.*, 194, 182, 1991.

484. **Delorme, E.,** Transformation of *Saccharomyces cerevisiae* by electroporation, *Appl. Environ. Microbiol.*, 55, 2242, 1989.

485. **Hill, D. E.,** Integrative transformation of yeast using electroporation, *Nucleic Acids Res.*, 17, 8011, 1989.

486. **Karube, I., Tamiya, E. and Matsuoka, H.,** Transformation of *Saccharomyces cerevisiae* spheroplasts by high electric pulse, *FEBS Lett.*, 182, 90, 1985.

487. **Grey, M. and Brendel, M.,** A ten-minute protocol for transforming *Saccharomyces cerevisiae* by electroporation, *Curr. Genet.*, 22, 335, 1992.

488. **Meilhoc, E., Masson, J.-M. and Teissié, J.,** High efficiency transformation of intact yeast cells by electric field pulses, *Bio/Technology*, 8, 223, 1990.

489. **Hashimoto, H., Morikawa, H., Yamada, Y. and Kimura, A.,** A novel method for transformation of intact yeast cells by electroinjection of plasmid DNA, *Appl. Microbiol. Biotechnol.*, 21, 336, 1985.

490. **Simon, J. R. and McEntee, K.,** A rapid and efficient procedure for transformation of intact *Saccharomyces cerevisiae* by electroporation, *Biochem. Biophys. Res. Commun.*, 164, 1157, 1989.

491. **Tarassov, I. A. and Entelis, N. S.,** Mitochondrially-imported cytoplasmic tRNA^{Lys}(CUU) of *Saccharomyces cerevisiae: in vivo* and *in vitro* targetting systems, *Nucleic Acids Res.*, 20, 1277, 1992.

492. **Hood, M. T. and Stachow, C.,** Transformation of *Schizosaccharomyces pombe* by electroporation, *Nucleic Acids Res.*, 18, 688, 1990.

493. **Furfine, E. S. and Wang, C. C.,** Transfection of the *Giardia lamblia* double-stranded RNA virus into *Giardia lamblia* by electroporation of a single-stranded RNA copy of the viral genome, *Mol. Cell. Biol.*, 10, 3659, 1990.

494. **Laban, A. and Wirth, D. F.,** Transfection of *Leishmania enriettii* and expression of chloramphenicol acetyltransferase gene, *Proc. Natl. Acad. Sci. U.S.A.*, 86, 9119, 1989.

495. **Brunk, C. F. and Navas, P.,** Transformation of *Tetrahymena thermophila* by electroporation and parameters affecting cell survival, *Exp. Cell Res.*, 174, 525, 1988.

496. **Kamdar, P., Von Allmen, G. and Finnerty, V.,** Transient expression of DNA in *Drosophila* via electroporation, *Nucleic Acids Res.*, 20, 3526, 1992.

497. **Mann, S. G. and King, L. A.,** Efficient transfection of insect cells with baculovirus DNA using electroporation, *J. Gen. Virol.*, 70, 3501, 1989.

498. **Inoue, K., Yamashita, S., Hata, J., Kabeno, S., Asada, S., Nagahisa, E. and Fujita, T.,** Electroporation as a new technique for producing transgenic fish, *Cell Diff. Devel.*, 29, 123, 1990.

499. **Hayasaka, K., Sato, M., Mitani, H. and Shima, A.,** Transfection of cultured fish cells RBCF-1 with exogeneous oncogene and their resistance to malignant transformation, *Comp. Biochem. Physiol.*, 96B, 349, 1990.

500. **Showe, M. K., Williams, D. L. and Showe, L. C.,** Quantitation of transient gene expression after electroporation, *Nucleic Acids Res.*, 20, 3153, 1992.

501. **Toneguzzo, F. and Keating, A.,** Stable expression of selectable genes introduced into human hematopoietic stem cells by electric field-mediated DNA transfer, *Proc. Natl. Acad. Sci. U.S.A.,* 83, 3496, 1986.

502. **Narayanan, R., Jastreboff, M. M., Chiu, C. F. and Bertino, J. R.,** *In vivo* expression of a nonselected gene transferred into murine hematopoietic stem cells by electroporation, *Biochem. Biophys. Res. Commun.,* 141, 1018, 1986.

503. **Saijo, Y., Perlaky, L., Valdez, B. C., Busch, R. K., Henning, D., Zhang, W. W. and Busch, H.,** The effect of antisense p120 construct on p120 expression and cell proliferation in human breast cancer MCF-7 cells, *Cancer Lett.,* 68, 95, 1993.

504. **Ziober, B. L., Willson, J. K. V., Hymphrey, L. E., Childress-Fields, K. and Brattain, M. G.,** Autocrine transforming growth factor-α is associated with progression of transformed properties in human colon cancer cells, *J. Biol. Chem.,* 268, 691, 1993.

505. **Chu, G., Hayakawa, H. and Berg, P.,** Electroporation for the efficient transfection of mammalian cells with DNA, *Nucleic Acids Res.,* 15, 1311, 1987.

506. **Hama-Inaba, H., Sato, K., Nishimoto, T., Ohtsubo, M. and Kasai, M.,** Establishment and application of a standard method of electroporation for introduction of plasmid and cosmid DNAs to mammalian cells, *Bioelectrochem. Bioenerg.,* 21, 355, 1989.

507. **Boggs, S. S., Gregg, R. G., Borenstein, N. and Smithies, O.,** Efficient transformation and frequent single-site, single-copy insertion of DNA can be obtained in mouse erythroleukemia cells transformed by electroporation, *Exp. Hematol.,* 14, 988, 1986.

508. **Potter, H., Weir, L. and Leder, P.,** Enhancer-dependent expression of human κ immunoglobulin genes introduced into mouse pre-B lymphocytes by electroporation, *Proc. Natl. Acad. Sci. U.S.A.,* 81, 7161, 1984.

509. **Andreason, G. L. and Evans, G. A.,** Introduction and expression of DNA molecules in eukaryotic cells by electroporation, *BioTechniques,* 6, 650, 1988.

510. **Yorifuji, T. and Mikawa, H.,** Co-transfer of restriction endonucleases and plasmid DNA into mammalian cells by electroporation: effects on stable transformation, *Mut. Res.,* 243, 121, 1990.

511. **Yorifuji, T., Tsuruta, S. and Mikawa, H.,** The effect of cell synchronization on the efficiency of stable gene transfer by electroporation, *FEBS Lett.,* 245, 201, 1989.

512. **Yang, J.-T., Wang, X.-Q. and Wu, M.,** Transfer of foreign genes by electroporation and their expression in mammalian cells, *Sci. China B.,* 32, 543, 1989.

513. **Ray, J. and Gage, F. H.,** Gene transfer into established and primary fibroblast cell lines: comparison of transfection methods and promoters, *BioTechniques,* 13, 598, 1992.

514. **Chen, L. S., Couwenhoven, R. I., Hsu, D., Luo, W. and Snead, M. L.,** Maintenance of amelogenin gene expression by transformed epithelial cells of mouse enamel organ, *Archs. Oral. Biol.,* 37, 771, 1992.

515. **Tur-Kaspa, R., Teicher, L., Levine, B. J., Skoultchi, A. I. and Shafritz, D. A.,** Use of electroporation to introduce biologically active foreign genes into primary rat hepatocytes, *Mol. Cell. Biol.,* 6, 716, 1986.

516. **Oshima, A., Itho, K., Nagao, Y., Sakuraba, H. and Suzuki, Y.,** β-Galactosidase-deficient human fibroblasts: uptake and processing of the exogeneous precursor enzyme expressed by stable transformant COS cells, *Hum. Genet.,* 85, 505, 1990.

517. **Kluxen, F.-W. and Lübbert, H.,** Maximal expression of recombinant cDNAs in COS cells for use in expression cloning, *Anal. Biochem.,* 208, 352, 1993.

518. **Doncheva, J., Georgiev, O., Milchev, G. and Tsoneva, I.,** Stable transfer of plasmid pSV3 *neo* in CV-1 cells by electroporation, *Bioelectrochem. Bioenerg.,* 26, 339, 1991.

519. **Aran, J. M. and Plagemann, P. G. W.,** Nucleoside transport-deficient mutants of PK-15 pig kidney cell line, *Biochim. Biophys. Acta,* 1110, 51, 1992.

520. **Toneguzzo, F., Hayday, A. C. and Keating, A.,** Electric field-mediated DNA transfer: transient and stable gene expression in human and mouse lymphoid cells, *Mol. Cell. Biol.,* 6, 703–706, 1986.

521. **McNally, M. A., Lebkowski, J. S., Okarma, T. B. and Lerch, L. B.,** Optimizing electroporation parameters for a variety of human hematopoietic cell lines, *BioTechniques*, 6, 882, 1988.
522. **Sudduth-Klinger, J. Schumann, M., Gardner, P. and Payan, D. G.,** Functional and immunological responses of Jurkat lymphocytes transfected with the substance P receptor, *Cell. Mol. Neurobiol.*, 12, 379, 1992.
523. **Thorn, J. T., Todd, A. V., Croaker, G. M. and Iland, H. J.,** Characterization of K562 cells following introduction of a mutant N-*RAS* gene, *Leuk. Res.*, 17, 23, 1993.
524. **Ulrich, M. J. and Ley, T. J.,** Function of normal and mutated γ-globin gene promoters in electroporated K562 erythroleukemia cells, *Blood*, 75, 990, 1990.
525. **Takahashi, M., Furukawa, T., Nikkuni, K., Aoki, A., Nomoto, N., Koike, T., Moriyama, Y., Shinada, S. and Shibata, A.,** Efficient introduction of a gene into hematopoietic cells in S-phase by electroporation, *Exp. Hematol.*, 19, 343, 1991.
526. **Tykocinski, M. L., Shu, H.-K., Ayers, D. J., Walter, E. I., Getty, R. R., Groger, R. K., Hauer, C. A. and Medof, M. E.,** Glycolipid reanchoring of T-lymphocyte surface antigen CD8 using the 3′end sequence of decay-accelerating factor's mRNA, *Proc. Natl. Acad. Sci. U.S.A.*, 85, 3555, 1988.
527. **Spandidos, D. A.,** Electric field-mediated gene transfer (electroporation) into mouse friend and human K562 erythroleukemic cells, *Gene. Anal. Techn.*, 4, 50, 1987.
528. **Teshigawara, K. and Katsura, Y.,** A simple and efficient mammalian gene expression system using an EBV-based vector transfected by electroporation in G2/M phase, *Nucleic Acids Res.*, 20, 2607, 1992.
529. **Ryan, J. J., Danish, R., Gottlieb, C. A. and Clarke, M. F.,** Cell cycle analysis of p53-induced cell death in murine erythroleukemia cells, *Mol. Cell. Biol.*, 13, 711, 1993.
530. **Dembic, Z., Haas, W., Zamoyska, R., Parnes, J., Steinmetz, M. and von Boehmer, H.,** Transfection of the CD8 gene enhances T-cell recognition, *Nature*, 326, 510, 1987.
531. **Rüker, F., Liegl, W., Mattanovich, D., Reiter, S., Himmler, G., Jungbauer, A. and Katinger, H.,** Electroporative gene transfer (Electrotransfection): a method for strain improvement of animal cells, *Bioelectrochem. Bioenerg.*, 17, 253, 1987.
532. **Uemura, N., Ozawa, K., Tojo, A., Takahashi, K., Okano, A., Karasuyama, H., Tani, K. and Asano, S.,** Acquisition of interleukin-3 independence in FDC-P2 cells after transfection with the activated c-H-*ras* gene using a bovine papilloma virus based plasmid vector, *Blood*, 80, 3198, 1992.
533. **Anderson, M. L. M., Spandidos, D. A. and Coggins, J. R.,** Electroporation of lymphoid cells: factors affecting the efficiency of transfection, *J. Biochem. Biophys. Meth.*, 22, 207, 1991.
534. **Bos, R. and Nieuwenhuizen, W.,** Enhanced transfection of a bacterial plasmid into hybridoma cells by electroporation: application for the selection of hybrid hybridoma (Quadroma) cell lines, *Hybridoma*, 11, 41, 1992.
535. **Evans, G. A., Ingraham, H. A., Lewis, K., Cunningham, K., Seki, T., Moriuchi, T., Chang, H. C., Silver, J. and Hyman, R.,** Expression of the Thy-1 glycoprotein gene by DNA-mediated gene transfer, *Proc. Natl. Acad. Sci. U.S.A.*, 81, 5532, 1984.
536. **Iannuzzi, M. C., Weber, J. L., Yankaskas, J., Boucher, R. and Collins, F. S.,** The introduction of biologically active foreign genes into human respiratory epithelial cells using electroporation, *Am. Rev. Respir. Disease*, 138, 965, 1988.
537. **Zerbib, D., Amalric, F. and Teissié, J.,** Electric field mediated transformation: isolation and characterization of a TK+ subclone, *Biochem Biophys. Res. Commun.*, 129, 611, 1985.
538. **Barsoum, J.,** Laboratory Methods. Introduction of stable high-copy-number DNA into Chinese hamster ovary cells by electroporation, *DNA Cell Biol.*, 9, 293, 1990.
539. **Winterbourne, D. J., Thomas, S., Hermon-Taylor, J., Hussain, I. and Johnstone, A. P.,** Electric shock-mediated transfection of cells, *Biochem. J.*, 251, 427, 1988.
540. **Nickoloff, J. A. and Reynolds, R. J.,** Electroporation-mediated gene transfer efficiency is reduced by linear plasmid carrier DNAs, *Anal. Biochem.*, 205, 237, 1992.

541. **Wolf, H., Rols, M.-P., Boldt, E., Neumann, E. and Teissié, J.,** Control by pulse parameters of electric field-mediated gene transfer in mammalian cells, *Biophys. J.*, 66, 524, 1994.

542. **Reiss, M., Jastreboff, M. M., Bertino, J. R. and Narayanan, R.,** DNA-mediated gene transfer into epidermal cells using electroporation, *Biochem. Biophys. Res. Commun.*, 137, 244, 1986.

543. **Titomirov, A. V., Sukharev, S. and Kistanova, E.,** *In vivo* electroporation and stable transformation of skin cells of newborn mice by plasmid DNA, *Biochim. Biophys. Acta*, 1088, 131, 1991.

544. **Schnabl, H. and Zimmermann, U.,** Immobilization of plant protoplasts, in *Biotechnology in Agriculture and Forestry, Vol. 8, Plant Protoplast and Genetic Engineering,* Bajaj, Y. P. S., Ed., Springer-Verlag, Berlin, 1989, 63.

545. **Goldberg, E. P.,** *Targeted Drugs*, John Wiley & Sons, New York, 1983.

546. **Orlowski, S. and Mir, L. M.,** Cell electropermeabilization: a new tool for biochemical and pharmacological studies, *Biochim. Biophys. Acta,* 1154, 51, 1993.

547. **Mir, L. M., Orlowski, S., Belehradek, J. Jr., and Paoletti, C.,** Electrochemotherapy: potentiation of antitumor effect of bleomycin by local electric pulses, *Eur. J. Cancer*, 27, 68, 1991.

548. **Mir, L. M., Belehradek, M., Domenge, C., Orlowski, S., Poddevin, B., Belehradek, J. Jr., Schwaab, G., Luboinski, B. and Paoletti, C.,** L'électrochimiothérapie, un nouveau traitement antitumoral: premier essai clinique, *C. R. Acad. Sci. Paris,* 313, 613, 1991.

549. **Belehradek, J. Jr., Orlowski, S., Poddevin, B., Paoletti, C. and Mir, L. M.,** Electrochemotherapy of spontaneous mammary tumours in mice, *Eur. J. Cancer*, 27, 73, 1991.

Chapter 2

ELECTROTRANSFORMATION OF BACTERIA

Akira Taketo

CONTENTS

I. Introduction 107

II. Conditions for Electrotransformation 109
  A. Culture Conditions 109
  B. Preparation of Recipient Cells.. 110
  C. Electric Field Parameters 111

III. Resealing and Recovery 114

IV. Transforming DNA 116

V. Potential Problems... 120

VI. Mechanisms of Bacterial Electropermeabilization 120

VII. Future Applications 128

References 129

I. INTRODUCTION

Since 1944 bacterial transformation has been extensively used for the characterization of biologically active DNA and the detailed genetic analysis of microbial cells.[1] Modern genetic engineering technology now depends heavily on transformation systems in prokaryotic, as well as in eukaryotic, cells. Various techniques have been developed for the introduction of foreign DNA into bacterial cells, most notably electropermeabilization (also termed electroporation, electroinjection, or electrotransformation — see Chapter 1). This technology is universally applicable to a wide range of genera/species, covering gram-positive as well as gram-negative cocci and bacilli. It is more efficient and reproducible than chemical (e.g., Ca^{2+}-dependent transfection) techniques and has consequently now become the method of choice for bacterial genetic manipulation.

Reversible membrane breakdown and electrically mediated transcellular ion flow in bacteria was first described more than 20 years ago.[2,3] However, electroinjection of intact bacteria for the purpose of DNA transfection awaited the development of the requisite recombinant DNA technology and was not reported until some 10 years later, in the mid 1980s. Electrotransformation (electric field-mediated DNA transformation) was first applied to *Bacillus cereus* protoplasts[4] and later to *Streptomyces lividans* protoplasts.[5]

Escherichia coli was the initial intact bacterial species electrotransformed,[6,7] and the technique was soon applied to other bacterial species including *Lactococcus lactis*.[8] The success of early investigators in the field and the ready availability of inexpensive instruments resulted in a virtual explosion in the use of electropermeabilization technology for bacterial transformation (see also References 9 and 10). Representatives of virtually all known genera have now been used for such experiments. The most remarkable organism to be successfully transformed is the extremely thermophilic archaeon *Sulfolobus*.[11]

Although the basic principles of reversible electric membrane breakdown discussed in Chapter 1 provided the impetus and early direction for bacterial electrotransformation, the field has been subject to an enormous amount of empiricism. This was due, in part, to the early recognition that electromanipulation of bacteria would present special challenges because of their very small size (radius of about 1 μm), lack of internal organelles, non-spherical shape, and their thick, complex envelope.* In addition, their turgor pressure is very high. These properties complicate the practical and theoretical aspects of electropermeabilization of these organisms.

As is the case for intact plant and yeast cells, the most important challenge to successful electromanipulation of bacteria is the existence of the bacterial cell envelope. The cell wall is often porous, highly conductive, and thus does not lend itself to the generation of sufficiently high electric potentials to permit breakdown (see Chapters 1 and 3). The application of extrinsic electric fields in this setting may also be associated with a number of complications, particularly if more conductive media are used. Empirically derived protocols have been developed using quite different strategies than those derived from the application of membrane biophysical principles (see Chapter 1). These demonstrated that, under appropriate circumstances, it was possible to perform electrotransfer in the presence of the bacterial cell envelope. However, it must be noted that such methods rely as heavily upon biological and biochemical manipulations of the cell envelope as they do upon electric field manipulation. These methods are generally based upon the application of single long duration (millisecond range) pulses, relatively low field strength pulses generated by widely available, and low cost electropermeabilization apparatuses. The protocols are usually performed in the presence of low conductivity solutions. It has been sufficient for most workers in the field to have protocols that "work,"

---

* The envelope comprises the cell wall and the cell membranes.

and there is little doubt that the current "user friendly" methods described below achieve this end.

II. CONDITIONS FOR ELECTROTRANSFORMATION

A. CULTURE CONDITIONS

The conditions under which bacteria are cultured have significant consequences for the state of the cell envelope as well as for other critical parameters. These would include the fraction of cells in cycle, susceptibility to selection media, spore formation, and others. Critical cell envelope factors probably include net charge, density (thickness), protein content, or other factors yet to be revealed. Taken together, such factors determine the "competence" of the bacteria for electrotransformation. In certain bacteria, complicated culture conditions are required for the induction of natural "competence" for plasmid transfer. However, to attain "electrocompetence,"* that is the ability of bacterial cells to be successfully electrotransformed, specified medium or conditioning is not usually necessary. This is not to state that moderate differences in electrotransformation efficiency may vary with culture conditions, even within the same strain. Nevertheless, bacterial cells grown in conventional liquid medium (e.g., L-broth) at optimal temperatures may be efficiently transformed upon application of high intensity electric fields. Although much less efficient, even cells grown on solid media such as agar have been successfully transformed by electropermeabilization without the prior removal of the cell wall.[12] This underscores the importance of biological and biochemical factors in electrotransformation (see also Chapter 1).

The bacterial growth phase has a significant effect on electrotransformation. Microbial cells harvested at early- or mid-logarithmic phase of growth are most commonly used for electrotransformation and are widely thought to be most suitable for this purpose. This is perhaps because the cell envelope is most permissive during this phase, although this has not been documented. Such differences might parallel the cell cycle dependence observed in eukaryotic cells (see Chapter 1) or have other undocumented effects upon the cell envelope. Occasionally, late logarithmic phase cells are used as recipient cells and for certain specific purposes, stationary phase bacteria are chosen.[13-18] The transformation efficiencies of *Streptococcus sanguis*,[19] *Brevibacterium sp.* R312,[20] and *Pasteurella multocida*,[21] have been reported to be independent of

---

* The term "electrocompetence" is an operational one and evolved from the concept of bacteria that were suitable for chemical transformation i.e., "competent cells." As noted in Chapter 1, it is possible to induce reversible breakdown in the membrane of any living cell, once the appropriate conditions have been discovered. However, as detailed in the text, the presence of the cell wall in bacteria brings into play a unique series of problems for electromanipulation that can be overcome, in part, by selection, cell culture, media additives, and so on. Such selected and/or prepared bacterial cells are thus termed "electrocompetent." It is of interest to note that the proportion of bacterial species that is at least somewhat susceptible to transformation by electric means is very large and certainly greater than when chemical means are used.

the growth phase. Since the content of lipoteichoic acid, hyaluronic acid, autolytic activity, or degradative enzymes varies in dependence on growth phase, it may in part contribute to phase differences in electrotransformation efficiency.

Growth temperature is one easily manipulated culture condition that influences the efficiency of electrotransformation. For example, *E. coli* grown at temperatures of 40–42°C may be more efficiently transformed by electropermeabilization than identical strains grown at 30–35°C.[22] Similar effects have been reported for *Acidiphilium* PW1.[23] It could be argued that the influence of the growth temperature on electrotransformation is related to the bacterial membrane fluidity, but this remains unproven.

A related strategy to enhance transformation efficiency involves supplementation of the culture medium with additives to "weaken" the cell envelope and thereby improve transformability by increasing the sensitivity of the cell envelope to electric fields. Such supplements vary with the species and strain to be transformed and include antibiotics (ampicillin,[24] or penicillin[25,26]), glycine,[13,27-32] isonicotinic acid hydrazide,[30,32] lysostaphin,[33] lysozyme,[34] sucrose,[25,35] threonine,[36-40] and Tween 80.[29] Pretreatment with hyaluronidase was effective in increasing the transformation efficiency of *Streptococcus pyogenes*.[41] Unlike Ca^{2+}-dependent transfection,[42,43] the efficiency of electrotransformation is not influenced by surface lipopolysaccharide[44] and may even be reduced by supplemental O-antigen.[45]

The recently described technology, cell electrorotation, is an exciting and potentially valuable technique for assessing the effect of various manipulations and supplements on the cell envelope (discussed in detail in Chapter 5). This noninvasive method can very easily be used to determine the surface conductance of the membrane and/or of the cell envelope. Such information might be invaluable both for elucidating the mechanisms underlying bacterial cell envelope electropermeabilization and also for optimizing the conditions for electrotransfer of macromolecules.

B. PREPARATION OF RECIPIENT CELLS

When low cost apparatuses are used, extensive washing of the recipient bacteria is necessary to remove medium components that can otherwise severely disrupt the bacterial electrotransformation process. In most cases, the cells are harvested, washed by centrifugation, first with chilled deionized and distilled water or 1 mM HEPES (pH 7), and finally with the buffer or solution to be used for electrotransformation. Low ionic strength sugar solutions are preferred for bacterial electrotransfer. Glycerol (10–15%) or sucrose (0.27–0.5 M) with or without supplementation of 1 mM $MgCl_2$, HEPES (1–7 mM), or phosphate (pH 7–8) have been used most successfully. In addition to the medium mentioned above, 10–30% polyethylene glycol 6000 (with or without 0.1 M mannitol supplementation), 0.5 M raffinose or sorbitol, or distilled water may also be used for electroinjection. Several bacteria such as *Methanococcus*[46] require special ionic conditions for their stabilization. The suspension density* of the bacteria is adjusted to 10^9-10^{11} cells ml^{-1}.

Even in conventional genera, certain strains release significant amounts of UV-absorbing material, when suspended in deionized water. This presumably results from the release of cytosolic contents as a result of a mild osmotic shock. For these strains, one should minimize the time they are kept in low ionic strength medium. The washed cells, once suspended in electrotransfer medium containing 10–15% glycerol can be cryopreserved for fairly long periods of time at –70 to –80°C. Suitable frozen "electrocompetent" *E. coli* strains are now commercially available in just such a format.

C. ELECTRIC FIELD PARAMETERS

The field parameters for cell membrane breakdown and electroinjection have been discussed in detail in Chapter 1. As is the case for electropermeabilization of eukaryotic cells, the critical electric parameters for bacterial electrotransformation are:

1. field strength,
2. pulse shape,
3. pulse duration.

The field strength calculation given in Equation 1 of Chapter 1 can be modified as necessary for application to non-spherical cells. Many bacterial species used for electrotransformation experiments are cylindrical in shape, i.e., "rods," rather than spherical cells. The shape of these bacteria also dictates that a "non-uniform" orientation within the field alters the calculations. For such rod bacteria, the equation is rewritten as follows (it must be noted that these calculations hold only for the bacterial membrane and **not** the cell envelope):

$$V_c = f_s E_c a \tag{1}$$

Here f_s is the "shape factor" which is about 1 for bacteria oriented with their long axis parallel to the external field and about 2 for bacteria aligned perpendicularly to the electric field lines.[3,47] As for eukaryotic cells, the reversible breakdown voltage, V_c, is about 1 V. As detailed in Chapter 1, the breakdown voltage is also pulse length and turgor pressure dependent. With millisecond duration pulses, the breakdown voltage may fall to the order of 0.5 V or lower.

A field strength of about 12–20 kV cm^{-1} is frequently reported as being satisfactory for bacterial transformation. As shown in Figure 1a, a field strength of 17.5 kV cm^{-1} is optimal for transformation of *E. coli* C suspended in 10% polyethylene glycol or in 10% glycerol.

---

*   At high cell densities, i.e., about 10^{11} cells ml^{-1}, bacterial "sludge" is noted that can interfere with the electrotransformation process. The lower end of the density range is thus preferred for most applications. Increased bacterial suspension viscosity may also be noted, a consequence of cell death and release of macromolecules including nucleic acids.

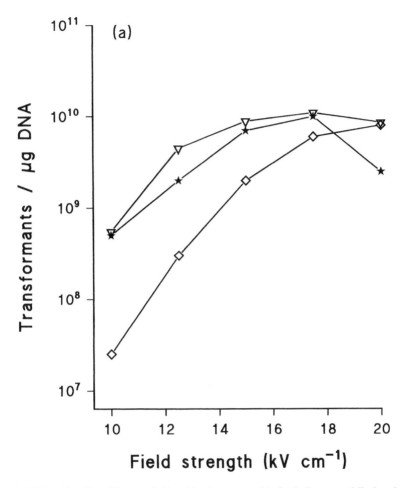

FIGURE 1. The effect of the strength (a) and the time constant (b) of a single exponentially decaying electric field pulse on bacterial transformation efficiency. The plasmid pBR322 (0.09 μg ml⁻¹) was electroinjected into *E. coli C* cells suspended in 10% polyethylene glycol (PEG) at a suspension density of (10¹¹ cells ml⁻¹) using a Gene Pulser (Bio-Rad, 25 μF capacitor). Pulse application was performed at room temperature using cells pre-cooled to 4°C. The pulsed cells were kept for an additional 2 min on ice and then transferred into complete (L-broth) medium and then incubated for an additional 30 min at 37°C to allow resealing. (a) Transformation efficiency as a function of the field strength for different settings of the pulse resistor (= different time constants of the exponentially decaying pulse): 200 Ω (◊), 600 Ω (★), and 800 Ω (▽). (b) Transformation efficiency as a function of the time constant for various field strengths: 10 kV cm⁻¹ (○), 12.5 kV cm⁻¹ (△), 15 kV cm⁻¹ (◊), and 20 kV cm⁻¹ (▼).

Virtually all bacterial electrotransfer has been performed with exponentially decaying pulses generated by capacitor discharge devices. The time constant of the exponentially decaying pulse reported in the literature varies from 4–12 ms.* At

---

*   The time constants often quoted in the literature are derived from capacitance and resistance
    settings or display panels of the electropermeabilization apparatus used. As such these
    "parameters" may not necessarily reflect the true time constants.

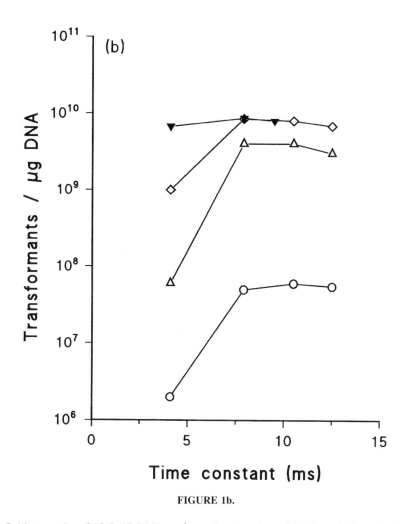

FIGURE 1b.

field strengths of 12.5–17.5 kV cm⁻¹, a pulse duration of 8–11 ms is found to be efficient for the transformation of *E. coli* (Figure 1b). It should be noted that the time constant can change during the discharging process as a result of pulse-mediated cellular disruption with a consequent increase in the media conductivity.

There are occasional reports of alternating current protocols.[48] Very few papers describe the successful use of short pulse durations (30–50 μs) applied by square-wave generating apparatuses[49,50] (for a more rigorous discussion, see Chapter 1). Unfortunately, it is difficult to compare these studies to the majority of studies which have used exponentially decaying pulses.

Application of a single exponential pulse is most widely used for bacterial transformation, except in a few cases where 3 to 10 pulses are applied in succession.[50,51] Indeed, repeated pulses have been found to reduce the fraction of cells surviving electrotransfer, without improving transformation efficiency. As noted above, long duration pulses (in the millisecond range) are not considered to be optimal for reversible membrane breakdown owing to their propensity to produce salt shifts, punch-through, local Joule heating, and other deleterious effects in more conductive solutions (see Chapters 1 and 6). Such conditions may occur if there is substantial bacterial cell death prior to or during the pulse application with release of organic ions into the media. However, as further discussed in Chapter 1 and below, the application of such long duration pulses may have the capacity to "open" or disrupt the cell envelope.

The electric parameters chosen for electrotransformation probably reflect a compromise between optimal reversible membrane breakdown and cell envelope disruption. As noted, they often reflect the field strength and pulse duration limitations imposed by widely distributed apparatuses. Thus field parameters derived from "trial and error" experiments have evolved and have come into common use.

III. RESEALING AND RECOVERY

The bacterial survival rate markedly increases if the electroinjected cells are post-incubated in liquid complete growth medium at high suspension density, at 37°C for 30 min (Figure 2). This time is very similar to that observed for eukaryotic cells. This is, nevertheless, relatively long with respect to the duration of the bacterial cell cycle. Recovery in complete growth medium is optimal for both transformant yield and cell viability. In contrast, recovery in pulse medium results in decreased yields. When a cell suspension electroinjected in 10% polyethylene glycol 6000 was incubated at 37°C for 30 min in this same solution, the transformant yield was reduced to nearly 0.4% of a control culture that was incubated in L-broth.

Neither heat shock at 42°C in the electropermeabilization medium nor preservation at 0°C in L-broth appears to be optimal for transfer or recovery. Also of interest is the apparent superiority of survival of bacterial cells grown first in suspension and then in the solid phase compared to immediate transfer to agar (Figure 2). This might be explained by diffusion deficiencies of nutrients essential for recovery in the agar.

Exceptions to the above protocol exist. This is due, in part, to varying conditions for phenotypic expression depending on bacterial strains used and selection markers. For example *Pseudomonas aeruginosa* cells were successfully electrotransformed with pLAFR1 after incubation in SOC[52] at 0°C for 30 min and then at 37°C for 2 h. It should be noted, however, that bacterial multiplication during a longer recovery time period can result in an overestimation of the efficiency of transformation.

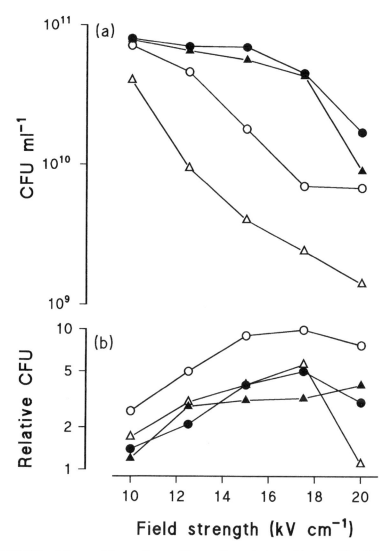

FIGURE 2. Influence of the resealing conditions on the survival rate of the pulsed *E. coli* C cells (in the presence of 0.09 µg ml^{-1} pBR322) determined by counting the colony forming units (CFU) ml^{-1}. (a) Total number of CFU after electropermeabilization in 10% PEG (circles) or 10% glycerol (triangles) media as a function of the field strength (capacitor of the pulser: 25 µF, resistor of the pulser 600 Ω). Open symbols (○, △): pulsed cells were immediately transferred to complete agar at 37°C without preceeding resealing in liquid complete L-broth medium. Closed symbols (●, ▲): CFU after 4 fold dilution of the pulsed sample with L-broth medium and resealing at 37°C for 30 min before the cells were plated on agar. (b) Relative number of CFU as a function of the field strength of the applied pulse. The relative number of CFU was determined by subjecting one part of the pulsed cells to resealing conditions in liquid L-broth medium at 37°C (see Figure 2a), whereas the other part was directly plated on solid agar. The relative number of CFU is defined as the ratio of CFUs obtained after pre-resealing in liquid medium to those obtained after direct transfer to agar. Pulsing was performed in PEG (●, ▲) or in glycerol solutions (○, △) for different pulser resistors: 200 Ω (▲), 600 Ω (○, ●), and 800 Ω (△).

IV. TRANSFORMING DNA

For electrotransformation of bacteria, double-stranded circular DNA with or without an insert derived from various organisms has been almost exclusively used. Transformation using relaxed or nicked and re-ligated forms gives an efficiency nearly equal to that obtained with supercoiled circular DNA. On the other hand, linearized double-stranded plasmid DNA has only feeble activity in the bacterial transformation system.[24,34,35,53-59] Whether this low activity is inherent to the linear form or to a residual circular DNA fraction that escapes nuclease attack is unknown. Intracellular degradation of the linear DNA by *E. coli rec* BC nuclease (exonuclease V)-like activity may partially explain the results obtained with the linear form. An exception has been observed for *Methanococcus voltae,* which may be successfully electrotransformed using chromosomal, but not plasmid DNA.[46] In *Brevibacterium lactofermentum,*[24] the transformation efficiency of pBRL is reduced by denaturation, from 8×10^5 transformants per µg DNA to 3.4×10^5 transformants per µg DNA. Similarly, the yield of transfectants in *E. coli* was higher for the double-stranded replicative form than the single-stranded phage DNA.[60] Even for the single-stranded DNA electropermeabilization is considerably more effective than Ca^{2+}-dependent transfection.

In general, the efficiency of electrotransformation decreases with an increase in plasmid size.[38,40,61-65] According to some authors,[59] an inverse relationship exists between the plasmid size and the transfection efficiency.

Nevertheless, plasmids ranging from 2.7 Kb (pUC18)[52] to 50 Kb (pHMC5)[64] are successfully introduced into bacteria, and efficient electropermeabilization of large plasmids (136–300 Kb) has been recently reported in *E. coli.*[63,66] Moreover, the transformation efficiency of certain bacteria is not significantly affected by plasmid size (4.4–26.5 Kb).[34,67] It should, however, be taken into consideration that these results have been obtained using plasmids differing not only in size, but also in DNA sequence. It is quite likely that many factors apart from plasmid size per se contribute to the final transformation efficiency.

For most bacterial species, a linear relationship exists between the number of the transformants and the amount of input DNA. As seen in Figure 3a, the yield of the transformants of *E. coli* was directly proportional to the concentration of pBR322 or pUC19, over a wide range. The "saturation level" differs significantly, depending on recipients and plasmids. Thus, the transformant yield reached a plateau at 1 ng pHY416 DNA in *Corynebacterium glutamicum* AS019[53] and 0.1 µg pRF29 DNA in *Rhodococcus fascians,*[68] whereas the saturation level was attained at 10 µg of pC194 in *Bacillus subtillis.*[69] In certain cases, a number of the transformants did not plateau, even at 10–15 µg of DNA.[70-72] Conventional plasmids, if properly prepared, may be used at a final concentration of 1–2 µg ml⁻¹ or close to the concentration that has proven successful for eukaryotic cells.

The efficiency of transformation (i.e., the number of transformants per μg DNA) increases parallel to the amount of DNA, until a plateau is reached. On the other hand, the yield of electro-transformants or -transfectants rises continuously, corresponding to the density of the recipient bacteria. As shown in Figure 3b, the saturation level in electropermeabilization is very high, as compared with transformation of naturally competent bacteria or in ordinal phage infection. This fact suggests absence of specific adsorption process in the recipient host for field-mediated uptake of DNA.

The transformation frequency (the number of transformants per total colony forming units) is related to the survival of the electropermeabilized bacteria.* Several bacterial species are exceptionally tolerant to high electric fields. No direct correlation between transformation and cell survival has been observed for these species.[68,73] If the amount of available DNA is limited, the efficiency will be more important than the frequency, whereas the frequency may be a critical factor when the selection method is inefficient. Conditions have been reported, under which 80–100% of *E. coli* cells can be transformed by electropermeabilization.[74] Such efficient methods are likely to be increasingly required as additional demands are placed upon the existing technology.

It has been reported that the efficiency of electropermeabilization-mediated transfection of mammalian cells is improved by the addition of carrier DNA.[75] However, at least one study has shown that supplementation with linearized DNA substantially reduced the efficiency of electrotransformation.[76] Such a so-called DNA "carrier effect" has not been observed with heterologous DNA (native and denatured) nor with RNA in bacterial transfection.[22] Similar ineffectiveness of the carrier DNA has been demonstrated in electrotransformation of *E. coli*.[77] On the other hand, contaminating nucleic acids seem to have little effect on the transformation frequency. This implies that it is not necessary to thoroughly purify DNA preparations prior to electropermeabilization of bacteria.

In recombinant DNA experiments, a ligation mixture is often used for transformation. Dialysis,[78] dilution,[79] ethanol precipitation,[80] or heat treatment[81] may be applied for removal of interfering salts or ligase in the mixture. Unless isogenic or "restrictionless" bacterial strains are used, there is a risk of endonucleolytic degradation of transfected DNA within the recipient cell. Unmodified plasmid DNA is especially vulnerable to restriction, even in recipients belonging to the same or closely related genera such as *E. coli* and *Salmonella typhimurium*.[44] In order to circumvent this difficulty, one can consider using plasmid DNA that has been propagated in the expected recipient (thus leading to autologous methylation *in vivo*), modification *in vitro* of the to be transfected DNA by methylase, or selection of a restrictionless host.

Electroinjection can also be used to efficiently release or extract plasmid molecules from bacterial cells.[72] This technique has been applied for direct transfer of plasmid DNA between different bacterial strains[17] or genera

---

* Transformation frequency is calculated by determining the colony count of bacteria on selection medium post-transfection divided by the number of bacterial colonies on non-selective medium.

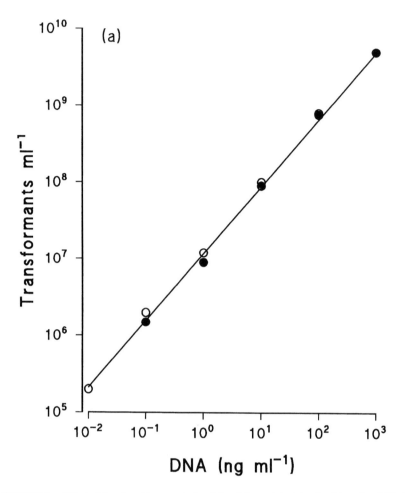

FIGURE 3.   Effect of the plasmid concentration (a) and the recipient density (b) on the yield of transformants. Field conditions: 17.5 kV cm⁻¹, capacitor of the pulser 25 μF, resistor of the pulser 600 Ω. The cells were pulsed in 10% PEG solutions and allowed to reseal in liquid complete medium before they were plated on agar (see Figures 1 and 2). (a) Suspension density of the *E. coli C* cells 1×10^{11} ml⁻¹; plasmids: pBR322 (○) and pUC19 (●). (b) Concentration of the plasmid pBR322: 0.9 μg ml⁻¹.

(e.g., *E. coli* versus *Lactococcus lactis*;[82] *Saccharomyces cerevisiae* versus *E. coli*[83]). Three variations have been reported:[84]

1.   After application of a field pulse, the donor bacterial preparation is centrifuged and the supernatant (containing the released plasmid) is added to the recipient bacteria and subjected to a second electropermeabilization (so-called "pulse-spin-pulse").

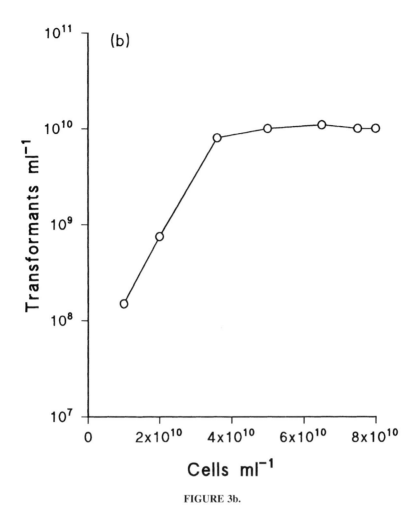

FIGURE 3b.

2. The electropermeabilized donor bacterial preparation is directly mixed with the recipient cells and then pulsed again ("sequential pulsing").
3. A mixture of the donor cells and the recipient bacteria is pulsed simultaneously ("simultaneous") allowing plasmid "exchange."

Although the efficiency of the transformation is relatively lower than with other methods, a tedious isolation procedure of plasmid DNA may be obviated.

A variation of plasmid release is the so-called "curing" of plasmid-bearing bacteria. This technique, which entails plasmid release followed by selection of plasmid-deleted colonies using appropriate selection media, has been accomplished in *E. coli*.[85] Interestingly, electropermeabilization did not increase the frequency of the plasmid "curing" in lactic streptococci.[34] Yet another variant of this technique using electropermeabilization to introduce a second plasmid into *E. coli*, without curing of the resident plasmid, has been reported.[86] As noted above, the mechanisms of the electrically-mediated DNA uptake/release are complex, and the efficiency of the plasmid curing is probably affected by a number of as yet poorly understood factors that might include plasmid stability and copy number.

V. POTENTIAL PROBLEMS

As listed in Tables 1–6, various bacteria have been genetically transformed by electroinjection. The electric field strength (kV cm^{-1}) and the time constant (ms) are given for each study. Other information (including culture conditions, washing, size of plasmid, expression time, and so on) will be found in the cited references. The reader will note that the application range is rapidly increasing and it is expected that genetic analysis of hitherto "untransformable" strains important for biomedical research or industrial applications might soon be possible.

Nevertheless, one must be aware of potential pitfalls and problems. Moderate Joule heating and other processes (see Chapter 1) are sometimes encountered during pulse application, especially with long pulse durations. High conductance media is likely to exacerbate this problem. This does not necessarily adversely affect the transformation efficiency, especially in situations where electropermeabilization of bacteria such as staphylococci is performed at lower temperatures, i.e., 20°C.[113] Under some circumstances this might create problems for transformation of temperature-sensitive bacteria.

The occurrence of voltage pulse-induced breakage of DNA has been noted.[136] Thus far, this has been of little practical importance.

The release of Al^{3+} from the aluminum electrodes common to some of the more popular inexpensive electropermeabilizators has also been observed.[137] In addition to the potential for direct toxicity of the Al^{3+}, aluminum electrodes may become oxidized, thus resulting in unpredictable field strengths (see Chapter 1). These potential problems may be avoided by using gold or platinum electrodes or at least by changing aluminum electrodes frequently.

VI. MECHANISMS OF BACTERIAL ELECTROPERMEABILIZATION

A rigorous treatment of the physical parameters of electric field effects on cell membranes has been given in Chapters 1 and 3. These physical parameters are equally valid for bacterial membranes. Not unexpectedly, a large fraction of cells are often killed by electropermeabilization (when using long pulses), but enough survive for selection, which is an adequate result in most cases.

But how is the impermeant bacterial cell envelope breached for DNA transfer without killing the surviving cells? The actual mechanism(s) by which

TABLE 1
Transformation of Aerobic Gram–Negative Bacteria

Recipient	DNA	Electric parameters		Efficiency	Frequency	Ref.
		kV cm^{-1}	ms			
Aquaspirillum dispar	RP4	12.5	12.0–16.0	3.0×10^4	—	87
Aquaspirillum itersonii	RP4	12.5	12.0–16.0	2.0×10^4	—	87
Azospirillum brasilense	pRK290	7.5	7.9	1.5×10^4	3.0×10^{-5}	88
Campylobacter jejuni	pILL512	13.0	2.0	1.2×10^6	—	71
Pseudomonas aeruginosa	pRO1614	12.5	~5.0	5.8×10^5	2.3×10^{-3}	89
Pseudomonas cepacia	pSP329	6.3	~18.0	3.6×10^4	—	57
Pseudomonas putida	pKK1	12.5	4.6	—	—	90
Pseudomonas stutzeri	pHSG298	25.0	5.0	2.0×10^7	—	91
Pseudomonas syringae	pMON7197	12.5	3.9	4.6×10^6	—	92
Xanthomonas campestris	pLAFR1	12.5	4.6?	$2.0–3.0 \times 10^5$	—	93
Xanthomonas campestris	pFUR027	12.5	7.5	2.0×10^8	—	94
Azotobacter vinelandii	pRK2501	3.8	29.0	2.6×10^4	—	95
Bradyrhizobium japonicum	PZB32	12.5	5.0–8.0	1.0×10^7	9.0×10^{-7}	96
Agrobacterium rhizogenes	pBI121	12.5	25.0	$10^7–10^8$	—	97
Agrobacterium tumefaciens	pBI121	12.5	8.0–12.0	8.0×10^5	—	98
Acetobacter xylinum	pUCD2	12.5	6.0–8.0	1.0×10^7	9.1×10^{-4}	99
Acidiphilium pWl	pL17-2	10.0–15.0	5.0–10.0	$2.0–6.0 \times 10^4$	—	23
Bordetella pertussis	pRK404ΔMCS	25.0	—	1.0×10^6	—	100
Brucella abortus	pSUP2021	12.5	10.0	—	$(1.8–2.0 \times 10^{-9})$	101
Methylobacterium extorquens	pLA2917	10.0[a]	0.3	8.0×10^3	—	50

[a] square pulse.
Note: ? Estimated indirectly

TABLE 2
Transformation of Facultatively Anaerobic Gram–Negative Rods

Recipient	DNA	Electric parameters		Efficiency	Frequency	Ref.
		kV cm^{-1}	ms			
Escherichia coli	pUC18	12.5	~4.0	10^9–10^{10}	~8×10^{-1}	52
Escherichia coli	pUC19	16.6	4.0	1.0–7.0×10^{10}	—	74
Escherichia coli	pUC18	12.0^a	>1.0	1.0×10^8	—	102
Escherichia coli	pBR322	4.0	5.0	3.0×10^8	—	103
Citrobacter freundii	pKT231	6.3	—	3.6×10^3	—	104
Enterobacter aerogenes	pKT321	6.3	—	5.0×10^1	—	104
Erwinia amylovora	pfdA31 Tn5	12.5–25.0	5.0–10.0	0.6–1.0×10^6	—	105
Klebsiella aerogenes	pBR322	6.3	—	7.9×10^4	—	106
Salmonella typhimurium	pBR322	12.5	—	10^8–10^9	—	44
Serratia marcescens	pKT231	6.3	—	4.0×10^3	—	104
Serratia polymuthica	pKT231	6.3	—	1.8×10^9	—	104
Yersinia enterocolitica	pSU2718	12.5	4.8	1.0×10^5	—	54
Yersinia pestis	pSU2718	12.5	4.8	1.0×10^7	—	54
Yersinia pseudotuberculosis	pSU2718	12.5	4.8	1.0×10^6	—	54
Vibrio alginolyticus	pACYC184	7.5	—	6.0×10^2	—	107
Vibrio cholerae	pBR322	12.1	3.9	1.5×10^5	—	64
Vibrio parahemolyticus	pACYC184	7.5	—	6.0×10^2	—	107
Pasteurella multocida	pVM109	12.5	10.0	1.3×10^7	—	21
Actinobacillus actinomycetemcomitans	pDL282	12.5	—	2.0×10^5	—	108
Haemophilus influenzae	pJ1–8	12.5	9.0–9.3	2.0×10^8	—	109
Haemophilus ducreyi	pLS88	16.2	—	1.0×10^6	—	59
Zymomonas mobilis	pZA22	10.0	4.5–5.0	2.5×10^6	—	110

^{a} square pulse.

TABLE 3
Transformation of Anaerobic Gram–Negative Bacteria

Recipient	DNA	Electric parameters		Efficiency	Frequency	Ref.
		kV cm^{-1}	ms			
Bacteroides fragilis	pBІІ91	12.5	5.0	8.8×10^6	—	65
Bacteroides ovatus	pFD288	12.5	5.0	1.1×10^4	—	65
Bacteroides ruminicola	pRR14	12.5	2.0–4.0	6.0×10^5	$0.8–1.5 \times 10^{-9}$	111
Bacteroides uniformis	pDP1	12.5	2.0–4.0	1.0×10^6	—	111
Porphyromonas gingivalis	pYT7	10.0	5.0	7.0×10^3	—	112

TABLE 4
Transformation of Gram–Positive Cocci

Recipient	DNA	Electric parameters		Efficiency	Frequency	Ref.
		kV cm⁻¹	ms			
Deinococcus radiodurans	pS28	5.3	~4.0	1.0×10^4	—	74
Staphylococcus aureus	pSK265	23.0	2.5	4.0×10^8	—	113
Staphylococcus carnosus	pC194	10.0	2.5	1.0×10^5	—	55
Staphylococcus epidermis	pC194	10.0	2.5	3.0×10^5	—	55
Staphylococcus staphylolyticus	pC194	10.0	2.5	5.0×10^3	—	55
Enterococcus faecalis	pGB354	12.5	4.7	2.8×10^6	—	114
Enterococcus faecium	pAM401	12.5	9.0–16.0	1.4×10^5	—	18
Enterococcus hirae	pAM401	12.5	9.0–16.0	1.2×10^3	—	18
Enterococcus malodoratus	pAM401	12.5	9.0–16.0	6.0×10^1	—	18
Enterococcus mundtii	pAM401	12.5	9.0–16.0	6.2×10^2	—	18
Lactococcus lactis	pGK12	12.5	4.5–5.0	5.7×10^7	—	28
Leuconostoc dextranicum	pGK12	6.3	—	1.1×10^4	—	61
Leuconostoc lactis	pGK12	6.3	—	7.7×10^3	—	61
Leuconostoc paramesenteroides	pNZ12	6.3	—	4.0×10^3	—	115
Pediococcus acidilactici	pGK12	12.5	—	$3.0–6.0 \times 10^1$	—	40
Streptococcus agalactiae	pDL414	6.3?	11.0–18.0	4.8×10^3	—	116
Streptococcus bovis	pVA838	12.5	5.0	1.5×10^1	—	117
Streptococcus cremoris	pGKV110	5.0	—	4.0×10^3	—	36
Streptococcus lactis	pMU1328	6.3	3.5–5.0	$\sim5.0 \times 10^5$	—	34
Streptococcus mutans	pII2607	6.3	5.3	3.0×10^2	—	118
Streptococcus pneumoniae	pSP2	6.3	—	1.4×10^3	$2.0–3.6 \times 10^{-5}$	119
Streptococcus pyogenes	pIL252	12.5	4.7	$\sim5.4 \times 10^7$	—	41
Streptococcus sanguis	pV1101	6.3?	—	3.0×10^5	—	37
Streptococcus thermophilus	pVA736	3.8–4.3	3.6	$10^3–10^4$	—	38
Ruminococcus albus	pSC22	12.5	4.5	3.0×10^5	—	120
Thiobacillus ferrooxidans	pKMZ51	12.5	3.5–4.5	1.2×10^2	—	121

Note: ? Estimated indirectly

TABLE 5
Transformation of Gram–Positive Rods

Recipient	DNA	Electric parameters		Efficiency	Frequency	Ref.
		kV cm⁻¹	ms			
Bacillus amyloliquefaciens	pUB110	7.5	8.0–10.0	1.0×10^5	—	35
Bacillus anthracis	pAB124	6.0	~4.7	2.6×10^4	5.3×10^{-4}	56
Bacillus brevis	pBAM101	7.5	5.0	2.6×10^4	—	15
Bacillus cereus	pC194	3.8	2.7	—	2.0×10^{-5}	122
Bacillus circulans	pMK4	6.3	—	8.0×10^3		123
Bacillus sphaericus	pUB110	12.5	4.0–8.0	3.0×10^6	7.5×10^{-4}	124
Bacillus subtilis	pUB110	7.0ᵃ	0.05	4.0×10^4	2.4×10^{-1}	49
Bacillus subtilis	pUB110	15.7	—	2.4×10^6	—	69
Bacillus thuringiensis	pBC16	3.3?	—	$0.2–1.0 \times 10^7$	—	125
Clostridium acetobutylicum	pMTL5OOE	5.0	4.7–6.9	$~2.9 \times 10^3$	—	126
Clostridium botulinum	pGK12	6.3	—	1.0×10^3	—	127
Clostridium perfringens	pHR106	6.3	7.5	3.0×10^5	2.5×10^{-4}	33
Lactobacillus acidophilus	pGK12	6.3	—	5.3×10^4	—	61
Lactobacillus acidophilus	pULA105E	6.3	—	7.4×10^6	—	62
Lactobacillus casei	pLZ15	5.0	—	8.5×10^4	—	128
Lactobacillus curvatus	p1L253	6.3	—	2.0×10^4	—	129
Lactobacillus delbrückii	pGK12	6.5	—	3.0×10^4	—	130
Lactobacillus fermentum	pGK12	6.3	—	9.7×10^5	—	61
Lactobacillus helveticus	pLHR	4.0	10.0–15.0	1.3×10^4	—	13
Lactobacillus plantarum	pGK12	12.5	10.0	5.0×10^6	—	131
Lactobacillus reuteri	pLUL634	12.5	—	3.0×10^7	—	132
Lactobacillus sake	pNZI2	6.3	—	1.0×10^2	—	129
Listeria innocua	pGK12	8.5	4.4	3.5×10^6	—	133
Listeria ivanovii	pGK12	8.5	4.4	1.3×10^6	—	133
Listeria monocytogenes	pGK12	8.5	4.4	3.9×10^6	—	133
Listeria seeligeri	pGK12	8.5	4.4	5.0×10^3	—	133

ᵃ square pulse.
Note: ? Estimated indirectly

TABLE 6
Transformation of Other Bacteria

Recipient	DNA	Electric parameters		Efficiency	Frequency	Ref.
		kV cm^{-1}	ms			
Brevibacterium ammoniagenes	pHY416	12.5	4.5–5.0	1.1×10^3	—	53
Brevibacterium flavum	pCGL7	12.5	3.0–5.0	7.4×10^5	—	73
Brevibacterium lactofermentum	pUL340	12.5	—	4.6×10^7	—	29
Brevibacterium stationis	pCRY3	12.5	—	1.7×10^3	—	26
Brevibacterium sp. R312	pSV73	11.8	2.4	1.0×10^8	—	20
Corynebacterium glutamicum	pCGL7	12.5	3.0–5.0	6.4×10^7	—	73
Clavibacter michiganense	pDM100	12.5	13.5	2.3×10^3	2.3×10^{-6}	58
Clavibacter xyli	pLAFR3	16.6	12.0	2.0×10^5	—	12
Helicobacter pylori	*H. Pylori*	12.5	—	2.0×10^5	—	134
Propionibacterium jensenii	pGK	6.3	—	3.2×10^1	—	61
Mycobacterium aurum	pAL8	12.5	3.5–4.5	2.3×10^4	—	30
Myobacterium parafortuitum	pJRD215	12.5	3.5–5.0	3.0×10^2	1.2×10^{-6}	31
Rhodococcus fascians	pRF29	12.0	4.1	10^5–10^7	—	68
Streptomyces lividans	pVE28	10.0[a]	1600	1.0×10^4	—	5
Caulobacter crescentus	pKT230	12.5	4.2	$\sim 1.5 \times 10^8$	—	135
Methanococcus voltae	*M.voltae*	2.0	1.5	2.5–8.0×10^2	—	46
Butyrivibrio fibrisolvens	pBS42	12.5	5.0	2.5×10^1	—	117

[a]   square pulse.

this occurs has/have been little studied and is/are therefore poorly understood. It is possible to derive several plausible hypotheses that might explain the success of this approach in breaching the cell envelope. These include:

1. the release of water after reversible breakdown of the cell membrane. This release is driven by the high turgor pressure of the bacterial cell and could result in splitting or "cracking" the cell envelope;
2. local heating of the membrane produced by the use of long duration pulses in the presence of a weakly conducting medium. The dense current flowing through a charged structure like the cell envelope could easily damage its integrity;
3. electrophoresis of DNA through electroleaks in the cell envelope;
4. electrointernalization (Chapter 1).

In fact it seems quite likely that none of these are mutually exclusive and that any or all might operate in parallel.

For example, it has been observed that DNA added after a voltage pulse results in only minimal transformation efficiency.[10,60,138] This suggests that DNA is taken up by bacteria primarily during the pulse application, favoring electrophoresis[139] or electroosmosis.[140] However, this finding could also be explained by electrointernalization, because cracks in the cell envelope may have diminished such that the DNA could not penetrate (see Chapter 1). By using freeze-fracture electron microscopy, one group[138] detected extensive blebbing immediately after electric pulse and observed the formation of large membrane "openings" 30–40 s after the pulse in cells of *E. coli*. These authors were, however, unable to relate the membrane changes with permeability to exogenous DNA. This finding would be compatible with disruption of the cell envelope by the collapse of the turgor pressure.

Based on transformation experiments using low-amplitude, low-frequency alternating electric fields (presumably below the breakdown voltage), some authors believe that plasmid DNA can enter *E. coli* without electric breakdown of the cell envelope.[48] In these experiments the maximal efficiency was only 1×10^5 transformants per μg of pUC18 DNA at a field strength of 200 V cm^{-1}.

Once electroinjected, the infectivity[60] or transformability[10] of the viral or bacterial DNA-recipient mixture promptly becomes resistant to DNase. Irrespective of this, a significant amount of free DNA remained in the medium after electroinjection.[22] This fact suggests that, unlike Ca^{2+}-dependent systems,[141] bulk adsorption of DNA onto the recipient surface does not take place in electroinjection. The significance of these findings in light of the foregoing discussion remains somewhat uncertain.

Small increases in the efficiency of transformation of *E. coli* appear to be induced by pretreatment with EDTA[142] or alkaline earth metal ions such as Mg^{2+}, Ba^{2+}, or Sr^{2+},[143] perhaps by binding DNA to the cell surface prior to envelope permeabilization. It remains to be elucidated whether DNA is directly driven by the electric field into bacterial cytoplasm or, as in the Ca^{2+}-dependent system,[141] forwarded after temporary interaction with the outer membrane. Our laboratory has shown that the efficiency of electrotransformation is much higher without such pretreatment (Taketo, A., manuscript in preparation). As previously noted, electrointernalization might also be playing a role here. It seems plausible that the transport of the smaller plasmids follows a one-step mechanism, whereas that of the large plasmids (100–300 Kb) is more complex.

In all, considerable work remains to be done to discover the mechanisms underlying the electrotransfer of DNA into bacterial cells. The ultimate "pay off" of such investigation should be the development of more cogent and efficient methods suitable for the next generation of electrotransformation applications.

VII. FUTURE APPLICATIONS

The successful introduction of large plasmids (136–300 Kb) into *E. coli* predicts that further refinement of the conditions for the electrotransformation will be of value in the construction of mammalian gene libraries using specialized bacterial systems. A practical size limit has yet to be demonstrated. Moreover, the high frequency of transformation in *E. coli* is a promising development for the direct screening of recombinant plasmids deficient in selection markers. Intact *E. coli* are easily transfected with phage RNA by electropermeabilization.[144] This result suggests applicability of the technique for introduction of different RNA species (messenger RNA, antisense RNA, or transfer RNA) into bacteria for a variety of experimental purposes. In other bacteria, however, neither the efficiency nor the frequency is high enough for these purposes. Substantial increases in the efficiency of the electrotransformation system will be required to expand applications for strains that are well-suited for the production of secretory proteins such as *Bacillus subtilis*.

The technique of bacterial electrotransfer need not, of course, be restricted to the introduction of nucleic acids. The method seems to have great potential for the analysis of influences of impermeable drugs, effectors, or nucleotides on microbial cells. Other future applications might include the introduction of enzymes, antibodies, and functional proteins into bacteria. Paralleling data obtained for eukaryotic cells, we have obtained preliminary results suggesting voltage-dependent penetration of pancreatic RNase into *E. coli*. Additionally, and as noted in Chapter 1, electropermeabilization has been used for import of low-molecular weight substances into animal cells.

Like other related fields, bacterial electrotransformation has recently entered a logarithmic growth phase. A myriad of refinements of existing techniques as well as new applications are expected in the coming years. Continued progress in this field will require a more detailed knowledge of the biophysical parameters involved in walled cell permeabilization and resealing as well as a better understanding of the molecular and biological events that underlie the processes of nucleic acid entry and transfection. The placement of this field at the intersection of the disciplines of molecular biology, microbiology, and biophysics makes this one of the most exciting and fertile areas of investigation.

REFERENCES

1. **Avery, O. T., MacLeod, C. M. and McCarty, M.,** Studies on the chemical nature of the substance inducing transformation of pneumococcal types. I. Induction of transformation by a desoxyribonucleic acid fraction isolated from *Pneumococcus* type III., *J. Exp. Med.,* 79, 137, 1944.
2. **Zimmermann, U., Schulz, J. and Pilwat, G.,** Transcellular ion flow in *Escherichia coli* B and electrical sizing of bacteria, *Biophys. J.,* 13, 1005, 1973.
3. **Zimmermann, U., Pilwat, G. and Riemann, F.,** Dielectric breakdown of cell membranes, *Biophys. J.,* 14, 881, 1974.
4. **Shivarova, N., Forster, W., Jacob, H. E. and Grigorova, R.,** Microbiological implications of electric field effects. VII. Stimulation of plasmid transformation of *Bacillus cereus* protoplasts by electric field pulses, *Z. Allg. Mikrobiol.,* 23, 595, 1983.
5. **MacNeil, D. J.,** Introduction of plasmid DNA into *Streptomyces lividans* by electroporation, *FEMS Microbiol. Lett.,* 42, 239, 1987.
6. **Potter, H., Weir, L. and Leder, P.,** Enhancer-dependent expression of human κ immunoglobulin genes introduced into mouse pre-B lymphocytes by electroporation, *Proc. Natl. Acad. Sci. U.S.A.,* 81, 7161, 1984.
7. **Dower, J.,** Electro-transformation of intact bacterial cells, *Molecular Biology Reports,* Bio-Rad Laboratories, 1, 5, 1987.
8. **Harlander, S. K.,** Transformation of *Streptococcus lactis* by electroporation, in *Streptococcal Genetics,* Ferretti, J. J. and Curtiss, R., III, Eds., American Society for Microbiology, Washington, D.C., 1987, 229.
9. **Shigekawa, K. and Dower, W. J.,** Electroporation of eukaryotes and prokaryotes: a general approach to the introduction of macromolecules into cells, *BioTechniques,* 6, 742, 1988.
10. **Chassy, B. M., Mercenier, A. and Flickinger, J.,** Transformation of bacteria by electroporation, *Trends Biotechnol.,* 6, 303, 1988.
11. **Schleper, C., Kubo, K. and Zillig, W.,** The particle SSV1 from the extremely thermophilic archaeon *Sulfolobus* is a virus: demonstration of infectivity and of transfection with viral DNA, *Proc. Natl. Acad. Sci. U.S.A.,* 89, 7645, 1992.
12. **Metzler, M. C., Zhang, Y. P. and Chen, T. A.,** Transformation of the gram-positive bacterium *Clavibacter xyli* subsp. *cynodontis* by electroporation with plasmids from the IncP incompatibility group, *J. Bacteriol.,* 174, 4500, 1992.
13. **Hashiba, H., Takiguchi, R., Ishii, S. and Aoyama, K.,** Transformation of *Lactobacillus helveticus* subsp. *jugurti* with plasmid pLHR by electroporation, *Agric. Biol. Chem.,* 54, 1537, 1990.
14. **McIntyre, D. A. and Harlander, S. K.,** Improved electroporation efficiency of intact *Lactococcus lactis* subsp. *lactis* cells grown in defined media, *Appl. Environ. Microbiol.,* 55, 2621, 1989.
15. **Takagi, H., Kagiyama, S., Kadowaki, K., Tsukagoshi, N. and Udaka, S.,** Genetic transformation of *Bacillus brevis* with plasmid DNA by electroporation, *Agric. Biol. Chem.,* 53, 3099, 1989.
16. **Phillips-Jones, M. K.,** Plasmid transformation of *Clostridium perfringens* by electroporation methods, *FEMS Microbiol. Lett.,* 66, 221, 1990.

17. **Summers, D. K. and Withers, H. L.,** Electrotransfer: direct transfer of bacterial plasmid DNA by electroporation, *Nucleic Acids Res.,* 18, 2192, 1990.

18. **Friesenegger, A., Fielder, S., Devriese, L. A. and Wirth, R.,** Genetic transformation of various species of *Enterococcus* by electroporation, *FEMS Microbiol. Lett.,* 79, 323, 1991.

19. **Somkuti, G. A. and Steinberg, D. H.,** Electrotransformation of *Streptococcus sanguis* Challis, *Current Microbiol.,* 19, 91, 1989.

20. **Chan Kuo Chion, C. K. N., Duran, R., Arnaud, A. and Galzy, P.,** Electrotransformation of whole cells of *Brevibacterium* sp. R312 a nitrile hydratase producing strain: construction of a cloning vector, *FEMS Microbiol. Lett.,* 81, 177, 1991.

21. **Jablonski, L., Sriranganathan, N., Boyle, S. M. and Carter, G. R.,** Conditions for transformation of *Pasteurella multocida* by electroporation, *Microb. Pathogen.,* 12, 63, 1992.

22. **Taketo, A.,** Properties of electroporation-mediated DNA transfer in *Escherichia coli., J. Biochem.,* 105, 813, 1989.

23. **Glenn, A. W., Roberto, F. F. and Ward, T. E.,** Transformation of *Acidiphilium* by electroporation and conjugation, *Can. J. Microbiol.,* 38, 387, 1992.

24. **Bonnassie, S., Burini, J. F., Oreglia, J., Trautwetter, A., Patte, J. C. and Sicard, A. M.,** Transfer of plasmid DNA to *Brevibacterium lactofermentum* by electrotransformation, *J. Gen. Microbiol.,* 136, 2107, 1990.

25. **Park, S. F. and Stewart, G. S. A. B.,** High-efficiency transformation of *Listeria monocytogenes* by electroporation of penicillin-treated cells, *Gene,* 94, 129, 1990.

26. **Kurusu, Y., Kainuma, M., Inui, M., Satoh, Y. and Yukawa, H.,** Electroporation-transformation system for coryneform bacteria by auxotrophic complementation, *Agric. Biol. Chem.,* 54, 443, 1990.

27. **Aukrust, T. and Nes, I. F.,** Transformation of *Lactobacillus plantarum* with the plasmid pTV1 by electroporation, *FEMS Microbiol. Lett.,* 52, 127, 1988.

28. **Holo, H. and Nes, I. F.,** High-frequency transformation, by electroporation, of *Lactococcus lactis* subsp. *cremoris* grown with glycine in osmotically stabilized media, *Appl. Environ. Microbiol.,* 55, 3119, 1989.

29. **Haynes, J. A. and Britz, M. L.,** Electrotransformation of *Brevibacterium lactofermentum* and *Corynebacterium glutamicum:* growth in tween 80 increases transformation frequencies, *FEMS Microbiol. Lett.,* 61, 329, 1989.

30. **Hermans, J., Boschloo, J. G. and de Bont, J. A. M.,** Transformation of *Mycobacterium aurum* by electroporation: the use of glycine, lysozyme and isonicotinic acid hydrazide in enhancing transformation efficiency, *FEMS Microbiol. Lett.,* 72, 221, 1990.

31. **Hermans, J., Suy, I. M. L. and deBont, J. A. M.,** Transformation of Gram-positive microorganisms with the Gram-negative broad-host-range cosmid vector pJRD215, *FEMS Microbiol. Lett.,* 108, 201, 1993.

32. **Haynes, J. A. and Britz, M. L.,** The effect of growth conditions of *Corynebacterium glutamicum* on the transformation frequency obtained by electroporation, *J. Gen. Microbiol.,* 136, 255, 1990.

33. **Scott, P. T. and Rood, J. I.,** Electroporation-mediated transformation of lysostaphin-treated *Clostridium perfringens, Gene,* 82, 327, 1989.

34. **Powell, I. B., Achen, M. G., Hillier, A. J. and Davidson, B. E.,** A simple and rapid method for genetic transformation of lactic streptococci by electroporation, *Appl. Environ. Microbiol.,* 54, 655, 1988.

35. **Vehmaanpera, J.,** Transformation of *Bacillus amyloliquefaciens* by electroporation, *FEMS Microbiol. Lett.,* 61, 165, 1989.

36. **van der Lelie, D., van der Vossen, J. M. B. M. and Venema, G.,** Effect of plasmid incompatibility on DNA transfer to *Streptococcus cremoris, Appl. Environ. Microbiol.,* 54, 865, 1988.

37. **Suvorov, A., Kok, J. and Venema, G.,** Transformation of group A streptococci by electroporation, *FEMS Microbiol. Lett.,* 56, 95, 1988.

38. **Somkuti, G. A. and Steinberg, D. H.,** Genetic transformation of *Streptococcus thermophilus* by electroporation, *Biochemie,* 70, 579, 1988.

39. **Dornan, S. and Collins, M. A.,** High efficiency electroporation of *Lactococcus lactis* LM0230 with plasmid pGB301, *Lett. Appl. Microbiol.,* 11, 62, 1990.

40. **Kim, W. J., Ray, B. and Johnson, M. C.,** Plasmid transfers by conjugation and electroporation in *Pediococcus acidilactici, J. Appl. Bacteriol.,* 72, 201, 1992.

41. **Simon, D. and Feretti, J.,** Electrotransformation of *Streptococcus pyogenes* with plasmid and linear DNA, *FEMS Microbiol. Lett.,* 82, 219, 1991.

42. **Taketo, A.,** Sensitivity of *Escherichia coli* to viral nucleic acid, XII, Ca^{2+}- or Ba^{2+}-facilitated transfection of cell envelope mutants, *Z. Naturforsch.,* 32c, 429, 1977.

43. **Taketo, A.,** Effect of mutations in surface lipopolysaccharide or outer membrane protein of *Escherichia coli* C on transfecting competence for microvirid phage DNA, *J. Gen. Appl. Microbiol.,* 24, 51, 1978.

44. **Binotto, J., MacLachlan, P. R. and Sanderson, K. E.,** Electrotransformation in *Salmonella typhimurium* LT2, *Can. J. Microbiol.,* 37, 474, 1991.

45. **Regué, M., Enfedaque, J., Camprubi, S. and Tomas, J. M.,** The O-antigen lipopolysaccharide is the major barrier to plasmid DNA uptake by *Klebsiella pneumoniae* during transformation by electroporation and osmotic shock, *J. Microbiol. Methods,* 15, 129, 1992.

46. **Micheletti, P. A., Sment, K. A. and Konisky, J.,** Isolation of a coenzyme-M-auxotrophic mutant and transformation by electroporation in *Methanococcus voltae, J. Bacteriol.,* 173, 3414, 1991.

47. **Grover, N. B., Naaman, J., Ben-Sasson, S. and Doljanski, F.,** Electrical sizing of particles in suspensions, *Biophys. J.,* 9, 1398, 1969.

48. **Xie, T. D. and Tsong, T. Y.,** Study of mechanisms of electric field induced DNA transfection, II. Transfection of low-amplitude, low-frequency alternating electric fields, *Biophys. J.,* 58, 897, 1990.

49. **Kusaoke, H., Hayashi, Y., Kadowaki, Y. and Kimoto, H.,** Optimum conditions for electric pulse-mediated gene transfer to *Bacillus subtilis* cells, *Agric. Biol. Chem.,* 53, 2441, 1989.

50. **Ueda, S., Matsumoto, S., Shimizu, S. and Yamane, T.,** Transformation of a methylotrophic bacterium, *Methylobacterium extorquens,* with a broad-host-range plasmid by electroporation, *Appl. Environ. Microbiol.,* 57, 924, 1991.

51. **Fiedler, S. and Wirth, R.,** Transformation of bacteria with plasmid DNA by electroporation, *Anal. Biochem.,* 170, 38, 1988.

52. **Dower, W. J., Miller, J. F. and Ragsdale, C. W.,** High efficiency transformation of *E. coli* by high voltage electroporation, *Nucleic Acids Res.,* 16, 6127, 1988.

53. **Dunikan, L. K. and Shivnan, E.,** High frequency transformation of whole cells of amino acid producing coryneform bacteria using high voltage electroporation, *Bio/Technology,* 7, 1067, 1989.

54. **Conchas, R. F. and Carniel, E.,** A highly efficient electroporation system for transformation of *Yersinia, Gene,* 87, 133, 1990.

55. **Augustin, J. and Gotz, F.,** Transformation of *Staphylococcus epidermidis* and other staphylococcal species with plasmid DNA by electroporation, *FEMS Microbiol. Lett.,* 66, 203, 1990.

56. **Quinn, C. P. and Dancer, B. N.,** Transformation of vegetative cells of *Bacillus anthracis* with plasmid DNA, *J. Gen. Microbiol.,* 136, 1211, 1990.

57. **Burns, J. L. and Hedin, L. A.,** Genetic transformation of *Pseudomonas cepacia* using electroporation, *J. Microbiol. Methods,* 13, 215, 1991.

58. **Meletzus, D. and Eichenlaub, R.,** Transformation of the phytopathogenic bacterium *Clavibacter michiganense* subsp. *michiganense* by electroporation and development of a cloning vector, *J. Bacteriol.,* 173, 184, 1991.

59. **Hansen, E. J., Latimer, J. L., Thomas, S. E., Helminen, M., Albritton, W. L. and Radolf, J. D.,** Use of electroporation to construct isogenic mutants of *Haemophilus ducreyi, J. Bacteriol.,* 174, 5442, 1992.

60. **Taketo, A.,** DNA transfection of *Escherichia coli* by electroporation, *Biochim. Biophys. Acta,* 949, 318, 1988.

61. **Luchansky, J. B., Muriana, P. M. and Klaenhammer, T. R.,** Application of electroporation for transfer of plasmid DNA to *Lactobacillus, Lactococcus, Leuconostoc, Listeria, Pediococcus, Bacillus, Staphylococcus, Enterococcus* and *Propionibacterium, Molec. Microbiol.,* 2, 637, 1988.

62. **Kanatani, K., Yoshida, K., Tahara, T., Yamada, K., Miura, H., Sakamoto, M. and Oshimura, M.,** Transformation of *Lactobacillus acidophilus* TK8912 by electroporation with pULA105E plasmid, *J. Ferment. Bioeng.,* 74, 358, 1992.

63. **Leonardo, E. D. and Sedivy, J. M.,** A new vector for cloning large eukaryotic DNA segments in *Escherichia coli, Bio/Technology,* 8, 841, 1990.

64. **Marcus, H., Ketley, J. M., Kaper, J. B. and Holmes, R. K.,** Effects of DNase production, plasmid size, and restriction barriers on transformation of *Vibrio cholerae* by electroporation and osmotic shock, *FEMS Microbiol. Lett.,* 68, 149, 1990.

65. **Smith, C. J., Parker, A. and Rogers, M. B.,** Plasmid transformation of *Bacteroides* spp. by electroporation, *Plasmid,* 24, 100, 1990.

66. **Shizuya, H., Birren, B., Kim, U. J., Mancino, V., Slepak, T., Tachiiri, Y. and Simon, M.,** Cloning and stable maintenance of 300-kilobase-pair fragments of human DNA in *Escherichia coli* using an F-factor-based vector, *Proc. Natl. Acad. Sci. U.S.A.,* 89, 8794, 1992.

67. **Allen, S. P. and Blaschek, H. P.,** Factors involved in the electroporation-induced transformation of *Clostridium perfringens, FEMS Microbiol. Lett.,* 70, 217, 1990.

68. **Desomer, J., Dhaese, P. and Van Montagu, M.,** Transformation of *Rhodococcus fascians* by high-voltage electroporation and development of *R. fascians* cloning vectors, *Appl. Environ. Microbiol.,* 56, 2818, 1990.

69. **Matsuno, Y., Ano, T. and Shoda, M.,** High-efficiency transformation of *Bacillus subtilis* NB22, an antifungal antibiotic iturin producer, by electroporation, *J. Ferment. Bioeng.,* 73, 261, 1992.

70. **Cymbalyuk, E. S., Chernomordik, L. V., Broude, N. E. and Chizmadzhev, Y. A.,** Electro-stimulated transformation of *E. coli* cells pretreated by EDTA solution, *FEBS Lett.,* 234, 203, 1988.

71. **Miller, J. F., Dower, W. J. and Tompkins, L. S.,** High-voltage electroporation of bacteria: genetic transformation of *Campylobacter jejuni* with plasmid DNA, *Proc. Natl. Acad. Sci. U.S.A.,* 85, 856, 1988.

72. **Setlow, J. K. and Albritton, W. L.,** Transformation of *Haemophilus influenzae* following electroporation with plasmid and chromosomal DNA, *Current Microbiol.,* 24, 97, 1992.

73. **Bonamy, C., Guyonvarch, A., Reyes, O., David, F. and Leblon, G.,** Interspecies electrotransformation in *Corynebacteria, FEMS Microbiol. Lett.,* 66, 263, 1990.

74. **Simith, M., Jessee, J., Landers, T. and Jordan, J.,** High efficiency bacterial electroporation: 1×10^{10} *E. coli* transformants/μg, *Focus* Life Technologies Inc., 12, 38, 1990.

75. **Chu, G., Hayakawa, H. and Berg, P.,** Electroporation for the efficient transfection of mammalian cells with DNA, *Nucleic Acids Res.,* 15, 1311, 1987.

76. **Nickoloff, J. A. and Reynolds, R. J.,** Electroporation-mediated gene transfer efficiency is reduced by linear plasmid carrier DNAs, *Anal. Biochem.,* 205, 237, 1992.

77. **Calvin, N. M. and Hanawalt, P. C.,** High-efficiency transformation of bacterial cells by electroporation, *J. Bacteriol.,* 170, 2796, 1988.

78. **Jacob, M., Wendt, S. and Stahl, U.,** High-efficiency electro-transformation of *Escherichia coli* with DNA from ligation mixtures, *Nucleic Acids Res.,* 18, 1653, 1990.

79. **Willson, T. A. and Gough, N. M.,** High voltage *E. coli* electro-transformation with DNA following ligation, *Nucleic Acids Res.,* 16, 11820, 1990.

80. **Zabarovsky, E. and Winberg, G.,** High efficiency electroporation of ligated DNA into bacteria, *Nucleic Acids Res.,* 18, 5912, 1990.

81. **Ymers, S.,** Heat inactivation of DNA ligase prior to electroporation increases transformation efficiency, *Nucleic Acids Res.,* 19, 6960, 1991.

82. **Ward, L. J. H. and Jarvis, A. W.,** Rapid electroporation-mediated plasmid transfer between *Lactococcus lactis* and *Escherichia coli* without the need for plasmid preparation, *Lett. Appl. Microbiol.,* 13, 278, 1991.

83. **Marcil, R. and Higgins, D. R.,** Direct transfer of plasmid DNA from yeast to *E. coli* by electroporation, *Nucleic Acids Res.,* 20, 917, 1992.

84. **Kilbane, J. J. II and Bielaga, B. A.,** Instantaneous gene transfer from donor to recipient microorganisms via electroporation, *BioTechniques,* 10, 354, 1991.

85. **Heery, D. M., Powell, R., Gannon, F. and Dunican, L. K.,** Curing of a plasmid from *E. coli* using high-voltage electroporation, *Nucleic Acids Res.,* 17, 19131, 1989.

86. **Li, S. J., Landers, T. A. and Smith, M. D.,** Electroporation of plasmids into plasmid-containing *Escherichia coli, Focus,* Life Technologies Inc., 12, 72, 1990.

87. **Eden, P. A. and Blakemore, R. P.,** Electroporation and conjugal plasmid transfer to members of the genus, *Aquaspirillum, Arch. Microbiol.,* 155, 449, 1991.

88. **Broek, A. V., van Gool, A. and Vanderleyden, J.,** Electroporation of *Azospirillum brasilense* with plasmid DNA, *FEMS Microbiol. Lett.,* 61, 177, 1989.

89. **Farinha, M. A. and Kropinski, A. M.,** High efficiency electroporation of *Pseudomonas aeruginosa* using frozen cell suspensions, *FEMS Microbiol. Lett.,* 70, 221, 1990.

90. **Trevors, J. T. and Starodub, M. E.,** Electroporation of pKK1 silver-resistance plasmid from *Pseudomonas stutzeri* AG259 into *Pseudomonas putida* CYM318, *Current Microbiol.,* 21, 103, 1990.

91. **Pemberton, J. M. and Penfold, R. J.,** High-frequency electroporation and maintenance of pUC- and pBR-based cloning vectors in *Pseudomonas stutzeri, Current Microbiol.,* 25, 25, 1992.

92. **Wendt-Potthoff, K., Niepold, F. and Backhaus, H.,** High-efficiency electro-transformation of the plant pathogen *Pseudomonas syringae* pv. *syringae* R32, *J. Microbiol. Meth.,* 16, 33, 1992.

93. **Wang, T. W. and Tseng, Y. H.,** Electrotransformation of *Xanthomonas campestris* by RF DNA of filamentous phage φ Lf., *Lett. Appl. Microbiol.,* 14, 65, 1992.

94. **Shaw, J. J. and Khan, I.,** Efficient transposon mutagenesis of *Xanthomonas campestris* pathovar *campestris* by high voltage electroporation, *BioTechniques,* 14, 556, 1993.

95. **Trevors, J. T. and Starodub, M. E.,** Electroporation and expression of the broad host-range plasmid pRK2501 in *Azotobacter vinelandii, Enzyme Microb. Technol.,* 12, 653, 1990.

96. **Hattermann, D. R. and Stacey, G.,** Efficient DNA transformation of *Bradyrhizobium japonicum* by electroporation, *Appl. Environ. Microbiol.,* 56, 833, 1990.

97. **Nagel, R., Elliott, A., Masel, A., Birch, R. G. and Manners, J. M.,** Electroporation of binary Ti plasmid vector into *Agrobacterium tumefaciens* and *Agrobacterium rhizogenes, FEMS Microbiol. Lett.,* 67, 325, 1990.

98. **Wen-jun, S. and Forde, B. G.,** Efficient transformation of *Agrobacterium* spp. by high voltage electroporation, *Nucleic Acids Res.,* 17, 8385, 1989.

99. **Hall, P. E., Anderson, S. M., Johnston, D. M. and Cannon, R. E.,** Transformation of *Acetobacter xylinum* with plasmid DNA by electroporation, *Plasmid,* 28, 194, 1992.

100. **Zealey, G., Dion, M., Loosmore, S., Yacob, R. and Klein, M.,** High frequency transformation of *Bordetella* by electroporation, *FEMS Microbiol. Lett.,* 56, 123, 1988.

101. **Lai, F., Schuring, G. G. and Boyle, S. M.,** Electroporation of a suicide plasmid bearing a transposon into *Brucella abortus, Microbiol. Pathogen,* 9, 363, 1990.

102. **Elvin, S. and Bingham, A. H. A.,** Electroporation-induced transformation of *Escherichia coli*: evaluation of a square waveform pulse, *Lett. Appl. Microbiol.,* 12, 39, 1991.

103. **Xie, T. D. and Tsong, T. Y.,** Study of mechanisms of electric field-induced DNA transfection. III. Electric parameters and other conditions for effective transfection, *Biophys. J.,* 63, 28, 1992.

104. **Fiedler, S., Friesenegger, A. and Wirth, R.,** Electroporation: a general method for transformation of Gram-negative bacteria, in *Genetic Transformation and Expression,* Butler, L. O., Harwood, C. and Moseley, B. E. D., Eds., Intercept, Andover, 1989, 65.

105. **Metzger, M., Bellermann, P., Schwartz, T. and Geider, K.,** Site-directed and transposon-mediated mutagenesis with pfd-plasmids by electroporation of *Erwinia amylovora* and *Escherichia coli* cells, *Nucleic Acids Res.,* 20, 2265, 1992.

106. **Trevors, J. T.,** Electroporation and expression of plasmid pBR322 in *Klebsiella aerogenes* NCTC418 and plasmid pRK2501 in *Pseudomonas putida* CYM318, *J. Basic Microbiol.,* 30, 57, 1990.

107. **Hamashima, H., Nakano, T., Tamura, S. and Arai, T.,** Genetic transformation of *Vibrio parahemolyticus, Vibrio alginolyticus* and *Vibrio cholerae* non 0–1 with plasmid DNA by electroporation, *Microbiol. Immunol.,* 34, 703, 1990.

108. **Sreenivasan, P. K., LeBlanc, D. J., Lee, L. N. and Fives-Taylor, P.,** Transformation of *Actinobacillus actinomycetemcomitans* by electroporation, utilizing constructed shuttle plasmids, *Infect. Immun.,* 59, 4621, 1991.

109. **Mitchell, M. A., Skowronek, K., Kauc, L. and Goodgal, S. H.,** Electroporation of *Haemophilus influenzae* is effective for transformation of plasmid but not chromosomal DNA, *Nucleic Acids Res.,* 19, 3625, 1991.

110. **Okamoto, T. and Nakamura, K.,** Simple and highly efficient transformation method for *Zymonas mobilis:* electroporation, *Biosci. Biotech. Biochem.,* 56, 833, 1992.

111. **Thomson, A. M. and Flint, H. J.,** Electroporation induced transformation of *Bacteroides ruminicola* and *Bacteroides uniformis* by plasmid DNA, *FEMS Microbiol. Lett.,* 61, 101, 1989.

112. **Yoshimoto, H., Takahashi, Y., Hamada, N. and Umemoto, T.,** Genetic transformation of *Porphyromonas gingivalis* by electroporation, *Oral Microbiol. Immunol.,* 8, 208, 1993.

113. **Schenk, S. and Laddaga, R. A.,** Improved method for electroporation of *Staphylococcus aureus, FEMS Microbiol. Lett.,* 94, 133, 1992.

114. **Cruz-Rodz, A. L. and Gilmore, M. S.,** High efficiency introduction of plasmid DNA into glycine treated *Enterococcus faecalis* by electroporation, *Mol. Gen. Genet.,* 224, 152, 1990.

115. **David, S., Simons, G. and de Vos, W. M.,** Plasmid transformation by electroporation of *Leuconostoc paramesenteroides* and its use in molecular cloning, *Appl. Environ. Microbiol.,* 55, 1483, 1989.

116. **Dunny, G. M., Lee, L. N. and LeBlanc, D. J.,** Improved electroporation and cloning vector system for Gram-positive bacteria, *Appl. Environ. Microbiol.,* 57, 1194, 1991.

117. **Whitehead, T. R.,** Genetic transformation of the ruminal bacteria *Butyrivibrio fibrisolvens* and *Streptococcus bovis* by electroporation, *Lett. Appl. Microbiol.,* 15, 186, 1992.

118. **Lee, S. F., Progulske-Fox, A., Erdos, G. W., Piacentini, D. A., Ayakawa, G. Y., Crowley, P. J. and Bleisweis, A. S.,** Construction and characterization of isogenic mutants of *Streptococcus mutans* deficient in major surface protein antigen PI (I/II), *Infec. Immun.,* 57, 3306, 1989.

119. **Bonnassie, S., Gasc, A. M. and Sicard, A. M.,** Transformation by electroporation of two Gram-positive bacteria: *Streptococcus pneumoniae* and *Brevibacterium lactofermentum,* in *Genetic Transformation and Expression,* Butler, L. O., Harwood, C. and Moseley, B. E. D., Eds., Intercept, Andover, 1989, 71.

120. **Cocconcelli, P. S., Ferrari, E., Rossi, F. and Bottazzi, V.,** Plasmid transformation of *Ruminococcus albus* by means of high-voltage electroporation, *FEMS Microbiol. Lett.,* 94, 203, 1992.

121. **Kusano, T., Sugawara, K., Inoue, C., Takeshima, T., Numata, M. and Shiratori, T.,** Electroporation of *Thiobacillus ferrooxidans* with plasmid containing a *mer* determinant, *J. Bacteriol.,* 174, 6617, 1992.

122. **Belliveau, B. and Trevors, J. T.,** Transformation of *Bacillus cereus* vegetative cells by electroporation, *Appl. Environ. Microbiol.,* 55, 1649, 1989.

123. **Aubert, E. and Davies, J.,** Transformation, by electroporation, of *Bacillus circulans* NRRLB3312, the producer of butirosin, in *Genetic Transformation and Expression,* Butler, L. O., Harwook, C. and Moseley, B. E. D., Eds., Intercept, Andover, 1989, 77.

124. **Taylor, L. D. and Burke, W. F. Jr.,** Transformation of an entomopathic strain of *Bacillus sphaericus* by high voltage electroporation, *FEMS Microbiol. Lett.,* 66, 125, 1990.

125. **Schurter, W., Geiser, M. and Mathé, D.,** Efficient transformation of *Bacillus thuringiensis* and *B. cereus* via electroporation: transformation of acrystalliferous strains with a cloned delta-endotoxin gene, *Mol. Gen. Genet.,* 218, 177, 1989.

126. **Oultram, J. D., Loughlin, M., Swinfield, T. J., Brehm, J. K., Thompson, D. E. and Minton, N. P.,** Introduction of plasmids into whole cells of *Clostridium acetobutylicum* by electroporation, *FEMS Microbiol. Lett.,* 56, 83, 1988.

127. **Zhou, Y. and Johnson, E. A.,** Genetic transformation of *Clostridium botulinum* Hall by electroporation, *Biotech. Lett.,* 15, 121, 1993.

128. **Chassy, B. M. and Flickinger, J. L.,** Transformation of *Lactobacillus casei* by electroporation, *FEMS Microbiol. Lett.,* 44, 173, 1987.

129. **Gaier, W., Vogel, R. F. and Hammes, W. P.,** Genetic transformation of intact cells of *Lactobacillus curvatus* Lc2-c and *Lact. sake* Ls2 by electroporation, *Lett. Appl. Microbiol.,* 11, 81, 1990.

130. **Zink, A., Klein, J. R. and Plapp, R.,** Transformation of *Lactobacillus delbrückii* ssp. *lactis* by electroporation and cloning of origins of replication by use of a positive selection vector, *FEMS Microbiol. Lett.,* 78, 207, 1991.

131. **Bringel, F. and Hubert, J. C.,** Optimized transformation by electroporation of *Lactobacillus plantarum* strains with plasmid vectors, *Appl. Microbiol. Biotechnol.,* 33, 664, 1990.

132. **Ahrné, S., Molin, G. and Axelsson, L.,** Transformation of *Lactobacillus reuteri* with electroporation: studies on the erythromycin resistance plasmid pLUL631, *Current Microbiol.,* 24, 199, 1992.

133. **Alexander, J. E., Andrew, P. W., Jones, D. and Roberts, I. S.,** Development of an optimized system for electroporation of *Listeria* species, *Lett. Appl. Microbiol.,* 10, 179, 1990.

134. **Segal, E. D. and Tompkins, L. S.,** Transformation of *Helicobacter pylori* by electroporation, *BioTechniques,* 14, 225, 1993.

135. **Gilchrist, A. and Smit, J.,** Transformation of freshwater and marine caulobacters by electroporation, *J. Bacteriol.,* 173, 921, 1991.

136. **Tamiya, E., Nakajima, Y., Kamioka, H., Suzuki, M. and Karube, I.,** DNA cleavage based on high voltage electric pulse, *FEBS Letters,* 234, 357, 1988.

137. **Loomis-Husselbee, J. W., Cullen, P. J., Irvine, R. F. and Dowson, A. P.,** Electroporation can cause artefacts due to solubilization of cations from the electrode plates. Aluminium ions enhance conversion of inositol 1,3,4,5-tetrakisphosphate into inositol 1,4,5-trisphosphate in electroporated L1210 cells, *Biochem. J.,* 277, 883, 1991.

138. **Sabelnikov, A. G., Cymbalyuk, E. S., Gongadge, G. and Borovyagin, V. L.,** *Escherichia coli* membranes during electrotransformation: an electron microscopy study, *Biochim. Biophys. Acta,* 1066, 21, 1991.

139. **Klenchin, V. A., Sukharev, S. I., Serov, S. M., Chernomordik, L. V. and Chizmadzhev, Y. A.,** Electrically induced DNA uptake by cells is a fast process involving DNA electrophoresis, *Biophys. J.,* 60, 804, 1991.

140. **Dimitrov, D. S. and Sowers, A. E.,** Membrane electroporation-fast molecular exchange by electroosmosis, *Biochim. Biophys. Acta,* 1922, 381, 1990.

141. **Taketo, A.,** Sensitivity of *Escherichia coli* to viral nucleic acid. XVI. Temperature conditions for Ca^{2+}-dependent DNA uptake in *Escherichia coli,* *Z. Naturforsch.,* 37c, 87, 1982.

142. **Taketo, A. and Kuno, S.,** Sensitivity of *Escherichia coli* to viral nucleic acid, I. Effect of lysozyme, EDTA, penicillin and osmotic shock treatment, *J. Biochem.,* 65, 361, 1969.

143. **Taketo, A.,** Sensitivity of *Escherichia coli* to viral nucleic acid. X. Ba^{2+}-induced competence for transfecting DNA, *Z. Naturforsch.,* 30c, 520, 1975.

144. **Taketo, A.,** RNA transfection of *Escherichia coli* by electroporation, *Biochim. Biophys. Acta,* 1007, 127, 1989.

Chapter 3

HIGH-INTENSITY FIELD EFFECTS ON ARTIFICIAL LIPID BILAYER MEMBRANES: THEORETICAL CONSIDERATIONS AND EXPERIMENTAL EVIDENCE

Mathias Winterhalter, Karl-Heinz Klotz, and Roland Benz

CONTENTS

I. Introduction 137

II. Lipid Membranes as Model Systems 138

III. Interaction of External Electric Fields
with Lipid Membranes 143
   A. Interaction of Electric Fields with Lipid Vesicles 143
      1. Electric Field Distribution and Maxwell Forces 143
      2. Coupling Between Maxwell and Mechanical Stresses 146
   B. Models of the Electric Breakdown of Lipid Membranes 147

IV. Irreversible Electric Breakdown of Planar Lipid
Bilayer Membranes 152
   A. Theoretical Considerations 152
   B. Experimental Observations 155

V. Reversible Electric Breakdown in
Planar Lipid Bilayer Membranes 161
   A. Theoretical Considerations 161
   B. Experimental Observations 162

VI. Conclusions 168

References 168

I. INTRODUCTION

Biological systems are exceedingly complex. It is often necessary to simplify such systems to a model that includes a minimal number of variables in order to gain understanding of their basic mechanisms. While limited in their scope, these systems offer enormous advantages of convenience and the possibility to formulate and test hypotheses experimentally. The biological sciences

0-8493-4476-X/96/$0.00+$.50

137

have been greatly advanced by the development and experimentation using such model systems. The study of high intensity field effects is one of the biological disciplines that has benefited from model systems. Although the biophysical effects of electric fields can be calculated and predicted with great rigor, it is necessary to verify the physical properties of the membrane at the bench.

The model systems described herein facilitate this process, but more importantly, lend themselves to experimentation. They have thus been indispensable for the development of new technologies as well as for the refinement and optimization of standard approaches. In this chapter we will discuss in detail those aspects of cell membrane modelling applicable to cell electromanipulation and summarize the state of the art with respect to electric fields on lipid membranes.

We will focus, in particular, on high electric field effects which we account for by including the Maxwell stresses due to the presence of electric field lines. The application of an external electric field has many effects on biological systems. These include coupling to internal fields, change in transport properties, phase transitions, and several others (see Chapter 6). Within this paradigm, external electric fields act as additional forces which are balanced by mechanical forces. Throughout this chapter we will use a continuum approach. This implies that the models are based upon the average over a large number of molecules, rather than at the level of individual molecules. The gain in simplicity of this is counterbalanced by the loss of molecular structure. For a more general discussion we refer to several recent text books.[1-3]

II. LIPID MEMBRANES AS MODEL SYSTEMS

This section is devoted to the introduction of some basic physical features of lipid membranes. In order to understand the action of external electric fields on membranes, we must first understand the mechanical behavior of the lipid membrane alone.

Many properties of lipid bilayer membranes can be well characterized within a viscoelastic model.[4,5] Such models describe membranes by one or more material properties depending upon their lipid composition. Models based upon these macroscopic properties are justified only if it is safe to neglect local properties on the molecular level. Below we will detail three applications of viscoelastic models, namely area stretching, shear deformation, and bending deformation of membranes.

Area Stretching

The first example deals with area stretching of a lipid vesicle or a planar lipid membrane. In Figure 1a we show a typical membrane deformation induced by the application of lateral stress τ.[4,5] The origin of the latter could be a consequence of osmotic swelling,[6,7] electric field stresses,[8,9] or manipulation with micropipettes.[5,9,10]

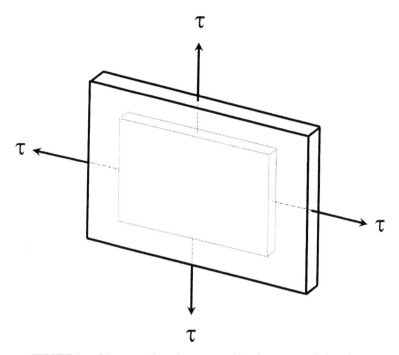

FIGURE 1a. Diagram to show the area stretching due to an applied tension τ.

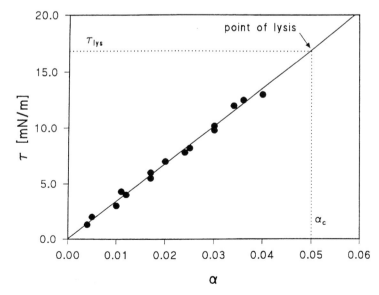

FIGURE 1b. Stress (τ) *versus* strain (α) plot for lipid vesicles made up of 1-stearoyl,2-oleoyl phosphatidylcholine/cholesterol mixtures (SOPC/Chol molar ratio 62/38).[10] The slope of the τ *versus* α plot is the elastic modulus of area stretching.

Indeed, experimental observations of such membranes show decidedly elastic behavior. Adopting Hook's law, which relates the tension τ to the relative change in area $\Delta A/A$, we see that:

$$\tau = k_E \cdot \frac{\Delta A}{A} = k_E \cdot \alpha \tag{1}$$

where k_E represents the elastic modulus of area stretching and $\alpha = \Delta A/A$ the relative change in area. A typical recording of an applied stress versus areal strain curve is shown in Figure 1b for lipid vesicles made from a mixture of 1-steroyl-2-oleoyl-phosphatidylcholine (SOPC)/cholesterol. The larger the applied stress, the larger the area extension. The observed linear behavior between tension and relative area increase is reversible up to the point of lysis,[10] a clear indication that the membranes follow Hook's law in a manner similar to that of an elastic spring.

This linear dependence of strain on stress has been observed for many lipids as well as for cell membranes.[5,10] The slope of the curve yields the elastic modulus of area dilatation k_E and depends only upon the lipid composition. This is well demonstrated for vesicles made up of various lipid/cholesterol mixtures (see Table 1). The data clearly indicate that the addition of cholesterol stiffens the membrane by an order of magnitude. At low cholesterol contents, the elastic modulus is that of a pure lipid itself. At 45 mol% cholesterol, the elastic modulus increases sharply to that of cholesterol alone. This is typical elastic behavior for composite materials and is well-known in material science.

Inspection of Table 1 allows us to draw the intriguing conclusion that lipid membranes are incompressible. Dividing the area elastic modulus k_E by a typical membrane thickness of about 4 nm yields $k_E/d \approx 10^7$ N m^{-2}. This value can be compared to the bulk compressibility modulus of alkane which is about 10^9 N m^{-2}, two orders of magnitude larger than the corresponding value obtained from the area stretching.[4,5] If we assume that the membrane interior must behave similarly to an alkane then, to the first approximation, application of tension causes the area to increase under the constraint of constant volume.

In Figure 1b a second material property, the critical stress τ_{lys} causing mechanical membrane rupture, is demonstrated.[5,10] In many reports it was thought that exceeding a critical area strain of 2% would lead to rupture. However, it has now been shown experimentally that, depending upon its composition, a membrane could resist rupture below 5% area dilatation.[10] These experimental data indicate that neither the elastic modulus, the relative area increase, nor the elastic energy stored is directly correlated to mechanical membrane rupture. This means that varying these parameters by changing the composition is not conclusive for the mechanical membrane stability.

TABLE 1
Cohesive Properties of
Vesicle Bilayer Membranes[10]

Lipid	Cholesterol [mol %]	k_E [mN m^{-1}]	τ_{lys} [mN m^{-1}]
SOPC[a]	0	190 ± 20	5.7 ± 0.2
	50	781 ± 45	19.7 ± 3.2
	78	1,286 ± 105	28.0 ± 2.2
BSM[b]	50	1,718 ± 484	23.2 ± 5.9
	80	1,732 ± 354	22.3 ± 3.5
DAPC[c]	0	57 ± 14	2.3 ± 0.6
	50	102 ± 24	4.5 ± 0.6
	80	67 ± 13	3.4 ± 1.7

Note: Material property parameters for lipid vesicles at 15°C composed of a variety of lipid types and lipid/cholesterol compositons. The elastic area compressibility, k_E and the critical tension causing lysis, τ_{lys}[10] are given.

[a] SOPC, 1-stearoyl,2-oleoyl phosphatidylcholine; [b] BSM, bovine sphingomyelin; [c] DAPC, diarachidonoyl phosphatidylcholine.

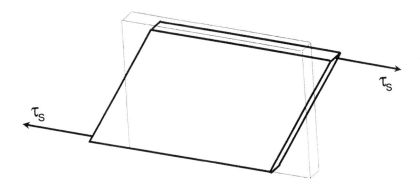

FIGURE 2. Deformation of a membrane subject to a shear stress τ_S.

Shear Deformation:

Figure 2 demonstrates the deformation of a membrane caused by application of shear stress τ_s. For frozen lipid membranes and also for cells such deformation has to be taken into account. In the following discussion we will focus only on lipids in the fluid phase (which behave like liquids, including the property of vanishing shear resistivity), hence deformation need not be considered.

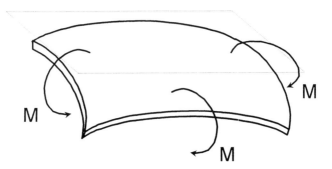

FIGURE 3.   A bending moment *M* acting on a planar membrane.

Bending Deformation:

Figure 3 demonstrates the third kind of deformation: membrane bending.[4,5,8,11] This property could be understood if we perform a "Gedanken-experiment" — what happens if we reverse area stretching, by say deflating, e.g., osmotically, the lipid vesicles. If we allow the vesicle to remain under tension causing it to assume a spherical shape and then deflate it by removing water from the inside, the tension relaxes to zero as the area strain vanishes. If we continue to remove water, the vesicle will have excess area. We would expect that entropy would cause the interface to mix with the aqueous phase. On the contrary, surprisingly well-defined structures can be observed. Inspection of the general linear elastic theory (a generalization of Hook's law) readily yields higher order elastic moduli, named the curvature elastic modulus or the bending rigidity k_C, for such a deformation. Lipid bilayer membranes resist bending, and the force induced by such bending is a linear function of the deformation. For a symmetric planar membrane the bending moment *M* due to the two principle curvatures R_1 and R_2 is given by:[4,5]

$$M = k_C \cdot \left(\frac{1}{R_1} + \frac{1}{R_2} \right) \qquad (2)$$

Membranes are asymmetrical in many cases (see Chapter 1). This can be easily taken into account by subtraction of the spontaneous curvature $1/R_0$ in Equation 2:

$$M = k_C \cdot \left(\frac{1}{R_1} + \frac{1}{R_2} - \frac{1}{R_0} \right) \qquad (3)$$

This approach provides a powerful tool for explaining the stability of the membrane against shape transformation for vesicles with excess area. Almost twenty years ago vesicle shapes were considered from the theoretical point of view on the basis of their bending elasticity.[11] As pointed out above, however, area dilatation would require much higher energies than bending. Taking this into account yields a second constraint: that of constant area. Once the volume and the area have been fixed, the energy is minimized with respect to curvature. Within this model, many shapes: spherical, discocytes, and stomatocytes, as well as the initial step of endocytosis/exocytosis, could be parameterized using spontaneous curvature, bending rigidity, and the constraint of the ratio of enclosed volume to area.

Recently, more refined calculations related to this model system have been performed.[12-17] A detailed phase diagram of shapes with the lowest energy as a function of the various parameters has been thoroughly documented. In theory, the spontaneous curvature has been modelled more realistically by taking into account possible differences between the outer and inner monolayers. Several of the predicted shapes have been observed and transitions between them could be triggered by slight variation of temperature.[14,15] However, it should be noted that within this model, the observed shapes depend strongly upon slight variations of the mechanical properties of the two monolayers. This could explain that the swelling of lipids in excess water causes an extreme variation in membrane shapes.

III. INTERACTION OF EXTERNAL ELECTRIC FIELDS WITH LIPID MEMBRANES

A. INTERACTION OF ELECTRIC FIELDS WITH LIPID VESICLES

The previous section outlined the stability of lipid membranes in the absence of external electric fields; now we will consider the effect of electric field forces on the lipid membrane. In this context we recall that application of an electric field across an aqueous solution causes an electric current to flow, and thereby Joule's heat. Theoretical models based on the minimization of energy must therefore be applied very carefully.[18] This problem is probably much more appropriately treated as a balance of forces.[19]

1. Electric Field Distribution and Maxwell Forces

External electric fields interact with charged membranes and cause electrophoretic movement (see Chapters 4–6). The force is linearly dependent upon the applied field strength. In the following we will restrict ourselves to an *a priori* uncharged membrane. In this case, external fields cause charge separation, which then will couple to the external electric field. This results in so-called Maxwell forces, which are quadratic functions of the applied field.[20]

In Figure 4a we show (schematically) a lipid vesicle in an external electric field. Application of a steady state external field would cause this vesicle to move due to field inhomogeneities. For a detailed discussion of this dielectrophoretic effect please refer to Chapters 4 and 5. In this context it is interesting to note that application of alternating current fields of about 1 kHz to vesicle suspensions avoids this effect.

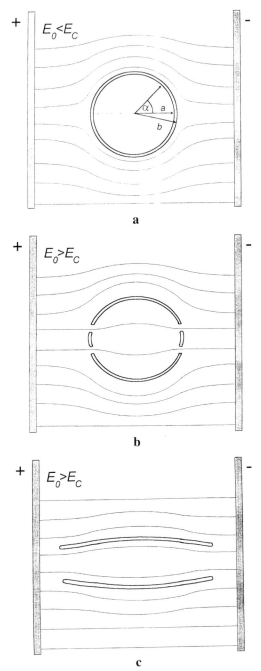

FIGURE 4. The action of external electric fields on vesicles. a. Field line distribution around a vesicle under stationary conditions and small membrane conductance (no breakdown; see also Chapter 1). b. Beyond the critical field strength, the vesicle is permeabilized at the pole areas. c. Still stronger fields could yield cylindrical structures.

The spatial distribution of the external electric field inside the chamber between the two electrodes is described by Maxwell's Equations.[20] In the absence of free charges all field lines start at the positive (+) pole and end at the negative (−) pole. Due to the difference in the relative dielectric permittivity of the lipid ($\varepsilon_l = 2$) and the aqueous phase ($\varepsilon_w = 80$), the perpendicular field lines will be discontinuous at the outer ($r = b$) water/lipid interface

$$\varepsilon_w \cdot \varepsilon_0 \cdot E_r^w(r = b) - \varepsilon_l \cdot \varepsilon_0 \cdot E_r^l(r = b) = \sigma(r = b) \tag{4}$$

with $\sigma(r = b)$ as the surface charge density at the outer interface and $\varepsilon_0 = 8.8 \cdot 10^{-12}$ AsV^{-1}m^{-1} is the dielectric permittivity of a vacuum. $E_r^w(r = b)$ and $E_r^l(r = b)$ stands for the radial field strength in the aqueous and the lipid medium. The boundary condition for the tangential fields strengths are:

$$E_\alpha^w(r = b) = E_\alpha^l(r = b) \tag{5}$$

Similar boundary conditions hold at the inner ($r = a$) membrane/water boundary. Because of the conductivity of the aqueous phase, ions start to flow along potential gradients. In addition to Maxwell's Equation, the continuity Equation for the charge flow

$$\kappa_w \cdot E_r^w(r = b) - \kappa_l \cdot E_r^l(r = b) = 0 \tag{6}$$

must be satisfied everywhere.

The difference in membrane conductivity κ_l and water conductivity κ_w (about 10 orders of magnitude) causes an accumulation of charges at the membrane/water interface until the boundary conditions satisfy Equations 4 and 6. Inspection of Maxwell's Equation yields for the induced charge density, σ, an angular dependency of the direction of the field, i.e., $\sigma \sim \cos\alpha$, where α is the angle between the electric field and the location on the surface of the vesicle (see Figure 4a). These induced charge densities σ of opposite sign at each water/lipid interface cause the external applied potential to drop across the lipid bilayer.

As schematically indicated in Figure 4a, the vesicle interior is screened by these induced charges and is (almost) field-free. For better understanding, we assume that the vesicle is in a homogeneous external electric field E_0. The induced charge densities on either side of the membrane in this case cause a very high screening field inside of about $E_r^l(b) = (3/2)(b/d)E_0\cos\alpha$ with d as the membrane thickness. This causes large Maxwell forces acting at the outer interface but directed towards the membrane of approximately

$$\Pi_{el}^l(r = b) = -\varepsilon_l \cdot \varepsilon_0 \cdot \frac{1}{2}(E_r^l(r = b))^2 = -\varepsilon_l \cdot \varepsilon_0 \cdot \frac{9}{8}\left(\frac{b}{d}\right)^2 E_0^2 \cos^2\alpha \tag{7}$$

and a similar force at the inner boundary by replacing b by the inner radius a. The two forces combine to give a pressure perpendicular to the membrane plane, which tends to thin the membrane.

The field lines "squeezed out" from the vesicle show a higher density at the equator (see Figure 4a) and give rise to a pressure gradient

$$\Pi_{el}^{w}(r = b) = -\left(\frac{9}{8}\right) \cdot \varepsilon_{w} \cdot \varepsilon_{0} \cdot E_{0}^{2} \cdot \sin^{2} \alpha \tag{8}$$

which causes elongation of the spherical shape of the vesicles.[8,20,21]

Detailed calculations can account for the conductivity and dielectric effects if it is assumed that a lipid vesicle is described by a single lipid shell.[20] These calculations are straightforward but result in rather lengthy expressions. Careful inspection of the detailed model yields three different regimes for the electric field induced forces on vesicles:[20,21]

> The *conducting regime* at low frequency in which only the conductivity ratio of the lipid to the aqueous phase need to be taken into account.
>
> A *high frequency regime* in which the deformational force of the electric field is determined by the ratio of the dielectric constants of lipid and water. Here the conductivity of both media can be neglected.
>
> An *intermediate regime* in which the conductivity of the aqueous phase and the dielectric constants have the main contributions.

The model presented above describes a lipid vesicle as a single lipid shell within an aqueous surrounding. In other approaches the lipid bilayer was modelled using many shells with different dielectric and conductive properties[22] (see also Chapter 5). With respect to the Maxwell forces, one could theoretically take such corrections into account. However, due to the small volume of each shell, no significant contribution from these forces would be expected. Other models oversimplified the presence of a lipid vesicles in an external electric field by assuming an infinitely thin shell.[23,24] Such approaches describe a homogeneous droplet in a surrounding of different electric properties.[24] In this case, the shielding effect of the lipid shell is neglected, and the potential drops across the entire vesicle, not only across the membranes themselves.

2. COUPLING BETWEEN MAXWELL AND MECHANICAL STRESSES

The Maxwell stresses due to electric field lines act as real forces at the membrane/water boundaries and have to be balanced by mechanical stresses. At low field strength, electric field effects are "hidden" by Brownian motion. If the field strength becomes larger, an orientation of nonspherical-shaped vesicles could be observed.[21] In the low frequency regime, flaccid vesicles become oriented parallel to the field lines, whereas in the high frequency regime an orientation perpendicular to the field is observed (see Chapter 4). Higher field strengths cause deformation of the original spherical vesicles to elongated ones[8,21] (see the schemes of Figure 4a–c). Recently electric fields have been used to smooth thermal undulations of the vesicle membrane.[8] Larger visible deformations can only be obtained for flaccid vesicles. Larger deformations influence the field distribution around the vesicles in such a way that the perturbations of the field lines due to their presence decrease.[25] This reduces the deformational force of the field. In extreme cases, even very high fields do not cause vesicle lysis.

Attempts have been made to include the effects of an external electric field[23,26,27] in the shape calculation described in Section II. The main difficulty is that a correction of the field distribution due to the vesicle deformation induced by the presence of external fields must be taken into account. Such corrections are of a higher order and require numerical solutions to the problem.

Caution must be exercised when considering conditions of low ionic strength. Thermal motion within the induced charges causes the ions to be smeared out across one Debye length. This tends to further reduce the net force of the applied electric field[8,21,28] (Chapter 5).

When the vesicle or cell lacks sufficient surface to counteract the Maxwell stresses by deformation of the shape, the electric field starts to stretch the membrane. This stress can exceed even the critical stress of lysis. Experimentally, this could result in single large pores or a sieve of pores on each pole (schematically represented in Figure 4b and 4c). This deformational power and the formation of large pores on either side of a lipid vesicle has been observed in the phase contrast microscope.[29] Larger field amplitudes caused a tube-like structure to form. The structure has been explained by a balance of bending energy and edge energy. A few years later it was shown experimentally that vesicles can be permeabilized by short electric field pulses.[30] The relation of applied field strength and subsequently the efflux of tracer molecules has been studied in detail.[30] High electric fields can also be used to produce large unilamellar vesicles.[31] Vesicles of small size were aligned with small alternating current fields (see Chapter 5) and fused with a series of pulses of higher amplitudes.

B. MODELS OF THE ELECTRIC BREAKDOWN OF LIPID MEMBRANES

Although it is clear that the breakdown of a lipid membrane can be caused either by mechanical or electric forces,[9] the exact process of mechanical membrane rupture is still a question of speculation. Nothing is known about the early stages of membrane defect* formation. Until now, it was not possible to

---

*   Due to the lack of knowledge about the structure of these initial steps we prefer the notation defect instead of pore. If these defects become large enough we use the expression pore (see also footnote on page 3 in Chapter 1).

resolve this process by any experimental approach. This probably indicates that the time resolution of the experiments was probably not good enough to distinguish between different possible processes.

Earlier investigations suggested an electromechanical compression as the most likely description of the electric breakdown[32,33] (see Chapter 1). In this model it is assumed that the membrane can be regarded as a capacitor filled with a homogeneous elastic dielectric material. The thickness of the dielectric material depends on the electric compressive forces (caused by the membrane potential) and counterbalanced by the mechanical forces of the lipids. It was shown that at high electric fields a critical thickness is reached at which the compressive forces (created by the electric field, see Equation 7) are no longer counterbalanced by the elastic restoring forces. At this point electric break-down occurs. The critical voltage is dependent upon the elastic compressive modulus perpendicular to the membrane plane, the relative dielectric constant of the membrane interior, and the membrane thickness in the absence of an electric field. However, this model requires partial thinning of the membrane by 30% of the original thickness. Taking into account the incompressibility of the lipid itself, this model would allow a very large area of expansion prior to lysis (which is in contrast to the experimental findings[5,9,10]). Based on this model, many refinements have been suggested.[33-35] Due to the lack of an observed large area increase at lysis, these models are restricted to the very small range beyond the linear regime which has not yet been investigated.

On the other hand, thinning of the membrane by electric compressive forces may cause the energy of the ions in the electric field to reach the Born charging energy, which is required to move ions from the high dielectric aqueous to the low dielectric membrane interior.[36] The energy gain for a monovalent ion moving through an electric potential of 1 V corresponds to $9.6 \cdot 10^4$ J mol^{-1}. The Born charging energy for a membrane with a hydrocarbon thickness of 4 nm and a relative dielectric constant of 2 is $1.5 \cdot 10^5$ J mol^{-1} when the image force is taken into account.[36,70] Assuming a dielectric constant of 3 for the membrane interior the Born energy decreases to $1 \cdot 10^5$ J mol^{-1}. This value is close to the energy gained by an ion when it falls through a potential difference of one volt. Artificial and biological membranes are not very homogeneous. Furthermore, the dielectric constant is probably a function of the membrane composition. This heterogeneity may lead to a further decrease of the energy needed for the injection of ions into the low dielectric membrane interior. This could also mean that a combination of electric compressive forces with the Born charging energy represents the primary event for defect formation in artificial and biological membranes during reversible electric breakdown and field-induced mechanical rupture.

Recently[9] the stability of vesicles during electric field pulses has been investigated. For these experiments, single vesicles are held under a fixed lateral mechanical tension τ using micropipettes. The influence of additional field pulses is seen in a decrease of the critical tension for lysis. The analysis of the experimental results suggests at a first glance that the electric field acts as an additional tension and leads to an effective total tension:

$$\tau_{eff} = \tau + \frac{3}{8} \cdot \varepsilon_l \cdot \varepsilon_0 \cdot \left(\frac{b}{d}\right)^2 \cdot E_0^2 \cdot d \tag{9}$$

When τ_{eff} exceeds the critical tension τ_{lys}, lysis was observed. However, according to Equation 7 one would expect an opposite sign for the electric field induced contribution to the tension. The Maxwell pressure (directed inwards) should thin the membrane and therefore release the energy stored by elastic area stretching. In other words this pressure acts against the mechanical stress. Under other experimental conditions, an increase in area attributed to electric field pulses was observed in free vesicles.[8] In fact, such an area increase has not been observed since the experiments have been performed at the point of lysis (at which any additional area increase is not visible). We would like to suggest the following explanation of this apparent discrepancy. Under the experimental conditions used,[9] the electric field pulses cause an increase of the area, which is reduced by the aspiration pressure in the micropipette. After removal of the field pulses, the ensuing increase of tension causes lysis.

Following our discussion we suggest instead of Equation 9 the effective tension

$$\tau_{eff} = \tau + \tau_{el} + \tau_{osm} \tag{10}$$

where τ is an external applied tension (e.g., by micropipettes), $\tau_{el} = -\frac{3}{8} \cdot \varepsilon_l \cdot \varepsilon_0 (\frac{b}{d})^2 \cdot E^2_0 \cdot d$ the electric tension and $\tau_{osm} = \Delta\Pi_{osm} \cdot \frac{b}{2}$ represents an additional tension due to the presence of an osmotic gradient across the membrane. Again, if the effective tension exceeds the critical point, vesicle lysis would be expected. In this context it should be noted that both models, the previous described electromechanical coupling model and the effective tension described by Equation 10, lead to identical results in the linear regime of stress versus area increase.

It should also be noted that these considerations depend on a direct effect of the electric field on the structure and mechanical properties of the membrane. Other models are based on previously existing and non-conducting small pores or defects within the membranes. These defects either fuse or grow in size as a consequence of the action of the electric field.[37-52]

A general model to describe the mechanical stability of a membrane assumes that a defect has been formed initially.[38] This could be due to thermal fluctuation or to other perturbation. Within such a model the exact process of the initial step does not matter. Once formed, the defect expands reversibly or irreversibly, both cases being separated by an energy barrier. For large defects, also called pores without external electric fields, this energy contains two opposite contributions:

$$E_{pore} = 2\pi \cdot a \cdot \Gamma - \pi \cdot a^2 \tau \qquad (11)$$

The first term represents the energy needed to build up the edge of a pore of radius α where Γ stands for the line tension. The second term is the energy which is gained by decrease of the membrane area due to the surface tension τ. Similar models are based on the monomer solvation energy or on statistical mechanics which yields the energy barrier for pore formation.[39-45] The mechanical lifetime is assumed to be given by a Boltzmann distribution with the energy barrier E_{pore} given for example by Equation 11 in the exponent.[6,34-52]

Several authors[37,46,47] accounted for electric field effects by adding the condenser energy of a lipid bilayer to Equation 11. The inside of a lipid pore was thought to be replaced with non-conducting water. Under the additional assumption that the electric potential is not altered after a pore is created, the electric field acts simply as a source of additional surface tension. This influence of the electric field has been modelled by assuming the existence, in the early stage of electric breakdown, of hydrophilic pores within the membrane.[48-52] However, even for hydrophobic pores, the electric potential will be altered.[49-54] Moreover, if the extreme difference in conductivity between the membrane and the ionic solution is taken into account, the electric field distribution will be substantially changed (see Figure 5). Many attempts have been made to overcome this deficiency by modelling detailed structures on the molecular level. During the early stage a so-called hydrophobic pore was postulated.[48-52] A simplified model shows that such effects cause an electric contribution to both the line tension and the surface tension.[54] If we let V_0 be the electric potential across the membrane, the electric line tension $\Gamma_{el} = -\varepsilon_w \cdot \varepsilon_0 \cdot V_0^2/2\pi$ tries to open the pore, and the contribution to the surface tension $\tau_{el} = -\varepsilon_l \cdot \varepsilon_0 \cdot V_0^2/2d$ emanating from the field along the membrane surface tends to close it again. Incorporating both contributions into Equation 11 by replacing the pure mechanical parameters with their effective ones we have:

$$E_{pore} = 2\pi \cdot a \cdot \Gamma_{eff} - \pi \cdot a^2 \tau_{eff} = 2\pi \cdot a \cdot \left(\Gamma - \varepsilon_w \cdot \varepsilon_0 \frac{V_0^2}{2\pi} \right) - \pi \cdot a^2 \cdot \left(\tau - \varepsilon_l \cdot \varepsilon_0 \frac{V_0^2}{2d} \right) \quad (12)$$

Here d denotes the membrane thickness and V_0 the applied voltage, whereas $\varepsilon_0 = 8.8 \cdot 10^{-12}$ A s V^{-1} m^{-1} is the permittivity of vacuum. Inspection of Equation 12 shows that two possible mechanisms of pore formation are possible. Using the applied potential V_0 as a parameter and increasing it continuously, in Equation 12 either $\Gamma_{eff} = 0$ or $\tau_{eff} = 0$. In the first case an enhancement of the field reduces the critical radius and promotes rupture. In the second case an increase of the applied external voltage V_0 stabilizes the membrane and causes stable pore formation at higher voltages. The mechanical properties of the membranes described in the first part of this chapter suggested the first process, which is described in more detail below. The second process has not

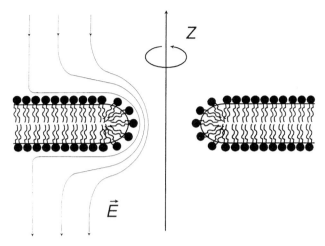

FIGURE 5.   The electric field lines close to a large pore.

yet been experimentally verified but might occur in the case of electro-permeabilization of flaccid vesicles or cells.

The stabilization of large pores by balancing the mechanical forces against electric field pulses has recently been demonstrated using the micropipette technique.[55] Single vesicles were held under fixed tension applied by micropipettes. In this series of experiments subcritical tensions were used. Careful permeabilization yields subcritical pores which are stabilized by the outflux of the internal volume. These measurements allowed an independent determination of the edge energy.[55]

The above models are based on a static view. Once a pore of a given radius, *a,* has formed, this model predicts whether the pore will close again or grow to infinite size causing membrane rupture. Nothing can be concluded about the dynamic process of pore widening or closing per se from these models. Due to the lack of experimental data about the very early stages of pore formation, none of the models could yet be disproved.

All of these models suggest an energy barrier E_{pore}. Within this framework the lifetime of a membrane is assumed to be distributed according a Boltzmann function:

$$t_{life} = t_0 \cdot e^{\frac{E_{pore}}{k \cdot T}} \tag{13}$$

Measurement of the lifetime allows an estimation of the pore energy.[51,56,57] The surface tension of a bilayer can be obtained by independent measurement. Variation of the external field strength and the successive variation in lifetimes have been used to fit the edge energy as an independent parameter. The observed Boltzmann distributions were very broad and required many measurements. However, the observed values (Table 2) were in reasonable agreement with those from other methods.

TABLE 2
Line Tension Γ and Surface Tension τ
for Various BLM Systems

Lipid	Γ[10⁻¹¹ N]	τ[mNm⁻¹]	Ref.
PE[a] in decane	1.60 ± 0.06	2.70 ± 0.30	56
PE in squalene	1.66 ± 0.07	2.40 ± 0.20	56
PE in squalene/LPC[b] (4.5 · 10⁻⁴ gl⁻¹)	0.41 ± 0.05	0.33 ± 0.10	56
PC[c] in decane	0.86 ± 0.04	1.00 ± 0.30	56
PC in decane/LPC (4 · 10⁻⁴ gl⁻¹)	0.33 ± 0.06	0.20 ± 0.10	56
EggPC[d]	2.00	—	29
EggPC[d]	2.00	0.22 ± 0.02	57
SOPC[e]	0.92 ± 0.07	—	55
SOPC/Cholesterol	3.05 ± 0.12	—	55

[a]   PE, phosphatidylethonolamine; [b] LPC, lysophosphatidylcholine; [c] PC, phosphatidylcholine; [d] EggPC, eggphosphatidylcholine, [e] SOPC, 1-stearoyl,2-oleoyl phosphatidylcholine.

IV. IRREVERSIBLE ELECTRIC BREAKDOWN OF PLANAR LIPID BILAYER MEMBRANES

The irreversible breakdown of thin films is a general phenomenon. Bursting of soap films is a typical example, wetting of a liquid on a solid or liquid support another.[58,59] Most investigations in this area have dealt with static properties; relatively few authors have considered the dynamics. The velocity of the phase boundary, or for the purposes of our discussion the velocity of the pore rim, can range from several m s⁻¹ down to μm s⁻¹.[58-64] The observed velocity represents the balance between the mechanical forces exceeding those driving the opening (or closing) of the pore as well as damping forces. The latter depend upon the velocity and are typically dominated by viscosity, or in fast processes, by inertia. Some years ago the bursting of soap films was carefully investigated.[64] Large soap films were spanned across a rim and defects were induced by electrical discharges. The kinetics were registered using a high speed camera.[64] Recently, in planar lipid membranes, defects have been induced with electric field pulses. The successive widening of the defects to give large aqueous pores was recorded electrically.[62,63] In this section we will discuss the theoretical basis upon which these experimental observations are based.

A. THEORETICAL CONSIDERATIONS

In Section III of this chapter a static picture of a pore was developed. Such a model elucidates the criteria of pore stability but says nothing of the dynamics involved.

An attempt to describe the dynamics of pore widening has been made based upon statistical mechanics.[60] Another procedure based upon the so-called defect model (Equation 12) and including the conductivity correction mentioned earlier has been proposed.[61] The observed experimental features of an irreversible breakdown were adjusted by several well-known membrane parameters. The kinetics of this model were due mainly to diffusion of the lipids and quantitatively described by the self diffusion coefficient. Unfortunately, such models seem to be in conflict with the experimental findings.[61]

Deriving the energy barrier of a defect (Equation 12) with respect to the pore radius yields a force which is either directed outward, causing further widening, or inward, causing resealing.[62] This force will cause a movement of the lipid which is damped by friction or inertia. In general the movement of a thin film is described by the Navier-Stokes Equation, which is quadratic with respect to flow velocity. The prediction of the pertinent dynamics requires knowledge of the experimental boundaries and the exact material flow profile. A general solution based on the Navier-Stokes Equation is not possible. However, taking into account experimental observations allows us to simplify the calculations and allows a special solution for the pore velocity in the observed time range.

If thermal fluctuations are neglected, then Equation 12 allows an irreversible breakdown only if the radius a becomes larger than the critical one $a > a^* = \Gamma_{eff}/\tau_{eff}$. Inserting typical values of the mechanical edge energies for pure lipids of about $\Gamma = 1 \cdot 10^{-11}$ N (Table 2) requires a voltage of about 400 mV for Γ_{eff} to vanish. Inspection of Equation 12 shows that the applied voltages $V_0 = 400$ mV together with a bilayer thickness $d = 5$ nm is responsible only for a relative small electric contribution to the surface tension $\tau_{el} = 0.3$ mN m^{-1} in comparison with the mechanical tension of roughly $\tau = 2 - 3$ mN m^{-1}. We may therefore neglect the electric contribution to the surface tension and take τ_{eff} to be independent of the applied voltage V_0 with reasonable accuracy. In the following discussion we assume that defects (or pores) are statistically formed (the exact mechanism is not critical for these purposes). For a given applied external electric field, the resulting forces will close the pore if $a < a^*$ or will cause further widening. The larger the pore becomes, the less important is the influence of the edge energy. After the initial pore forming, the opening process is driven by the mechanical surface tension alone.

In order to understand the presented experimental data we have to model the flow of the lipid material during the irreversible mechanical breakdown. Due to its finite surface tension, the lipid is homogeneously stretched like a soap film. Furthermore, we assume that the pore is of cylindrical shape and large enough to neglect effects due to finite thickness.

During the "opening process" of the pore, lipid flows radially away from the center to the torus of the membrane. As mentioned above, an exact solution of this membrane material flow requires knowledge of the local stress distribution after breakdown. Whatever the local stress distribution, surface tension remains the driving force for the opening process. As a first approximation one would expect that the membrane viscosity would determine the time course of the opening. As recently shown,[62] a damping force acting at the boundary and working against the movement of the pore rim must be added to the mechanical forces derived from Equation 12. Neglecting the edge energy contribution for larger pores yields an exponential increase of the pore radius with time:

$$a(t) = a_0 e^{\left(\frac{\tau}{4\eta}\right)t} \tag{14}$$

The decay time $t_{vis} = 4\eta/\tau$ is about $2 \cdot 10^{-6}$ s for reasonable values[52,53] of η ($\approx 10^{-9}$ Ns m^{-1}) and of τ ($= 2 \cdot 10^{-3}$ N m^{-1}). The pore conductivity was recorded in the breakdown experiments over much longer times than t_{vis}. This means that an exponential increase in the pore conductivity would have been clearly detected. Its absence indicates that it is probably controlled by a different process with a larger degree of damping. In this context an earlier report[66] using a voltage clamp on bilayer membranes in a 1 M KCl solution should be mentioned. Within the first few microseconds after the initiation of the breakdown, an exponential increase attributable to viscosity was measured.

A completely different process was proposed some years ago to explain the bursting of soap films.[64] After initiating the irreversible breakdown in a soap film with a spark, the kinetics of the opening were recorded with a high speed camera. As a first approximation it was assumed that the film itself stays immobile while only the lipid material of the hole moves (with constant velocity). The dynamics of the pore widening were obtained by bringing the kinetic energy term into Equation 11:

$$E_c = \pi \cdot a^2 d\rho \cdot (a)^2 \tag{15}$$

where ρ is the lipid density. Again, neglecting the electrostatic contribution to the surface tension, as well as the edge effect, we obtain the time dependency of the pore radius[62,64]

$$a(t) = a_0 + \sqrt{\frac{\Phi \cdot \tau}{d\rho}} \cdot t = a_0 + u \cdot t \tag{16}$$

where ϕ is a parameter depending on the unknown material flow, as well as on dissipation effects. In the following section we will qualitatively describe the behavior of the opening velocity u by specific variation of the effective mass of the membrane. In our analysis we attributed the measured conductivity to a single pore. Experimental observation (shown in the following subsection) suggests that the observed defects are rather large with respect to the thickness of the membrane. The conductivity is then simply related to the so called access resistor (R_{pore}) of a single pore:[62,65]

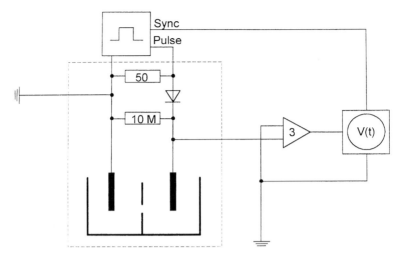

FIGURE 6.   Schematic diagram of the charge-pulse instrumentation used for the measurement of electric field-induced irreversible breakdown of lipid bilayer membranes.

$$G(t) = \frac{1}{R_{pore}} = 2\kappa \cdot a(t) = 2\kappa \cdot (a_0 + u \cdot t) \tag{17}$$

and contributions stemming from the area of the pore are negligible.

B. EXPERIMENTAL OBSERVATIONS

The conductivity of planar lipid bilayers has been studied for many years. Two different methods (or combinations of both) are most commonly used to study the membrane electric properties: voltage clamp or charge-pulse relaxation. Each has its own advantages and disadvantages. The first one uses large currents under breakdown conditions causing a poorly defined voltage drop at the pore edge. The second is charge-pulse relaxation studies which have the advantage of being much faster and allowing investigation within the nano- to millisecond ranges. In this section we will outline recently reported investigations on irreversible breakdown using both of these methods.

The kinetics of pore formation followed by mechanical rupture of lipid bilayers were investigated in detail using the charge-pulse relaxation technique. Figure 6 shows the electric circuit used for investigation of lipid bilayer membranes under charge-pulse conditions. For the purposes of this discussion, the membrane can be regarded as a plate condensor (of specific capacitance C_m between 0.3 and 0.7 μF cm^{-2}), and a resistor, R_m, (corresponding to a specific resistance between 10^6 and 10^8 Ω cm^2) connected in parallel and switched in series to a resistance R_E of electrodes (with electrolytes on both sides of the membrane). In charge pulse experiments the membrane capacitance, C_m, is connected to a pulse generator. After the pulse, with a duration of 100 ns to a few μs, the resistance of the external circuit is made very large. Hence the charge on the membrane capacitance can decay only by charge leakage across the membrane or through an external resistor, R_{par}, introduced in parallel to the membrane to obtain a defined $R_{par} \cdot C_m$ (discharge) time constant.

From the exponential voltage decay at small membrane potentials it is possible to determine the relaxation time $t_{RC} = R_{par} \cdot C_m$ of the membrane. Figure 7a shows charge pulse experiments on planar lipid bilayers consisting of diphytanoyl phosphatidylcholine/n-decane. One of these membranes was previously polarized to a voltage of 12 mV with a charge pulse of about 20 µs duration (trace 1). Because of the high specific resistance of the unmodified lipid bilayer, the membrane discharges exclusively through the external resistor (10 MΩ). The same membrane was subsequently charged to about 500 mV (trace 2). At this voltage, mechanical breakdown of the membrane is observed, i.e., the membrane is irreversibly destroyed. The membrane voltage drops to zero within about 100 ms, because of the increase of the pore diameter. It is noteworthy that the voltage decay depends upon the specific conductivity of the bulk aqueous phase and is much slower in 10 mM KCl. Figure 7a also shows the irreversible breakdown of other membranes initiated by the same high electric fields (traces 3 – 5). Although the onset of the mechanical breakdown occurred at different times after the application of the electric fields to the membranes, the time course of the voltage decays (i.e., that of the irreversible breakdown) is very similar for the different membranes (Figure 7a).

Figure 7b shows the analysis of the voltage decay during irreversible electrical breakdown of lipid bilayer membranes. The instantaneous membrane conductance, $G(t)$, (which corresponds to that of the widening pore caused by the high electric field) was calculated according to:

$$G(t) = \frac{I(t)}{V(t)} = -\frac{C \cdot dV(t)}{dt} \cdot \frac{1}{V(t)} \tag{18}$$

This approximation neglects the variation of the capacity due to widening of the pore and the resulting reduction in membrane area. This is justified whenever the pore area is much smaller than the total area. Figure 7b shows the corresponding conductivity plots of Figure 7a. A linear increase of the conductivity with time is clearly evident. This is further discussed below.

Single or Multiple Pores During Irreversible Breakdown?

Several experimental observations suggest that irreversible breakdown is a result of widening of a single pore. These observations include:

- The statistically broad distribution of lifetimes. A large number of pores would yield a rather narrow distribution of lifetimes.
- After raising the applied voltage to a critical level, a series of narrow distributions of opening velocities was observed.

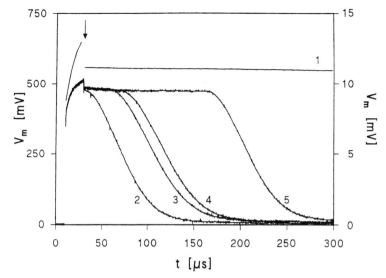

FIGURE 7a. Trace 1 shows the exponential decay of membrane voltage $V_m(t)$ during the discharge process through an external resistor (see text). Traces 2–5 exhibit the time course of membrane voltage $V_m(t)$ during electric field-induced irreversible breakdown of four membranes formed of diphytanoyl phosphatidylcholine (DPhPC). The length of the charge pulse was 20 μs. The aqueous phase contained 100 mM KCl; T = 20°C. Note: whereas for trace 1 the right y-axis is valid, the traces 2–5 relate to the left y-axis. The arrow indicates the end of the charge pulse.

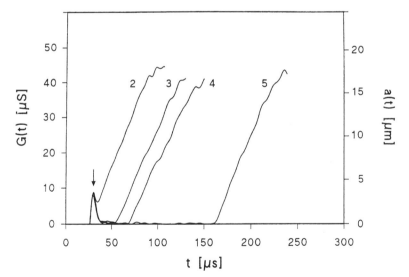

FIGURE 7b. Conductance $G(t)$ *versus* time curves during irreversible breakdown of the four different membranes of Figure 7a (traces 2–5). All curves exhibit a slope of 0.7 S s^{-1}. $G(t)$ was calculated according to Equation 18 by using the capacitance of the individual membranes and the velocity of the pore rim by using Equation 17. The length of the charge pulses was 20 μs and the amplitude 520 mV. Because of the considerable scatter of the signal, the data have been averaged by standard methods.[62] The aqueous phase contained 100 mM KCl; T = 20°C. The arrow indicates the end of the charge pulse.

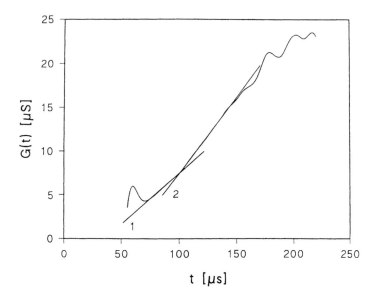

FIGURE 8. Conductance $G(t)$ *versus* time curve during irreversible breakdown of a mixed DPhPC/CPCl membrane bathed in 100 mM KCl. $G(t)$ was calculated according to Equation 18 by using the capacitance of the membrane and the velocity of the pore rim by using Equation 17. The length of the charge pulse was 20 μs; T = 20°C. Note that the initial slope of the curve was 0.1 S s^{-1} at first, and changed after about 30 ms to 0.2 S s^{-1}, due to the opening of a second pore.

- In a few experiments, the initial slope suddenly doubled (Figure 8). This was interpreted as the occurrence of a second pore. These findings contradict theoretical predictions based upon the coalescence of many pores.[61] However, it should be noted that despite the observations above, large supercritical field pulses may create many pores.

How Large are Pores During Irreversible Breakdown?

As an extension of arguments that single pores are responsible for the measured conductivity changes one can easily estimate the effective radii of these pores. Inspection of Figure 7b suggests that the actual pores have radii of the order of μm. In this case, a membrane thickness of a few nanometers is negligible and the conductivity is given by the access resistance of the pore.[65] One should note that this yields a linear dependence on a and not a quadratic one as with small pores. In Figure 7b the corresponding effective radii are included on the right hand axis.

Kinetics of Pore Formation

Theoretically, two distinct damping mechanisms are possible. The first is the viscosity of the membrane itself, which yields an exponential opening with time. In all of the lipids investigated to date, a clear linear opening was observed suggesting that inertia becomes more important as the pore grows fast enough.

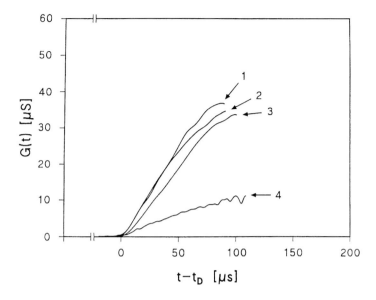

FIGURE 9. Conductance $G(t)$ *versus* time curves during irreversible breakdown of membranes made of: (1) DPhPC, curve slope 0.52 S s^{-1} (0.21 ms^{-1}); (2) DPhPC and 0.3 mg ml^{-1} PAA added to the aqueous phase, curve slope 0.52 S s^{-1} (0.21 ms^{-1}); (3) DPhPC and 0.2 mg ml^{-1} Pluronic F-68 added to the aqueous phase, curve slope 0.42 Ss^{-1} (0.17 ms^{-1}); (4) mixed diphytanoyl phosphatidylcholine/cetylpyridinium chloride (DPhPC/CPCl) bilayer, curve slope 0.20 S s^{-1} (0.08 ms^{-1}); Note: the significant part we fitted covers the first 50 μs. $G(t)$ was calculated according to Equation 18 by using the capacitance of the individual membranes and the velocity of the pore rim (see Equation 17). The length of the charge pulses was 20 μs. Due to the considerable scatter of the signal, the data have been averaged by standard methods.[62] The aqueous phase contained 100 mM KCl; T = 20°C.

A series of investigations devoted to the study of parameters affecting the kinetics of the irreversible breakdown have been performed. Variation of the lipid composition varies several parameters simultaneously, including membrane thickness, surface tension, edge energy, viscosity, etc. Modification of the lipid solvent, e.g., use of hexadecane instead of decane, results in only a slight variation of the bilayer thickness which is far too small to allow proof of a square root dependency in thickness.[67]

Large increase in effective thickness (i.e., an increase of the mass of the membrane) is expected when polymers adsorb to the membranes.[68] Addition of high molecular weight polymers (Pluronic F-68) resulted in a visible decrease of the pore opening velocity (Figure 9). Due to the unknown quantity of adsorbed polymer, it is not possible to quantitate this measurement.[63] A control experiment using polyacrylamide (PAA) (a polymer presumed not to interact with the membrane) showed no change in the opening velocity.

Quantitative measurements of the thickness effect were made in a series of experiments using a defined mixture of diphytanoyl phosphatidylcholine (DPhPC) with distearoyl phosphatidylethanolamine-polyethylene glycol

TABLE 3

**Dynamics of Pore Increase During Irreversible Breakdown
of Lipid Bilayer Membranes Made of Pure and Stealth®-
Endowed EggPC:Cholesterol**

System	R_f/[nm][a]	1 mol% u [ms⁻¹]	5 mol% u [ms⁻¹]	10 mol% u [ms⁻¹]
pure		0.10 ± 0.03	—	—
DSPE-PEG-350[b]	1.2	0.10 ± 0.02 (0.10)	0.08 ± 0.02 (0.09)	0.06 ± 0.01 (0.09)
DSPE-PEG-750	1.9	0.10 ± 0.02 (0.09)	0.09 ± 0.03 (0.08)	0.06 ± 0.02 (0.06)
DSPE-PEG-1900	3.4	0.10 ± 0.02 (0.07)	0.07 ± 0.02 (0.05)	—
DSPE-PEG-5000	6.1	0.03 ± 0.01 (0.03)	—	—

Note: Quantitative measurement of the opening velocity u of specific lipid/DSPE-
PEG mixtures. The numbers in brackets correspond to the theoretically
expected values calculated from scaling theory.

[a] R_f, Flory radius; [b] DSPE-PEG-X, distearoyl phosphatidylethonolamine-
polyethyleneglycol (X molecular mass).

(DSPE-PEG Stealth®-lipid). DSPE-PEG lipids consist of a fully saturated
DSPE lipid with a covalently bonded PEG of fixed molecular weight. These
lipids allow a precise control of the lipid composition and hence the creation
of lipid films of a defined mass per unit area. The experimentally measured
pore-opening velocity of the non-PEG lipid alone was used as the starting point
for the theoretical prediction. Unknown dissipative effects should not be changed
by the presence of the small amount of PEG-polymer; their effects will be
included in the opening velocity of the lipid alone. The additional increase in
effective mass density due to the presence of the PEG-polymer moiety was
then calculated using scaling theory.[69] The latter yields the Flory radius[69]
representing the size of the polymer coil and is given in the first column in
Table 3. It was assumed that interstitial water has to follow the movement of
the polymer coil and therefore accounts for the effective thickness.

In Table 3 the measured opening velocity is shown. Increasing the polymer
length from PEG-350 to PEG-5000 causes the mass per unit area to increase,
and consequently the velocity to decrease. The same was found when the molar
fraction of the added polymer was increased. Most likely the polymer exists in
a coiled state and the interstitial water follows the movement of the lipid.
Measurements of the surface tension using a film balance showed that addition
of DSPE-PEG-lipids to DPhPC had no influence [unpublished data]. We
concluded from these data that the addition of DSPE-PEG will not influence
the contact angle between the membrane and the Teflon chamber, and thus will
not influence the surface tension of the bilayer. The expected theoretical values

for the breakdown velocities are given in brackets below the corresponding experimental ones. It will be noted that these show remarkable agreement.

In a separate series of experiments, 10 mg ml⁻¹ CPCl (cetyl pyridinium chloride, MW 358 Da) was added as surfactant. Capacity measurements showed no reduction in membrane thickness. It is expected that CPCl reduces the contact angle and therefore reduces the surface tension of the bilayer spanned over the Teflon rim. This results in a smaller force driving the widening, in accordance with which the opening velocity of the pore shown in Figure 9 is clearly reduced. Again a quantitative study of this observation is lacking.

V. REVERSIBLE ELECTRIC BREAKDOWN IN PLANAR LIPID BILAYER MEMBRANES

A. THEORETICAL CONSIDERATIONS

The results described in the previous section indicate that voltages around 500 mV induce mechanical rupture of lipid bilayer membranes. However, in some studies it has been demonstrated that bilayer membranes made from oxidized cholesterol can also exhibit reversible electric breakdown (i.e., increase of the membrane conductance) without mechanical rupture.[70,71] More recently, similar phenomena have also been observed on azolectin membranes treated with UO_2^{2+}.[72]

To date no theoretical model has been able to explain the experimental features of reversible electric breakdown. A recent attempt to introduce a general model was based on fusion of small pores with a time dependence determined by the self-diffusion constant of the lipid molecules.[61] Several models dealing with the initial steps of pore formation were discussed in the preceding section. In the following paragraph we will outline general arguments for a possible underlying mechanism of reversible electric breakdown.

According to the aforementioned viscoelastic model, the opening of pores in membranes is driven by the mechanical surface tension. For vesicles, the latter depends upon the enclosed volume to area ratio. Obviously, whenever inner volume flows out, the mechanical tension starts to relax and decreases to almost zero (see Figure 1b). The mechanical tension of the vesicle depends directly on the flow rate as well as on the pore radii. In contrast to planar bilayer experiments (as described in the previous section) the tension is determined by the contact angle at the rim. In all geometries, these forces work against the edge energy, which promotes resealing. As already pointed out, experiments suggest that electric fields lower the edge energy. Once the expanding pore is sufficiently large, the contribution stemming from the edge becomes negligible with respect to the energy gain due to surface tension. As shown for irreversible breakdown, the resulting force vector between these two forces is balanced by a dynamic force, either viscosity or inertia. When the applied electric field pulse lasts long enough to allow the kinetics to overcome the critical mechanical pore radius a^*, irreversible breakdown will occur. When this process is too slow to overcome the critical pore size during pulse application, the pore will reseal. Inspection of the underlying dynamics of the resealing process shows that the edge energy is the driving force. Due to its linear dependency on the pore radius a (Equation 11), this is a slow process in which viscosity should be taken into account. As in the previous section, similar arguments regarding the time dependency of the pore radius can also be made. The time dependency for the resealing of a pore with radius a is readily obtained:

$$a(t) = a_0 \cdot \left(1 - e^{\frac{\tau}{\eta} \cdot t} \right) \qquad (19)$$

Such a conclusion is in agreement with the observation of the resealing process of long lived pores (visible over many seconds). In contrast, the final resealing process is finished within a few milliseconds.[55]

B. EXPERIMENTAL OBSERVATIONS

For a better understanding of these events, we should discuss the effect of electric field pulses on vesicles. Application of the theory of irreversible breakdown to vesicles readily shows the potential for reversibility. We again expect that surface tension is the driving force for pore opening. However, the creation of pores allows the surface tension to relax through efflux of internal volume. As mentioned above, large pores were induced in giant vesicles and observed under the phase contrast microscope many years ago.[29] The pores were stable as long as the alternating current-field was applied and resealed after switch-off. Recently, electric field-induced pores have been stabilized by the application of additional mechanical tension.[55] The reduction in tension due to a flux through the pores was measured as the applied mechanical tension was increased. If the tension was reduced below a critical threshold the pore began to close. Although the pores were stable for several seconds, the resealing process itself was finished within a few milliseconds.

Similar results were found with artificial lipid membranes comprised of oxidized cholesterol/n-decane.[70,71] Figure 10 illustrates the typical time course of such a system during a charge pulse experiment. The membrane was charged to substantially higher voltages within 500 ns. In the first experiment (relaxation 1) the membrane was polarized to about 1 V. The charging process (upper trace of Figure 10) was followed by a very rapid discharge of the membrane, which was caused by a high conductance of the membrane initiated by the high electric field. Since electric breakdown of membranes formed of oxidized cholesterol/n-decane is reversible, the membrane remains mechanically stable, so that breakdown experiments can be readily repeated. Furthermore the decay of the membrane voltage to 100 mV indicates that the membrane had resealed rapidly. In the second experiment (relaxation 2) the length of the charge pulse was 1 μs. The voltage at the output of the charge-pulse generator shows a small "hump," which indicates that electric breakdown

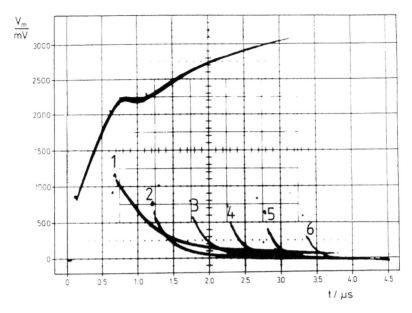

FIGURE 10. Oscillographic records of charge-pulse experiments performed on a membrane made of oxidized cholesterol/n-decane. Six charge-pulses of increasing duration (0.5 to 3 μs) were applied to the same membrane. T = 25°C; 1 M KCl. The upper trace corresponds to the voltage at the output of the charge-pulse generator and represents the superposition of 6 pulses. The lower traces represent the discharge of the membrane following the charge-pulses. (From Benz, R., Beckers, F. and Zimmermann, U., *J. Membrane Biology*, 48 , 181, 1979. With permission.)

occurred during the charging process. The membrane conductance at this point is so large that the membrane is only polarized to a voltage of about 600 mV, followed by a rapid discharge. The application of four further charge pulses of 1.5, 2, 2.5, and 3 μs results in an even higher membrane conductance, corresponding to a smaller voltage across the membrane. The tracings of the six charge pulses coincide, indicating excellent reproducibility of the breakdown experiments. The rapid discharge processes of the membrane in these experiments can be explained by a rapid transient resistance change of the membrane, (which could be as large as eight to nine orders of magnitude from 10^7–10^8 Ω cm² down to 0.3 Ω cm²).

Because of the electric breakdown of the membrane, the voltage of about 1 V cannot be exceeded at room temperature, (even when the amplitude of the charge pulse is raised several fold). Rather, the use of supercritical charge pulses results in electric breakdown during the charging process, so that the initial voltage drops as a result of the high conductivity of the membrane (relaxations 3 to 6). By extrapolating the relaxation curve to the end of the charge pulse, it is possible to estimate from this and similar experiments that reversible electric breakdown occurs within 10 ns. Since breakdown is a very rapid event, longer charging times lead to current flow through the membrane which, in turn, can lead to secondary reactions. This result provides an explanation for the observation (see above) that, with longer pulse applications or higher field intensities, electric breakdown causes increasingly irreversible changes in the cell and membrane (see Chapter 1). Studies on artificial lipid bilayers have shown that the currents flowing through the membrane can reach substantial values.

Similar studies have been performed on UO_2^{2+}-modified azolectin membranes as well as with oxidized cholesterol using voltage clamp techniques.[72-78] It should be noted that this method is less well suited to the study of fast processes. In addition breakdown studies using voltage clamp techniques necessarily deal with large current at the edge of the pore (complicating the analysis). The results for oxidized cholesterol were in agreement with those obtained using charge pulses. UO_2^{2+}-modified azolectin membranes seem to slow down the dynamics of the pore formation. In this case pulse length of several milliseconds could be applied without causing irreversible breakdown.

As in biological membranes, the regeneration of aqueous pores in lipid bilayer membranes after reversible electric breakdown is a very rapid process that is complete within a few microseconds. This can be shown in lipid bilayer experiments using the double-pulse method. A current pulse of low amplitude is applied simultaneously with the pulse which is responsible for initiating electric breakdown. By measuring current and voltage simultaneously, it is possible to monitor the closure of the electrically induced pores as a function of time. An experiment of this type is shown in Figure 11, in which four double pulses were applied to a membrane comprised of oxidized cholesterol/n-decane. The short (intense) pulses used an output voltage of 8 V in all experiments, while the duration of the pulses was varied between 300 ns (trace 1) and 500 ns (trace 4). In all cases the long (weaker) pulses had an amplitude of 65 mV and a duration of 50 μs. The upper traces of Figure 11 represent the voltage across the membrane and the lower traces the current density. Trace 1 indicate that the membrane conductance did not increase appreciably during the intense pulse. In the second experiment, the membrane conductance is so high that the voltage across the membrane drops below 65 mV but recovers after resealing of the membrane has been completed. The traces 3 and 4 indicate an even higher membrane conductance. Figure 12 shows semilogarithmic plots of the specific membrane conductance vs. time derived from experiments 2 to 4 of Figure 11. The decrease of the specific conductance (reciprocal of the specific resistance) of all three curves follows an exponential curve with time constants of about 3 μs. The initial specific conductance of the membrane (i.e., the specific conductance extrapolated to the end of the high pulse) can be as large as 0.76 Scm^{-2} (trace 4). In other experiments initial membrane conductances up to 3.3 Scm^{-2} have been observed. This is the largest membrane conductance which has been measured in experiments with artificial lipid bilayer membranes.

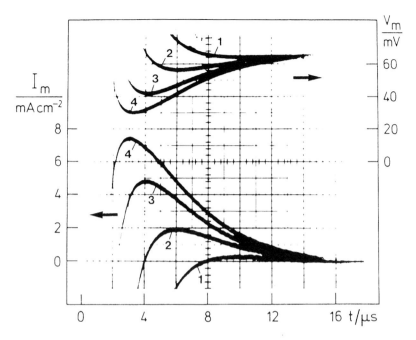

FIGURE 11. Double-pulse experiments performed on a membrane made of oxidized choles-
terol/n-decane. The duration of the intense pulse (output voltage 8 V) ranged between 300 ns (trace
1) and 500 ns (trace 4), whereas the duration of the long pulse was 50 ms (output voltage 65 mV)
in all cases. T = 10°C; 1 M KCl. (From Benz, R. and Zimmermann, U., *Biochim. Biophys. Acta,*
640, 169, 1981. With permission.)

The breakdown voltage in experiments similar to those given in Figure 10
is approximately 1 V. However, such a value was only observed for the
breakdown voltage when the charging time of the membrane was approximately 500 ns. For longer or shorter charging times the critical voltage V_c
differed considerably from 1 V. Figure 13 shows the dependence of the critical
voltage on the pulse-length of the charge pulses. The data were derived from
experiments on membranes from oxidized cholesterol/n-decane, which exhibit
reversible electric breakdown.[79] For a pulse of 10 ns duration, the critical
voltage is approximately 1.2 V at 25°C. It remains approximately constant as
the pulse length is increased up to 200 ns, but then decreases to about 400 mV
at a pulse-length of 10 μs (25°C). Even longer pulses resulted in irreversible
electric breakdown of the oxidized cholesterol/n-decane membranes, possibly
because the current through the membrane during breakdown was too high,
(leading to an increase of the pore size above that of the critical value).

The pulse-length dependence of the critical voltage suggests that it is an
intrinsic property of a single membrane. Such a relationship is consistent with
the requirements of the models described previously. No adequate theoretical
model for the kinetics of the reversible breakdown is yet available.

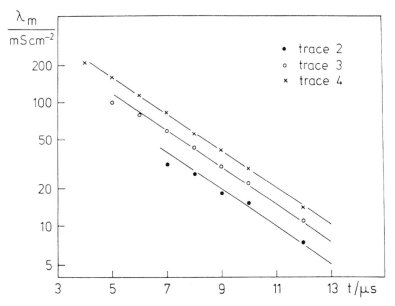

FIGURE 12. Semilogarithmic plot of membrane conductance *versus* time of the data given in Figure 11. The time constant of the exponential decay of the membrane conductance during the resealing process was about 3 ms in all cases. The initial specific conductance (at the end of the intense pulse) ranged between 0.37 S cm^{-2} and 0.76 S cm^{-2} (trace 4). (From Benz, R. and Zimmermann, U., *Biochim. Biophys. Acta,* 640, 169, 1981. With permission.)

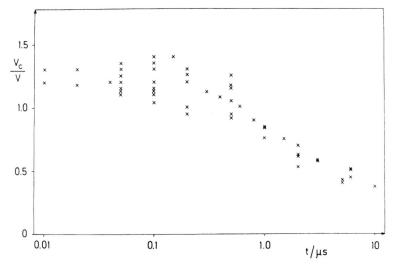

FIGURE 13. Breakdown voltage, V_c, of planar lipid bilayers made from oxidized cholesterol/n-decane as a function of the charging time in charge pulse experiments. V_c is defined as the maximum voltage the bilayers can be charged to at a given pulse length; T = 25 °C. The aqueous phase contained 1 M KCl. (From Benz, R. and Zimmermann, U., *Biochim. Biophys. Acta,* 597, 637, 1980. With permission.)

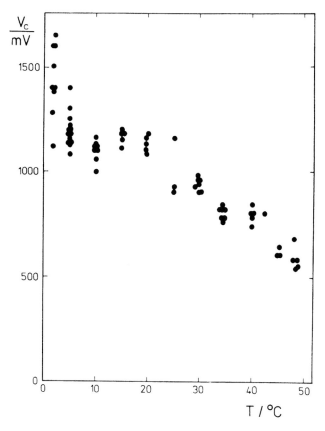

FIGURE 14.   Breakdown voltage, V_c, as a function of temperature. V_c is defined as the maximum voltage to which the bilayers can be charged by a charge-pulse of 400 ns duration. The membranes were formed from oxidized cholesterol/n-decane. The aqueous phase contained 1 M KCl. (From Benz, R., Beckers, F. and Zimmermann, U., *J. Membrane Biology*, 48, 181, 1979. With permission.)

Figure 14 shows the dependence of the critical voltage on temperature.[70] The experiments were performed with membranes comprised of oxidized cholesterol/n-decane, which exhibit reversible electric breakdown. The breakdown voltage was approximately 1.2 V between 2°C and 20°C (pulse-length 400 ns). At higher temperatures the critical voltage decreased, becoming about 600 mV at 48°C. This result may also be explained by the viscoelastic model, if the elastic restoring forces are dependent on the mechanical properties of the lipid molecules. Higher temperature thus induces a greater fluidity of the lipids, which may result in faster-growing defects.

VI. CONCLUSIONS

We have outlined different viscoelastic models for the effect of electric fields on reversible or irreversible breakdown. We assume that thermal fluctuations cause statistical defects in the lipid membrane. If the radii of pores formed in this case are smaller than a critical value ($a < a^*$) the membrane will reseal. Larger radii will lead to irreversible breakdown of planar lipid films. Due to their different structures, vesicles behave differently. The surface tension (the driving force for the opening) relaxes as the pore widens, and so vesicles can close due to the finite edge energy.

The process of membrane rupturing can be initiated either by lowering the edge energy or by increasing the surface tension. Addition of "edge active" molecules like lysolecithin or surfactants can promote the former, while osmotic or mechanical stresses favor the latter. Electric fields act by lowering the edge energy, whereby the applied voltage gives rise to only a small change in the mechanical surface tension. Depending on the experimental boundary condition, two distinct scenarios can be distinguished:

1. voltage clamp wherein the electric forces are present during the whole opening process;
2. charge pulse relaxation wherein the external field acts only temporarily.

In the latter case the field helps to overcome the energy barrier and will lower the critical pore radius. The kinetics at later stages are determined entirely by the mechanical properties of the membrane. If such an applied pulse is short enough, the opening process will lead to a pore smaller than the critical radius for mechanical breakdown. This pore will reseal because the resealing process is driven by the edge energy and shows a very different time dependency.

REFERENCES

1. **Cevc, G. and Marsh, D.,** *The Phospholipid Bilayer*, Wiley, New York, 1985.
2. **Israelachvili, J. N.,** *The Intermolecular and Surface Forces*, Academic Press, New York, 1985.
3. **Lasic, D. D.,** *Liposomes: From Physics to Applications*, Elsevier, Amsterdam, 1993.
4. **Bloom, M., Evans, E. and Mouritsen, O. G.,** Physical properties of the fluid lipid-bilayer component of cell membranes, *Quart. Rev. Biophys.*, 24, 293, 1991.
5. **Evans, E. and Needham, D.,** Physical properties of surfactant bilayer membranes, *J. Phys. Chem.*, 91, 4219, 1987.
6. **Taupin, C., Dvolaitzky, M. and Sauterey, C.,** Osmotic pressure induced pores in phospholipid vesicles, *Biochem.*, 14, 4771, 1981.
7. **Mui, B. L. S., Cullis, P. R., Evans, E. A. and Madden, T. D.,** Osmotic properties of large unilamellar vesicles prepared by extrusion, *Biophys. J.*, 64, 443, 1993.
8. **Kummrow, M. and Helfrich, W.,** Deformation of giant lipid vesicles by electric fields, *Phys. Rev. A.*, 44, 8356, 1991.

9. **Needham, D. and Hochmuth, R. M.,** Electro-mechanical permeabilization of lipid vesicles, *Biophys. J.,* 55, 1001, 1989.

10. **Needham, D. and Nunn, R. S.,** Elastic deformation and failure of lipid bilayer membranes, *Biophys. J.,* 58, 997, 1990.

11. **Deuling, H. J. and Helfrich, W.,** Red blood cell shapes as explained on the basis of curvature elasticity, *Biophys. J.,* 16, 861, 1976.

12. **Sackmann, E., Duwe, H. P. and Engelhardt, H.,** Membrane bending elasticity and its role for shape fluctuations and shape transformations of cells and vesicles, *Faraday Discuss. Chem. Soc.,* 81, 281, 1986.

13. **Svetina, S. and Zeks, B.,** Membrane bending energy and shape determination of phospholipid vesicles and red blood cells, *Europ. Biophys. J.,*17, 101,1989.

14. **Berndl, K., Käs, J., Lipowsky, R., Sackmann, E. and Seifert, U.,** Shape transformation of giant vesicles, *Europhys. Letters,* 13, 659, 1990.

15. **Nezil, F. A., Bayerl, S. and Bloom, M.,** Temperature reversible eruptions of vesicles in model membranes studied by nmr, *Biophys. J.,* 61, 1413, 1992.

16. **Lipowsky, R.,** The conformation of membranes, *Nature,* 349, 475, 1991.

17. **Wiese, W., Harbich, W. and Helfrich, W.,** Budding of lipid bilayer vesicles and flat membranes, *J. Phys.,* C 4, 1647, 1992.

18. **Schwarz, G.,** General equation for the mean electrical energy of a dielectric body in an alternating electrical field, *J. Chem. Phys.,* 39, 2387, 1963.

19. **Sauer, F. A.,** Interaction-forces between microscopic particles in an external electromagnetic field. In *Interactions between Electromagnetic Fields and Cells.* Chiabreara, A., Nicolini, C. and Schwan, H. P., Eds., Plenum Press, New York, 1986, 181.

20. **Winterhalter, M. and Helfrich, W.,** Deformation of spherical vesicles by external fields, *J. Coll. Interface Sci.,* 122, 583, 1988.

21. **Mitov, M. D., Méléard, P., Winterhalter, M. and Bothorel, P.,** Electric-field dependent thermal fluctuation of giant vesicles, *Phys. Rev.,* E 48, 628, 1993.

22. **Asami, K. and Irimajiri, A.,** Dielectric analysis of mitochondria isolated from rat liver, *Biochim. Biophys. Acta,* 778, 570, 1984.

23. **Hyuga, H., Kinoshita, K. and Wakabayashi, N.,** Steady state deformation of a vesicle in alternating fields, *Bioelectrochem. Bioenerg.,* 32, 15, 1993.

24. **Torza, S., Cox, R. G. and Mason, S. G.,** Electrohydrodynamic deformation and burst of liquid drops, *Phil. Trans. R. Soc. (London)* A269, 295, 1971.

25. **Bryant, G. and Wolfe, J.,** Electromechanical stresses produced in the plasma membranes of suspended cells by applied electric fields, *J. Membrane Biolology,* 96, 129, 1987.

26. **Zeks, B., Svetina, S. and Pastushenko, V. F.,** The shape of phospolipid vesicles in an external electric field, *Studia biophysica,* 138, 137,1990.

27. **Zeks, B., Svetina, S. and Pastushenko, V. F.,** Theoretical analysis of lipid shape in constant electric field, *Biol. Membr.,* 8, 430, 1991.

28. **Grosse, C. and Barchini, R.,** On the permittivity of a suspension of charged particles in an electrolyte, *J. Phys.,* D 19, 1113, 1986.

29. **Harbich, W. and Helfrich, W.,** Alignment and opening of giant vesicles by electric fields, *Z. Naturforsch.,* 34 A, 1063, 1979.

30. **Teissié, J. and Tsong, T. Y.,** Electric field induced transient pores in phospholipid bilayer vesicles, *Biochem.,* 20, 1548, 1981.

31. **Büschl, R., Ringsdorf, H. and Zimmermann, U.,** Electric field induced fusion of large liposomes from natural and polymerizable lipids, *FEBS Lett.,* 150, 38, 1982.

32. **Crowley, J. M.,** Electrical breakdown of bimolecular lipid membranes as an electromechanical instability, *Biophys. J.,* 13, 711, 1973.

33. **Zimmermann, U., Pilwat, G., Péqueux, A. and Gilles, R.,** Electro-mechanical properties of human erythrocyte membranes, *J. Membrane Biology,* 54, 103, 1980.

34. **Dimitrov, S. D.,** Electric field induced breakdown of lipid bilayer and cell membranes, *J. Membrane Biology,* 78, 53, 1984.

35. **Dimitrov, S. D. and Jain, R. K.,** Membrane stability, *Biochim. Biophys. Acta,* 779, 437, 1984.
36. **Parsegian, V. A.,** Energy of an ion crossing a low dielectric membrane. Solution to four relevant electrostatic problems, *Nature (London),* 221, 844, 1969.
37. **Powell, K. T. and Weaver, J. C.,** Transient aqueous pores in bilayer membranes, *Bioelectrochem. Bioenerg.,* 15, 211, 1986.
38. **Deryagin, B. V. and Gutop, Y. V.,** Theory of the breakdown of free films, *Coll.J.* (USSR), 24, 370, 1962.
39. **Bivas, I.,** Molecular theory of the lifetime of black lipid membranes, *J. Coll. Interface Sci.,* 144, 63, 1991.
40. **Bivas, I.,** Pores and their number in bilayer lipid membranes, *J. Phys. Lett. (Paris),* 46, L 513, 1985.
41. **Sugar, I. P. and Neumann, E.,** Stochastic model for the electric field-induced membrane pores, *Biophys. Chem.,* 19, 211, 1984.
42. **Sugar, I. P., Förster, W. and Neumann, E.,** Model of cell electrofusion, *Biophys. Chem.,* 26, 321, 1987.
43. **Deryagin, B. V. and Prokhorov, A. V.,** On the theory of the rupture of black films, *J. Coll. Interface Sci.,* 81, 108, 1981.
44. **Prokhorov, A. V. and Derjaguin, B. V.,** On the generalized theory of the bilayer film rupture, *J. Coll. Interface Sci.,* 125, 111, 1988.
45. **Kashchiev, D. and Exerova, D.,** Bilayer lipid membrane permeation and rupture due to hole formation, *Biochim. Biophys. Acta,* 732, 133, 1983.
46. **Abidor, I. G., Arakelyan, V. B., Chernomordik, L. V., Chizmadzhev, Y. A., Pastushenko, V. F. and Tarasevich, M. R.,** Electrical breakdown of bilayer lipid membranes, *Bioelectrochem. Bioenerg.,* 6, 37, 1979.
47. **Weaver, J. C. and Mintzer, R. A.,** Decreased bilayer stability due to transmembrane potentials, *Phys. Lett.,* 86 A, 57, 1981.
48. **Petrov, A. G., Mitov, M. D. and Derzhanski, A.,** Edge energy and pore stability in bilayer membranes. In *Advances in Liquid Crystal Research and Applications.* Bata, L., Ed., Pergamon Oxford, 1980, 695, 2.
49. **Pastushenko, V. F. and Petrov, A. G.,** Electro-mechanical mechanism of pore formation in bilayer lipid membranes, *Proc. 7th School on Biophysics of Membrane Transport,* Poland, 1984.
50. **Pastushenko, V. F. and Chizmadhev, Y. A.,** Stabilisation of conducting pores in blm by electric current, *Gen. Physiol. Biophys.,* 1, 43, 1985.
51. **Leikin, S. L., Glaser, R. W. and Chernomordik, L. V.,** Mechanism of pore formation under electrical breakdown of membranes, *Biol. Membr.,* 6, 944, 1986.
52. **Chizmadhev, Y. A. and Pastushenko, V. F.,** Electrical stability of biological and model membranes, *Biol. Membr.,* 6, 1013, 1989.
53. **Barnett, A.,** The current voltage relation of an aqueous pore in a lipid bilayer membrane, *Biochim. Biophys. Acta,* 1025, 10, 1990.
54. **Winterhalter, M. and Helfrich, W.,** Effect of voltage on pores in membranes, *Phys. Rev.,* 36A, 5874, 1987.
55. **Zhelev, D. and Needham, D.,** Tension stabilized pores in giant vesicles, *Biochim. Biophys. Acta,* 1147, 89, 1993.
56. **Chernomordik, L. V., Kozlov, M. M., Melikyan, G. B., Abidor, I. G., Markin, V. S. and Chizmadhev, Y. A.,** The shape of lipid molecules and monolayer membrane fusion, *Biochim. Biophys. Acta,* 812, 643, 1985.
57. **Genco, I., Gliozzi, A., Relini, A., Robello, M. and Scalas, E.,** Electroporation in symmetric and asymmetric membranes, *Biochim. Biophys. Acta,* 1149, 10, 1993.
58. **DeGennes, P. G.,** Wetting, *Rev. Mod. Phys.,* 57, 827, 1985.
59. **Redon, C., Brochard, F. and Rondelez, F.,** Dynamics of dewetting, *Phys. Rev. Lett.,* 66, 715, 1991.

60. **Sugar, I. P.,** Stochastic model of electric field induced membrane pores. In *Electroporation and Electrofusion in Cell Biology,* Neumann, E., Sowers, A. and Jordan, C. A., Eds., Plenum Press, New York, 1989, Chapter 6, 97.

61. **Barnett, A. and Weaver, J. C.,** Electroporation, *Bioelectrochem. Bioenerg.,* 25, 163, 1991.

62. **Wilhelm, C., Winterhalter, M., Zimmermann, U. and Benz, R.,** Kinetics of pore size during irreversible electrical breakdown of lipid bilayer membranes, *Biophys. J.,* 64, 121, 1993.

63. **Klotz, K.-H., Winterhalter, M. and Benz, R.,** Use of irreversible electrical breakdown for the study of interaction with surface active molecules, *Biochim. Biophys. Acta,* 1147, 161, 1993.

64. **Frankel, S. and Mysels, K. J.,** The bursting of soap films, *J. Phys. Chem.,* 73, 3028, 1969.

65. **Hille, B.,** *Ionic Channels in Excitable Membranes.* Sinauer Associates, Sunderland, MA, 1984.

66. **Sukharev, S. I., Arakelyan, V. V., Abidor, I. G., Chernomordik, L. V. and Pastushenko, V. F.,** BLM destruction as a result of electrical breakdown, *Biofizika,* 28, 756, 1982.

67. **Dilger, J. P. and Benz, R.,** Optical and electrical properties of thin monoolein membranes, *J. Membrane Biol.,* 85, 181, 1985.

68. **King, A. T., Davey, M. R., Mellor, I. R., Mulligan, B. J. and Lowe, K. C.,** Surfactant effects on yeast cells, *Enzyme Microb. Technol.,* 13, 148, 1991.

69. **DeGennes, P. G.,** *Scaling Concepts in Polymer Physics.* Cornell University Press, Ithaca, 1985.

70. **Benz, R., Beckers, F. and Zimmermann, U.,** Reversible electrical breakdown of lipid bilayer membranes, *J. Membrane Biology,* 48, 181, 1979.

71. **Benz, R. and Zimmermann, U.,** The resealing process of lipid bilayer membranes after reversible electrical breakdown, *Biochim. Biophys. Acta,* 640, 169, 1981.

72. **Abidor, I. G., Chernomordik, L. V., Chizmadhev, Y. A. and Sukharev, S. I.,** Reversible electrical breakdown of lipid bilayer membranes modified by uranyl ions, *Sov. Electrochem.,* 17, 1543, 1981 and *Bioelectrochem. Bioenerg.,* 9, 141, 1982.

73. **Chernomordik, L. V., Abidor, I. G., Chizmadhev, Y. A. and Sukharev, S. I.,** Mechanism of reversible electrical breakdown of lipid bilayer membranes in the presence of UO_2^{2+}, *Sov. Electrochem.,* 18, 88, 1982.

74. **Chernomordik, L. V., Abidor, I. G., Kushnev, V. V. and Sukharev, S. I.,** Pore development during the reversible electric breakdown of lipid bilayer membranes, *Sov. Electrochem.,* 19, 1097, 1983.

75. **Pastushenko, V. F. and Chizmadhev, Y. A.,** Electrical breakdown of vesicles, *Biofizika,* 28, 1036, 1983.

76. **Pastushenko, V. F. and Chizmadhev, Y. A.,** Breakdown of lipid vesicles by an external field, *Biol. Membr.,* 2, 1116, 1985.

77. **Chernomordik, L. V., Abidor, I. G., Chizmadhev, Y. A. and Sukharev, S. I.,** Reversible breakdown of lipid bilayer membranes in the electric field, *Biol. Membr.,* 1, 229, 1984.

78. **Glaser, R. W., Leikin, S. L., Sokirko, A. I., Chernomordik, L. V. and Pastushenko, V. F.,** Reversible electrical breakdown of lipid bilayer membranes, *Biochim. Biophys. Acta,* 940, 275, 1988.

79. **Benz, R. and Zimmermann, U.,** Pulse-length dependence of the electrical breakdown in lipid bilayer membranes, *Biochim. Biophys. Acta,* 597, 637, 1980.

Chapter 4

ELECTROFUSION OF CELLS: STATE OF THE ART AND FUTURE DIRECTIONS

Ulrich Zimmermann

CONTENTS

I. Introduction 174

II. Cell Apposition and Membrane Contact 177
   A. Electrofusion of Cells which are
      in Natural Membrane Contact 177
   B. Electrofusion in Cell Pellets or
      in Chemically Aggregated Cells 178
   C. Cell and Membrane Contact by Physical Forces 181
      1. Mechanically Induced Contact 181
      2. Ultrasonically and Magnetically
         Mediated Cell Alignment 182
      3. Electrically Induced Cell-Cell Contact 182
         a. Factors Controlling Dielectrophoresis 183
         b. Beneficial Aspects of Dielectrophoresis 187
         c. Pitfalls in Electrofusion of
            Dielectrophoretically Aligned Cells 190

III. Electrode Assembly and Power Supply 193
   A. Electrode Assembly 193
   B. Power Supply 196

IV. Fusion Conditions and Constraints:
   Biochemical and Biophysical Factors 197
   A. Chemical Facilitators 197
   B. Media Ingredients 200
   C. Fusion Pulse Parameters 203
      1. Temperature 203
      2. Field Strength 203
      3. Shape, Duration, and Number of Fusion Pulses 207

V. The Membrane Intermingling Process 208

0-8493-4476-X/96/$0.00+$.50
© 1996 by CRC Press Inc.

VI. Applications of Electrofusion 215
    A. Basic Research 215
    B. Bacterial, Fungal, and Yeast Biotechnology 219
    C. Plant Breeding.. 219
    D. Cloning of Embryos.. 223
    E. Hybridoma Technology 227

VII. Conclusions 230

References 231

I. INTRODUCTION

One of the most significant advances in bioscience during the last three decades has been the development of efficient *in vitro* cell-cell fusion techniques for somatic hybridization of yeast, plant, mammalian, and blastomere cells. This fascinating field of research has opened new vistas in basic science, plant breeding, agriculture, biomedicine, and biotechnology. Recent improvements in fusion technology have facilitated the production of a great variety of cell hybrids capable of proliferating indefinitely in culture and expressing combinations of desirable traits. The best known application of cell-cell fusion is hybridoma technology which permits the routine production of both murine and human monoclonal antibodies, which are now indispensable in both research and clinical medicine. Expanding knowledge and further refinements in the technologies of human antibody generation from small cell numbers will allow the introduction of new therapeutic applications (see Chapter 1 and below).

Cell-cell fusion is a process in which two or more cells integrate their structures (including membranes, cytosol, nuclear membranes, and nucleoplasm) into a new viable cell capable of proliferation.[1] Fusion can occur between cells of the same species (intraspecies) or of different species (interspecies). When identical cells are fused, a homokaryon is produced. When cells of different origins are fused, heterokaryons are obtained. In homo- and heterokaryons (so-called "cybrids"), only the cell membranes and cytoplasm undergo fusion. Synkaryons, i.e., hybrid cells, are formed when the nuclear membranes, the nucleoplasm, and the chromosomes have become integrated as well.

It is now known that membrane fusion is an ubiquitous event in cell biology.[2] In fact, every biomembrane has the potential to fuse. However, certain membranes exhibit a greater proclivity to fuse than others. Whereas in most intracellular membranes, including those of the endoplasmic reticulum, the lysosomes, and the Golgi system, fusion events take place continuously,[3,4] spontaneous fusion between the plasma membrane of most cultured cells occurs only occasionally (for exceptions, see Reference 5).

Early observations of cell-cell fusion date from the 19th century, when pathologists reported the occurrence of multinucleated cells in various human diseases.[6-9] At the beginning of this century cell-cell fusion was also observed *in vitro*.[10-13] However, cell-cell fusion remained a curiosity[14] until the publication of a report in which hybridization between distinct cell types in mixed cultures was described.[15]

Teleological arguments suggest that it is desirable to maintain the integrity of cell membranes in most cell types. It is thus not surprising to find that spontaneous cell-cell fusion is not commonly observed. The relatively weak potential of spontaneous plasma membrane fusion can be explained by the physical and chemical properties of the plasma membrane. The presence of glycoproteins, microvilli, and cytoskeletal elements in the outer half of the plasma membrane, as well as the electrostatic repulsion of the "double layers" between two cells, normally inhibit cell-cell contact and fusion[2,16] (for a more rigorous discussion, see Chapter 6). Furthermore, enormous hydration forces (resulting from the energy needed to remove water from the polar groups stabilizing the membrane) dominate most bilayer membrane interactions at distances less than 2 or 3 nm. These forces hinder the very close juxtaposition of membranes that must precede any fusion event between suspended cells.[17]

Given the foregoing, it would be expected that bilayer dehydrating agents, such as polyethylene glycol (PEG) or Ca^{2+}-ions, allow membrane apposition and molecular contact.[18-20] PEG (MW 6000, usually 35 to 50% w/w) has been exploited extensively as a fusogen[18-20] and in particular for the production of hybrid cells. Due to the high osmotic pressure exerted by the added PEG, intracellular water is depleted and the cell volume collapses. Collapsed cells aggregate in clumps, with their membranes in tight contact. Phospholipid vesicles fuse at this stage, yet cells apparently remain unfused. Detectable cell-cell fusion is induced by dilution or removal of the PEG. Membrane and cytoskeletal proteins may have kept the cells unfused until this stage, when the osmotic swelling further stretches and weakens the membranes in contact. PEG has several advantages: it exhibits a relatively low degree of cytotoxicity, is inexpensive, commonly available, water-soluble, and relatively easy to remove by washing. The fusion and hybridization rates are, however, often very poor because PEG does not agglutinate cells as effectively as do some of the other methods described below.

The use of fusogenic viruses provides another *in vitro* approach to overcome the plasma membrane's reticence to fuse.[19-21] Such viruses have been observed to induce syncytia during *in vivo* infections. They appear to disturb the organization of membrane components and/or their aqueous environment, leading to a structurally unstable membrane with fusogenic capacity. Moreover, by binding to receptors on the cell surface these viruses are capable of binding (or agglutinating) cells, further enhancing the fusion process. Virally induced fusion reactions (including those of Sendai, influenza viruses, as well as other viruses, and certain retroviruses) have been extensively studied.[19-21]

Although a very valuable tool for study of the interactions between viruses and biomembranes, virus-induced fusion techniques have in recent years been supplanted by chemical fusogens for most applications. A number of short-comings limit the general application of this technique. For example, many viruses are species-specific. Some cell types (including many human cells) are not readily fused by animal viruses. The use of human viruses for the fusion of human cells poses additional problems of pathogen containment. Where viral fusion is effective, preparation and purification of the large quantities of virus required for fusion is very time-consuming and enormously expensive. In addition, the integration of the viral genome and components into the host cells may complicate the analysis of the results as well as biomedical and biotech-nological applications. It is also very difficult to control.

A major step in improving hybridization frequencies resulted from the observation that administration of a short field pulse of high intensity to red blood cells at high suspension densities induced cell fusion[22] (see also References 23–27). The systematic investigation of the potential of electrofusion was stimulated by the discovery that electropermeabilization of suspended cells after pre-alignment in an inhomogeneous alternating electric field improved the electric generation of cell hybrids manyfold.[23,24,28–46]

Electrofusion has become an increasingly popular method for cell fusion for several reasons. Fusion occurs in a more or less synchronized manner that allows the study of dynamic fusion intermediates. Such analysis is impossible with chemically or virally induced fusion. Biological variability with respect to "fusibility" of various cell types or subpopulations is less of a problem with electrofusion than with other methods. Other significant advantages of electrofusion include predictability of the fusion parameters, application to all biological and artificial systems, enhanced control and monitoring of the fusion process microscopically, gentle fusion conditions, high reproducibility, very high yields, and the ability to produce hybrids when input cell numbers are limited (see below). Electrofusion has dramatically increased the spectrum of feasible fusion experiments. The widespread popularity of electrofusion argues that these advantages have outweighed the one major disadvantage of the technology — the relatively high initial cost for a suitable power supply.

Electrofusion is based upon the application of well known biophysical principles as well as a thorough understanding of the biological systems involved. It is not surprising, therefore, that the success of this method (as well as other cell electromanipulation procedures) in biology, genetics, and immu-nology, as well as in plant and animal agriculture, has benefitted from a mutually productive interplay between physicists and biologists. Many of the physical phenomena of the electrofusion process are understood[27-46] and must be taken into account if the technology is to be successfully applied, particu-larly when approaching difficult problems. Unfortunately, as noted for chemi-cally or virally induced fusion, the molecular mechanisms of electrofusion are still poorly understood. Moreover, cell fusion involves too many changing and unidentified parameters, too many biochemical "triggers," and too many other factors capable of influencing cell membrane organization (see Chapter 6) to allow the construction of a purely biophysical model with any confidence.

In this chapter, I will, therefore, focus on the conceptual framework of the electrofusion process that currently prevails in the field and the facts that support these concepts. In addition, the current state-of-the-art and of the potential of electrofusion is reviewed, especially the areas of biotechnology and genetic engineering. I will also outline some of the significant technical advances in recent years that have led to enormous improvement of suspension electrofusion technology. I will also highlight some of the more significant aspects of the considerable body of knowledge of electric field interactions with biological and artificial lipid membranes that has accumulated from electrofusion research (see also Chapters 3 and 5).

II. CELL APPOSITION AND MEMBRANE CONTACT

Electrofusion techniques combine controlled field-induced permeabilization of plasma membrane areas (Chapter 1) with methods to enhance contact between cells. This cell-cell contact must proceed to intermembrane (molecular) contact associated with (partial) dehydration of the membrane surface in order to ensure fusion by the breakdown pulse. As mentioned above, the hydration forces become significant as the separation of the membrane surfaces becomes small. With decreasing distance, the hydration forces increase steeply and overwhelm both the van der Waals attraction and electrostatic repulsion forces. Hydration repulsion alone, without regard to the additive electrostatic or undulation repulsion, poses the major barrier to the establishment of direct intermembrane, molecular contact between aggregated cells. Considerable energy must be expended to remove the intervening water and bring the two apposed membranes into contact. If this is accomplished electrically, the energy expended during the high intensity field pulse is, in principle, sufficient to induce membrane fusion with the proviso that the cell-cell contact is achieved.

However, experience has shown that high fusion and hybridization rates can only be obtained if a very tight membrane contact is established before field pulse application (for exceptions, see below). This suggests that the approach used to induce cell-cell contact probably also contributes to the creation of a "fusogenic" membrane state. A number of different strategies have been used to achieve this purpose.

A. ELECTROFUSION OF CELLS WHICH ARE IN NATURAL MEMBRANE CONTACT

Several workers have used the "natural" intercellular contact between anchorage-dependent cells without additional alignment prior to electrofusion. Contact-inhibited mammalian cells grown to confluence in monolayer cultures,[47-54] on microcarriers,[55] or on porous membranes,[51,56] have thus been electrofused with high strength electric field pulses of appropriate pulse duration and direction. For generation of heterokaryons, co-cultures must either be plated and grown to confluence[52,57,58] or a second cell layer must be formed or overlaid on the first confluent one.[51,56] Hybridization frequencies have been obtained that were equal or significantly higher than those observed with PEG. However, the frequencies were generally far lower than those usually obtained after electric alignment with an inhomogeneous alternating field (see below).

The use of cells grown to confluence offers a number of advantages including the feasibility of using relatively conductive solutions[51,56] (comparable to those employed in electroinjection of foreign molecules into cultured cells, see Chapter 1). The requirement of using adherent cell lines is a major limitation, however. An additional disadvantage is the preferential formation of multi-nucleated homo- or heterokaryons containing three or more nuclei. While not desirable for most cell fusion applications, these multinucleated cells are interesting for some basic research purposes (with the proviso that they can be produced in a reproducible and regular manner, see below).

Fusion has also been successfully induced by electrical treatment of cells that form a natural contact during development such as *Dictyostelium*[59] or blastomeres* of 2-, 4-, and 8-cell embryos within the zona pellucida.[60-71] Electrofusion of zona-intact blastomeres is used to create enlarged, viable chimeric (mouse, rabbit, etc.) embryos which have proved useful in studying nucleo-cytoplasmic interactions in oogenesis and early embryogenesis (see below).

B. ELECTROFUSION IN CELL PELLETS OR IN CHEMICALLY AGGREGATED CELLS

Mammalian cells or plant protoplasts kept at high density[22,29,32,46,72-79] or cells pelleted by sedimentation/centrifugation[37,51,80] can be readily electrofused.** Under these conditions (which resemble those used very often in electrotransformation of bacteria, see Chapter 2) the separation between the cells is reduced to a very small value. Thus, application of a high intensity field pulse results in a strong and highly non-uniform field[29] and in cellular dipole interaction pressures[27,81,82] leading to cell motion and contact (mutual dielectrophoresis, see also below).

Fusion has also been reported for mammalian cells brought into contact by a specially designed device allowing an intense field pulse to be applied to the cells during centrifugation.[83,84] The fusion frequency was very high and increased with increasing centripetal acceleration. When the pulses were injected

---

* However, in many cases, an additionally induced membrane contact, e.g., induced by dielectrophoresis (see below), was beneficial. In addition, high fusion frequencies require that the cleavage plane is oriented perpendicularly to the field lines.[61,63-65] Therefore, preceding dielectrophoresis is also useful for appropriate positioning and alignment of the eggs.

** By analogy to the fusion of blastomeres additional application of an alternating field very often increased the yield of fusion products[73,75,77] indicating that — at non-optimal cell density — dielectrophoretic movement during the pulse was not sufficient to induce the required membrane contact.

after centrifugation the fusion yield decreased dramatically. This apparently indicates that swelling of the permeabilized cells within the pellet due to the colloid-osmotic pressure caused an additional compression of the pellet during centrifugation that was beneficial for fusion. Consistently, electromicrographs showed[84] that after field treatment the cells were tightly packed: the extracellular space was almost completely obliterated, and the intermembrane contact appeared more intimate. The extracellular water was apparently adsorbed into the cell by the colloid-osmotic pressure. In contrast, in the presence of sucrose or albumin (minimizing the colloid-osmotic pressure, see below and Chapter 1) these effects were not observed. Although this work clearly showed that the cells must be in close contact before field treatment, there are other reports[72,85,86] in which fusion was also observed after centrifugation of field-pretreated cells into a pellet. Most surprisingly, rat hepatocytes suspended in media without Ca^{2+} fused during centrifugation without making use of either a field pulse or any fusogenic additive.[87] The effect was explained by the assumption that proteins spanning the hepatocyte membranes may play a role in inducing local and transient destabilization of the plasma membrane during centrifugation. However, an alternative (and more likely) explanation is that high gravitational forces induce electric breakdown via the resting membrane potential. As discussed in Chapter 1, the breakdown voltage decreases exponentially when the membrane is precompressed by mechanical forces or hydrostatic pressure gradients. Breakdown occurs if the resting membrane potential difference in the compressed membrane exceeds the breakdown voltage.

Some authors have also used chemical additives (such as PEG, lectins, dextran, or spermine) for the establishment of cell-cell contact prior to field pulse treatment.[26,78,88-92] This approach as well as the others described above have yet to find wide application in hybrid production. Although of interest for basic research, several disadvantages must be overcome to increase their utility. For example, fusion occurs randomly between chemically agglutinated or pelleted cells. The investigator can exert little control over the size of the cell aggregates, and undesired polykaryons are frequently produced. Reproducibility is also less than satisfactory.

Thus, pending further refinement, these applications remain specialized. However, chemical pre-agglutination may be very useful in the egg-egg or blastomere-egg electrofusion because the relatively large single eggs (after removal of the zona pellucida) can be transferred as pairs to droplets of solutions containing agglutinating additives[93-96] (however, see Reference 66). This procedure has also proven useful in electrofusion for pronuclear transplantation of eggs.[97,98]

More selective and controlled cell-cell contact can be achieved chemically, when the two cells that are to be fused are bridged by receptor-mediated binding before field pulse administration.[25,99-104] An interesting example is the avidin-biotin system, which was used for myeloma-splenocyte electrofusion. Five steps are involved in the basic design:

1. chemically cross-linking avidin to the antigen by using various reagents[100]
2. mixing the avidin-antigen complex with B cells (the avidin-antigen complex will preferentially bind to the desired B cells through antigen-receptor interaction);
3. biotinylation of myeloma cells;
4. mixing the B cell preparation with the biotinylated myeloma cells (only B cells with membrane-bound avidin-antigen complexes will pair with the myeloma cells through avidin-biotin interactions);
5. electrofusion of the B cell-myeloma pairs.

High frequencies of antibody-producing hybridomas have been reported.[99,100,104] However, there are some technical problems (e.g., non-specific binding of avidin, which is highly positively charged, to negatively charged lipids, DNA, and hydrophobic surfaces) and other disadvantages[25,38] which make the avidin-biotin bridging method less than ideal. These problems can be partly circumvented by replacing avidin with streptavidin.[25,100-103] Streptavidin has an isoelectric point near neutral pH and does not bind nonspecifically to hydrophobic surfaces. In this approach, an antigen is cross-linked to biotin and then to streptavidin. The B cell-antigen-biotin-streptavidin conjugate is then linked to a biotinylated myeloma cell. The complete chain is B cell-antigen-biotin-streptavidin-biotin-myeloma. There are other approaches to link a myeloma cell with a B cell (e.g., through the Fab fragment of antibodies against myeloma cell surface antigen). A special variant of this technique is peptide-mediated electrofusion.[105] Specific peptides can be used to preferentially select the required B cell clone from the entire B cell repertoire. This seems to be a valuable approach where amino acid sequence data are available, as in the case with many potyviruses.

The most obvious advantage of these receptor-mediated electrofusion methods is the elimination of the antibody screening step for hybridomas and the potential of preparing monoclonal antibodies of high affinity. The methods were designed originally for hybridoma production; the concept could, however, be applied equally to other areas where targeting of cells for chemical treatment is desired. A very interesting example was published recently using a modified flow cytometer for selective electrofusion of conjugated cells in flow.[106] The two mammalian cell types were stained with two different fluorescent membrane probes to facilitate optical recognition, and then coupled through an avidin-biotin bridge. In the flow cytometer, the hydrodynamically focused cells and cell pairs were first optically analyzed in a normal flow channel and then forced through an orifice of a particle analyzer (see Chapter 1). If the optical analysis indicated that a cell pair was present, a fusion pulse was applied across the orifice.

C. CELL AND MEMBRANE CONTACT BY PHYSICAL FORCES

The use of physical forces for establishment of intercellular and intermembrane contact before electrofusion offers the great advantage that cell processing, apposition, fusion, and post-alignment can be controlled using the vectorial character of these forces. The initiation of the fusion process by the application of a breakdown pulse is restricted to the contact area. Thus, merging of the two apposed membranes occurs over a very small membrane domain without affecting the remaining membrane surface — at least at the first instant. This ensures very gentle but highly efficient fusion conditions with high rates of viable fusion and hybrid products.

1. Mechanically Induced Contact

Mechanical establishment of intimate contact between two cells, each held on a micropipette, has been used since the beginning of this century[13] and has also been used successfully in electric field-mediated fusion of plant protoplasts.[78,107,108] This method is very time-consuming and requires relatively large cells. However, it has some advantages for the processing of small numbers of cells. An added advantage is the ability to select hybrids morphologically, obviating the need for selection media.

A special but important variant of the micropipette technique is used in nuclear transplantation experiments[109-122] (see below). A blastomere from a donor embryo is aspirated into a pipette and then transferred to the perivitelline space of an enucleated mature oocyte. This creates a membrane-bound karyoplast within the perivitelline space of the enucleated oocyte. Fusion is accomplished by injection of a field pulse of sufficient strength. By analogy to the fusion of blastomeres within the zona pellucida, the alignment of the karyoplast/oocyte is critical. The fusing membranes must be parallel to the electrodes. Because of this, several workers have used additional dielectrophoresis for positioning and alignment.[114,115,117-122] A similar procedure is used for pronuclear transplantation of non-enucleated eggs[97,98,123,124] and for artificially induced fusion between eggs and sperm microinjected into the perivitelline space.[125] In the latter case, when injected sperm were not in visible contact with the egg membrane prior to pulsing, sperm-egg membrane contact was provided through gentle pressure of the microinjection needle against the zona pellucida.

In general, the use of mechanical pressure (in combination with dielectrophoresis) is a fascinating alternative to other (physical) cell alignment techniques. An example of this is the fusion of cells to tissue monolayers (so-called "electro-mechanical fusion").[126-129] In one protocol this is accomplished by layering the cells onto Millipore filters by centrifugation; then placing the filter with the cell side down, onto the tissue layer. Mechanical pressure is applied with specially designed electrodes (with adjustable spacing) to obtain juxtaposition between the cultured cells and tissue-immobilized cells prior to application of the breakdown pulse. This "sandwich" method can be also applied to the fusion of cultured cells of different origins when one or two filters are pre-layered with the respective cell line.[37,56,129] Fluorimetric assays for quantitating cell-tissue electrofusion have been described which measure the number of cells fused to intact tissue.[130]

Instead of mechanical pressure, ordered pair formation of the two fusion partners can also be obtained by differential sedimentation resulting in two distinct monolayers.[131] This technique, however, only leads to high yields of heterokaryons when the sedimented cells were pre-aligned by an alternating electric field. The application of the technique is rather limited because the two fusion partners must differ significantly in density. This is for example the case for fusion of vacuolated with evacuolated plant protoplasts.

2. Ultrasonically and Magnetically Mediated Cell Alignment

Several workers have recently shown that cells can be brought into intimate contact by ultrasonic radiation forces prior to administering an electrofusion pulse (so-called "electro-acoustic fusion").[38,132-136] In an ultrasonic field, forces are exerted on cells in a manner somewhat analogous to those exerted in an inhomogeneous alternating electric (or magnetic) field (see below). Cell-cell contact in an ultrasonic standing-wave depends upon the size and density of the particle (cell), the sound velocity in the particle as well as sound velocity in the suspending medium, and its density.[135,137] If the sound wavelength is much smaller than that of the fusion chamber (e.g., 1 mm corresponding to a frequency of 1 MHz), "pearl chaining" of cells is observed. A concentration of the cells occurs at the standing-wave-pressure maximum. The effect is so powerful that cells are often completely absent from other regions of the standing-wave pattern. The ultrasonic technique of bringing cells together has the advantage that the efficiency of cell banding can be observed by eye in the fusion chamber. Since the sound velocity of the cell and of the suspending phase are not strongly dependent upon the ionic composition of the medium, cell-cell contact is achievable in physiological saline solutions.[134,136]

Inhomogeneous magnetic fields can similarly be used for cell alignment and contact (so-called "electro-magneto-electrofusion").[138] This technique allows for viewing of the fusion process under the light microscope but has the interesting requirement that the cells be "pre-magnetized" by adsorbing small magnetic particles onto the external surface of the plasma membrane. Adsorbed magnetic particles are likely to be incorporated into the cell during the fusion process (presumably due to the removal of a part of the plasma membrane in forms of vesicles,[139] see below) and, therefore, can not be removed by subsequent washing. This provides an interesting (but not yet systematically explored) means of separating the resulting hybrids from nonfused cells or other desirable fusion products with magnets rather than with selection media.

3. Electrically Induced Cell-Cell Contact

Although there are still some vocal scientists who doubt the advantages of electric field-moderated cell juxtaposition and membrane contact as compared with the other previously mentioned techniques for cell alignment, most work-ers in the field exclusively use non-uniform, alternating electric fields for cell and membrane contact.[23,24,28-46] The electric field-mediated contact results from a unidirectional motion of (neutral) polarizable particles and cells in a non-uniform electric field resulting in the formation of "pearl-chains"[140-153] (Figure 1). This phenomenon was discovered at the beginning of this century[154-156] and was termed dielectrophoresis around 1960.[143,144,148-151]

a. Factors Controlling Dielectrophoresis

The net dielectrophoretic force exerted on a cell (or more precisely on a dielectric particle) in an inhomogeneous field depends on several parameters. Force and migration increase with the square of the field strength and with field divergence (non-uniformity), but contrary to the belief of some authors[157] are not dependent on the current. The dielectrophoretic force is proportional to the volume of the particle. Thus, the square of the threshold field strength for cell chain formation is inversely proportional to the volume of the particle.[145] Consistently, a double logarithmic plot of the experimentally determined thresh-old field strength versus the diameter yields a straight line with a negative slope.[152,158] In other words, whereas alignment of particles (cells) with a diameter of about 10 μm requires field strengths of the order of 3 kV cm^{-1}, field strengths of 100 kV cm^{-1} must be applied to align particles with a diameter of only 10 nm. This can only be achieved by using microstructures (see Chapter 5).

Because of the square-dependence of the force on the field strength, cell migration can occur both in non-uniform d.c. and a.c. fields. The direction of cell movement is dictated by the difference of the complex dielectric constants between the particle and the medium (see Chapter 5). If the real part of the complex dielectric constant of the particle is higher than that of the medium, the particle will move in the direction of maximum field strength (i.e., to the electrodes, so-called positive dielectrophoresis, Figure 1). In the opposite case (Figure 2), particles will be repelled from the electrodes and migration will be observed in the direction of the region of weakest field strength (i.e., away from the electrodes and/or out of the electrode space, so-called negative dielectrophoresis).*

The polarization properties of a cell are strongly dependent upon the fre-quency of the field (see Chapter 5). Thus, cells can exhibit either positive or negative dielectrophoresis over the frequency spectrum attainable with a few electrofusion apparatuses, and changes of the conductivity of the medium can alter the direction of cell dielectrophoresis at a given frequency. Dielectrophoresis

---

* Recent investigations, in which interdigitated castellated electrodes (large electrode areas in combination with a small electrode space) were used,[159] showed that cells can be collected at electrodes under the influence of both positive and negative dielectrophoretic forces due to the contribution of surface charge properties to the apparent low-frequency dielectrophoretic behavior of cells and colloidal particles. The general definition of dielectrophoresis given above should, therefore, be extended to take into account polarization associated with particle surface conductivity and electrical "double-layer" dynamics (see also Chapter 5).

electrical wires

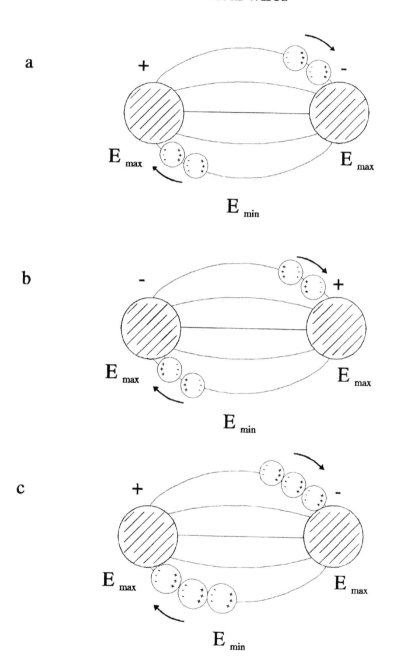

FIGURE 1 a-c.

plate electrodes

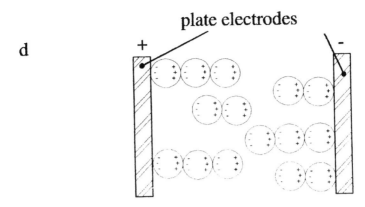

FIGURE 1. Schematic diagram of positive dielectrophoresis of cells in an inhomogeneous, alternating electric field. The field is generated between two cylindrical electrodes (200 μm apart) glued parallel on a microslide. A cross section of the electrode arrangement is shown. Owing to the external field, a dipole is created. The induced positive and negative charges are equal, but the field strength on the two sides of the cell are different. This gives rise to a net force, which pulls the cells into the region of the highest field intensity (a), provided that the complex dielectric constant of the cells is higher than that of the medium. If the field direction is reversed (b), the charge separation within the cell is opposite to the conditions described above; however, the force still acts toward the electrodes. If the cells come close together, the dipoles will attract each other (mutual dielectrophoresis) and make intimate membrane contact between the cells (c). Pearl chain formation can also be observed in a "homogeneous" field between two flat platinum electrodes (d) provided that the suspension density is high enough to create the required distortion of the field lines. *Note — only net induced charge is shown (see also Chapter 5, Figure 2).

and cell alignment can be performed, in principle, in solutions of high ionic strength.[33,37,160-162] However, the current density is proportional to the specific conductivity of the medium. High conductivity gives large currents, increased heating, and increased conductivity. This may disrupt membrane contact in the cell chains, and conductive solutions[37,44,157] are not, therefore, recommended for electrofusion, unless integrated circuits (microstructures) with rapid heat dissipation properties are used[163-165] (see also Chapter 5).

In conductive solutions (>1 S cm^{-1}) living cells always show negative dielectrophoresis.[165] In contrast, the prediction of optimum frequency ranges for the positive or negative dielectrophoresis of cells is difficult for weakly conductive solutions (<0.1 S cm^{-1}). Several qualitative and semi-quantitative analyses of both loss-free and lossy particles (including conducting media if the radius of the particle is large compared to the Debye-Hückel-length of the counter-ion cloud) have been published[142,144-146,148,151,155,166-171] (see Chapter 5). The most rigorous analysis of the migration of particles in an external electric field is given in Reference 171.

Theoretical explanations of the trajectories, relative motion, and the approach of two identical particles in the field accord with experimental results.[171] Generally speaking, a particle (cell) approaching another polarizable

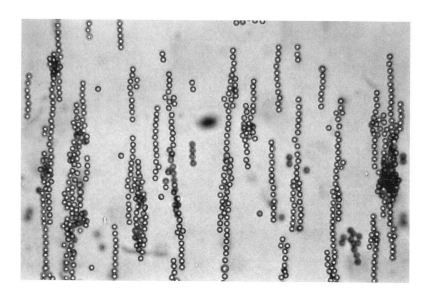

FIGURE 2. Negative dielectrophoresis of polystyrene beads (diameter 4 μm) using the electrode assembly of Figure 1 (distance 200 μm, electrodes are not shown). If the complex dielectric constant of cells or particles is lower than that of the medium, the charge separation is opposite to that shown in Figure 1a and b, respectively. The cells or particles will be repelled from the electrodes and migration will be observed in the direction of the region of weakest field strength (i.e., into the center of the electrode gap or out of the electrode space). (From Mehrle, W., Hampp, R., Zimmermann, U., and Schwan, H. P., *Biochim. Biophys. Acta,* 939, 561, 1988. With permission.)

particle (cell) as it moves toward the regions of highest or lowest field intensity, will encounter an enhanced local field divergence in its immediate environment due to the second polarized particle. Consequently, the cells or particles will form "pearl-chain"-like aggregates with point-to-point membrane contact (so-called "mutual dielectrophoresis"[151]).

In electrofusion of cells, weakly conductive solutions in combination with positive dielectrophoresis are normally used for electrically induced cell alignment.* Usually, sinusoidal electric fields of about 200 kHz to 2 MHz are applied[23,24,29,172-188] (for a description of pulsed d.c. or a.c. fields, see Chapter 1). These frequencies are high enough to "mask" electrophoretically induced motion of the (normally) negatively charged cells as well as avoiding electrode polarization and the development of toxic electrolysis products which disappear with increasing field frequency.

Low frequency alignment has been advocated by some authors. However, work in which frequencies of only 60 Hz were used for alignment[157,189-194] is very difficult to interpret because possible influence of the d.c. component has not been definitely excluded by control experiments. It is quite conceivable that

---

* Electrofusion in highly conductive (physiological) media have not been reported. Electrofusion of cells aligned in microstructures failed when highly conductive solutions were used (unpublished results). This is theoretically expected because breakdown will not occur in the contact zone of the aligned cell (see Chapters 1 and 5).

at least some of the effects (e.g., field-induced mobility of fluorescent dyes within the membrane) which were attributed by the authors to dielectrophoresis and/or to the fusion pulse[157,189-194] may have resulted from reactions induced by electrolysis products and/or electrophoresis effects within the membrane (see Chapters 1 and 6).

Frequencies around 2 to 4 MHz have the added advantage that intracellular dielectrophoresis of the nuclei can be induced under some circumstances, resulting in a positioning of the nuclei of the two attached cells close to the contact zone.[43,160] As outlined in Chapter 1 (Equations 6 and 7), the membrane capacity becomes (partly) short-circuit above 1 to 2 MHz; thus the cell interior experiences a non-uniform field in these circumstances. Field-induced motion of the nuclei into the cell contact zone is particularly efficient during post-pulse alignment when the cytoskeleton has been dissolved (see below). There is evidence to support facilitated nuclear fusion after nuclear alignment occurs, an event that would be expected to influence the somatic cell hybridization substantially as well as having considerable intrinsic interest from the basic research point of view.

Depending on the cell radius, the peak-to-peak amplitude of the a.c. field required for the formation of cell-cell contacts within about 10 to 30 s ranges between about 10 V cm^{-1} and 100 V cm^{-1}. If the field strength is increased slightly after point-to-point contact, the two attached cells are observed to "flatten" in the contact zone (due to the compression dipole forces[27]) and to form two adjacent parallel boundary layers,[29,30,195] presumably by some kind of disturbance of the hydration layer. In this configuration the two apposed membranes fuse very readily at the contact point when an intense field pulse is administered. There is evidence from the measurement of human erythrocyte electrofusion kinetics[196] that a very close dielectrophoretically induced membrane amplifies membrane destabilization caused by the high field pulse. The reason for this may be simply the enhancement of the field strength at and within the contact zone which increases with increasing cell contact (see further below). However, there is other evidence which suggests that changes in the membrane assembly occur upon tight dielectrophoretically induced contact. It has been reported[197] that, with increasing amplitude of the a.c. field (up to the critical field strength for breakdown), the membrane boundaries between the aligned (and deformed) cells, as viewed under the light microscope, disappear (i.e., the field-stretched cell chain becomes completely transparent). These uniform-appearing aggregates remain when the field is switched off, provided that weakly conductive solutions are used. However, if electrolytes are added, the membrane boundaries become visible again and the "uniform" chains disaggregate into unattached cells.[29]

b. *Beneficial Aspects of Dielectrophoresis*

Dielectrophoresis has been and, nowadays, is used for a variety of experimental applications (see also Chapters 5 and 6) including the separation and levitation of cells (or particles) as well as for the investigation of the dielectric properties of biological and non-biological materials.[151,152,166,198-208] This work contains an impressive amount of information and may form the foundation for the construction of future guidelines for improvement of electric field-mediated cell contact when new electrode assemblies becomes available (for examples, see References 159, 163-165).

For electrofusion, the phenomenon of dielectrophoresis has enormous potential in terms of practicality, flexibility, application spectrum, and hybrid yield. Electric field-mediated cell contact is more efficient and selective than any other method described above (particularly if fluid integrated circuits are considered, see Chapter 5). If the cell contact and the subsequent fusion process are appropriately performed,[28-46,209] optimal conditions can be found for each cell-cell system which facilitate reproducible production of a high number of viable hybrids or giant cells.

An important aspect of electric alignment is the probability of attaining two-cell fusions (versus polyploid homo- or heterokaryotic cells). Even though it is common for dielectrophoretically aligned "cell chains" to contain 4 to 15 members (depending upon the distance between the electrodes), experimental observation has shown a preponderance of binucleated homo- or heterokaryons in the fusion product. The preferential fusion between two homologous cells in a chain* was investigated quantitatively for Muntjac- and V79-EAT cells[42,160] by determination of the partial fusion index, F_n. F_n is defined as the ratio of the number of cells containing n nuclei to the total number of cells. According to this definition, F_2 represents a measure of two-cell fusion, F_3 of three-cell fusion, and so on. Experimentally it was found[160] that the partial fusion index decreased with increasing number of nuclei in the fused product. When the F_n-values were plotted versus the number of the nuclei, n, in a semilogarithmic plot, straight lines were obtained for the investigated cell systems (Figure 3). Such a relationship is expected if membrane breakdown occurs only in the contact zone of adherent cells in a chain and if the probability, P, for fusion of a cell pair after electropermeabilization is constant over the entire chain, i.e., P < 1 = constant. In this case, the experimental results can be described by Equation 1:

$$\log F_n = \log P \cdot (n - 1) \tag{1}$$

For P < 1, F_2 is equal to P, F_3 is equal to P × P, and so on. P was determined[160] to be 0.4 for V79-EAT and 0.35 for Muntjac-EAT fusion as calculated from the experimental curves (Figure 3). Assuming that the probability, P, is generally <0.5, it is immediately obvious that fusion of three cells has a probability less than 25%, fusion of six cells has less than about 1% probability, and so on.

---

* "Cell Fusion in an Electric Field," a movie available on a commercial basis from the "Institut für den Wissenschaftlichen Film," Göttingen, Germany, shows this cell-cell fusion event in a "cell chain" in a very impressive way.

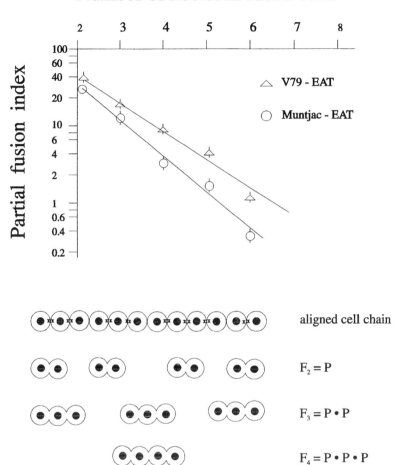

FIGURE 3. A semi-logarithmic plot of the percentage partial fusion index F_n of two different cell lines in dependence on the number of nuclei counted in the electrofused cells. Electrofusion was performed between (adherent) cells of an Ehrlich ascites tumor (EAT) line with cells of the attaching cell line V 79-S181 (triangles) and with Muntjac cells (circles). For experimental conditions, see Reference 160. The partial fusion index is defined by Equation 1. P is the probability for fusion between two aligned cells in a multi-cell chain generated by dielectrophoresis. P is assumed to be constant. Corresponding to this definition, F_2 is equal to P, F_3 to P × P, F_4 to P × P × P (and so on) as indicated in the lower part of the figure. The solid lines through the experimental data points represent a fit to Equation 1. From the straight lines P is calculated to be 0.35 and 0.4 for the Muntjac/EAT and V79/EAT fusion system, respectively. (From Bertsche, U., Mader, A., and Zimmermann, U., *Biochim. Biophys. Acta,* 939, 509, 1988. With permission.)

These considerations hold only for reasonable cell suspension densities in the fusion chamber (see below). At higher suspension densities, where adjacent cell chains come into lateral contact, Equation 1 is no longer valid. Under these conditions multinucleated, giant cells will result.

Equation 1 should also be valid for other heterologous fusion systems with the proviso that cells of approximately equal size and physiological state are fused. Otherwise, the fusion probability, P, may not be completely constant throughout the entire cell chain. From this point of view, therefore, there is no need for special manipulations which "select" for the pairing of heterologous cells.[99-104,181,183,210] However, as pointed out above and below, such procedures may have a great advantage in handling the cells and/or in identifying heterologous fusion products without use of selection media.

c. Pitfalls in Electrofusion of Dielectrophoretically Aligned Cells

Alignment is a critical component of the electrofusion process for both two- and multi-cell fusions. During pulsing, the alternating field must be switched off in order to avoid electric interferences between the high intensity field pulse and the alternating field. The pulse usually induces translational and rotational cell movement in the chain[29,196,211] because of electrophoretic and dielectrophoretic effects as well as of local fluid turbulences. Therefore, high fusion and hybrid yields are dependent on the maintenance of the pearl chain structures after pulsing[29-46,196] (Figure 4). Thus, in most electrofusion protocols, the alternating field is switched on immediately after the last fusion pulse.

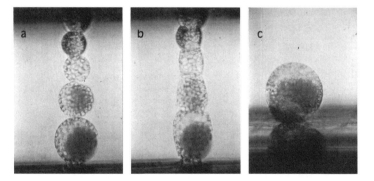

FIGURE 4.   Electrofusion of oat mesophyll protoplasts. The vacuoles of some protoplasts had been stained with neutral red. Protoplasts were aligned between two cylindrical electrodes (see Figure 1) by application of a sinusoidal electric field of 4 V peak amplitude and 500 kHz frequency (a). Administration of a square pulse of 750 V cm^{-1} strength and 20 μs duration induced fusion (b). Fusion (including fusion of the vacuoles) was completed after about 7 min (c). Note that the field strength values given here and in other parts of this chapter represent average values which were calculated by dividing the applied voltage through the electrode distance. (From Zimmermann, U. and Scheurich, P., *Biochim. Biophys. Acta,* 641, 160, 1981. With permission.)

Tight dielectrophoretic contact may fail when considerable ion leakage into the fusion medium has occurred because of inappropriate intensity of the fusion pulse.[29,196] After about 30 s the fusion process has proceeded far enough to allow the alternating field to be switched off without irreversible interruption of the fusion process. At this stage, if the membrane of one fusion partner had been pre-stained, diffusion of the dye over almost the entire surface of the fusion product can be observed.[34] Further application of the alternating field is not recommended because current flow through the intercellular interior may create adverse side effects.

After the post-alignment phase, mammalian cells are transferred into nutritional medium, wherein the mixture of the membrane components and the cytosolic content will be completed.

Since non-physiological solutions are used during alignment and fusion, the experimental regimes and the timing must be carried out with the greatest care in order to avoid cell death.

The entire fusion process (from the transfer of the cells into fusion medium to re-transfer to culture medium after the post-alignment phase) should be completed within about 10 to 15 min if high viabilities of the fused products are to result. Experience has shown that mammalian cells can be kept for about half to one hour in low-ionic solutions without significant detrimental effects on cellular function and integrity.[212] Yeast and plant protoplasts can be kept longer in low-ionic media without cell death.

Problems may arise if parental cell types dramatically different in size and/or density are aligned dielectrophoretically for fusion.[213-215] Because of the volume-dependence of the dielectrophoretic force and of differential sedimentation, separation between the two fusion partners can occur during the alignment phase (normally in about 30 s). This results in a preferential alignment of the same cell type and, in turn, homofusions are favored over heterofusions. Indeed, if cells of different size or density are fused under microgravity conditions (i.e., in parabolic flight or in orbit, thus eliminating sedimentation and convection), fusion efficiency and hybrid yields are markedly increased.[212,216-222]

Under ordinary laboratory conditions, the segregation of cells of disparate density or size can be avoided, by using an alternating field of modulated rather than constant amplitude.[215,223,224] With this approach, an advantage is gained by the very rapid motion of cells at relatively high field strengths. At field strengths of about 500 V cm^{-1} (which are still well below the breakdown threshold, see Chapter 1) cell chain formation is observed within a few seconds, so minimizing volume and density-dependent segregation effects. These field strengths are potentially lethal for the cells within a very short time (about 5 to 10 s). Therefore, the field strength must be reduced to a level just sufficient to keep the cells in an aligned position for a fraction of a second after high-speed alignment is initiated in order to prevent cell death. Experiments with cells of different sizes and density have shown that, under these conditions, the fusion and hybrid yield can be increased by more than an order of magnitude.[215,223,224]

High-density solutions have been used to avoid sedimentation as an alternative to modulated amplitude fields. In one experiment,[225] the addition of Ficoll (20% w/v) to the fusion medium resulted in a significant increase of the viability of mammalian cells (determined by trypan blue exclusion assay), but did not significantly enhance the fusion yield. The higher survival rate is not surprising (see Chapter 1) because the osmotic pressure exerted by the high polymer substance after electropermeabilization (reflection coefficient close to 1) would be expected to approximately balance the colloid-osmotic pressure of the intracellular proteins. Thus cell swelling is largely avoided.[84] Such additives are, therefore, advantageous if one wishes to study the fusion process after exclusion of secondary osmotic processes. Unfortunately, however, the fusion rate per number of input cells is proportionately reduced because of the lack of the osmotic pressure-induced processes which seem to play an important part in driving the fusion process of the two apposed cells after the breakdown pulse (see below). Thus, high density solutions are expected to have no net beneficial effect on the process. An additional argument against the use of high polymer substances for elimination of sedimentation is the concern that such substances may be toxic and are likely to be taken up during the fusion process (Chapter 1 and below).

Difficulties may also arise when non-spherical cells are fused. When non-spherical (or polar) cells, organelles, particles, etc. are suspended in a medium with a different complex dielectric constant, they will be subjected to additional "orienting" forces in the alternating field.[29,32,40,226-240] Because of the frequency-dependence of the generated membrane potential (see Equations 6 and 7 in Chapter 1), orientation and corresponding alignment will change with the frequencies applied (so-called "electro-orientation"). At certain frequencies a "turnover" from one stable orientation to the other is observed.[29,32,38,39,145,147,230-233,236,237,240] When non-spherical (ellipsoidal) cells are electrofused, sufficient membrane contact may not be achieved unless frequencies are used at which the shortest semi-axis of the cells are oriented in field direction.[32]

Besides their effects on cell orientation, alternating fields can induce other cellular events such as deformation,[29,197,241-246] elongation, rotation, protoplasmic streaming, and bleb formation[197,237,243,245] and, with eggs, induction of polarity.[234] These effects are frequency-, field strength-, and time-dependent. Appropriate dielectrophoretic field conditions must be selected (and are always found) where these and other effects on membrane transport properties (see Chapter 6) are minimized, in order to avoid possible adverse side effects on the fusion process and on the viability of the fused products.

Problems can also arise when cells attached to electrodes are fused.[29,196] Under these circumstances, fused cells can become adsorbed tightly to the electrodes resulting in mechanical rupture upon removal.[180] The material, geometry, and surface structure of the electrodes as well as the field and media conditions used are factors in determining the likelihood of those adverse effects. Fewer problems are likely if the electrodes are made of platinum, gold, or stainless steel (which have the added advantage that these materials are relatively inert during pulsing, see References 247, 248, and also Chapter 1).

III. ELECTRODE ASSEMBLY AND POWER SUPPLY

A. ELECTRODE ASSEMBLY

Because of the additional requirement of cell and membrane contact prior to pulse application, fusion chambers must meet several criteria. A number of engineering solutions have been constructed for this purpose. The factors which dictate electrode configurations are:

1. the method used for inducing cell-cell contact (centrifugal, mechanical, acoustic magnetic, or electrical);
2. the cell fusion system (single or mass cell-pair, cell-tissue, or cell-monolayer);
3. the objective of fusion (hybridization or formation of giant cells).

It is beyond the scope of this review to discuss the great variety of specialized electrode assemblies developed for electrofusion although some of them have interesting features (for examples, see References 247–250).

Evaluation of fusion chamber types is hampered by the lack of direct comparisons and by a paucity of solid objective data. For example, in some situations, chambers have been only used for homologous fusion; in others, the viability of the fused products was tested by trypan blue exclusion alone (see Chapter 1). Furthermore, the development of fusion chambers based upon microstructures and fluid integrated circuits (see Chapter 5), seems likely to revolutionize the field and may render existing technology obsolete (but see Footnote on page 186). The interested reader is directed to the cited publications describing the various fusion chambers in the relevant literature mentioned above and below. I will focus here on a description of the best described chambers currently in use. All of these chambers have been successfully used for fusion and hybridization of dielectrophoretically aligned cultured cells and suspended plant and yeast protoplasts.

The simplest electrode assembly is illustrated in Figure 1. It consists of two wires or metal strips mounted parallel to each other on a glass microscope slide.[172-180,251-257] The distance between the electrodes is usually adjusted to 200 µm to 500 µm, depending on the cell size and voltage supply. For large oocytes and embryos, electrode distances of from 500 µm up to 2 mm may be required.[67,186] The chamber is simple, cheap, and can be constructed without specialized equipment. Owing to its configuration, the chamber (and the fusion process) can be viewed "in real time" under the light microscope. The cell suspension is pipetted into the electrode space. Cell alignment occurs at both electrodes due to the distribution of the field lines between the two parallel cylindrical electrodes (Figure 1c). However, depending upon the width of the electrode gap and the suspension density, the cells also tend to form "pearl chains" without attaching to the electrodes.[175,247] In order to limit evaporation and resulting changes of the osmolality of the fusion medium, the electrode assembly can be covered by vaseline or a cover slip.[183,253]

This open chamber is best suited to the study of the fusion process and for the establishment of optimal fusion parameters for a particular cell population. Although the conditions are not always identical to those in larger "scale up" chambers, they are sufficiently close to provide a first approximation in most situations.

For electrofusion on the two-cell level (microfusion) it is (sometimes) advantageous that cells be aligned on the bottom of the electrode assembly after pretreatment of the microslide surface with a mixture of poly(L)-lysine and/or pronase.[183,258] This approach increases the adhesion between the glass surface and the cells without significant hindrance of cell motion during dielectrophoresis. Such procedures are not required when large, dense oocytes or embryos are electrofused owing to their tendency to fall quickly to the bottom of the electrode slot.[259] Fusion on the bottom of the chamber generally has the advantage that stable cell chains can also be obtained in the presence of physiological ionic solutions.[161] Fusion in microdroplets between two electrodes also allows defined fusion between two cells[183] and has been successfully used in plant protoplast fusion and hybridization.[181,260-264] Using special arrangements, many fusions of cell pairs can be performed in a very short time.

With minor modifications, the two-electrode chamber can also be used as a "flow-through" system for the fusion of large batches of protoplasts or cells.[32,180,247] Large-scale fusion can also be achieved by gluing six or more wires (or strips) parallel to each other on the same microslide or by using movable multi-electrode assemblies[265-270] as well as by arrangement of the electrodes in a "meander" configuration.[30,209,271-273]

Functionally similar, but more sophisticated and expensive, chambers consisting of a glass or plexiglass carrier plate onto which electrodes have been previously vacuum-evaporated are described in the literature.[40,45,257,274] Of the various possible configurations, assemblies in which electrode pairs (two electrodes running in parallel at a distance of about 125 μm) radiate from the center of a circular carrier disk have proved suitable for mass- (macro-) fusion and hybrid generation.[37,40] They are also quite suitable for fusion of single cell-pairs (micro-fusion). The chamber can be sealed with a lid allowing completion of the entire fusion process under sterile conditions.

The helical chamber (Figure 5) has proven to be the most versatile and useful of all of the chambers currently in use. Its configuration allows easy and gentle manipulation of the electrofusion process under both sterile and well-defined conditions combined with high fusion efficiency and hybrid yield. Volumes (and thus input cell numbers) can be varied over a surprisingly wide range. The chamber consists of a central cylindrical perspex core helically wrapped with two wires of about 1 m length on a pitch of, say, 200 μm.[31,37,41,46,267,275] The two wires are connected to two protruding electrodes

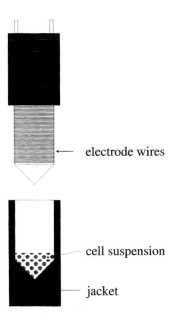

FIGURE 5.   Helical chamber designed for large-scale hybrid production. The chamber consists of a hollow Perspex tube around which are wound two cylindrical platinum wires 200 μm apart. A total of 1 m of electrode wire is used. The wrapped electrode is introduced into a cylindrical Perspex "jacket" to which the cells had been added. As the electrode is introduced into the tight-fitting jacket, the cell suspension is forced to rise between the wires.

which are then attached to the power supply. This electrode assembly fits into a hollow chamber or "jacket" (giving a volume of about 200 μl to 350 μl) into which the cells are introduced. The cell suspension rises up through the intervening space as the electrode is inserted. The cells are then dielectrophoretically collected in the electrode gaps by administering the alignment field.

For small input numbers of parental cells, the helical chamber can only be used when the volume is considerably reduced. This can be achieved by partial filling of the chamber with a cell suspension and/or by elimination of the dead volume at the bottom of the receptacle by pre-adding high density fusion medium (e.g., by addition of albumin or Ficoll). The cells are then overlayered onto the surface of this cushion.[215]

The ability to work with a liquid spacer allows the use of the same helical chamber for both large-scale and small-scale fusion. This chamber has been used successfully for the production of various yeast, mammalian, and plant hybrids, including the generation of human hybridoma cells from a small input number of B cells (see below). It allows easy adaptation and optimization for many specific applications.

Despite the many advantages of the helical electrode assembly, the newly developed "adjustable plate microchamber" (Figure 1d) is superior for handling very small cell numbers.[32,46,276] This chamber consists of two parallel flat or disk-shaped electrodes on an adjustable micrometer. In contrast to the chambers used for electroinjection (see Chapter 1), the distance between the electrodes can be variably adjusted between 10 μm and 100 μm, using a micrometer screw. In chambers consisting of two plate electrodes the electric field is homogeneous, but such a field becomes inhomogeneous when the cell suspension density exceeds a certain value (see Chapter 1). It is erroneous to believe[73] that such electrode arrangements ensure an almost homogeneous electric field which allows the precise calculation of the external field strength (see also below). The advantage of the "adjustable plate microchamber" is that a small total number of cells is required to reach the threshold density, whereas fusion in "electroinjection" chambers only takes place if the absolute cell number is very high.[22,73,75-79,277,278]

A further advantage of the "adjustable plate microchamber" over the helical chamber is that all cells are exposed to the electric field. The cell suspension is contained by surface tension in a droplet fixed between the two electrode plates. In contrast to the helical chamber, fusion does not occur under sterile conditions. As with the two-parallel wire chamber, the fusion must be performed in a sterile hood.

An alternative electrode assembly can be created by separating parallel electrode plates or disks with an insulating (Teflon or polyvinyl chloride) spacer. This results in either an enclosed fusion chamber volume (for microfusion) or, when the electrode plates are equipped with an inlet and an outlet, in a continuous-flow electrofusion system (for macrofusion).[131,247,248,279-282] In the latter case, variable electrode distances are achieved by using spacers of different thickness.

Electrofusion of bacterial protoplasts or of organelles, such as chloroplasts or mitochondria, usually requires a specialized approach dictated by their small size. Here it is advantageous to employ very non-uniform fields (such as those created by a pin-plate electrode assembly) or to pellet organelles by centrifugal forces[37,84,283-285] in order to bring them into close membrane contact. Pin-plate electrode systems are also very efficient for the production of giant cells because of their ability to simultaneously fuse thousands of cells.

B. POWER SUPPLY

Dielectrophoretic alignment requires a function generator appropriately connected to the electrodes.[44,172,249,286] The generator must also be capable of injecting square wave pulses into the cell suspension to induce breakdown in the cell contact area. Because of the relatively small distance between the electrodes, voltages of only about 100–400 V are needed to generate the required field strengths. Electrofusion instruments which combine the various electric requirements for fusion of suspended cells are commercially available from a number of sources, with a wide range of price, performance, and sophistication. Functional devices can also be constructed by electronically literate investigators.

IV. FUSION CONDITIONS AND CONSTRAINTS: BIOCHEMICAL AND BIOPHYSICAL FACTORS

A. CHEMICAL FACILITATORS

As is the case for the successful electroinjection of exogenous molecules into cells (see Chapter 1), one must take into account and optimize a number of variables in order to maximize electrofusion and hybridization efficiency. Considerable controversy exists in the literature and it is, therefore, important to point out that factors which increase the fusion efficiency do not necessarily increase the number of viable and dividing hybrids obtained from a given fusion protocol. Thus, even when enhanced fusion efficiency is claimed by the addition of chemical additives, one must take into account the net effect of said additive(s) on cell viability in the post-fusion period.

There are several reports that show that some chemicals, such as PEG, dextran, spermine, spermidine, lysophosphatidylcholine, sodium deoxycholate, dimethyl sulfoxide, S-methyl-L-cysteine, or prostaglandins facilitate the frequency of electrofusion of plant or mammalian cells.[252,271,287-295] Although this seems to contradict the above, it was also reported that compounds such as ethanol,[49,50] lysolipids, or anaesthetics[296] depress the fusion rate.

Enzymes* such as pronase, trypsin, dispase, driselase, or neuraminidase are very potent facilitators of electrofusion of mammalian cells.[28-30,81,175,176,225,251,280,282,291,297-306]

The beneficial effect of enzymatic pretreatment on the fusion frequency of (mammalian) cells may be due to differing reasons. First, such treatment could remove the glycocalyx and surface charges, allowing for closer cell-cell approach and, in turn, apparently tighter membrane-membrane interactions.[34] Work on human erythrocytes has shown that, after pronase treatment, adjacent cells can approach within 5 to 10 nm during dielectrophoresis in contrast to the 15 to 25 nm distance observed with untreated cells.[27,304] It is also well-known[40,314] that adhesion of cells to a hydrocarbon interface is much improved in solutions of low ionic strength after enzymatic treatment. However, electron microscopy of cross fractures of enzymatically treated erythrocytes frozen very rapidly during dielectrophoresis has revealed[304] that the cells were still separated by a well-defined aqueous boundary which was only disrupted by the subsequent application of the breakdown pulse.

Secondly, while proteins are crucial for the modulation of membrane fusion, it is the lipid portion of the membrane that actually fuses.[29,175] This assumption

---

* For electrofusion of blastomeres of different origin, oocytes, bacterial, yeast, algal, and plant protoplasts, the cells must be treated enzymatically in order to remove the mucin coat, the zona pellucida, or the cell wall, respectively.[91,94,95,117,174,180,274,307-313] Therefore, it seems reasonable to assert that the susceptibility of eggs and protoplasts to electrofusion as compared to that of mammalian cells is due to the enzymatic removal of the cell wall or the zona pellucida preceding fusion. However, in special cases[313] addition of pronase may also enhance electrofusion of protoplasts because of its effect on the stability of the cells against high electric field strengths.

is supported by the finding that the susceptibility of human erythrocytes to electrofusion was greatly enhanced when ATP-depleted cells were used.[175] Under these conditions giant cells could be produced by multi-cell fusion without enzymatic pre-treatment. Electron-microscopic studies have shown[175] that ATP-depletion leads to the emergence of intramembrane particle-free lipid domains. Consistently, extracellular application of phospholipase A_2 or C considerably reduces the fusion yield of mammalian cells.[303] Isolated vacuoles[316] (which contain presumably a smaller percentage of proteins than plasma membranes) and liposomes[49,317] fuse very rapidly. Although the high fusion rates may be partly due to the absence of a cytoskeleton, these results could also be interpreted as evidence that the intermingling process is facilitated when intramembrane particle-denuded regions are subjected to electrofusion.

Enzymatic treatment presumably enhances the emergence of regions free of intramembrane proteins. Digestive enzymes degrade intrinsic membrane proteins to some extent (depending on the exposure time to the enzyme). For example, it has been shown[304] that the band III protein of pronase-treated human erythrocyte membranes was degraded into two fragments as revealed by SDS gel electrophoresis. Degradation of intrinsic membrane proteins presumably increases the mobility of other membrane components which may result in the emergence of protein-free lipid domains, particularly under the influence of an alternating electric field.[30] Such areas, which are inherently more "fusogenic" than protein-containing regions, may also occur after treatment with enzyme molecules that neutralize charged groups on the outer membrane surface by adsorption. These charges may be permanently present and/or created by the enzymatic digestion. Such alterations of the membrane surface are suggested by measurements of the electrophoretic mobility of pronase-treated mouse L-cells (using the Free Flow Electrophoresis technique). Low concentrations of Ca^{2+}-free pronase (about 0.1 mg ml^{-1}) reduced the electrophoretic mobility of the cells, whereas addition of progressively higher concentrations (up to 1 mg ml^{-1} as usually used in electrofusion studies) correspondingly enhanced the electrophoretic mobility (unpublished data). This and staining experiments show that adsorption of enzyme molecules to the outer membrane surface does occur. This may be the reason that, despite a reduction in distance, a well-defined aqueous boundary is still visible between the aligned cells as mentioned above.

Evidence that patches of the plasma membrane are clear of intramembrane particles (proteins) after enzymatic digestion is also suggested by the increased stability of the cells against high intense field pulses.[29,175,251,297,315] As detailed in Chapter 1, the breakdown voltage increases as the protein content of the membranes decreases.

Another way that enzyme treatment might exert a possible beneficial effect on cell fusion may be via the presence of contaminating Ca^{2+}-ions in commercial enzyme preparations.[251,300] This however appears somewhat unlikely because:

1. The intermingling process of the membrane can proceed at very low concentrations of Ca^{2+}-ions (although the fusing aggregate does not pass through the so-called "dumb-bell shape" stage).[29,36]
2. The use of heat-inactivated pronase* was not effective.[300]
3. The addition of protease inhibitors abrogated the enhancement of fusion by trypsin.[300]

As mentioned, divalent cations have a dramatic stimulative effect on the viability of the electrofused cells[32-46,187,253,295,298,300-303,318-322] and, therefore, on the hybrid frequency when added at concentrations around 0.1 mM. Lower or higher concentrations of these ions result in a dramatically decreased hybrid yield.[253,323] This critical level of Ca^{2+}-ion concentrations in the fusion medium presumably arises from the interaction of this divalent cation with intracellular and membrane components (after membranes have been electropermeabilized). Intracellular biochemical processes require very low levels[324] of Ca^{2+}-ions of less than 10^{-7} M (see also Chapter 1). In contrast, extracellular Ca^{2+}-concentrations of about 1 mM are generally required to stabilize membrane structures.[20,325] A Ca^{2+}-concentration of about 0.1 mM in the electrofusion medium apparently balances these two extreme requirements for cell survival.

It may also be possible that the effect of Ca^{2+}-ions is due to the involvement of Ca^{2+}-binding sites in the electrofusion process. Because calmodulin inhibitors partially inhibit electrofusion of lymphoma cells[303] and plant protoplasts,[326] this Ca^{2+}-binding protein might be involved. This explanation is also supported by the finding that a 30-min stimulation of mouse embryo fibroblasts with relatively low field strengths of 10 V cm^{-1} disrupted the cytoskeletal stress fibers and that this event was correlated with a field-induced Ca^{2+}-influx.[327] There was evidence that these processes (which were also associated with cell shape and orientation changes) occurred via a calmodulin-mediated mechanism. Furthermore, Ca^{2+}, at different concentrations, is known to have a dual effect on the function of the microfilaments: enhancement of the gelling of actin and resolution of actin gels.[328] There are indeed some findings[329] that the transient increase in intracellular Ca^{2+} after the fusion pulse may activate a microfilament-associated factor which promotes spherulation (i.e., rounding up) of the fusing cells (see also below).

Furthermore, resealing of disrupted cell membranes generally requires external Ca^{2+} and can be antagonized by elevated concentrations of Mg^{2+}-ions.[330] The Ca^{2+}-dependent reaction pattern for cell membrane resealing may also involve vesicle delivery seen both under electroinjection (Chapter 1) and electrofusion conditions[139] (see also below). It is also likely that the presence of Ca^{2+} in the fusion medium may also modulate the first steps in the

---

* Heat inactivation of enzymes leads to precipitation of part of the enzymes and thus to a loss of activity in the supernatant. In addition, Ca^{2+} is partly removed by binding to the insoluble proteins. These two effects were not considered in that particular experiment and call for a reinvestigation.

intermingling process of the apposed membranes by binding to the polar groups of the anionic phospholipids[16,331] leading to removal of water from the hydration layer.[332]

These considerations demonstrate that Ca^{2+} presumably assumes a multi-functional role in the entire electrofusion process. However, other ions such as Mg^{2+} must also be considered in any discussion of fusion supplements.[44,47,51,318,319] Optimum results in hybridization are only obtained when about 0.5 mM Mg^{2+}-ions are simultaneously added to the fusion medium. Although some authors have used the chloride salts of Mg^{2+} (and also of Ca^{2+}),[48,187,253,302,323,333] the use of acetate salts generally results in better hybridization rates, particularly when a train of field pulses is administered.[318,319] A possible reason for this is that electrolysis products (e.g., hypochlorite) are produced by low-frequency fields (see above), as well as by use of a high number of field pulses (say 5 to 10) of high intensity and/or long duration.[334-336] Such products are very toxic and may change the membrane properties (e.g., by lipid peroxidation[337,338]). The reactions induced might then be mistaken as direct field effects.

Work on yeast protoplasts has shown that Ca^{2+}-ions can be substituted by approximately equivalent concentrations of Zn^{2+}-ions.[320,339] A limited uptake of Zn^{2+}-ions into the cells during fusion seems to have no detrimental side effects (this divalent cation is required as a cofactor for many enzymes). Zn^{2+} is also involved in the control of lateral mobility of lipid molecules within the membrane as well as in the maintenance of membrane structure and function.[340-342] In some cell species, the addition of Zn^{2+} led to hybrid yields manyfold higher than the yields obtained with Ca^{2+}-ions.[339] There is also a report on mammalian cells[303] which showed that Ca^{2+} can be substituted by Mn^{2+}, whereas Mg^{2+}, Co^{2+}, Ni^{2+}, and Ba^{2+} were considerably less effective. As expected, La^{3+}, a Ca^{2+} antagonist, inhibited electrofusion of plant protoplasts.[326,343]

After reviewing the vast electrofusion literature, it can be concluded that it is unnecessary to add any of the chemicals or enzymes mentioned above to obtain high-efficiency hybridization frequencies, with the proviso that the electric fusion and medium parameters are properly selected (see also below). An exception would be made for divalent cations which should be added in optimal concentration.[38-46,248,305] However, supplemental enzymatic pre-treatment may prove to be advantageous in screening for the optimum fusion conditions with new cell populations (with unknown electric properties) and with fibroblastic cells which frequently resist fusion.[56]

B. MEDIA INGREDIENTS

Besides Ca^{2+} and Mg^{2+}, the media used for dielectrophoretic alignment and subsequent field pulse-mediated fusion normally contains albumin (0.1 to 0.2%), buffer substances, and sugars as required to adjust the osmolarity (or more precisely the osmolality). Albumin was found to be very beneficial in generating a larger number of viable hybrids.[39,44,46] The effect of albumin on the electrofusion process is poorly understood.[192] However, it is likely that insertion of albumin (or other macromolecules) into the permeabilized membrane (apposed to the contact zone) may prevent high current flow through the intermingling cells at the time of the square pulse application and the subsequent post-alignment phase (see Chapter 1 and below). The buffering capacity of albumin is usually sufficient for this application. However, the addition of 1 to 10 mM phosphate buffer or 1 to 10 mM histidine will improve the buffering capacity without significantly changing the conductivity of the medium.[39,40,44,46,210,236,251,273,332] Histidine, in particular, has proven beneficial because of its isoelectric point of 7. There is one study[344] in which it is reported that a significant fusion yield was only obtained when the phosphate buffer concentration was raised to 30 mM. Experimental variations might explain some of the inconsistencies present in the literature with respect to media composition.

A pH of approximately 7 is most favorable for electrofusion. Fusion yield decreases dramatically when the pH decreases to 6, but surprisingly is not as adversely affected[345] when the pH is increased to 9.

High fusion and hybrid rates are normally obtained[28-46,333] when the fusion medium contain no or low amounts of K^+ and/or Na^+ (see above).

The osmolality of the fusion medium is a very critical parameter. Sorbitol, inositol, and sucrose are usually used to adjust the osmolality of the medium because these sugars do not change the water structure or the activities of enzymes or other membrane and cell components. Highly purified preparations of these sugars are commercially available and one is thus able to avoid potential complications of heavy metal contaminants (see Chapter 1).

Performance of electric alignment and fusion in hypo-osmolar[51,346] and particularly in strongly hypo-osmolar media (75 to 90 mOsmol)[44-46,215,276,347-349] greatly enhances fusion efficiency and hybrid yield of mammalian cells as compared to iso-osmolar conditions (Figure 6). Strongly hypo-osmolar conditions are particularly useful in fusion of adherent cell lines. Treatment of such cells with hypo-osmolar solutions leads to gentle dissolution of the cells from the confluent layer (unpublished data). Enzymatic digestion or mechanical manipulations are not required. The suspended cells are immediately susceptible for electrofusion. Plant and yeast protoplast[223,224,331] as well as two-cell embryo[350] fusion is also improved when slightly hypo-osmolar conditions (usually around 450 to 600 mOsmol) are used. As a general rule, hypo-osmolar conditions are optimal when the osmolality is adjusted 10 to 30 mOsmol above the sugar concentration at which cell bursting occurs. It is interesting to note that strongly hypo-osmolar conditions also increase the fusion yield in electro-acoustic fusion.[136]

Pre-swelling of the cells prior to injection of the breakdown pulse seems to dissolve the cytoskeleton and enhance the mobility of membrane components. Strongly hypo-osmolar conditions will therefore not be useful for cell lines which withstand swelling because of a tight cytoskeleton.[49,351] However, such cell lines are exceptional and it may be conceivable that, in these cases, electrofusion can be facilitated by a further (gradual?) decrease in osmolality. It is also important to note the beneficial effect of low osmolality on fusion and

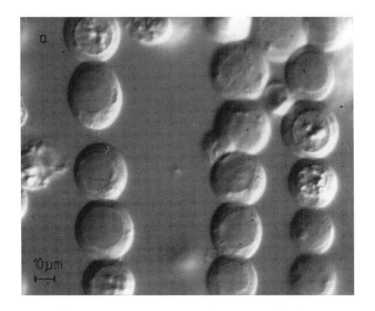

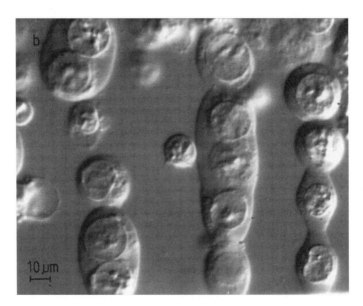

FIGURE 6. Electrofusion of cells of a heteromyeloma strain with human lymphocytes under strongly hypo-osmolar conditions (75 mOsmol sorbitol, 0.1 mM Ca^{2+}-acetate, 0.5 mM Mg^{2+}-acetate, and 1 mg ml^{-1} bovine serum albumin). Fusion was performed within two cylindrical electrode wires as shown in Figure 1a. (a) shows alignment of the cells in an alternating field of 250 V cm^{-1} strength and 1.5 MHz frequency and (b) fusion after injection of an electric breakdown pulse of 1.25 kV cm^{-1}. (Photographs courtesy of Schnettler, R.)

hybrid yield is not as apparent when the field strength or the breakdown pulse is not properly adjusted, that is by taking into account the increase in cell volume[289,352,353] (for a more detailed discussion, see Chapter 1).

The discovery of hypo-osmolar electrofusion solved a major dilemma facing research in this field for some time.[354-358] This was the inability to efficiently and reproducibly produce stable, immunoglobulin-secreting human hybridomas as readily as their murine counterparts. Even though the hybrid yield was significantly higher with iso-osmolar electrofusion than with chemical or viral methods,[359] the fusion frequency was too small and too unreproducible for reliable immortalization of the small number of lymphocytes usually available from the peripheral blood or from biopsy material (see below).

Electrofusion in strongly hypo-osmolar solutions is at least efficient as receptor-mediated electrofusion[25,99-105] or more so, but much easier to perform. Additionally, in combination with the Free-Flow Electrophoresis technique, the generated hybridoma cells can be separated from parental cells and homologous fusion products.[360] In this way, the use of selection media is eliminated which may lead to a further increase in hybrid yield.

C. FUSION PULSE PARAMETERS

1. Temperature

In contrast to electroinjection conditions (see Chapter 1) optimum fusion is achieved at temperatures between 20 to 30°C.[28-39,44,46,303,359] Outside this temperature range the fusion and hybrid yield decreases significantly.[253,359] Higher temperatures seem to favor resealing of the individual membranes, thus preventing the intermingling of the two attached membranes. Conversely, lower temperatures favor loss of intracellular (ionic) substances because of the delayed resealing of the membrane and the unphysiological environment. This may lead to increased heating in the close vicinity of the aligned cells and resulting turbulence which can disturb cell apposition during the post-alignment phase. Therefore, it is not surprising that electrofusion of mammalian cells is completely inhibited at temperatures below 10°C.[303]

2. Field Strength

Fusion and hybridization rates also depend strongly upon the field strength of the rectangular breakdown pulse, the pulse duration, and the number of administered pulses. Empirically it has been found that, as a general rule, optimum fusion and high yield of viable hybrids are obtained (using noble metal electrodes) when pulses equal to or twofold larger than that calculated by Equation 1 in Chapter 1 are used.[33] The field strengths for intercellular exchange of cytoplasmic material between aligned cells are apparently much lower than those required for field-mediated incorporation of DNA into freely suspended cells.[29-46,361] Increasing the field strength of the pulse increases the fusion speed. At high field strengths, intermingling of the membranes is observed within a fraction of a second and rounding up is completed within about 1 minute. However, the increase in fusion speed occurs at the expense of increased membrane damage, which usually leads to a dramatic decrease in viable hybrids.

The usefulness of Equation 1 of Chapter 1 for estimation of the voltage to be applied across the electrode gap for optimal electrofusion is *a priori* not straightforward (if the electrode distance is small in relation to the cell sizes and numbers). Due to the special features of the chambers required for dielectrophoretic alignment of the cells, the field strength will differ considerably in different parts of the chamber, as shown theoretically.[151,152,362] The field strength will, of course, be highest in close proximity to the electrode surface. In particular, if graphite electrodes are used,[284,285] cells or organelles may be exposed to very high field strengths, when they accumulate partly within the porous structures of the electrodes. Furthermore, as mentioned above, the field distribution will change dramatically after injection of the cells and subsequent dielectrophoretic alignment, even in chambers consisting of parallel plate electrodes if the distance between them is relatively small. This can be experimentally shown by mapping the field distribution around dielectrophoretically aligned plant protoplasts using small particles as field probes.[158] Biological probes (such as bacteria or chloroplasts which show positive dielectrophoresis) will gather at the poles of the terminal protoplasts and close to the contact zones of adherent protoplasts in the alignment chains (Figure 7). In contrast, artificial beads (such as polystyrene or glass beads which exhibit negative dielectrophoresis) form concentric rings around the individual aligned protoplasts (Figure 8). After application of the fusion pulse the two concentric polysterene rings around adherent cells move and merge into each other at the former membrane contact zone during the fusion process.* The position of these concentric rings as well as of the bacteria assembled at the contact zone and terminal ends, depends upon the volume of the aligned cells and their location within the chain. The field inhomogeneity increases with increasing diameter of the cells and should, therefore, be very significant with plant protoplasts, with mammalian cells under strongly hypo-osmolar fusion conditions, and with oocytes. Furthermore, if small cells are aligned with large cells in a chain, it is possible that the field strength is enhanced in the entire space around the small cell because of the pronounced field distortion created by the large adjacent cells (Figure 7b).

The spatial accumulation of field probes occurs at much lower field strengths than those required for the alignment of the probes themselves (i.e., in the absence of aligned mammalian cells or plant protoplasts[158,230]). This strongly suggests that the magnitude of the field strength at and within the contact zone

---

* Similar events can be observed in well-pigmented eggs of the sea urchin species *Paracentrotus lividus.*[274] Capping of the pigment granules occurs in the membrane contact zone and at the pole areas of terminal eggs under dielectrophoresis. After application of the fusion pulse the concentrated pigment spots increased in diameter forming a ring-like band in the waist of the fusing aggregate. "Low-field," spatial accumulation of biological particles such as (bacteria and granules) in the proximity of the contact zone apparently also contradicts recent theoretical considerations that the field applied to the membrane is significantly less than predicted by Equation 1 in Chapter 1.[363]

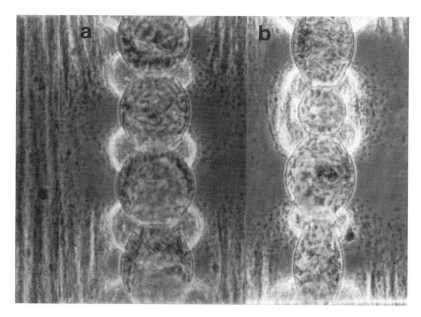

FIGURE 7. Demonstration of the distortion of the field lines around dielectrophoretically aligned oat mesophyll protoplasts by using bacteria as field probes. Accumulation of bacteria occurs at the poles and at, and within, the contact zone provided that the cells were of comparable size (a). If the chains contain differently sized cells bacteria accumulation occurs over the entire surface of the small cell (b). Accumulation is observed at very low field strengths (around 70 V cm⁻¹ peak strength). For technical reasons the photographs were taken at a field strength of 300 V cm⁻¹. At this field strength the protoplasts become slightly elongated. Simultaneously, formation of chains of bacteria is induced in the remaining electrode space because the threshold field strength for alignment of these tiny cells is exceeded. (From Mehrle, W., Hampp, R., Zimmermann, U., and Schwan, H. P., *Biochim. Biophys. Acta*, 939, 561, 1988. With permission.)

of aligned protoplasts is considerably higher and, correspondingly, is lower in the external space than that calculated from the applied voltage across the electrode gap. This may have several consequences:

1. due to the high field inhomogeneity, redistribution of membrane molecules can be induced which may be associated with the emergence of protein-free lipid domains;[30]
2. depending on the field strength of the fusion pulse, breakdown will be more or less restricted to the proximity of the contact zone or to the poles of the terminal cells in the chain;
3. the field strength within the apposed membranes can considerably exceed the threshold value for reversible breakdown in non-aligned cells.

It is likely that the field strength may be high enough to reach even the critical value for an irreversible, mechanical breakdown of the membrane under some circumstances (see Chapters 1 and 3 as well as the next section).

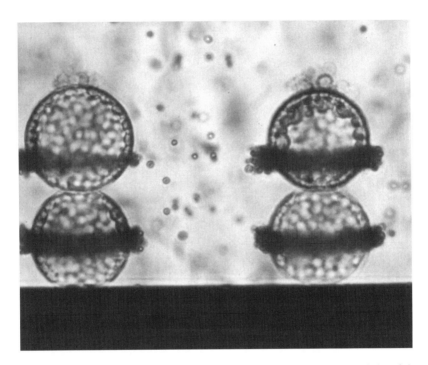

FIGURE 8. Illustration of the extremely low electric field strength in the proximity of the equatorial plane of aligned oat mesophyll protoplasts by using polystyrene beads (diameter 4 μm). In contrast to Figure 2 these beads form concentric rings perpendicular to the protoplast chains indicating that the field strength in the equatorial plane of the cells is much less than in the remaining electrode gap. (From Mehrle, W., Hampp, R., Zimmermann, U., and Schwan, H. P., *Biochim. Biophys. Acta*, 939, 561, 1988. With permission.)

This may be one of the reasons why electrofusion of cells pre-treated with hypo-osmolar conditions requires the injection of only one pulse, whereas administration of two pulses is detrimental to the cells. The field enhancement at and within the contact zone can also explain the result that fusion was occasionally observed between pronase-treated human erythrocytes when subjected to an alternating field of only about 0.8 kV cm⁻¹ strength[304] although the critical field strength for breakdown of non-aligned erythrocytes[24] is about 3 kV cm⁻¹.

The actual field strength and distribution around and within the contact zone is difficult to predict. The use of field probes for monitoring the field distribution is helpful, but one cannot exclude completely the possibility that the introduction of the probes induce additional disturbances in the original field distribution because of their size. The complexity of the problem is best demonstrated by the finding that the breakdown of the membrane of plant protoplasts under alignment conditions appears to be asymmetric[211,364] although both membranes in the contact zone were permeabilized (presumably because the above field enhancement effect in the contact zone cannot occur in freely suspended cells, see Chapter 1). The asymmetry of electropermeabilization was suggested by the finding that dye uptake was more pronounced through the hemisphere oriented to the anode than through the other one.[211] It is conceivable (Figures 7a and 7b) that asymmetric breakdown did not occur directly in the contact zone, but rather in the proximity of the contact zone (in particular in the case of small cells), where the field strength is still very high. Among other things, one can take advantage of this effect if differently sized cells are subjected to electrofusion.[211] Fusion was greatly enhanced if the smaller fusion partner of a cell pair was oriented towards the anode.[364]

Asymmetry of breakdown under alignment conditions was not reported for other cells. This may have technical causes (e.g., the lack of suitable-sized probes) or may reflect other peculiar (electric) properties of plant protoplasts which may be partly responsible for asymmetric breakdown. Except for mer-istematic cells, plant cells usually possess a vacuole which occupies a large part of the volume. Two membranes, i.e., the plasmalemma and the tonoplast are arranged in series. Both sides of the tonoplast (particularly the vacuolar side) are bathed in highly conductive solutions. This may be the reason that an external field pulse causes primarily breakdown of the plasmalemma without successive breakdown of the tonoplast. In addition, the tonoplast seems to exhibit a significant higher breakdown voltage than the plasmalemma.[211] This assumption is supported by the finding that fusion of the plasmalemma occurs immediately, but fusion of vacuoles about 1 min or even considerably later[33,365,366] after pulse administration (Figure 4). Recordings of the time course of fusion with a video recorder, which reveal the sequence of events in slow motion (see Footnote on page 188), showed an ellipsoid cell union is formed in the first stage after intermingling of the plasma membranes. At the interface between the two unfused vacuoles, chloroplasts are pushed outward, to form a ring. The subsequent fusion of the vacuoles is a sudden event, during which the chloro-plasts distribute themselves evenly over the entire cytoplasmic space, and the hybrid cell becomes rounded in shape. The fusion of the vacuoles seems to be triggered by tension in the plasma membrane of the fused product thereby squeezing the vacuoles together.

These considerations show that there is still much to be learned about the fusion process in plant protoplasts.

3. Shape, Duration, and Number of Fusion Pulses

In most electrofusion studies, pulse duration times of 10 to 100 μs were used depending on the size of the fusion partner (for exceptions, see References 49, 50, 75, 107, 361). For fusion of oocytes (or blastomeres within the zona pellucida or for pronuclear transplantation), because of their large size, rela-tively long pulse durations are sometimes needed[61,63,65,97,98,312,367,368] (for details, see Chapter 1). Under iso-osmolar conditions, a train of two or three pulses (at intervals of about 1 s) is normally injected to obtain optimal fusion rates.[29,30,44,175,369-371] In contrast to "electroinjection" protocols, rectangular pulses are more beneficial than exponentially decaying pulses. A possible reason for these relatively "harsh" field conditions is the dissolution of the cytoskeleton which must precede the fusion of the two cells (see below). In agreement with this explanation, a single breakdown pulse is generally sufficient when electrofusion is performed in strongly hypo-osmolar solutions. Injection of more than two pulses usually leads to cell bursting.

For the production of giant (multinucleated) cells, thousands of input cells are required to induce fusion within the chain and also between the chains. This is simply achieved by using high suspension density conditions which facilitate the formation of many "cell chains" and by subsequent injection of several pulses of high intensity.[28,29,33,173,175,253,284] Due to the angular dependence of the generated membrane voltage and, in particular, to the constriction of the field lines at the contact zones, multiple fusions occur with the administration of a few pulses. Giant cells can fuse further if they are brought into contact immediately after formation because of the relatively long duration of the fusogenic state of the membrane (see Chapter 1 and below). The formation of giant cells depends not only on the suspension density and the (super-critical) field strength, but also on the fusogenic susceptibility of the parental cells. Because hypo-osmolar conditions enhance the fusogenity of the input (particularly of mammalian) cells, multiple fusion is observed at somewhat lower suspension densities than is the case for iso-osmolar conditions.[215]

If optimum fusion pulse as well as appropriate media and alignment conditions are selected, high yields of viable fusion products are always obtained independent of the species. However, it must be noted that after the post-alignment phase (see above) and the transfer of the hybrids into nutrition (selection) medium, restoration and regeneration in the fused cell product should have begun.[27] These processes may last some time,[213-215] 2 hours or more (see also Chapter 1). The cells should be left undisturbed during this process in order to avoid yield reduction.

V. THE MEMBRANE INTERMINGLING PROCESS

The molecular mechanisms that govern electric field-mediated membrane fusion and post-fusion cytosolic mixing have not yet been elucidated. Consequently they are the subject of numerous investigations and speculations. There is little doubt that electric field-induced cell-cell fusion shares many properties with chemically and virally induced cell-cell fusion and also with the fusion of lipid vesicles (see References 19, 20). This is not surprising because once triggered by the breakdown pulse, the ensuing cascade of events that culminates in membrane fusion parallels the processes that occur in the cell membrane with other fusion methods. Membrane repair processes and cytosolic rearrangements are most likely independent of the nature of the lesions in the membrane.[330]

Most scientists in the field of electrofusion accept that only the lipid matrix is involved in early fusion events.[29,153,175,195,196,372,373] As discussed above, the facilitation of electrofusion by pre-treatment with digestive enzymes or by other manipulations leading to protein-free membrane areas is taken as favoring this hypothesis. Even without enzymatic pre-treatment, removal of mobile charged molecules in the membrane contact zone is expected owing to the surface potential changes evoked by close approach of the cells and by the high field inhomogeneity[16,29,195] (see also Chapter 6).

If emergence of pure lipid membrane domains is enhanced by close membrane contact we would expect [in contrast to the breakdown of freely suspended cells (see Chapter 1)] to find many similarities between the breakdown events in pure lipid membranes (Chapter 3) and those in the contact zone of aligned cells. This means that the model of pore generation for electrofusion is more adequate than for electropermeabilization of non-aligned cells. Since reversible breakdown experiments on planar lipid membranes have shown that numerous "electropores" are generated (Chapter 3), a similar event may happen in the protein-free lipid domains in the contact zone of appropriately aligned cells.[29,175,317] This process is schematically illustrated in Figure 9a. If the (temperature-dependent) healing processes of the individual cells are not too fast, the randomly oriented lipid molecules within the electropores will be expected to form bridges between the two bilayers. Intercellular, hydrophilic (co-axial) channels between the two adherent cells will be created in a sieve-like configuration.[29,175,374-378] The high curvature of such structures renders them thermodynamically unstable with the result that fusion leads ultimately to the more stable spherical shape.[29]

Ultrastructural studies of the early events in electrofusion have given evidence for a multi-defect event in regions of the membranes free of proteins as depicted in Figure 9a. Electron microscopy of freeze fractures of pronase-pretreated human erythrocytes have revealed that within 100 ms following pulse application the fracture faces exhibited discontinuous areas which were predominantly free of intramembranous particles.[304] At 2 s after the pulse, transient point defects attributed to intercellular contact appeared in the same membrane areas and replaced the discontinuous areas. At 10 s after the pulse, the majority of the discontinuous areas and the point defects disappeared as the intercellular distance returned to approximately 20 nm except at sites of cytoplasmic bridge formation. Electronmicroscopic examinations of the fusion process of plant protoplasts (sycamore tissue culture cells) have also shown[246] that "aggregation" of two dielectrophoretically aligned protoplasts takes place by a series of very close points of contact along the surface of the plasmalemmas. Ultrastructural studies of the electrofusion process in soybean protoplasts[379] have also given indications of membrane fragmentation in the contact zones which were apparently the areas of the onset of fusion.

The fusion points in the work on sycamore tissue culture cells[246] were separated by vesicles that later disperse in the fused cytoplasm. Similarly, formation of large, microscopically visible vesicles (occurring some time after pulse administration) had been discovered in the contact zone of various plant protoplasts[33,139] (Figure 10), but could also be seen in fused mammalian cells

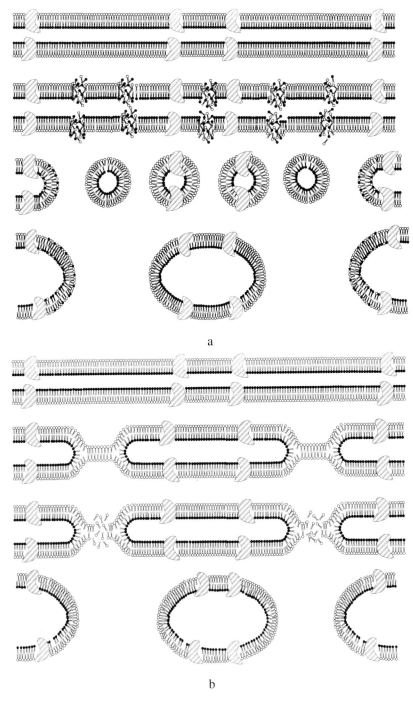

FIGURE 9.   Schematic diagrams of the mechanism of electrofusion (breakdown of a bilayer (a) and of a trilaminar (b) structure). For a detailed discussion, see text.

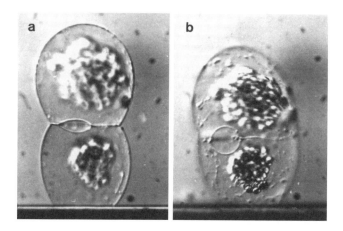

FIGURE 10. Formation of large vesicles in the contact zone of electrofusing mesophyll proto-plasts of *Kalanchoe daigremontiana.* The cells were aligned in a sinusoidal electric field of 100 V cm^{-1} strength and 1 MHz frequency. Fusion was induced by administration of a square pulse of 1 kV cm^{-1} strength and 15 μs duration. The two photographs were taken 1 min (a) and 6 min (b) after pulse application.

such as erythrocyte ghosts,[34,378] myeloma cells (unpublished observations), and mitochondrial inner membranes.[283] At least in plant protoplasts and mammalian cells, they remain visible a couple of hours after the fusion process has been completed.*

As mentioned above, Ca^{2+}-dependent mechanisms for membrane resealing are generally involved in vesicle delivery, but also in docking and fusion, similar to the exocytosis of neurotransmitters.[330] Therefore, field-induced Ca^{2+}-uptake may also trigger mechanisms, similar to exo- and endocytosis, which rapidly seal the membranes and ensure cell survival under electrofusion conditions. Vesicle formation is expected and obviously facilitated by the electrofusion process depicted in Figure 9a. According to the "bilayer-bridging" model numerous small vesicles must be formed immediately after reversible breakdown. There is little data which allow speculation about the putative composition of the tiny vesicles. According to the proposed mechanism in Figure 9a "outside-in" vesicles should be formed. However, a field-induced "flip-flop" of the phospholipids may occur in the membrane contact zone (presumably induced by Ca^{2+}-uptake)[380] (see also Chapter 1). As detailed in Chapter 1, electropulsing of non-aligned cells also results in the occurrence of many microscopically visible "endocytotic" vesicles distributed over the whole cell interior and over the entire inner membrane surface. They were observed after re-establishment of iso-osmolality of the external medium. The similarities of the vesicles seen under electrofusion and electropermeabilization conditions suggest a common mechanism for vesicle formation and composition although in aligned cells only a few visible vesicles are formed which are exclusively restricted to the contact zone. However, this can easily be ex-

* There is, however, one report wherein the formation of such vesicles is denied.[279]

plained by the field constriction at and within the contact zone of aligned cells as discussed above. In contrast to electropermeabilization of non-aligned cells, breakdown is restricted only to the contact area under electrofusion conditions. In Chapter 1, I gave some evidence that the large vesicles consist partly of "excess" membrane material either formed after the field-induced osmotic swelling in iso-osmolar solutions, or else during pre-swelling of the cells in hypo-osmolar solutions (to partly counter tensile forces). This is also likely to happen under electrofusion. Therefore, on the line of the model in Figure 9a the process of vesicle formation during electrofusion can be apparently divided into two parts. In the first step, the tiny vesicles (which are formed immediately after the "multi-point-bridging process" of the two apposed membranes) apparently fuse together as a result of their proximity in the contact zone and of Brownian collisions. The fused vesicles serve then as "nuclei" for incorporation of further "intrinsic" and also of "excess" membrane material because of the on-going surface reduction of the fused product. As a result of continuous growth and fusion a few large vesicles will be formed. If this interpretation is basically correct the occurrence of large vesicles reflects ultimately the multipoint fusion events which take place after pulsing.

Although the findings discussed above give circumstantial support to the concept that sieve-like "electropores" are generated in the lipid domains of the contact zone of aligned cells by reversible breakdown,[29,175,376-378] another (modified) mechanism seems more likely to me. A critical point of the fusion model depicted in Figure 9a is that the intermediate stage, in which the hydrophobic tails of the lipid molecules face the aqueous phase, is energetically unfavorable.[195] From the energy standpoint, pore formation is more likely[195] when a trilaminar or "trilayer" structure in the contact region is formed after dielectrophoretic alignment (Figure 9b). Breakdown in the single bilayer region of such a configuration would result in the immediate formation of a single hydrophilic channel. Experimental support for the intermediate existence of a trilaminar configuration comes from electric breakdown experiments on lipid bilayer membranes.[195,381] The microscopical and ultrastructural data discussed above are not in contradiction to the assumption of breakdown in a trilaminar bilayer structure. Within the framework of this model, one can simply assume that numerous trilaminar structures (as shown in Figure 9b) are formed in the plane of the contact zone of dielectrophoretically aligned cells. Under this condition many hydrophilic channels are formed at the instant of the membrane breakdown, so forming the tiny vesicles in the model of Figure 9a.

An alternative and more straightforward assumption is that only a single or very few trilaminar structures are formed in the membrane contact zone. In this case, the field strength within the lipid of these structures will be approximately double that in the rest of the contact zone. This can be expected to lead to a more severe breakdown in the area of the trilaminar structure, resembling the mechanical (irreversible) breakdown in planar lipid membranes. In contrast, the parallel breakdown in the double bilayer parts of the contact zone will tend to be more reversible. Studies of the mechanical breakdown of planar lipid membranes have presented evidence that very few, usually only one (or two) pore(s) are generated (see Chapter 3). These expand very rapidly to a radius of 300 μm at which point mechanical rupture of the planar membrane usually occurs.[382] Because of these features of mechanical breakdown, it appears to me that this phenomenon is the crucial event in fusion because it would readily explain all the available, and partly controversial data. The rapidly expanding pore(s) will immediately open a (or several) lumen(s) between the two apposed cells. Parallel to the growth of these few pores the "reversible breakdown points" (i.e., electropores) will disappear at the expense of the expanding "mechanical breakdown pore(s)." This supposes that the majority of the transient defects and discontinuous areas seen in electron microscopy reflect apparently the "reversible breakdown points" (as already assumed in the model of Figure 9a). The assumption of a single or a few rapidly expanding mechanical pores is also consistent with the finding of an immediate onset of mixing of the cytosols of the two attached cells after pulsing as monitored by fluorescence assays or by optical means.[196,212,355,359,383]

On the line of these arguments the membrane material released from the mechanical pores would represent the "nucleus" for the formation of the few large visible vesicles. In other words, in this case, the occurrence of a few visible vesicles can be taken as evidence for the formation of a few mechanical breakdown pores.

Conditions, which prevent the formation of a trilaminar bilayer configuration in the contact zone and/or the rapid expansion of the mechanical breakdown pore (e.g., "loose" contact between the aligned cells, or elevated temperatures and/or inappropriate osmolality of the medium) may induce a quite different process. In these cases, application of a critical field pulse will only lead to the formation of "reversible electro-leaks" in the contact zone. Membrane intermingling allowing diffusion of membrane molecules from one cell to the other might be predicted. However, because of the absence of a "mechanical breakdown pore" mixing of the cytosolic contents of the two cells should be prevented.

Indeed, membrane fusion without cytoplasmic fusion (so-called hemi-fusion) has been observed in carefully conducted electrofusion experiments on erythrocytes using different fluorescent membrane and cytoplasmic probes.[251] As expected, the osmolality and temperature of the fusion medium greatly influenced the hemi-fusion rate versus the cytoplasmic fusion rate.[251] The hemi-fusion yield increased with increasing temperature. This is presumed to be a result of the favoring of resealing by higher temperatures. The enhanced resealing prevents the mechanical pores from reaching the critical size for expansion. Hemi-fusion also increased with increasing osmolality. This is expected because lower osmolarities will favor the competitive generation and expansion of mechanical breakdown pores and, therefore, cytosolic exchange. Due to the osmotically induced (partial) dissolution of the membrane skeleton (spectrin) structural changes within the membrane are pre-formed which will facilitate the expansion of the generated mechanical pores.

The finding[251] that hemi-fused erythrocytes occasionally fused completely upon heating to 50°C cannot be taken as evidence that electrofusion of cells is normally mediated via a transient, hemi-fused state. By analogy to the effect of osmolality on hemi-fusion it suggests rather (in the framework of the electrofusion model discussed here) that the cytoskeleton, together with osmotic swelling, control the rate of the few expanding mechanical pores which lead ultimately to formation of permanent lumina between fusing cells.

It has been known for many years that osmotic processes are involved in electrofusion (and also in chemically and virally induced fusion — see References 254, 372, 373, 384). However, the overall contribution of osmotic forces to electrofusion remained controversial, mainly because of the interference of membrane and cytoskeleton proteins with the fusion and post-fusion reorganization reaction pattern.[53,84,161,187,385,386] The mechanism of electrofusion discussed here suggests that osmotic processes are generally involved, but that they contribute only indirectly to electrofusion by dissolution of the membrane and cytoplasmic skeleton. There is considerable evidence from electrofusion work[53,84,161,187,326,327,332,378,385-387] that the cytoskeleton (or spectrin in red blood cells) poses an elastic and viscous resistance to local fusion events. Aggregation of nuclei in the fusing cells (essential for hybridization) in the heterokaryon also seems to depend on the formation of parallel microtubule bundles between the nuclei.[388] The cytoskeleton apparently hinders the enlargement of primary fusion sites, i.e., the pore generated by a "mechanical" breakdown event. The opposing influence of the cytoskeleton can be overcome by the hydrostatic pressure inside the cells resulting from the colloid-osmotic pressure difference and/or by direct field-mediated dissolution of the cytoskeleton.[161] In the absence of a colloid-osmotic pressure gradient (such as in erythrocyte ghost cells), the field strength must be sufficiently high to dissolve the cytoskeleton before mechanical rupture occurs in the contact zone.[378] ATP depletion, Ca^{2+}, or ionic strength may also partially influence electrofusion in this way[175,389] (see above).

That the cytoskeleton must be considered as an important factor in the electrofusion of cells is also suggested by experiments with myeloma cells first subjected to strong hypo-osmolar and then returned to iso-osmolar conditions before being subjected to electrofusion. This treatment leads to a significant increase in fusion and hybrid yield compared to fusion under iso-osmolar conditions[348] and it is presumed that much of the enhanced efficiency is a consequence of cytoskeleton dissolution by osmotic "shock."

From the foregoing discussion we can conclude that, in contrast to non-aligned cells, mechanical breakdown does not necessarily lead to an irreversible destruction of the cells. This can be explained as follows: first, the postulated mechanical disruption of the membrane is restricted to the contact zone; most of the remaining surface apparently not being exposed to the field.*

---

* This is true except for the pole area of the terminal cells in the chain which are exposed to relatively high field strengths. This would explain why large yields of viable hybrids are only obtained with suspension densities leading to the formation of at least three- to four-cell chains.

Second, the cytoskeleton and the membrane proteins will hinder the opening of the "mechanical" pore (depending upon the deterioration induced by the electric field and the subsequent osmotic processes). This is not a factor in planar lipid membranes or liposomes. Third, the cells are kept in very close contact due to the post-alignment field forcing the membranes together and thus avoiding continuous pore opening which could lead to rupture.

The main contribution of the dielectrophoretic force during pre-alignment to electrofusion is apparently that, due to membrane alterations in the contact zone of aligned cells, experimental conditions which favor reversible break-down in non-aligned cells result in a mechanical, but locally very restricted, breakdown. Therefore, it is clear to me that pre-alignment and then pulsing must be more efficient[29-46,196] than pre-pulsing followed by establishment of membrane contact.[85,86,378,390] Under the latter conditions, the breakdown events are of more "reversible shape" because of the lack of the field strength enhancement in the contact zone. Fusion can apparently be only achieved when high (super-critical) field strengths or unfavorable frequency conditions are used which ensue large perturbations in the membrane and, therefore, a rela-tively long fusogenic state.[36,157,190,391]

These considerations also show how important it is to distinguish carefully between reversible and irreversible breakdown as well as "punch through effects" (see Chapters 1 and 3). Although these field effects on the membrane undoubtedly exist as a continuum, discrete events can be classified according to defined biological effects. This is analogous to the classification of electro-magnetic waves into discrete wavelength bands, i.e., X-ray, microwaves, vis-ible light, etc. because of their distinct interactions with material (see also Chapter 3). Unfortunately considerable confusion has been introduced into the literature by a number of authors who neglected these distinctions.

VI. APPLICATIONS OF ELECTROFUSION

A. BASIC RESEARCH

Most of the current basic research work is directed to the elucidation of the various processes involved in electrofusion of cells as well as in the subsequent structural rearrangements of cellular and membrane components in the fused products. From the many reports published in the field and discussed partly above, it is obvious that the numerous possibilities for manipulation of the electrofusion conditions can provide the future framework for clarifying the molecular events which play a pivotal role in many dynamic cellular and membrane processes.

Apart from its ability to reveal the mechanism of numerous intra- and intercellular processes, the potential of electrofusion is far from exhausted. In the opinion of the author, electrofusion represents a very valuable tool for answering many other (interdisciplinary) questions confronting biologists at present. However, the various beneficial aspects of electrofusion for many other fields have so far received less attention than may be expected from the potential of this technology. A few interesting examples are given in this section in order to enhance the interdisciplinary appeal of the electrofusion technique for future applications.

Electrofusion is the method of choice for "engineering" cell lines with a specific desired trait in order to study biochemical reactions, receptor-mediated processes, reorganization on the ultrastructural level, genetics, immunology, transcription of parental nucleolus, nuclei organization, mitochondrial synthesis, etc.[52,58,265,321,392-395]

One other application of electrofusion is the induction of premature chromosome condensation. When an interphase cell is fused with a cell in mitosis, factors generated by the mitotic cell induce condensation of the chromatin in the interphase nucleus into discernible chromosomes. This phenomenon has been termed premature chromosome condensation (PCC).[396] The morphology of the prematurely condensed chromosomes varies depending upon the stage of the cell cycle at which the interphase cells are fused. PCC analysis is an elegant tool to answer questions about the regulation of mitosis, the morphology of interphase chromatin, the requirements for entry into S-phase as well as the differential sensitivity of interphase chromatin to mutagenic agents (such as ionizing radiation or clastogenic chemicals), and the repair of chromosomal lesions.[52,397-399] Electrofusion has eliminated certain limitations and disadvantages of chemical and viral cell-cell fusion including cytotoxicity and impairment of cell viability.[399] In addition, the possibility of monitoring the fusion process under the microscope has been the major factor in obtaining good quality PCCs. Although high yields of heterokaryons were reported,[398] electrofusion did not always result in higher fusion rates than chemical or viral cell-cell fusion.[399] Electrofusion of metaphase cells of an Ehrlich ascites tumor cell line with interphase cells of a Muntjac cell line or of a Chinese Hamster subline showed a selectivity for PCC in G2 cells.[399] A possible reason for this is that the concentration and the transfer rate of the chromosome condensation factor (mitosis-inducing protein, MIP) from the metaphase to the interphase cell are presumably limited by the localized electropermeabilization of the membranes and the dominance of two-cell fusion events (see above). If this interpretation is correct, determination of the MIP transfer rate (measured by subsequent PCC) should allow estimation of the time-dependent size changes of the electric field-induced perturbations (mechanical breakdown versus reversible breakdown) in the contact zone of the two adherent cells.

The controlled formation of larger entities (giant cells) from tiny parental cells by multi-cell electrofusion is a further example of the considerable value of the field pulse technique in basic research. Giant cells allow measurement of the passive and active electric properties of the membranes (Figure 11) by using intracellular microelectrodes in combination with conventional electrophysiological techniques (e.g., current-, voltage-clamp, charge pulse technique, patch clamp, etc., see Chapters 3 and 6). Mechanical stabilization of the giant cells or of zona-free blastomeres can be achieved by immobilization of

a

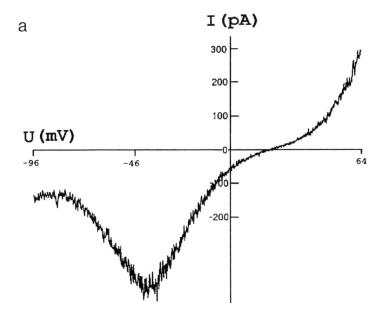

FIGURE 11. An example for patch-clamp studies on a three-cell fusion product generated by homologous fusion of three guard cell protoplasts of *Vicia faba.* Electrofusion was carried out between two parallel electrode wires (see Figure 1) attached to the patch chamber. Measurements were performed three minutes after application of the fusion pulse. (a): current-voltage (I-U) relationship of whole-cell anion fluxes, (b): single channel activity of an outside-out membrane patch measured at different clamped membrane potentials. The external solution contained: 40 mM $CaCl_2$, 2 mM $MgCl_2$, 10 mM MES-Tris, pH 5.6; the pipette solution contained: 150 mM TEAC1, 2 mM $MgCl_2$, 0.1 mM EGTA, 10 mM MgATP, 10 mM Na_2GTP, and 10 mM HEPES-Tris, pH 7.2. Note that the peak current amplitude recorded in the whole-cell patch was with –40 mV identical to that found for guard cell anion channels in unfused protoplasts. In addition, the slow capacitance value of the fusion product was with 19.7 pF approximately three-fold higher compared to the unfused protoplasts. The single channel activity was also not altered by fusion. For further details, see Reference 553. This finding suggests that electrofusion is a valuable tool for "enlargement" of tiny cells which are otherwise not accessible for electrophysiological studies (unpublished data obtained from Hedrich, R.).

the fused cells in purified alginate cross-linked by Ca^{2+}- or (more effectively) Ba^{2+}-ions.[28,95,96,400-404] By addition of polybrene (or polyimine) to the alginate solution prior to the immobilization process,[405] the mechanical stability of the artificial envelope can be even further enhanced if required. This would allow study of turgor-pressure-dependent processes in wall-less cells subjected to osmotic stress; these are measurements which are urgently needed to elucidate turgor-regulation and turgor-dependent growth phenomena in single plant cells and higher plants.[406]

It is equally likely that giant yeast cells have great potential for electrophysiological and membrane transport studies (see Chapter 6), because yeast is an ideal system for cloning of carriers and channels of other species. Transformation of yeast and subsequent production of giant cells offers, therefore, the

b

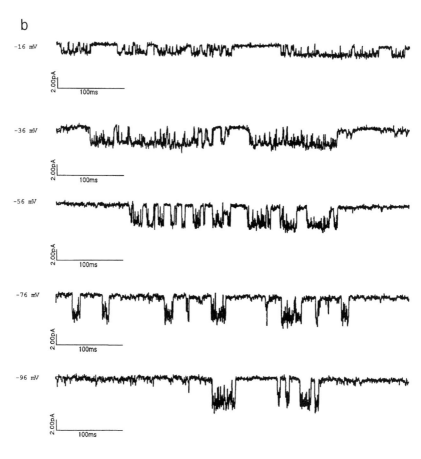

FIGURE 11b

possibility to study the biophysical features of these transporters on a completely new level.

Despite the tremendous potential of this technology for electrophysiology and the study of the water relations of cells, only two reports have been published dealing with this subject matter. In one of these studies[407] the D-glucose- and K[+]-induced changes in membrane potential of electrofused insulin-producing pancreatic islet cells of mice were investigated using impaled microelectrodes. Fused plant protoplasts were used to measure the hydraulic conductivity of the membrane barrier of fused entities.[307]

Fused giant cells can also be used in the study of highly specialized biochemical processes. Examples are the stimulation of hemoglobin synthesis in the presence of dimethyl sulfoxide in electrofused Friend cells[175] and investigations of the biochemical reaction patterns in interkingdom fusion products.[179,255,313,408,409]

Although the "giant cell" technique is still in its infancy, it can be expected that this technique will continue to develop rapidly and will soon be applied to other biological systems because optimum electrofusion conditions for fusing protozoa,[187,265,410] bacteria,[90,411,412] fungi,[108,413-417] yeast,[37,164,177,318,331,339,418-426] insect cells,[426] algae,[309-311] plant protoplasts,[172,180,181,184,223,248,289,316,427,428] (including moss[260,261,429] and conifer[430] protoplasts) and mammalian cells (see below) have been established in the recent years.

B. BACTERIAL, FUNGAL, AND YEAST BIOTECHNOLOGY

The interspecific and intergeneric hybridization of species by electrofusion of bacterial[431-437] or fungal[272,416,438-441] protoplasts represent interesting possibilities for biotechnological exploitation (production of antibiotics, etc.). Careful selection of strains on the basis of compatibility and biochemical similarity, with respect to secondary metabolism, could be productive in yielding hybrid recombinant progeny which synthesize a novel metabolite.

Much interest has also been devoted to the electrofusion of yeast protoplasts to generate interspecific and intergeneric hybrids.[308,339,418,419,421,442-444] The ultimate objective of this work is to improve the brewing process by generation of stable hybrid strains which are characterized by improved oxygen requirements and flocculation properties, as well as by enhanced ethanol tolerance and resistance against killer toxin. Recently, the baker's yeast industry has also become active in the construction of novel yeast strains with the ability to utilize melibiose, maltose, and/or lactose rapidly as well as increased osmo- and temperature-tolerance.[444,445]

Good industrial yeast strains are often polyploid or aneuploid, and as a consequence, do not possess a mating type, have a low degree of sporulation, and have a poor spore viability. While such strains are genetically stable, enabling the manufacture of a consistent product, their superior character could not be genetically analyzed or combined sexually. Protoplast fusion techniques, particularly electrofusion, provide a means of overcoming the genetic barrier of polyploidy in yeast because sexual mating is not required for the formation of fusion products. However, the fusion of two genomes is, generally speaking, never a predictable procedure, with fusion products often being very different from either parent. In contrast to recombinant DNA technology (see Chapter 1), it is very difficult to selectively introduce a single trait into a brewer's yeast using this technique. In addition and in contrast to laboratory strains, industrial yeast strains with genetic markers are usually not available which makes it difficult to select the desired hybrid strain.

As a closing note on this subject, the value of cell-cell fusion (including electrofusion) for bacterial, fungal, and yeast biotechnology remains to be determined. However, it is clear that electrofusion will find its place in the technological armory of geneticists because the efficiency of this technique very often provides definite answers to problems where conventional fusion techniques could not be used to reach firm conclusions.

C. PLANT BREEDING

Plant cells are totipotent and have the capacity to unfold their morphogenetic potential to develop into whole plants. Likewise, isolated protoplasts in culture regenerate a cell wall around themselves to reconstitute a cell and undergo repeated divisions to form a callus. By manipulation of the nutritional and physiological conditions, this cultured tissue may be induced to regenerate into an intact plant. Therefore, genetic modification of mesophyll, suspension, and/or callus protoplasts by fusion overcomes the sexual barriers. Somatic hybridization opens up new avenues and a host of scientifically interesting prospects for the improvement of crop plants (high protein and high seed production, disease resistance as well as insect, drought, frost and salt tolerance, etc.). However, at present the regeneration of complete plants from (fused) protoplasts is restricted to only a few species. Furthermore, the possibility of spontaneous or induced elimination of chromosomal material from a hybrid cell (callus or plant) is an important feature and has to be taken into account for transfer of desirable traits in breeding.[75,446-450] Somatic hybrid plants that combine the nuclear genes of two cross incompatible species are very often (but not always) sterile. This could be the result of polyploidy and aneuploidy which are frequently found in somatic hybrids. Several recent studies have shown[447] that fertility of somatic hybrid plants can be improved if one protoplast type prior to fusion is treated by X- or γ-irradiation. The rationale of this approach is that the fertility of the somatic hybrid plant is increased when the hybrid plants have received only one or a few donor chromosomes in addition to the complete genome of the recipient species (Figure 12, see also below).

Problems also exist with respect to selection systems because of the lack of appropriate markers. Recovery of heterokaryons from a mixture of fused and unfused parental protoplasts is, therefore, not as straightforward as in bacteria, yeast, or cultured animal cells. The latter have an inherently low level of biological competence which makes them ideal objects for the production and isolation of auxotrophic, temperature-sensitive, or drug-resistant mutants. The lack of suitable markers often represents a serious limitation in somatic hybridization of plant protoplasts, particularly if no visual markers (such as albinism, suspension/mesophyll fusion partners) are available or other strategies (such as vigor analysis of the fusion products) can be used.[448,451,452] Recent advances in transformation of plants with plasmids carrying antibiotic-resistance genes, in the development of selectable-markers genes and in the selection of mutants has improved the ability to isolate somatic hybrid cells from a population that consists of unfused protoplasts, homo- and heterokaryons.[76,77,224,246,266,295,370,371,446,447,453-457] Other, more physically based schemes for identifying and selecting heterokaryons (particularly at low fusion frequencies) comprise: manual isolation (of fluorescent-labelled hybrids) using pipettes and micromanipulators[75,267,323,458-460] and use of microfusion (microdroplet) assemblies (combined with optical control of the fusion process),[181,260-262,461] of flow cytometry (which also allowed determination of

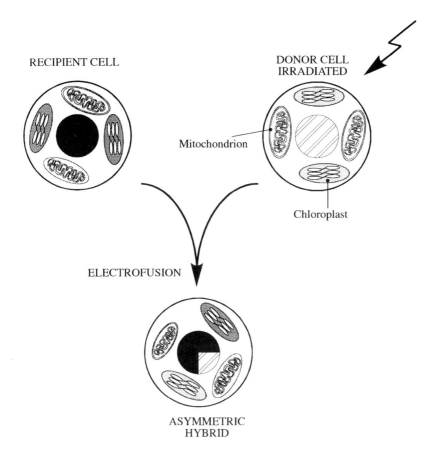

RECIPIENT CELL

DONOR CELL
IRRADIATED

Mitochondrion

Chloroplast

ELECTROFUSION

ASYMMETRIC
HYBRID

FIGURE 12.   Schematic diagram for the formation of asymmetric somatic hybrids. If no irradiation is applied, the hybrids obtained after electrofusion have most of the nuclear genes of both fusion partners. In contrast, irradiation inactivates the nucleus of the donor cell without affecting the transmission of organelles. Therefore, hybrids are recovered that contain the complete nuclear genome of the recipient cell and some nuclear DNA from the donor. The chloroplasts segregate so that only either the chloroplast set of the recipient or donor cells remain. In contrast, the mitochondria undergo genetic recombination, thereby producing a novel mitochondria genome that differs from both parental species. If the chloroplasts of the recipient cells are additionally inactivated with iodoacetate, cybrid plants are obtained that contain the nucleus of the recipient and the chloroplasts of the donor cells. (From Bates, G. W., *Guide to Electroporation and Electrofusion,* Academic Press, San Diego, 1992, 249. With permission.)

the polyploidy level),[223,462-464] or of cell electrophoretic techniques.[465] A further alternative and very interesting selection scheme for mass isolation of heterofusants is to fuse evacuolated (or meristematic) protoplasts with vacuole-containing proto-plasts.[131,213,214,223,224,364,466,467] Because of significant differences in density between the parental cells and hybrids, fractions enriched with heterokaryons can be obtained by density centrifugation (see above). Additional introduction of an antibiotic resistance gene into the evacuolated partner and use of appropriate selection media after density separation results in the complete elimination of the parental cells.[224] On the other hand, when purified parental plant protoplasts (exhibiting a distinct range of buoyant density) are used, almost pure fractions of heterokaryons can be obtained.[467]

Recent years have shown that the combination of the above heterokaryon selection schemes with the electrofusion technique were of great and immediate value for plant breeding (see literature cited in this section). There is a bulk of evidence that high yields of fusion products can be obtained by using the field pulse technique for intra/interspecific, intra/intergeneric protoplast fusion, very often eliminating a requirement for elaborate heterokaryon selection schemes. If needed, the electrofusion frequencies can be even further increased in many cases by adding chemical facilitators (although they may markedly reduce protoplast viability, see above). Most workers in this field have reported that electrofusion was found to be efficient, technically easy, and also quicker than chemical fusion.[185,447,452,454,458,468-470] Contrary findings which have been occasionally reported[471] resulted most probably from inappropriate field and media conditions. Following developmental work on the optimization of the electrofusion technique, many reports are published nowadays in which the successful regeneration of calli and/or complete (symmetric and asymmetric) hybrid plants from electrofused protoplasts is described.[76,77,92,224,262,266-270,278,295,298,370,371,427,447-450,452,453,455-457,459-461,464,467,470,472-485] The hybrid character of the plants (obtained by using both large-scale or microelectrofusion/microculture techniques) were identified by investigation of the morphology, cytological analysis, Southern hybridization, and/or isoenzyme analysis of the putative hybrids. Members of the nightshade family, particularly species of *Nicotiana* and *Solanum,* have been worked with most frequently because protoplasts from these species are easy to culture. Field tests of such hybrid plants are underway in order to study their performance in various agronomic traits.[448,460]

Recent success in producing symmetric or asymmetric somatic hybrids between other (closely or distantly related) species (such as *Rudbeckia,*[278] *Oryza,*[456,470,474,486] *Medicago,*[267,452,484] *Brassica,*[487] *Lactuca,*[472] *Citrus,*[479] *Cucumis,*[450] *Helianthus,*[483] *Raphanus,*[487] *Pisum,*[469] *Lycopersicon,*[457] *Lotus,*[76] and *Saccharum*[75] spp. as well as forage grass[478] and desiccation-tolerant grass species,[488,489] etc.) are prime indications that electrofusion is not restricted to protoplasts of plants that are easy to grow in tissue culture.

Protoplast fusion also represents a powerful approach to organelle genetics in higher plants. The genes for many agriculturally important traits mentioned above seem to be nuclear encoded. However, other traits, which for example control the sensitivity to herbicides or to cytoplasmic male sterility, are encoded in the chloroplast and mitochondria, respectively. Following fusion, the chloroplasts of the two species sort out (probably by random segregation) during subsequent cell divisions. The recovery of plants with mixed populations of chloroplasts or genetically recombined chloroplasts is therefore very rare.[447,448] By contrast, mitochondrial recombination is more frequent than segregation. Thus somatic hybrid plants may contain a novel mitochondrial genome that differs from both parental species.

The controlled transmission of organelles and nuclear genes can be performed in different ways using the electric field technique (see Figure 12). One possibility is to inactivate the chloroplasts of the recipient fusion partner with iodoacetate prior to (electro-) fusion.[76,472,478] This results in somatic hybrids that contain the chloroplasts of the untreated donor fusion partner. If the donor protoplasts are pre-irradiated, cybrid plants are obtained that contain the nucleus of the unirradiated recipient cells and the organelles of the donor.[482,486]

Another approach for organelle transfer is electrofusion of experimentally manipulated cell types such as cytoplasts (enucleated protoplasts) and karyoplasts (nuclei surrounded by plasma membrane).[447] Recent studies have shown[263,298,490] that micro-electrofusion of preselected pairs of manipulated cells in combination with individual cell culture techniques offers a very elegant and efficient method for cybridization because recovery of colonies of undesired origin is excluded.

Electrofusion is a powerful technique, not only for fusion of somatic protoplasts and for organelle transfer, but also for fusion of isolated gametes (i.e., an egg cell with single sperm cells). Gametes are predetermined to form zygotes and consequently to develop into embryos. These progenitor cells are, therefore, the most useful source for studying cytoplasmic inheritance, the processes involved in fertilization, and for regeneration of plants. As shown recently for maize,[491] egg cell protoplasts obtained by enzymatic treatment fused rapidly with sperm cells released from pollen grains after osmotic rupture. After application of the electric field pulses the sperm nucleus was integrated within the egg cell protoplast, migrated toward the egg cell nucleus, and fused with it within 1 hour, as demonstrated by electron microscopy.[492] Regeneration of plants occurred via embryogenesis and occasionally by polyembryony and organogenesis.[493] It has also been shown[494] that electrofusion-mediated *in vitro* fertilization, using isolated, single gametes, is a suitable and efficient system for transmitting chloroplasts and mitochondria via cytoplasts through the fertilization process.

D. CLONING OF EMBRYOS

The commercial cloning of domestic animals may have many important applications in basic research and agriculture.[109-121,495] The management of a herd of genomically identical individuals would probably be easier than a herd of non-identical animals, as they should all respond identically to environmental effects such as nutrition, housing, and drugs.

The number of clones produced by splitting the embryo at a preimplantation stage is limited to two or four.[119] In contrast, nuclear transplantation, in combination with serial nuclear transfers, could yield a large number of identical "clonal" offspring.[109-122,495] The method is based on the fact that nuclei of a multicellular embryo contain the same genetic material and, therefore, have the potential of giving rise to individuals with the same genotype.* Developmental genes are activated and subsequently inactivated during early embryonic growth, and nuclei become functionally restricted during differentiation. In order for nuclei from advanced embryos to direct the development of identical individuals, they must be "reprogrammed" to express these developmental genes at the appropriate time. Nuclei of embryonic mammalian cells can be reprogrammed by transferring them to enucleated mature mammalian oocytes.

Nuclear transfer, therefore, offers the possibility of multiplication of genotypes of superior economic value. In addition, nuclear transfer (like fusion-mediated generation of polyploid embryonic cells[60-71,93,94,186,259] and pronuclear transplantation[97,98,123,124]) also provides a method to study the mechanisms controlling cell differentiation as well as for understanding the processes and conditions by which the oocyte cytoplasm accomplishes nuclear de-differentiation and allows re-differentiation.[110,114,115,117,312,368,498-502]

The surgical method for transferring nuclei from early embryos in enucleated recipient oocytes requires that the plasma membranes of the cells be penetrated, which is difficult to accomplish in mammals. A non-surgical procedure using cell fusion to induce nuclear transfer is more efficient. The current procedure for nuclear transplantation involves isolation of blastomeres from preimplantation-stage embryos and their insertion as nuclear donors into the perivitelline space of enucleated oocytes on a 1:1 basis (Figure 13). The acceptor oocytes will have been matured and fertilized *in vivo* or *in vitro* for 24 to 48 h (nuclear recipient). Enucleation of the oocyte entails removing the first polar body and a portion of membrane bounded cytoplasm using a bevelled and sharpened pipette. The transfer of the blastomere creates a membrane-bound karyoplast within the zona pellucida as mentioned above. After fusion and oocyte activation, nuclear transfer embryos are cultured either *in vitro* or *in vivo* (in a ligated oviduct of an intermediate recipient) to the morula or blastocyst stage. Afterwards, the nuclear transfer embryos are transferred to the uteri of foster mothers for development to term or are used as donor embryos for a subsequent generation of nuclear transfer.

In most of the current work (see literature quoted in this section) nuclear transfer and fusion of embryonic cells is accomplished by the use of electrofusion, because of its simplicity and consistency compared to other fusion techniques. However, it must be noted that the rate of cell fusion depends critically on the alignment of the karyoplast-oocyte assembly, on the relative size and on the stage of development of the blastomere. Many studies have shown that embry-

---

* However, offspring produced by nuclear transfer may not be true clones, since the majority of the mitochondria present are from recipient cytoplasm. Although mitochondrial genomes with various species are thought to be similar,[496] differences have been detected.[497] Until the mitochondrial genomes within and between the various breeds of cattle are shown to be similar, it must be assumed that these embryos are not true clones.[110]

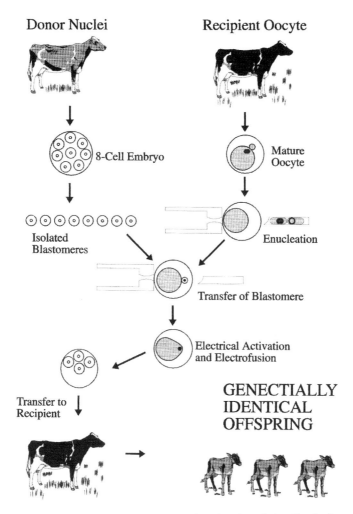

FIGURE 13. Schematic diagram of nuclear transfer using electrofusion. For details, see text.

onic development is hardly affected by field-mediated nuclear transplantation with the proviso that optimum fusion conditions were used.*

The degree to which nuclei are reprogrammed after transfer and activation of the recipient oocyte depends upon their degree of differentiation. In domestic animals donor nuclei from the 2- through the 32-cell stage, could be reprogrammed after transfer depending on the species. There were indications that the cytoplasm of the recipient was more crucial than the stage of the introduced nuclei for the development of reconstituted embryos to the blastocyst stage *in vitro* (provided that nuclei up to the 16- and 32-stage were electrically transferred)[122,499,500] and restricted by the source of recipient cyto-

---

* Most protocols are based on results obtained from research on electrofusion of sea urchin[274] and *Xenopus*[367] eggs as well as of blastomeres.[60-66,71,95,96,256,350,503]

plasm. These observations also indicate the importance of oocyte preparation for obtaining high-quality enucleated oocyte cytoplasm.

Apart from the profound effects of nuclear-cytoplasmic interactions on the development of reconstituted embryos, oocyte activation is also considered to affect the development of nuclear transplant embryos critically. Activation of the oocyte is also required for donor nuclear reprogramming. In the majority of experiments using an electric field pulse to cause cell fusion, the field pulse serves as a parthenogenetic activation factor, and it is considered that this type of treatment is sufficient to activate the oocytes of different species.[110,121,368,501,504-511] Optimal conditions for inducing efficient electrofusion coincided with optimal activating conditions. Activation occurs presumably by field-mediated Ca^{2+}-exchange[509,512] (see also Chapter 1).

This and other work have paved the way for successful cloning of domestic animals. In recent years many authors have reported that nuclear transfer embryos of rabbits, sheep, goats, pigs, and calves after electrofusion and implantation in foster mothers resulted in pregnancy and birth.[109-111,115,116,118-120,122,501,502,513] This is encouraging, but the overall efficiency of the nuclear transplantation procedure is still relatively low. In addition, the cloned animals (e.g., calves) were generally overweight compared to the controls (for an exception, see Reference 120).

Further progress is expected from improvements of the current enucleation procedures as well as in the maturation, fertilization, and culture conditions of the recipient oocyte *in vitro* before (and after) nuclear transplantation. Lack of sufficient information of these and other associated parameters (such as recipient oocyte age and cytoplasmic environment as well as the age and type of donor embryos) may account for a substantial portion of the present low efficiency of nuclear transfer.[368,500,501,505,514]

To develop a program for production of identical livestock, it is also important to develop methods for long-term storage of donor embryos: one will want to create a few animals for evaluation before producing multiple clones.[113] In addition, excess embryos will sometimes be available that need to be stored for some period of time prior to nuclear transfer. Thus the need for using frozen-thawed donor embryos arises. Several workers[109,515] in the field have recently addressed this objective and have shown that blastomeres from frozen-thawed donors, or from donors that were themselves the product of nuclear transfer, can in principle be used for serial recloning. Thus, the production of multiple identical offspring with superior traits by electrofusion-mediated nuclear transfer seems to be a powerful technology that offers the potential for tremendous genetic and economic benefits in animal agriculture.

The recent advances in the electrofusion technology have apparently made it also possible to produce multiple copies of individuals from other animals such as oysters, mussels, fish, etc.[516-519] However, even though the idea of "tailored" animals is appealing to meet the food demand of the world population, cloning by field-mediated nuclear transfer is not without problems. The success of this approach will certainly depend most strongly on the acceptance by consumers. It is true that researchers in this field are "not making copies of adults; in contrary, embryos are cloned which are genetically unknown entities".[113] With regard to cloning technology, many ethical and other questions remain which still must be answered.

E. HYBRIDOMA TECHNOLOGY

Electrofusion seems to be the method of choice for the cheap and efficient generation of large numbers of monoclonal antibody-(MAbs)-secreting hybridoma cells. Hybridoma cells are immortalized by fusion of non-secreting myeloma cell lines to IgG- or IgM-secreting B lymphocytes resulting in continuously growing, specific MAb-secreting somatic cell hybrid clones.[1,520] To assure growth of heterokaryons formed following fusion, but not of homokaryons, enzyme-deficient parental myeloma mutant cells and appropriate selective media are employed.[1]

The impact of MAbs is enormous. Today, hundreds of research and diagnostic procedures are based upon this technology and murine monoclonal antibodies specific for almost any antigen can be generated by electrically (or chemically) induced fusion, setting the stage for many different applications.

Most workers in the field have reported that electrofusion is significantly superior to chemical-mediated fusion for the production of B cell hybridomas[40,44,46,104,280-282,319,521-531] (and T cell hybridomas[532,533]). As mentioned above, the method is particularly efficient under hypo-osmolar conditions.[44,215,276,348,354,355]

Electrofusion has recently reached its full potential in human hybridoma technology. While the ultimate aim is trying to control and eventually eradicate human diseases through immunomodulation and to improve the ability to diagnose abnormalities, it is believed that human monoclonal antibodies (HMAbs) will prove far superior to those of murine origin. HMAbs should not only be less immunogenic, but should be more effective because of their ability to interact more appropriately with human complement or human effector cells. Early results from trials with HMAbs have shown excellent promise that these reagents will prove to be powerful weapons in the immunoregulation of human disease.

The principle obstacle to the ready production of HMAbs was the difficulty in obtaining sufficient numbers of antigen-responsive B cells for cell fusion. The conventional chemical fusion techniques required at least 10^7 to 10^8 B cells. Due to the recent innovations in the apparatus and technique protocols of electrofusion (see above), it is now possible to obtain reproducible fusion and hybrid yields at least 3 to 4 orders of magnitude higher than those attainable with chemical fusion methods. Thus, only a very small input number of B cells is required for fusion. A complimentary advantage is that the production of relevant antigen-specific B cells from different sources (including biopsy material) is greatly enhanced. The sources of B cells are lymph nodes, tonsils, spleen, bone marrow, and, in particular, peripheral blood as well as tumor-infiltrating lymphocytes (Figure 14). The source of B cells is significant since the immunology of the microenvironment of each source may be uniquely different.[534]

Human B cell (and T cell) hybridomas are produced in essentially the same way as murine hybridomas[176,306,321,349,355-358,531-543] (see also above). However, some key differences remain. The B cells must be "activated" (or "stimulated") for immunoglobulin synthesis prior to electrofusion. It is well-known (although notable exceptions are published[531]) that small, unstimulated lymphocytes are usually very difficult to fuse, whereas actively proliferating or stimulated cells have much higher fusion efficiencies. In traditional murine MAb technology, B cell activation is usually performed by subjecting the mouse to extensive hyperimmunization programs with highly immunogenic compounds. However, for ethical reasons, human subjects may only be immunized with a very limited number of immunogens (mostly after vaccinations or naturally occurring infections). Consequently, in many instances only relatively small numbers of lymphocytes are available, and the frequency of antigen-specific B cells is rather low.

The recent development of effective *in vitro* immunization protocols has bypassed most of the concerns about the safety and tolerance of critical antigens. Complimentary advantages of *in vitro* immunization are that only small amounts of antigen are required and that culture conditions can be manipulated to obtain the HMAb of desired specificity and isotype. *In vitro* activated B cells, which are highly suitable for electric field-mediated hybridization, can be obtained by stimulation with polyclonal activators (such as pokeweed mitogen, phytohemagglutinin, and bacterial lipopolysaccharides). It has also been reported[541,542,544] that the human B cell pool can be expanded by infection with Epstein-Barr virus (EBV), and that this leads to hybridization efficiencies comparable to those observed for murine cells. However, frequently encountered problems in using EBV for the production of human monoclonal antibodies are transient activation of B cells without subsequent transformation and the genetic instability of transformed B cells.

On the other hand, a recent report[545] of the stimulation of the CD40 receptor on the surface of human B lymphocytes, thereby allowing them to survive long-term culture, could have a significant impact on the production of HMAbs, particularly if microstructures for activation and subsequent electrofusion are used (see Chapter 5).

Another alternative and attractive method is the clonal expansion of single B cells in the presence of human T cell supernatant and irradiated murine thymoma helper cells.[546-548] This approach seems to be very efficient when applied to a

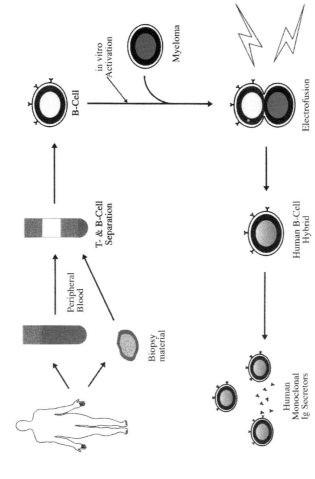

FIGURE 14. Schematic diagram for the production of human monoclonal antibody-secreting hybridoma cells by immortalization of lymphocytes taken from peripheral blood samples or from biopsy material.

number of antigen-specific B cells pre-selected by panning on antigen-coated dishes and rosetting with antigen-coupled paramagnetic beads.

Subsequent to antigen-specific B cell activation, the B cell must be rendered immortal through the somatic fusion with a selected fusion partner. Although this was a long-term obstacle in human (but not in mouse) hybridoma technology, numerous human-mouse heteromyeloma cell lines have been generated in the meantime, which lead to genetically stable, HMAb-producing hybridoma cells after electrofusion.[355-359,535] These cell lines have the properties of non-immunoglobulin secretion and selectivity.

Using the electrofusion technology and appropriate heteromyeloma strains, many (functional) human antibodies have been generated.[356-358,542,548-552] Most interestingly, electric field-mediated immortalization of B cells isolated from stomach carcinoma biopsy material of a patient resulted in two stable, HMAb-secreting clones.[357,358] The HMAbs exhibited functional activity against the autologous tumor cells (inhibition of cell adhesion and immunofluorescence staining of the membrane). This shows that a variety of hitherto inaccessible B cell populations from other human organ biopsies can be immortalized by the advanced electrofusion technique.

A pre-requisite for patient treatment with HMAbs is that gram quantities can be produced and that the HMAb will ultimately have to be purified to the specifications required for an injectable reagent intended for human use. The scale-up production of HMAbs is still a problem which is not solved satisfactorily. However, preliminary results in the laboratory of the author have shown that (serum-free) production of HMAbs in large quantities is possible. This field is currently in flux, and we can expect that appropriate technology for routine mass production of these reagents will be available soon.

VII. CONCLUSIONS

Several direct comparisons between the PEG and electrofusion methods have been made. From the data gathered so far, the efficiency of electrofusion seems to be up to three orders of magnitude better than that of PEG-induced fusion. This is not surprising in view of the many possibilities for fine tuning the electrofusion procedure. As shown in this chapter, electrofusion has an enormous potential for basic research, biotechnology, biomedicine, and plant agriculture, which have only been partly explored at the present. The tremendous advances that have been made in the relatively short time since the first disclosure of electric field-mediated fusion to the scientific world in 1978 indicate that the future holds profound and apparently unlimited perspectives. Needless to say, the possibilities that exist for the use of the electrofusion technology in improving microbial strains, mammalian cell lines, and higher organisms for industrial applications are manyfold and exciting. The coming years will undoubtedly see the emergence of "novel" cell strains for production of pharmaceutical and/or food proteins as well as "novel" crop plants and "novel" domestic animals. As detailed above, the electrofusion technique has almost reached the targeted goals in human hybridoma technology.

In all, electrofusion provides a powerful key to unlock the secrets of membrane and cellular processes involved in endocytosis, exocytosis, membrane biogenesis and transport, viral infectivity, fertilization, etc. Furthermore, continuing advances in electrofusion technology are expected to lead to a better understanding of possible thermal and direct effects of pulsed and sinusoidal electric fields on biological systems.

REFERENCES

1. **Bartal, A. H. and Hirshaut, Y.**, *Methods of Hybridoma Formation,* Humana Press, Clifton, New Jersey, 1987.
2. **Verkleij, A. J.,** Role of nonbilayer lipids in membrane fusion, in *Membrane Fusion,* Wilschut, J. and Hoekstra, D., Eds., Marcel Dekker, New York, 1991, 155.
3. **Verkleij, A. J., Leunissen-Bijvelt, J., de Kruijff, B., Hope, M. and Cullis, P. R.,** Nonbilayer structures in membrane fusion, in *Cell Fusion,* Pethica, B. A., Ed., Pitman Books, London (Ciba Foundation Symposium 103), 1984, 45.
4. **Morré, D. J., Kartenbeck, J. and Franke, W. W.,** Membrane flow and interconversions among endomembranes, *Biochim. Biophys. Acta,* 559, 71, 1979.
5. **Roos, D. S. and Choppin, P. W.,** Biochemical studies on cell fusion. II. Control of fusion response by lipid alteration, *J. Cell Biol.,* 101, 1591, 1985.
6. **Robin, M. Ch.,** Sur l'existence de deux espèces nouvelles d'éléments anatomiques qui se trouvent dans le canal médullaire des os, *C.R. Séanc. Soc. Biol.,* 149, 1849.
7. **Virchow, R.,** Reizung und Reizbarkeit, *Virchow's Arch. Path. Anat. Physiol.,* 14, 1, 1858.
8. **Langhans, Th.,** Über Riesenzellen mit wandständigen Kernen in Tuberkeln und die fibröse Form des Tuberkels, *Virchow's Arch. Path. Anat. Physiol.,* 42, 382, 1868.
9. **Krauss, E.,** Beiträge zur Riesenzellenbildung in epithelialen Geweben, *Virchow's Arch. Path. Anat. Physiol.,* 95, 249, 1884.
10. **Küster, E.,** Über die Verschmelzung nackter Protoplasten, *Ber. Deut. Bot. Ges.,* 27, 589, 1909.
11. **Lambert, R. A.,** The production of foreign body giant cells *in vitro, J. Exp. Med.,* 15, 510, 1912.
12. **Lewis, W. H.,** The formation of giant cells in tissue cultures and their similarity to those in tuberculous lesions, *Am. Rev. Tuberc. Pulm. Dis.,* 15, 616, 1927.
13. **Michel, W.,** Über die experimentelle Fusion pflanzlicher Protoplasten, *Arch. Exp. Zellforsch.,* 20, 230, 1937.
14. **Hofmeister, L.,** Mikrurgische Untersuchung über die geringe Fusionsneigung plasmolysierter, nackter Pflanzenprotoplasten, *Protoplasma,* 43, 278, 1954.
15. **Barski, G., Sorieul, S. and Cornefert, F.,** Production dans des culture *in vitro* de deux souches cellulaires en association, de cellules de caratère "hybride," *C.R. Acad. Sci. (Paris),* 251, 1825, 1960.
16. **Cecv, G.,** Membrane electrostatics, *Biochim. Biophys. Acta,* 1031, 311, 1990.

17. **Parsegian, V. A., Rand, R. P. and Gingell, D.,** Lessons for the study of membrane fusion from membrane interactions in phospholipid systems, in *Cell Fusion,* Pethica, B. A., Ed., Pitman Books, London (Ciba Foundation Symposium 103), 1984, 9.
18. **Kao, K. N. and Michayluk, M. R. A.,** Method for high-frequency intergeneric fusion of plant protoplasts, *Planta,* 115, 355, 1974.
19. **Wilschut, J. and Hoekstra, D.,** *Membrane Fusion,* Marcel Dekker, New York, 1991.
20. **Poste, G. and Nicolson, G. L.,** *Membrane Fusion,* North-Holland Publishing, Amsterdam, 1978.
21. **Okada, Y. and Tadokoro, J.,** Analysis of giant polynuclear cell formation caused by HVJ virus from Ehrlich's tumor cells. II. Quantitative analysis of giant polynuclear cell formation, *Exp. Cell Res.,* 26, 108, 1962.
22. **Zimmermann, U. and Pilwat, G.,** The relevance of electric field induced changes in the membrane structure to basic membrane research and clinical therapeutics and diagnosis, in *Abstract IV-19-(H) of the 6th International Biophysics Congress,* Kyoto, Japan, 140, 1978.
23. **Zimmermann, U., Vienken, J. and Scheurich, P.,** Electric field induced fusion of biological cells, *Biophys. Struct. Mech.,* 6 (Suppl.), 86, 1980.
24. **Zimmermann, U., Vienken, J. and Pilwat, G.,** Development of drug carrier systems: electrical field induced effects in cell membranes, *Bioelectrochem. Bioenerg.,* 7, 553, 1980.
25. **Tsong, T. Y. and Tomita, M.,** Selective B lymphocyte-myeloma cell fusion, *Meth. Enzym.,* 220, 238, 1993.
26. **Weber, H., Förster, W., Jacob, H.-E. and Berg, H.,** Enhancement of yeast protoplast fusion by electric field effects, in *Current Developments in Yeast Research: Advances in Biotechnology,* Stewart, G. G. and Russel, I., Eds., Pergamon Press, Toronto, 1981, 219.
27. **Hui, S. W. and Stenger, D. A.,** Electrofusion of cells: hybridoma production by electrofusion and polyethylene glycol, *Meth. Enzym.,* 220, 212, 1993.
28. **Zimmermann, U., Scheurich, P., Pilwat, G. and Benz, R.,** Cells with manipulated functions: new perspectives for cell biology, medicine, and technology, *Angewandte Chemie, Int. Ed. Engl.,* 20(4), 325, 1981.
29. **Zimmermann, U.,** Electric field-mediated fusion and related electrical phenomena, *Biochim. Biophys. Acta,* 694, 227, 1982.
30. **Zimmermann, U. and Vienken, J.,** Electric field-induced cell-to-cell fusion, *J. Membrane Biol.,* 67, 165, 1982.
31. **Zimmermann, U.,** Electrofusion of cells: principles and industrial potential, *Trends in Biotechnology,* 1(5), 149, 1983.
32. **Zimmermann, U., Vienken, J. and Pilwat, G.,** Electrofusion of cells, in *Investigative Microtechniques in Medicine and Biology,* Vol. 1, Chayen, J. and Bitensky, L., Eds., Marcel Dekker, New York, 1984, 89.
33. **Zimmermann, U. and Vienken, J.,** Electric field-mediated cell-to-cell fusion, in *Cell Fusion: Gene Transfer and Transformation,* Beers, R. F. and Basset, E. G., Eds., Raven Press, New York, 1984, 171.
34. **Zimmermann, U., Vienken, J., Pilwat, G. and Arnold, W. M.,** Electrofusion of cells: principles and potential for the future, in *Cell Fusion,* Pethica, B. A., Ed., Pitmann Books, London (Ciba Foundation Symposium 103), 1984, 60.
35. **Arnold, W. M. and Zimmermann, U.,** Electric field-induced fusion and rotation of cells, in *Biological Membranes,* Vol. 5, Chapman, D., Ed., Academic Press, London, 1984, 389.
36. **Zimmermann, U., Büchner, K.-H. and Arnold, W. M.,** Electrofusion of cells: Recent developments and relevance for evolution, in *Charge and Field Effects in Biosystems,* Allen, M. J. and Usherwood, P. N. R., Eds., Abacus Press, Turnbridge Wells, 1984, 293.
37. **Zimmermann, U., Vienken, J., Halfmann, J. and Emeis, C. C.,** Electrofusion: A novel hybridization technique, in *Advances in Biotechnological Processes,* Vol. 4, Mizrahi, A. and van Wenzel, A., Eds., Alan R. Liss, New York, 1985, 79.

38. **Zimmermann, U.,** Electrical breakdown, electropermeabilization and electrofusion, *Rev. Physiol. Biochem. Pharmacol.,* 105, 175, 1986.

39. **Zimmermann, U. and Urnovitz, H. B.,** Principles of electrofusion and electropermeabilization, *Meth. Enzym.,* 151, 194, 1987.

40. **Zimmermann, U.,** Electrofusion of cells, in *Methods of Hybridoma Formation,* Bartal, A. H. and Hirshaut, Y., Eds., Humana Press, Clifton, NJ, 1987, 97.

41. **Zimmermann, U. and Stopper, H.,** Electrofusion and electropermeabilization, in *Biomembrane and Receptor Mechanisms,* Vol. 7, Bertoli, E., Chapman, D., Cambria, A., and Scapagnini, U., Eds., Fidia Research Series, Liviana Press and Springer Verlag, Padova and Berlin, 1987, 371.

42. **Zimmermann, U.,** Electrofusion and electrotransfection of cells, in *Molecular Mechanisms of Membrane Fusion,* Ohki, S., Doyle, D., Flanagan, T. D., Hui, S. W. and Mayhew, E., Eds., Plenum Press, New York, 1988, 209.

43. **Zimmermann, U., Arnold, W. M. and Mehrle, W.,** Biophysics of electroinjection and electrofusion, *J. Electrostat.,* 21, 309, 1988.

44. **Zimmermann, U., Gessner, P., Wander, M. and Foung, S. K. H.,** Electroinjection and electrofusion in hypo-osmolar solution, in *Electromanipulation in Hybridoma Technology,* Borrebaeck, C. A. K. and Hagen, I., Eds., Stockton Press, New York, 1990, 1.

45. **Zimmermann, U.,** Electrofusion and electropermeabilization in genetic engineering, in *Membrane Fusion,* Wilschut, J. and Hoekstra, D., Eds., Marcel Dekker, New York, 1991, 665.

46. **Neil, G. A. and Zimmermann, U.,** Electrofusion, *Meth. Enzym.,* 220, 174, 1993.

47. **Teissié, J., Knutson, V. P., Tsong, T. Y. and Lane, M. D.,** Electric pulse-induced fusion of 3T3 cells in monolayer culture, *Science,* 216, 537, 1982.

48. **Blangero, C. and Teissié, J.,** Homokaryon production by electrofusion: a convenient way to produce a large number of viable mammalian fused cells, *Biochem. Biophys. Res. Comm.,* 114, 663, 1983.

49. **Teissié, J., Rols, M.-P. and Blangero, C.,** Electrofusion of mammalian cells and giant unilamellar vesicles, in *Electroporation and Electrofusion in Cell Biology,* Neumann, E., Sowers, A. E., and Jordan, C. A., Eds., Plenum Press, New York, 1990, 203.

50. **Orgambide, G., Blangero, C. and Teissié, J.,** Electrofusion of Chinese hamster ovary cells after ethanol incubation, *Biochim. Biophys. Acta,* 820, 58, 1985.

51. **Sukharev, S. I., Bandrina, I. N., Barbul, A. I., Abidor, I. G. and Zelenin, A. V.,** Electrofusion of fibroblast-like cells, *Studia biophysica,* 110, 45, 1987.

52. **Ishizaki, K., Chang, H. R., Eguchi, T. and Ikenaga, M.,** High voltage electric pulses efficiently induce fusion of cells in monolayer culture, *Cell Struct. Funct.,* 14, 173, 1989.

53. **Blangero, C., Rols, M.-P. and Teissié, J.,** Cytoskeletal reorganization during electric-field-induced fusion of Chinese hamster ovary cells grown in monolayers, *Biochim. Biophys. Acta,* 981, 295, 1989.

54. **Teissié, J. and Blangero, C.,** Direct experimental evidence of the vectorial character of the interaction between electric pulses and cells in cell electrofusion, *Biochim. Biophys. Acta,* 775, 446, 1984.

55. **Teissié, J. and Conte, P.,** Electrofusion of large volumes of cells in culture. Part I: Anchorage-dependent strains, *Bioelectrochem. Bioenerg.,* 19, 49, 1988.

56. **Sukharev, S. I., Bandrina, I. N., Barbul, A. I., Fedorova, L. I., Abidor, I. G. and Zelenin, A. V.,** Electrofusion of fibroblasts on the porous membrane, *Biochim. Biophys. Acta,* 1034, 125, 1990.

57. **Finaz, C., Lefevre, A. and Teissié, J.,** Electrofusion: a new, highly efficient technique for generating somatic cell hybrids, *Exp. Cell Res.,* 150, 477, 1984.

58. **Gallagher, T., Chignol, M. C., Chardonnet, Y., Guerret, S. and Schmitt, D.,** Ultrastructure of hybrids derived from electric pulse fusion of human HeLa cells and murine 3T3.4E cells, *J. Submicrosc. Cytol. Pathol.,* 22, 147, 1990.

59. **Neumann, E., Gerisch, G. and Opatz, K.,** Cell fusion induced by high electric impulses applied to *Dictyostelium, Naturwissenschaften,* 67, 414, 1980.

60. **Berg, H.,** Biological implications of electric field effects. Part V. Fusion of blastomeres and blastocysts of mouse embryos, *Bioelectrochem. Bioenerg.,* 9, 223, 1982.

61. **Kubiak, J. Z. and Tarkowski, A. K.,** Electrofusion of mouse blastomeres, *Exp. Cell Res.,* 157, 561, 1985.

62. **Winkel, G. K. and Nuccitelli, R.,** An electrofusion technique for simultaneous fusion of mouse blastomeres, *J. Cell Biol.,* 103, 87a, 1986.

63. **Ozil, J.-P. and Modlinski, J. A.,** Effects of electric field on fusion rate and survival of 2-cell rabbit embryos, *J. Embryol. Exp. Morph.,* 96, 211, 1986.

64. **Clement-Sengewald, A. and Brem, G.,** Electrofusion parameters for mouse two-cell embryos, *Theriogenology,* 32, 159, 1989.

65. **Clement-Sengewald, A.,** *Elektrofusion und Enukleation von Blastomeren bei Säugetierembryonen,* Dissertation, Institut für Tierzucht und Tierhygiene der Tierärztlichen Fakultät der Ludwig-Maximilians-Universität München, 1988.

66. **Winkel, G. K. and Nuccitelli, R.,** Octaploid mouse embryos produced by electrofusion polarize and cavitate at the same time as normal embryos, *Gamete Res.,* 24, 93, 1989.

67. **Henery, C. C. and Kaufman, M. H.,** Cleavage rates of diploid and tetraploid mouse embryos during the preimplantation period, *J. Exp. Zool.,* 259, 371, 1991.

68. **Kaufman, M. H.,** Histochemical identification of primordial germ cells and differentiation of the gonads in homozygous tetraploid mouse embryos, *J. Anat.,* 179, 169, 1991.

69. **Henery, C. C. and Kaufman, M. H.,** Relationship between cell size and nuclear volume in nucleated red blood cells of developmentally matched diploid and tetraploid mouse embryos, *J. Exp. Zool.,* 261, 472, 1992.

70. **James, R. M., Kaufman, M. H., Webb, S. and West, J. D.,** Electrofusion of mouse embryos results in uniform tetraploidy and not tetraploid/diploid mosaicism, *Genet. Res. Camb.,* 60, 185, 1992.

71. **Henery, C. C., Bard, J. B. L. and Kaufman, M. H.,** Tetraploidy in mice, embryonic cell number, and the grain of the developmental map, *Develop. Biol.,* 152, 233, 1992.

72. **Teissié, J. and Rols, M.-P.,** Electrofusion of large volumes of cells in culture. Part II: Cells growing in suspension, *Bioelectrochem. Bioenerg.,* 19, 59, 1988.

73. **Watts, J. W. and King, J. M.,** A simple method for large-scale electrofusion and culture of plant protoplasts, *Biosci. Rep.,* 4, 335, 1984.

74. **Morikawa, H., Sugino, K., Hayashi, Y., Takeda, J., Senda, M., Hirai, A. and Yamada, Y.,** Interspecific plant hybridization by electrofusion in *Nicotiana, Bio/Technology,* 4, 57, 1986.

75. **Chen, W. H., Sotak, R., Power, J. B., Cocking, E. C. and Davey, M. R.,** Electrofusion of sugarcane-petunia protoplasts and co-cultivation of heterokaryons, *Rep. Taiwan Sugar Res. Inst.,* 117, 1, 1987.

76. **Aziz, M. A., Chand, P. K., Power, J. B. and Davey, M. R.,** Somatic hybrids between the forage legumes *Lotus corniculatus* L. and *L. tenuis* Waldst et Kit., *J. Exp. Bot.,* 41, 471, 1990.

77. **Hamill, J. D., Watts, J. W. and King, J. M.,** Somatic hybridization between *Nicotiana tabacum* and *Nicotiana plumbaginifolia* by electrofusion of mesophyll protoplasts, *J. Plant Physiol.,* 129, 111, 1987.

78. **Senda, M., Morikawa, H. and Takeda, J.,** Electrical induction of cell fusion of plant protoplasts, *Plant Tissue Culture,* 615, 1982.

79. **Teissié, J., Reynaud, J. A. and Nicolau, C.,** Electric-field-induced morphological alterations and fusion of hepatocytes, *Bioelectrochem. Bioenerg.,* 17, 9, 1987.

80. **Abidor, I. G., Barbul, A. I., Zhelev, D., Sukharev, S. I., Kusmin, P. I., Pastushenko, V. F. and Zelenin, A. V.,** Electrofusion and electrical properties of cell pellets in centrifuge, *Biol. Membrany,* 6, 527, 1989.

81. **Stenger, D. A., Kaler, K. V. I. S. and Hui, S. W.,** Dipole interactions in electrofusion: contributions of membrane potential and effective dipole interaction pressures, *Biophys. J.,* 59, 1074, 1991.

82. **Hui, S. W. and Stenger, D. A.,** Effects of intercellular forces on electrofusion, in *Guide to Electroporation and Electrofusion,* Chang, D. C., Chassy, B. M., Saunders, J. A., and Sowers, A. E., Eds., Academic Press, San Diego, 1992, 167.

83. **Sukharev. S. I.,** On the role of intermembrane contact in cell electrofusion, *Bioelectrochem. Bioenerg.,* 21, 179, 1989.

84. **Abidor, I. G., Barbul, A. I., Zhelev, D. V., Doinov, P., Bandrina, I. N., Osipova, E. M. and Sukharev, S. I.,** Electrical properties of cell pellets and cell electrofusion in a centrifuge, *Biochim. Biophys. Acta,* 1152, 207, 1993.

85. **Teissié, J. and Rols, M.-P.,** Fusion of mammalian cells in culture is obtained by creating the contact between cells after their electropermeabilization, *Biochem. Biophys. Res. Commun.,* 140, 258, 1986.

86. **Montané, M.-H., Dupille, E., Alibert, G. and Teissié, J.,** Induction of a long-lived fusogenic state in viable plant protoplasts permeabilized by electric fields, *Biochim. Biophys. Acta,* 1024, 203, 1990.

87. **Reynaud, J. A., Caballero, M. and Mouneimne, Y.,** A comparison between fusion of rat hepatocytes induced by electric and by centrifugal fields, *Bioelectrochem. Bioenerg.,* 24, 33, 1990.

88. **Weber, H., Förster, W., Berg, H. and Jacob, H.-E.,** Parasexual hybridization of yeasts by electric field stimulated fusion of protoplasts, *Curr. Gen.,* 4, 165, 1981.

89. **Weber, H., Förster, W., Jacob, H.-E. and Berg, H.,** Microbiological implications of electric field effects. III. Stimulation of yeast protoplast fusion by electric field pulses, *Z. Allg. Mikrobiol.,* 21, 555, 1981.

90. **Shivarova, N., Grigorova, R., Förster, W., Jacob, H.-E. and Berg, H.,** Microbiological implications of electric field effects. Part VIII: Fusion of *Bacillus Thuringiensis* protoplasts by high electric field pulses, *Bioelectrochem. Bioenerg.,* 11, 181, 1983.

91. **Chapel, M., Teissié, J. and Alibert, G.,** Electrofusion of spermine-treated plant proto-plasts, *FEBS Lett.,* 173, 331, 1984.

92. **Chapel, M., Montane, M.-H., Ranty, B., Teissié, J. and Alibert, G.,** Viable somatic hybrids are obtained by direct current electrofusion of chemically aggregated plant proto-plasts, *FEBS Lett.,* 196, 79, 1986.

93. **Petzoldt, U.,** Developmental profile of glucose phosphate isomerase allozymes in parthe-nogenetic and tetraploid mouse embryos, *Development,* 112, 471, 1991.

94. **Eid, R. and Petzoldt, U.,** Stage-specific gene expression in asynchronous tetraploid mouse embryos formed by fusion of blastomeres and fertilized eggs, *Roux's Arch. Dev. Biol.,* 202, 198, 1993.

95. **Taniguchi, T., Cheong, H. T. and Kanagawa, H.,** Fusion and development rates of single blastomere pairs of mouse two- and four-cell embryos using the electrofusion method, *Theriogenology,* 36, 645, 1991.

96. **Taniguchi, T. and Kanagawa, H.,** Development of reconstituted mouse embryos pro-duced from the cytoplast of bisected oocytes or pronuclear-stage embryos and single blastomeres of 2-cell stage embryos, *Theriogenology,* 38, 921, 1992.

97. **Tsunoda, Y., Kato, Y. and Shioda, Y.,** Electrofusion for the pronuclear transplantation of mouse eggs, *Gamete Res.,* 17, 15, 1987.

98. **Kono, T. and Tsunoda, Y.,** Effects of induction current and other factors on large-scale electrofusion for pronuclear transplantation of mouse eggs, *Gamete Res.,* 19, 349, 1988.

99. **Lo, M. M. S., Tsong, T. Y., Conrad, M. K., Strittmatter, S. M., Hester, L. D. and Snyder, S. H.,** Monoclonal antibody production by receptor-mediated electrically induced cell fusion, *Nature,* 310, 792, 1984.

100. **Tsong, T. Y., Tomita, M. and Lo, M. M. S.,** Pre-selection of B-lymphocytes by antigen for fusion to myeloma cells by pulsed electric field (PEF) method, in *Molecular Mechanisms of Membrane Fusion,* Ohki, S., Doyle, D., Flanagan, T. D., Hui, S. W., and Mayhew, E., Eds., Plenum Press, New York, 1988, 223.

101. **Wojchowski, D. M. and Sytkowski, A. J.,** Hybridoma production by simplified avidin-mediated electrofusion, *J. Immunol. Meth.,* 90, 173, 1986.

102. **Hewish, D. R. and Werkmeister, J. A.,** The use of an electroporation apparatus for the production of murine hybridomas, *J. Immunol. Meth.,* 120, 285, 1989.

103. **Tomita, M. and Tsong, T. Y.,** Selective production of hybridoma cells: antigenic-based pre-selection of B lymphocytes for electrofusion with myeloma cells, *Biochim. Biophys. Acta,* 1055, 199, 1990.

104. **Abel, H., Stolley, P., Dreyer, G., Walther, I., Handschack, W., Schwarz, K., Platzer, C., Gröbel, C., Karsten, U. and Micheel, B.,** Hybridoma production by means of electrofusion — an experimental report, *Bioelectrochem. Bioenerg.,* 19, 173, 1988.

105. **Werkmeister, J. A., Tebb, T. A., Kirkpatrick, A. and Shukla, D. D.,** The use of peptide-mediated electrofusion to select monoclonal antibodies directed against specific and homologous regions of the potyvirus coat protein, *J. Immunol. Meth.,* 143, 151, 1991.

106. **Bakker Schut, T. C., Kraan, Y. M., Barlag, W., de Leij, L., de Grooth, B. G. and Greve, J.,** Selective electrofusion of conjugated cells in flow, *Biophys. J.,* 65, 568, 1993.

107. **Senda, M., Takeda, J., Abe, S. and Nakamura, T.,** Induction of cell fusion of plant protoplasts by electric stimulation, *Plant Cell Physiol.,* 20, 1441, 1979.

108. **Magae, Y., Kashiwagi, Y., Senda, M. and Sasaki, T.,** Electrofusion of giant protoplasts of *Pleurotus cornucopiae, Appl. Microbiol. Biotechnol.,* 24, 509, 1986.

109. **Willadsen, S. M.,** Nuclear transplantation in sheep embryos, *Nature,* 320, 63, 1986.

110. **Prather, R. S., Barnes, F. L., Sims, M. M., Robl, J. M., Eyestone, W. H. and First, N. L.,** Nuclear transplantation in the bovine embryo: assessment of donor nuclei and recipient oocyte, *Biol. Reprod.,* 37, 859, 1987.

111. **Barnes, F. L., Prather, R. S., Robl, J. M. and First, N. L.,** Multiplication of bovine embryos, *Theriogenology,* 27, 209, 1987.

112. **Sviridova, T. A., Chailakhyan, L. M., Nikitin, V. A., Veprintsev, B. N.,** A method of local electrofusion of pronuclei with an enucleated zygote, *Dokl. Akad. Nauk SSSR,* 295, 241, 1987.

113. **Marx, J. L.,** Cloning sheep and cattle embryos, *Science,* 239, 463, 1988.

114. **Landa, V.,** Cytoplasts from two-cell embryos for nuclear transplantation in the mouse, *Folia Biologica (Praha),* 35, 353, 1989.

115. **Smith, L. C. and Wilmut, I.,** Influence of nuclear and cytoplasmic activity on the development *in vivo* of sheep embryos after nuclear transplantation, *Biol. Reprod.,* 40, 1027, 1989.

116. **Prather, R. S. and First, N. L.,** Cloning of embryos, *J. Reprod. Fert. Suppl.,* 40, 227, 1990.

117. **Yang, X., Zhang, L., Kovács, A., Tobback, C. and Foote, R. H.,** Potential of hypertonic medium treatment for embryo micromanipulation: II. assessment of nuclear transplantation methodology, isolation, subzona insertion and electrofusion of blastomeres to intact or functionally enucleated oocytes in rabbits, *Mol. Reprod. Dev.,* 27, 118, 1990.

118. **McLaughlin, K. J., Davies, L. and Seamark, R. F.,** *In vitro* embryo culture in the production of identical merino lambs by nuclear transplantation, *Reprod. Fertil. Dev.,* 2, 619, 1990.

119. **Bondioli, K. R., Westhusin, M. E. and Looney, C. R.,** Production of identical bovine offspring by nuclear transfer, *Theriogenology,* 33, 165, 1990.

120. **Todorov, J., Todorova, T., Zimmermann, U., Arnold, W. M., Leiding, C., Hahn, R. and Hahn, J.,** Geklontes Kalb in Neustadt geboren, *Reprod. Dom. Anim.,* 27, 307, 1992.

121. **Todorov, J., Todorova, T., Zimmermann, U., Arnold, W. M., Leiding, C., Hahn, R. and Hahn, J.,** The influence of a prolonged *in vitro* maturation time and of a repeated field-pulse sequence upon the development rates of the oocytes in bovine nuclear transfer, 12th International Congress on Acanimal Reproduction, Congress Proceedings, Vol. 2., 754, 1992.

122. **Zhang, Y., Wang, J., Qian, J. and Hao, Z.,** Nuclear transplantation in goat embryos, *Sci. Agric. Sin.,* 24, 1, 1991.

123. **Henery, C. C. and Kaufman, M. H.,** The cleavage rate of digynic triploid mouse embryos during the preimplantation period, *Mol. Reprod. Dev.,* 34, 272, 1993.

124. **Henery, C. C. and Kaufman, M. H.,** Cleavage rate of diandric triploid mouse embryos during the preimplantation period, *Mol. Reprod. Dev.,* 32, 251, 1992.

125. **Rickords, L. F., White, K. L. and Wiltbank J. N.,** Effect of microinjection and two types of electrical stimuli on bovine sperm-hamster egg penetration, *Mol. Reprod. Dev.,* 27, 163, 1990.

126. **Grasso, R. J., Heller, R., Cooley, J. C. and Haller, E. M.,** Electrofusion of individual animal cells directly to intact corneal epithelial tissue, *Biochim. Biophys. Acta,* 980, 9, 1989.

127. **Heller, R. and Grasso, R. J.,** Transfer of human membrane surface components by incorporating human cells into intact animal tissue by cell-tissue electrofusion *in vivo, Biochim. Biophys. Acta,* 1024, 185, 1990.

128. **Heller, R. and Grasso, R. J.,** Reproducible layering of tissue culture cells onto electro-statically charged membranes, *J. Tissue Culture Methods,* 13, 25, 1991.

129. **Heller, R. and Gilbert, R.,** Development of cell-tissue electrofusion for biological applications, in *Guide to Electroporation and Electrofusion,* Chang, D. C., Chassy, B., Saunders, J. A., and Sowers, A. E., Eds., Academic Press, San Diego, 1992, 393.

130. **Heller, R.,** Spectrofluorometric assay for the quantitation of cell-tissue electrofusion, *Anal. Biochem.,* 202, 286, 1992.

131. **Naton, B., Hoffmann, E. M., Hampp, R. and Vasil, I. K.,** Improved electrofusion of protoplasts of varied fusibility by selective pairing: application of asymmetric breakdown of plasma membranes, *Plant Sci.,* 75, 93, 1991.

132. **Vienken, J., Zimmermann, U., Zenner, H. P., Coakley, W. T. and Gould, R. K.,** Electro-acoustic fusion of erythrocytes and of myeloma cells, *Biochim. Biophys. Acta,* 820, 259, 1985.

133. **Vienken, J., Zimmermann, U., Zenner, H. P., Coakley, W. T. and Gould, R. K.,** Electro-acoustic fusion of cells, *Naturwissenschaften,* 72, 441, 1985.

134. **Bardsley, D. W., Coakley, W. T., Jones, G. and Liddell, J. E.,** Electroacoustic fusion of millilitre volumes of cells in physiological medium, *J. Biochem. Biophys. Meth.,* 19, 339, 1989.

135. **Coakley, W. T., Bardsley, D. W., Grundy, M. A., Zamani, F. and Clarke, D. J.,** Cell manipulation in ultrasonic standing wave fields, *J. Chem. Tech. Biotechnol.,* 44, 43, 1989.

136. **Bardsley, D. W., Liddell, J. E., Coakley, W. T. and Clarke, D. J.,** Electroacoustic production of murine hybridomas, *J. Immun. Meth.,* 129, 41, 1990.

137. **Nyborg, W. L.,** Physical principles of ultrasound, in *Ultrasound: Its Applications in Medicine and Biology,* Vol. 3, Part I, Fry, F. J., Ed., Elsevier, Amsterdam, 1978, 1.

138. **Kramer, I., Vienken, K., Vienken, J. and Zimmermann, U.,** Magneto-electrofusion of human erythrocytes, *Biochim. Biophys. Acta,* 772, 407, 1984.

139. **Vienken, J., Zimmermann, U., Ganser, R. and Hampp, R.,** Vesicle formation during electrofusion of mesophyll protoplasts of *Kalanchoe daigremontiana, Planta,* 157, 331, 1983.

140. **Schwan, H. P.,** Biophysics of diathermy, in *Therapeutic Heat,* Licht, S., Ed., Waverly Press, Baltimore, 1958, 80.

141. **Wildervanck, A., Wakim, K. G., Herrick, J. F. and Krusen, F. H.,** Certain experimental observations on a pulsed diathermy machine, *Arch. Phys. Med.,* 40, 45, 1959.

142. **Saito, M. and Schwan, H. P.,** The time constants of pearl-chain formation, in *Biological Effects of Microwave Radiation,* Peyton, M. F., Ed., Proc. of the 1960 Conference, Plenum Press, New York, 1961, 85.

143. **Crane, J. S. and Pohl, H. A.,** A study of living and dead yeast cells using dielectrophoresis, *J. Electrochem. Soc.,* 115, 584, 1968.

144. **Sher, L. D.,** Dielectrophoresis in lossy dielectric media, *Nature,* 220, 695, 1968.

145. **Schwan, H. P. and Sher, L. D.,** Alternating-current field-induced forces and their biological implications, *J. Electrochem. Soc.,* 116, 22C, 1969.

146. **Sher, L. D., Kresch, E. and Schwan, H. P.,** On the possibility of nonthermal biological effects of pulsed electromagnetic radiation, *Biophys. J.,* 10, 970, 1970.

147. **Griffin, J. L. and Ferris, C. D.,** Pearl chain formation across radio frequency fields, *Nature,* 226, 152, 1970.

148. **Crane, J. S. and Pohl, H. A.,** Theoretical models of cellular dielectrophoresis, *J. Theor. Biol.,* 37, 15, 1972.

149. **Pohl, H. A. and Pethig, R.,** Dielectric measurements using non-uniform electric field (dielectrophoretic) effects, *J. Physics,* 10, 190, 1977.

150. **Pohl, H. A., Pollock, K. and Crane, J. S.,** Dielectrophoretic force, *J. Biol. Phys.,* 6, 133, 1978.

151. **Pohl, H. A.,** *Dielectrophoresis,* Cambridge University Press, Cambridge, 1978.

152. **Pethig, R.,** *Dielectric and Electronic Properties of Biological Materials,* John Wiley & Sons, Chichester, 1979.

153. **Hofmann, G. A. and Evans, G. A.,** Electronic genetic — physical and biological aspects of cellular electromanipulation, *IEEE Eng. Med. Biol.,* 5, 6, 1986.

154. **Muth, E.,** Über die Erscheinung der Perlschnurkettenbildung von Emulsionspartikelchen unter Einwirkung eines Wechselfeldes, *Kolloid. Z.,* 41, 97, 1927.

155. **Krasny-Ergen, W.,** Nicht-thermische Wirkungen elektrischer Schwingungen auf Kolloide, *Hochfrequenztechn. u. Elektroak.,* 40, 126, 1936.

156. **Liebesny, P.,** Athermic short wave therapy, *Arch. Phys. Ther.,* 19, 736, 1939.

157. **Sowers, A. E. and Kapoor, V.,** The mechanism of erythrocyte ghost fusion by electric field pulses, in *Molecular Mechanisms of Membrane Fusion,* Ohki, S., Doyle, D., Flanagan, T. D., Hui, S. W. and Mayhew, E., Eds., Plenum Press, New York, 1988, 237.

158. **Mehrle, W., Hampp, R., Zimmermann, U. and Schwan, H. P.,** Mapping of the field distribution around dielectrophoretically aligned cells by means of small particles as field probes, *Biochim. Biophys. Acta,* 939, 561, 1988.

159. **Pethig, R., Huang, Y., Wang, X.-B. and Burt, J. P. H.,** Positive and negative dielectrophoretic collection of colloidal particles using interdigitated castellated microelectrodes, *J. Phys. D: Appl. Phys.,* 24, 881, 1992.

160. **Bertsche, U., Mader, A. and Zimmermann, U.,** Nuclear membrane fusion in electrofused mammalian cells, *Biochim. Biophys. Acta,* 939, 509, 1988.

161. **Glaser, R. W. and Donath, E.,** Hindrance of red cell electrofusion by the cytoskeleton, *Studia biophysica,* 121, 37, 1987.

162. **Rickords, L. F. and White, K. L.,** Effect of electrofusion pulse in either electrolyte or nonelectrolyte fusion medium on subsequent murine embryonic development, *Mol. Reprod. Dev.,* 32, 259, 1992.

163. **Masuda, S., Washizu, M. and Nanba, T.,** Novel method of cell fusion in field constriction area in fluid integrated circuit, *IEEE Transact. Ind. Appl.,* 25, 732, 1989.

164. **Urano, N., Kamimura, M., Nanba, T., Okada, M., Fujimoto, M. and Washizu, M.,** Construction of a yeast cell fusion system using a fluid integrated circuit, *J. Biotechnol.,* 20, 109, 1991.

165. **Fuhr, G., Müller, T., Schnelle, Th., Hagedorn, R., Voigt, A., Fiedler, S., Arnold, W. M., Zimmermann, U., Wagner, B. and Heuberger, A.,** Radio-frequency microtools for particle and live cell manipulation, *Naturwissenschaften,* 81, 528, 1994.

166. **Jones, T. B.,** Dielectrophoretic force calculation, *J. Electrostat.,* 6, 69, 1979.

167. **Hu, C. J. and Barnes, F. S.,** A simplified theory of pearl chain effects, *Radiat. Environ. Biophys.,* 12, 71, 1975.
168. **Benguigui, L. and Lin, I. J.,** More about the dielectrophoretic force, *J. Appl. Phys.,* 53, 1141, 1982.
169. **Barnaby, E., Bryant, G. and Wolfe, J.,** What is "dielectrophoresis"?, *Bioelectrochem. Bioenerg.,* 19, 347, 1988.
170. **Marszalek, P., Zielinski, J. J. and Fikus, M.,** Experimental verification of a theoretical treatment of the mechanism of dielectrophoresis, *Bioelectrochem. Bioenerg.,* 22, 289, 1989.
171. **Sauer, F. A.,** Interaction-forces between microscopic particles in an external electromagnetic field, in *Interactions between Electromagnetic Fields and Cells,* Chiabrera, A., Nicolini, C., and Schwan, H. P., Eds., Plenum Press, New York, 1985, 181.
172. **Zimmermann, U. and Scheurich, P.,** High frequency fusion of plant protoplasts by electric fields, *Planta,* 151, 26, 1981.
173. **Scheurich, P. and Zimmermann, U.,** Giant human erythrocytes by electric-field-induced cell-to-cell fusion, *Naturwissenschaften,* 68, 45, 1981.
174. **Scheurich, P., Zimmermann, U. and Schnabl, H.,** Electrically stimulated fusion of different plant cell protoplasts, *Plant Physiol.,* 67, 849, 1981.
175. **Pilwat, G., Richter, H.-P. and Zimmermann, U.,** Giant culture cells by electric field-induced fusion, *FEBS Lett.,* 133, 169, 1981.
176. **Bischoff, R., Eisert, R. M., Schedel, I., Vienken, J. and Zimmermann, U.,** Human hybridoma cells produced by electrofusion, *FEBS Lett.,* 147, 64, 1982.
177. **Halfmann, H. J., Röcken, W., Emeis, C. C. and Zimmermann, U.,** Transfer of mitochondrial function into a cytoplasmic respiratory-deficient mutant of *Saccharomyces* yeast by electrofusion, *Curr. Gen.,* 6, 25, 1982.
178. **Zimmermann, U., Pilwat, G. and Pohl, H. A.,** Electric field-mediated cell fusion, *J. Biol. Phys.,* 10, 43, 1982.
179. **Zimmermann, U., Vienken, J. and Pilwat, G.,** Electric-field-induced fusion of cells, *Studia biophysica,* 90, 177, 1982.
180. **Bates, G. W., Gaynor, J. J. and Shekhawat, N. S.,** Fusion of plant protoplasts by electric fields, *Plant Physiol.,* 72, 1110, 1983.
181. **Koop, H. U., Dirk, J., Wolff, D. and Schweiger, H. G.,** Somatic hybridization of two selected single cells, *Cell Biol. Inter. Rep.,* 7, 1123, 1983.
182. **Jacob, H.-E., Siegemund, F. and Bauer, E.,** Fusion von pflanzlichen Protoplasten durch elektrischen Feldimpuls nach Dielektrophorese, *Biol. Zbl.,* 103, 77, 1984.
183. **Vienken, J., Zimmermann, U., Fouchard, M. and Zagury, D.,** Electrofusion of myeloma cells on the single cell level: fusion under sterile conditions without proteolytic enzyme treatment, *FEBS Lett.,* 163, 54, 1983.
184. **Tempelaar, M. J. and Jones, M. G. K.,** Fusion characteristics of plant protoplasts in electric fields, *Planta,* 165, 205, 1985.
185. **Fish, N., Karp, A. and Jones, M. G. K.,** Production of somatic hybrids by electrofusion in *Solanum, Theor. Appl. Genet.,* 76, 260, 1988.
186. **Iwasaki, S., Kono, T., Fukatsu, H. and Nakahara, T.,** Production of bovine tetraploid embryos by electrofusion and their developmental capability *in vitro, Gamete Res.,* 24, 261, 1989.
187. **Gaertig, J. and Iftode, F.,** Rearrangement of the cytoskeleton and nuclear transfer in *Tetrahymena thermophila* cells fused by electric field, *J. Cell Sci.,* 93, 691, 1989.
188. **Saunders, J. A., Roskos, L. A., Mischke, S., Aly, M. A. M. and Owens, L. D.,** Behavior and viability of tobacco protoplasts in response to electrofusion parameters, *Plant Physiol.,* 80, 117, 1986.
189. **Sowers, A. E.,** Characterization of electric field-induced fusion in erythrocyte ghost membranes, *J. Cell Biol.,* 99, 1989, 1984.
190. **Sowers, A. E.,** A long-lived fusogenic state is induced in erythrocyte ghosts by electric pulses, *J. Cell Biol.,* 102, 1358, 1986.

191. **Sowers, A. E.,** Evidence that electrofusion yield is controlled by biologically relevant membrane factors, *Biochim. Biophys. Acta,* 985, 334, 1989.

192. **Sowers, A. E.,** Low concentrations of macromolecular solutes significantly affect electrofusion yield in erythrocyte ghosts, *Biochim. Biophys. Acta,* 1025, 247, 1990.

193. **Sowers, A. E.,** Movement of a fluorescent lipid label from a labeled erythrocyte membrane to an unlabeled erythrocyte membrane following electric-field-induced fusion, *Biophys. J.,* 47, 519, 1985.

194. **Sowers, A. E.,** Fusion events and nonfusion contents mixing events induced in erythrocyte ghosts by an electric pulse, *Biophys. J.,* 54, 619, 1988.

195. **Chernomordik, L. V., Melikyan, G. B. and Chizmadzhev, Y. A.,** Biomembrane fusion: a new concept derived from model studies using two interacting planar lipid bilayers, *Biochim. Biophys. Acta,* 906, 309, 1987.

196. **Stenger, D. A. and Hui, S. W.,** Human erythrocyte electrofusion kinetics monitored by aqueous contents mixing, *Biophys. J.,* 53, 833, 1988.

197. **Scheurich, P., Zimmermann, U., Mischel, M. and Lamprecht, I.,** Membrane fusion and deformation of red blood cells by electric fields, *Z. Naturforsch.,* 35c, 1081, 1980.

198. **Jones, T. B. and Kallio, G. A.,** Dielectrophoretic levitation of spheres and shells, *J. Electrostatics,* 6, 207, 1979.

199. **Glaser, R., Pescheck, Ch., Krause, G., Schmidt, K. P. and Teuscher, L.,** Dielektrophorese als Grundlage für ein neues Verfahren zur präparativen Zelltrennung, *A. Allg. Mikrobiol.,* 19, 601, 1979.

200. **Jones, T. B.,** Cusped electrostatic fields for dielectrophoretic levitation, *J. Electrostatics,* 11, 85, 1981.

201. **Krause, G., Schade, W., Glaser, R. and Gräger, B.,** Anwendung der Dielektrophorese für die präparative Trennung thermotoleranter Hefezellen, *Z. Allg. Mikrobiol.,* 22, 175, 1982.

202. **Lin, I. J. and Benguigui, L.,** High-intensity, high-gradient electric separation and dielectric filtration of particulate and granular materials, *J. Electrostat.,* 13, 257, 1982.

203. **Jones, T. B. and Loomans, L. W.,** Size effect in dielectrophoretic levitation, *J. Electrostatics,* 14, 269, 1983.

204. **Shalom, A. L. and Lin, I. J.,** Principles of high-gradient dielectrophoresis with a rod matrix and determination of particle trajectories, *J. Electrostatics,* 18, 39, 1986.

205. **Stoicheva, N. and Dimitrov, D. S.,** Frequency effects in protoplast dielectrophoresis, *Electrophoresis,* 7, 339, 1986.

206. **Price, J. A. R., Burt, J. P. H. and Pethig, R.,** Applications of a new optical technique for measuring the dielectrophoretic behavior of micro-organisms, *Biochim. Biophys. Acta,* 964, 221, 1988.

207. **Burt, J. P. H., Al-Ameen, T. A. K. and Pethig, R.,** An optical dielectrophoresis spectrometer for low-frequency measurements on colloidal suspension, *J. Phys. E. Sci. Instrum.,* 22, 952, 1989.

208. **Tombs, T. N. and Jones, T. B.,** Digital dielectrophoretic levitation, *Rev. Sci. Instrum.,* 62, 1072, 1991.

209. **Förster, E. and Emeis, C. C.,** Quantitative studies on the viability of yeast protoplasts following dielectrophoresis, *FEMS Microbiol. Lett.,* 26, 65, 1985.

210. **Vienken, J. and Zimmermann, U.,** Electric field-induced fusion: electro-hydraulic procedure for production of heterokaryon cells in high yield, *FEBS Lett.,* 137, 11, 1982.

211. **Mehrle, W., Hampp, R. and Zimmermann, U.,** Electric pulse induced membrane permeabilisation. Spatial orientation and kinetics of solute efflux in freely suspended and dielectrophoretically aligned plant mesophyll protoplasts, *Biochim. Biophys. Acta,* 978, 267, 1989.

212. **Schnettler, R., Gessner, P., Zimmermann, U., Neil, G. A., Urnovitz, H. B. and Sammons, D. W.,** Increased efficiency of mammalian somatic cell hybrid production under microgravity conditions during ballistic rocket flight, *Appl. Microgravity Tech. (Microgravity sci. technol.),* II 1, 3, 1989.

213. **Hampp, R., Steingraber, M., Mehrle, W. and Zimmermann, U.,** Electric-field-induced fusion of evacuolated mesophyll protoplasts of oat, *Naturwissenschaften,* 72, 91, 1985.
214. **Naton, B., Mehrle, W., Hampp, R. and Zimmermann, U.,** Mass electrofusion and mass selection of functional hybrids from vacuolate × evacuolate protoplasts, *Plant Cell Rep.,* 5, 419, 1986.
215. **Zimmermann, U., Gessner, P., Schnettler, R., Perkins, S. and Foung, S. K. H.,** Efficient hybridization of mouse-human cell lines by means of hypo-osmolar electrofusion, *J. Immun. Meth.,* 134, 43, 1990.
216. **Zimmermann, U., Büchner, K.-H. and Schnettler, R.,** Elektrofusion von Hefeprotoplasten, in TEXUS 11/12, *Final Campain Report,* Bundesministerium für Forschung und Technologie, 1985, 39.
217. **Mehrle, W., Naton, B., Hampp, R. and Zimmermann, U.,** Electrofusion of plant protoplasts under µg conditions, in *Biological Sciences in Space,* Watanabe, S., Mitarai, G. and Mori, S., Eds., International Symposium on Biological Sciences in Space, Nagoja, Japan, MYU Research, Tokyo, 1987, 377.
218. **Zimmermann, U., Schnettler, R. and Hannig, K.,** Biotechnology in space: Potentials and perspectives, in *Biotechnology,* Vol. 6b, Rehm, H.-J. and Reed, G., Eds., Verlag Chemie Weinheim, 1988, 637.
219. **Mehrle, W., Hampp, R., Naton, B. and Grothe, D.,** Effects of microgravitation on electrofusion of plant cell protoplasts, *Plant Physiol.,* 89, 1172, 1989.
220. **Hampp, R., Naton, B., Hoffmann, E. and Mehrle, W.,** Suspensions of plant cells in microgravity, *Microgravity Sci. Technol.,* III, 3, 168, 1990.
221. **Hampp, R., Naton, B., Hoffmann, E., Mehrle, W., Schönherr, K. and Hemmersbach-Krause, R.,** Hybrid formation and metabolism of plant cell protoplasts under microgravity, *The Physiologist,* 35, 27, 1992.
222. **Baumann, T., Kreis, W., Mehrle, W., Hampp, R. and Reinhard, E.,** Regeneration and characterization of protoplast-derived cell lines from *Digitalis lanata* EHRH and *Digitalis purpurea* L. suspension cultures after electrofusion under microgravity conditions, *Proceedings of the Fourth European Symposium on Life Sciences Research in Space,* held in Trieste, Italy, ESA SP-307, 405, 1990.
223. **Klöck, G. and Zimmermann, U.,** Facilitated electrofusion of vacuolated × evacuolated oat mesophyll protoplasts in hypo-osmolar media after alignment with an alternating field of modulated strength, *Biochim. Biophys. Acta,* 1025, 87, 1990.
224. **Pfister, K., Klöck, G. and Zimmermann, U.,** Selection of hybrid plants obtained by electrofusion of vacuolated × evacuolated plant protoplasts in hypo-osmolar solution, *Biochim. Biophys. Acta,* 1062, 13, 1991.
225. **Herzog, R., Müller-Wellensiek, A. and Voelter, W.,** Usefulness of Ficoll in electric field-mediated cell fusion, *Life Sci.,* 39, 2279, 1986.
226. **Ludloff, K.,** Untersuchungen über den Galvanotropismus, *Pflügers Arch.,* 59, 525, 1895.
227. **Verworn, M.,** Untersuchungen über die polare Erregung der lebendigen Substanz durch den constanten Strom, *Pflügers Arch.,* 62, 415, 1896.
228. **Füredi, A. A. and Ohad, I.,** Effects of high-frequency electric fields on the living cell. I. Behavior of human erythrocytes in high-frequency electric fields and its relation to their age, *Biochim. Biophys. Acta,* 79, 1, 1964.
229. **Heller, J. H. and Teixeira-Pinto, A. A.,** A new physical method of creating chromosomal aberrations, *Nature,* 183, 905, 1959.
230. **Teixeira-Pinto, A. A., Nejelski, L. L., Jr., Cutler, J. L. and Heller, J. H.,** The behavior of unicellular organisms in an electromagnetic field, *Exp. Cell Res.,* 20, 548, 1960.
231. **Füredi, A. A. and Valentine, R. C.,** Factors involved in the orientation of microscopic particles in suspensions influenced by radio frequency fields, *Biochim. Biophys. Acta,* 56, 33, 1962.
232. **Schwarz, G., Saito, M. and Schwan, H. P.,** On the orientation of nonspherical particles in an alternating electrical field, *J. Chem. Phys.,* 43, 3562, 1965.

233. **Griffin, J. L.,** Orientation of human and avian erythrocytes in radio-frequency fields, *Exp. Cell Res.,* 61, 113, 1970.
234. **Novák, B. and Bentrup, F. W.,** Orientation of *Fucus* egg polarity by electric a.c. and d.c. fields, *Biophysik,* 9, 253, 1973.
235. **Gagliano, A. G., Geacintov, N. E. and Breton, J.,** Orientation and linear dichroism of chloroplasts and sub-chloroplast fragments oriented in an electric field, *Biochim. Biophys. Acta,* 461, 460, 1977.
236. **Vienken, J., Zimmermann, U., Alonso, A. and Chapman, D.,** Orientation of sickle red blood cells in an alternating electric field, *Naturwissenschaften,* 71, 158, 1984.
237. **Schwan, H. P.,** EM-field induced force effects, in *Interactions between Electromagnetic Fields and Cells,* Chiabrera, A., Nicolini, C., and Schwan, H. P., Eds., Plenum Press, New York, 1985, 371.
238. **Mishima, K. and Morimoto, T.,** Electric field-induced orientation of myelin figures of phosphatidylcholine, *Biochim. Biophys. Acta,* 985, 351, 1989.
239. **Miller, R. D. and Jones, Th. B.,** Electro-orientation of ellipsoidal erythrocytes: Theory and experiment, *Biophys. J.,* 64, 1588, 1993.
240. **Saito, M., Schwan, H. P. and Schwarz, G.,** Response of nonspherical biological particles to alternating electric fields, *Biophys. J.,* 6, 313, 1966.
241. **Pawlowski, P. and Fikus, M.,** Bioelectrorheological model of the cell. 4. Analysis of the extensil deformation of cellular membrane in alternating electric field, *Biophys. J.,* 65, 535, 1993.
242. **Pawlowski, P., Szutowicz, I., Marszalek, P. and Fikus, M.,** Bioelectrorheological model of the cell. 5. Electrodestruction of cellular membrane in alternating electric field, *Biophys. J.,* 65, 541, 1993.
243. **Takashima, S., Chang, S. and Asakura, T.,** The effects of pulsed RF fields on the shape and volume of normal and sickled erythrocytes, in *Interactions between Electromagnetic Fields and Cells,* Chiabrera, A., Nicolini, C., and Schwan, H. P., Eds., Plenum Press, New York, 1985, 391.
244. **Takashima, S., Chang, S. and Asakura, T.,** Shape change of sickled erythrocytes induced by pulsed rf electrical fields, *Proc. Natl. Acad. Sci., U.S.A.,* 82, 6860, 1985.
245. **Friend, A. W., Finch, E. D. and Schwan, H. P.,** Low frequency electric field induced changes in the shape and motility of amoebas, *Science,* 187, 357, 1975.
246. **Burlion, N. and Northcote, D. H.,** Electrical fusion of protoplasts from *Sycamore* mutants and hybrid selection on hypoxanthine-aminopterin-thymidine (HAT) medium, *Protoplasma,* 138, 23, 1987.
247. **Hibi, T., Kano, H., Sugiura, M., Kazami, T. and Kimura, S.,** High-speed electro-fusion and electro-transfection of plant protoplasts by a continuous flow electro-manipulator, *Plant Cell Rep.,* 7, 153, 1988.
248. **Hibi, T.,** Plant protoplast fusion by electromanipulation, *Adv. Cell Cult.,* 7, 147, 1989.
249. **Mischke, S., Saunders, J. A. and Owens, L.,** A versatile low-cost apparatus for cell electrofusion and other electrophysiological treatments, *J. Biochem. Biophys. Meth.,* 13, 65, 1986.
250. **Sato, M., Noguchi, Y., Inoue, Y. and Sadakata, M.,** Production of somatic hybrid plant by means of a new electrofusion technique in a convergent electric field, *Kagaku Kogaku Ronbunshu,* 15, 1126, 1989.
251. **Song, L., Ahkong, Q. F., Georgescauld, D. and Lucy, J. A.,** Membrane fusion without cytoplasmic fusion (hemi-fusion) in erythrocytes that are subjected to electrical breakdown, *Biochim. Biophys. Acta,* 1065, 54, 1991.
252. **Nea, L. J., Bates, G. W. and Gilmer, P. J.,** Facilitation of electrofusion of plant protoplasts by membrane-active agents, *Biochim. Biophys. Acta,* 897, 293, 1987.
253. **Zheng, Q. and Zhao, N.-M.,** Electrofusion of IBRS2 cells and the study of their fusion process, *Science in China (Ser. B),* 32, 303, 1989.
254. **Ahkong, Q. F. and Lucy, J. A.,** Osmotic forces in artificially induced cell fusion, *Biochim. Biophys. Acta,* 858, 206, 1986.

255. **Neumann, V., Siegemund, F. and Baeßler, B.,** Electrically induced fusion of different human cell types, *J. Neurol.,* 233, 153, 1986.

256. **Cheong, H. T., Taniguchi, T., Takahashi, Y. and Kanagawa, H.,** Influences of electrode geometry and field strength on the fusion and subsequent development of mouse two-cell embryos, *Animal Reprod. Sci.,* 29, 307, 1992.

257. **Rákosy-Tican, L., Lucaciu, C. M., Turcu, I., Cachita-Cosma, D., Varga, P., Stana, D. and Morariu, V. V.,** Viability of wheat mesophyll protoplasts and homokaryons in response to electrofusion parameters, *Rev. Roum. Biol. Ser. Biol. Veg.,* 33, 121, 1988.

258. **Sowers, A. E. and Lieber, M. R.,** Electropore diameters, lifetimes, numbers, and locations in individual erythrocyte ghosts, *FEBS Lett.,* 205, 179, 1986.

259. **Kaufman, M. H. and Webb, S.,** Postimplantation development of tetraploid mouse embryos produced by electrofusion, *Development,* 110, 1121, 1990.

260. **Hansen, S., Koop, H.-U. and Abel, W. O.,** Electrofusion of two selected single moss protoplasts, *Mitt. Inst. Allg. Bot. Hamburg.,* 22, 29, 1988.

261. **Mejía, A., Spangenberg, G., Koop, H.-U. and Bopp, M.,** Microculture and electrofusion of defined protoplasts of the moss *Funaria hygrometrica, Bot. Acta,* 101, 166, 1988.

262. **Koop, H.-U. and Schweiger, H.-G.,** Regeneration of plants after electrofusion of selected pairs of protoplasts, *Eur. J. Cell Biol.,* 39, 46, 1985.

263. **Eigel, L., Oelmüller, R. and Koop, H.-U.,** Transfer of defined numbers of chloroplasts into albino protoplasts using an improved subprotoplast/protoplast microfusion procedure: transfer of only two chloroplasts leads to variegated progeny, *Mol. Gen. Genet.,* 227, 446, 1991.

264. **Kirichenko, I. V. and Gleba, Y. Y.,** Individual cultivation of preselected plant protoplasts and of their fusion products in microdrops, *Biopolim. Kletka,* 7, 6, 1991.

265. **Eisenschlos, C. D., Paladini, A. A., Molina, Y., Vedia, L., Torres, H. N. and Flawia, M. M.,** Evidence for the existence of an N_S-type regulatory protein in *Trypanosoma cruzi* membranes, *Biochem. J.,* 237, 913, 1986.

266. **Puite, K. J., van Wikselaar, P. and Verhoeven, H.,** Electrofusion, a simple and reproducible technique in somatic hybridization of *Nicotiana plumbaginifolia* mutants, *Plant Cell Rep.,* 4, 274, 1985.

267. **Damiani, F., Pezzotti, M. and Arcioni, S.,** Electric field mediated fusion of protoplasts of *Medicago sativa* L. and *Medicago arborea* L., *J. Plant Physiol.,* 132, 474, 1988.

268. **Sihachakr, D., Haicour, R., Serraf, I., Barrientos, E., Herbreteau, C., Ducreux, G., Rossignol, L. and Souvannavong, V.,** Electrofusion for the production of somatic hybrid plants of *Solanum melongena* L. and *Solanum khasianum* C.B. Clark, *Plant Sci.,* 57, 215, 1988.

269. **Sihachakr, D., Haicour, R., Chaput, M.-H., Barrientos, E., Ducreux, G. and Rossignol, L.,** Somatic hybrid plants produced by electrofusion between *Solanum melongena* L. and *Solanum torvum* Sw., *Theor. Appl. Genet.,* 77, 1, 1989.

270. **Möllers, C. and Wenzel, G.,** Somatic hybridization of dihaploid potato protoplasts as a tool for potato breeding, *Bot. Acta,* 105, 133, 1992.

271. **Zhang, L., Fiedler, U. and Berg, H.,** Modification of electrofusion of barley protoplast by membrane-active agents, *Bioelectrochem. Bioenerg.,* 26, 87, 1991.

272. **Katenkamp, U., Jacob, H.-E., Kerns, G. and Dalchow, E.,** Hybridization of *Trichoderma reesei* protoplasts by electrofusion, *Bioelectrochem. Bioenerg.,* 22, 57, 1989.

273. **Zimmermann, U., Vienken, J. and Greyson, J.,** Electrofusion: a novel hybridization technique, *In Biotech 84 Europe,* 1, 231, 1984.

274. **Richter, H.-P., Scheurich, P. and Zimmermann, U.,** Electric field-induced fusion of sea urchin eggs, *Develop., Growth and Differ.,* 23, 479, 1981.

275. **Kuta, A. E., Rhine, R. S. and Heebner, G. M.,** Electrofusion: a new tool for biotechnology, *Am. Biotechnol. Lab.,* 3, 31, 1985.

276. **Zimmermann, U., Klöck, G., Gessner, P., Sammons, D. W. and Neil, G. A.,** Microscale production of hybridomas by hypo-osmolar electrofusion, *Hum. Antibod. Hybridomas,* 3, 14, 1992.

277. **Tempelaar, M. J. and Jones, M. G. K.,** Directed electrofusion between protoplasts with different responses in a mass fusion system, *Plant Cell Rep.,* 4, 92, 1985.

278. **Al-Atabee, J. S., Mulligan, B. J. and Power, J. B.,** Interspecific somatic hybrids of *Rudbeckia hirta* and *R. laciniata* (Compositae), *Plant Cell Rep.,* 8, 517, 1990.

279. **Zachrisson, A. and Bornman, C. H.,** Application of electric field fusion in plant tissue culture, *Physiol. Plant.,* 61, 314, 1984.

280. **Karsten, U., Stolley, P., Walther, I., Papsdorf, G., Weber, S., Conrad, K., Pasternak, L. and Kopp, J.,** Direct comparison of electric field-mediated and PEG-mediated cell fusion for the generation of antibody producing hybridomas, *Hybridoma,* 7, 627, 1988.

281. **van Duijn, G., Langedijk, J. P. M., de Boer, M. and Tager, J. M.,** High yields of specific hybridomas obtained by electrofusion of murine lymphocytes immunized *in vivo* or *in vitro, Exp. Cell Res.,* 183, 463, 1989.

282. **Karsten, U., Stolley, P. and Seidel, B.,** Comparison of PEG-induced and electric field-mediated cell fusion in the generation of monoclonal antibodies against a variety of soluble and cellular antigens, in *Guide to Electroporation and Electrofusion,* Chang, D. C., Chassy, B. M., Saunders, J. A., and Sowers, A. E., Eds., Academic Press, San Diego, 1992, 363.

283. **Sowers, A. E.,** Fusion of mitochondrial inner membranes by electric fields produces inside-out vesicles. Visualization by freeze-fracture electron microscopy, *Biochim. Biophys. Acta,* 735, 426, 1983.

284. **Güveli, D. E.,** Fusion of mitochondria, *J. Biochem.,* 102, 1329, 1987.

285. **Reynaud, J. A., Labbe, H., Lequoc, K., Lequoc, D. and Nicolau, C.,** Electric field-induced fusion of mitochondria, *FEBS Lett.,* 247, 106, 1989.

286. **Buckley, D. J., Lefebvre, M., Meijer, E. G. M. and Brown, D. C. W.,** A signal generator for electrofusion of plant protoplasts, *Comput. Electron. Agric.,* 5, 179, 1990.

287. **Lackney, V. K., Spanswick, R. M., Hirasuna, T. J. and Shuler, M. L.,** PEG-enhanced, electric field-induced fusion of tonoplast and plasmalemma of grape protoplasts, *Plant Cell, Tissue and Organ Culture,* 23, 107, 1990.

288. **Ruzin, S. E. and McCarthy, S. C.,** The effect of chemical facilitators on the frequency of electrofusion of tobacco mesophyll protoplast, *Plant Cell Rep.,* 5, 342, 1986.

289. **Tempelaar, M. J., Duyst, A., de Vlas, S. Y., Krol, G., Symmonds, C. and Jones, M. G. K.,** Modulation and direction of the electrofusion response in plant protoplasts, *Plant Sci.,* 48, 99, 1987.

290. **Christov, A. M. and Vaklinova, S. G.,** Effects of prostaglandins E_2 and $F_{2\alpha}$ on the electrofusion of pea mesophyll protoplasts, *Plant Physiol.,* 83, 500, 1987.

291. **Nea, L. J. and Bates, G. W.,** Factors affecting protoplast electrofusion efficiency, *Plant Cell Rep.,* 6, 337, 1987.

292. **Tsuzuki, K. and Sugiyama, H.,** Studies on the electrofusion of the dissociated cells obtained from planarian tissues II, *Bull. Nippon Vet. Anim. Sci. Univ.,* 41, 97, 1992.

293. **Pancheva, R., Naplatarova, M., Antonov, P. and Ovtscharoff, W.,** Electrofusion of cells from Ehrlich ascit tumour after treatment with sodium deoxycholate, *C.R. Acad. Bulg. Sci.,* 44, 89, 1991.

294. **Walton, P. D. and Brown, D. C. W.,** Electrofusion of protoplasts and heterokaryon survival in the genus *Medicago, Plant Breeding,* 101, 137, 1988.

295. **de Vries, S. E., Jacobsen, E., Jones, M. G. K., Loonen, A. E. H. M., Tempelaar, M. J., Wijbrandi, J. and Feenstra, W. J.,** Somatic hybridization of amino acid analogue-resistant cell lines of potato (*Solanum tuberosum* L.) by electrofusion, *Theor. Appl. Genet.,* 73, 451, 1987.

296. **Glaser, A. and Donath, E.,** Influence of membrane modification on electrofusion of oat protoplasts, *Studia biophysica,* 119, 93, 1987.

297. **Zimmermann, U., Pilwat, G. and Richter, H.-P.,** Electric-field-stimulated fusion: increased field stability of cells induced by pronase, *Naturwissenschaften,* 68, 577, 1981.

298. **Spangenberg, G. and Schweiger, H.-G.,** Controlled electrofusion of different types of protoplasts including cell reconstitution in *Brassica napus* L., *Eur. J. Cell Biol.,* 42, 51, 1986.

299. **Donath, E. and Arndt, R.,** Electric-field-induced fusion of enzyme-treated human red cells: kinetics of intermembrane protein exchange, *Gen. Physiol. Biophys.,* 3, 239, 1984.

300. **Ohno-Shosaku, T. and Okada, Y.,** Facilitation of electrofusion of mouse lymphoma cells by the proteolytic action of proteases, *Biochem. Biophys. Res. Commun.,* 120, 138, 1984.

301. **Ohno-Shosaku, T., Hama-Inaba, H. and Okada, Y.,** Somatic hybridization between human and mouse lymphoblast cells produced by an electric pulse-induced fusion technique, *Cell Struct. Funct.,* 9, 193, 1984.

302. **Okada, Y., Ohno-Shosaku, T. and Oiki, S.,** Ca^{2+} is prerequisite for cell fusion induced by electric pulses, *Biomed. Res.,* 5, 511, 1984.

303. **Ohno-Shosaku, T. and Okada, Y.,** Electric pulse-induced fusion of mouse lymphoma cells: roles of divalent cations and membrane lipid domains, *J. Membrane Biol.,* 85, 269, 1985.

304. **Stenger, D. A. and Hui, S. W.,** Kinetics of ultrastructural changes during electrically induced fusion of human erythrocytes, *J. Membrane Biol.,* 93, 43, 1986.

305. **Chang, D. C., Hunt, J. R., Zheng, Q. and Gao, P.-Q.,** Electroporation and electrofusion using a pulsed radio-frequency electric field, in *Guide to Electroporation and Electrofusion,* Chang, D. C., Chassy, B. M., Saunders, J. A., and Sowers, A. E., Eds., Academic Press, San Diego, 1992, 303.

306. **Mally, M. I., Mc Knight, M. E. and Glassy, M. C.,** Protocols of electroporation and electrofusion for producing human hybridomas, in *Guide to Electroporation and Electrofusion,* Chang, D. C., Chassy, B. M., Saunders, J. A., and Sowers, A. E., Eds., Academic Press, San Diego, 1992, 507.

307. **Salhani, N., Schnabl, H., Küppers, G. and Zimmermann, U.,** The hydraulic conductivity as a criterion for the membrane integrity of protoplasts fused by an electric field pulse, *Planta,* 155, 140, 1982.

308. **Avram, D., Petcu, I., Radu, M., Dan, F. and Stan, R.,** Electrically induced protoplast fusion for ergosterol-producing yeast strain improvement, *J. Basic Microbiol.,* 32, 369, 1992.

309. **Mizukami, Y., Okauchi, M. and Kito, H.,** Effects of cell wall-lytic enzymes on the electrofusion efficiency of protoplasts from *Porphyra yezoensis, Aquaculture,* 108, 193, 1992.

310. **Reddy, C. R. K., Iima, M. and Fujita, Y.,** Induction of fast-growing and morphologically different strains through intergeneric protoplast fusion of *Ulva* and *Enteromorpha* (Ulvales, Chlorophyta), *J. Appl. Phycol.,* 4, 57, 1992.

311. **Changsheng, C.,** Electrofusion of protoplasts from *Porphyra haitanensis* and *P. yezoensis* Thalli (Rhodophyta), *Chin. J. Biotechnol.,* 8, 33, 1992.

312. **Mattioli, M., Galeati, G., Bacci, M. L. and Barboni, B.,** Changes in maturation-promoting activity in the cytoplasm of pig oocytes throughout maturation, *Mol. Reprod. Dev.,* 30, 119, 1991.

313. **Lynch, P. T., Isaac, S. and Collin, H. A.,** Electrofusion of protoplasts from celery (*Apium graveolens* L.) with protoplasts from the filamentous fungus *Aspergillus nidulans, Planta,* 178, 207, 1989.

314. **Donath, E. and Gingell, D.,** A sharp cell surface conformational transition at low ionic strength changes the nature of the adhesion of enzyme-treated red blood cells to a hydrocarbon interface, *J. Cell Sci.,* 63, 113, 1983.

315. **Brown, S. M., Ahkong, Q. F., Sage, A. D. and Lucy, J. A.,** Electrically induced cell fusion in the production of monoclonal antibodies, *Biochem. Soc. Trans.,* 14, 298, 1986.

316. **Vienken, J., Ganser, R., Hampp, R. and Zimmermann, U.,** Electric field-induced fusion of isolated vacuoles and protoplasts of different developmental and metabolic provenience, *Physiol. Plant.,* 53, 64, 1981.

317. **Büschl, R., Ringsdorf, H. and Zimmermann, U.,** Electric field-induced fusion of large liposomes from natural and polymerizable lipids, *FEBS Lett.,* 150, 38, 1982.

318. **Schnettler, R. and Zimmermann, U.,** Influence of the composition of the fusion medium on the yield of electrofused yeast hybrids, *FEMS Microbiol. Lett.,* 27, 195, 1985.

319. **Vienken, J. and Zimmermann, U.,** An improved electrofusion technique for production of mouse hybridoma cells, *FEBS Lett.,* 182, 278, 1985.

320. **Tsoneva, I., Tomov, T., Panova, I. and Doncheva, J.,** Electrofusion in the presence of divalent cations, *Bioelectrochem. Bioenerg.,* 27, 37, 1992.

321. **Yoshida, N., Radbruch, A. and Rajewsky, K.,** Ig gene rearrangement and expression in the progeny of B-cell progenitors in the course of clonal expansion in bone marrow cultures, *The EMBO J.,* 6, 2735, 1987.

322. **Noda, K., Togawa, Y. and Yamada, Y.,** Quantification of physical and cyto-physiological conditions for the electrofusion of *Saccharomyces cerevisiae, Agric. Biol. Chem.,* 54, 2023, 1990.

323. **Chand, P. K., Davey, M. R., Power, J. B. and Cocking, E. C.,** Amphidiploid somatic hybrids between *Solanum viarum* and *S. dulcamara* by micromanipulation of heterokaryons following protoplast electrofusion, *The Nucleus,* 34, 53, 1991.

324. **Carafoli, E. and Semenza, G.,** *Membrane Biochemistry,* Springer Verlag, Berlin, 1979.

325. **Glaser, A. and Donath, E.,** Extracellular multivalent cations are necessary for plasma membrane expansion in oat protoplasts, *Studia biophysica,* 131, 35, 1989.

326. **Abe, S. and Takeda, J.,** Possible involvement of calmodulin and the cytoskeleton in electrofusion of plant protoplasts, *Plant Physiol.,* 81, 1151, 1986.

327. **Onuma, E. K. and Hui, S.-W.,** The effects of calcium on electric field-induced cell shape changes and preferential orientation, *Ionic Curr. Devel.,* 319, 1986.

328. **Schliwa, M.,** Action of cytochalasin D on cytoskeletal networks, *J. Cell Biol.,* 92, 79, 1982.

329. **Abe, S. and Takeda, J.,** Promotion by calcium ions and cytochalasin D of the rounding up process (spherulation) of electrofused barley protoplasts and its relation to the cytoskeleton, *J. Exp. Bot.,* 40, 819, 1989.

330. **Steinhardt, R. A., Bi, G. and Alderton, J. M.,** Cell membrane resealing by a vesicular mechanism similar to neurotransmitter release, *Science,* 263, 390, 1994.

331. **Urano, N., Kamimura, M. and Washizu, M.,** Physical parameters affecting electrofusion of yeasts: Zeta-Potential on the surface of yeast protoplasts and osmotic pressure of the solution, *J. Biotechnol.,* 18, 213, 1991.

332. **Song, L. Y., Baldwin, J. M., O'Reilly, R. and Lucy, J. A.,** Relationships between the surface exposure of acidic phospholipids and cell fusion in erythrocytes subjected to electrical breakdown, *Biochim. Biophys. Acta,* 1104, 1, 1992.

333. **Blangero, C. and Teissié, J.,** Ionic modulation of electrically induced fusion of mammalian cells, *J. Membrane Biol.,* 86, 247, 1985.

334. **Zimmermann, U., Beckers, F. and Coster, H. G. L.,** The effect of pressure on the electrical breakdown in the membranes of *Valonia utricularis, Biochim. Biophys. Acta,* 464, 399, 1977.

335. **Hülsheger, H. and Niemann, E.-G.,** Lethal effects of high-voltage pulses on *E. coli* K12, *Radiat. Environ. Biophys.,* 18, 281, 1980.

336. **Hülsheger, H., Potel, J. and Niemann, E.-G.,** Killing of bacteria with electric pulses of high field strength, *Radiat. Environ. Biophys.,* 20, 53, 1981.

337. **Biedinger, U., Youngman, R. J. and Schnabl, H.,** Differential effects of electrofusion and electropermeabilization parameters on the membrane integrity of plant protoplasts, *Planta,* 180, 598, 1990.

338. **Biedinger, U. and Schnabl, H.,** Ethane production as an indicator of the regeneration potential of electrically fused sunflower protoplasts (*Helianthus annuus* L.), *J. Plant Physiol.,* 138, 417, 1991.

339. **Schnettler, R. and Zimmermann, U.,** Zinc ions stimulate electrofusion of *Hansenula polymorpha* protoplasts, *FEMS Microbiol. Lett.,* 106, 47, 1993.

340. **Pasternak, C. A.,** A novel role of Ca^{2+} and Zn^{2+}: protection of cells against membrane damage, *Biosci. Rep.,* 8, 579, 1988.

341. **Chvapil, M.,** New aspects in the biological role of zinc: A stabilizer of macromolecules and biological membranes, *Life Sci.,* 13, 1041, 1973.

342. **Rygol, J., Arnold, W. M. and Zimmermann, U.,** Zinc and salinity effects on membrane transport in *Chara connivens, Plant Cell Environ.,* 15, 11, 1992.

343. **Abe, S. and Takeda, J.,** Effects of La^{3+} on surface charges, dielectrophoresis, and electrofusion of barley protoplasts, *Plant Physiol.,* 87, 389, 1988.

344. **Sowers, A. E.,** Electrofusion of dissimilar membrane fusion partners depends on additive contributions from each of the two different membranes, *Biochim. Biophys. Acta,* 985, 339, 1989.

345. **Chang, D. C., Hunt, J. R. and Gao, P.-Q.,** Effects of pH on cell fusion induced by electric fields, *Cell Biophysics,* 14, 231, 1989.

346. **Brown, S. M., Ahkong, Q. F. and Lucy, J. A.,** Osmotic pressure and the electrofusion of myeloma cells, *Biochem. Soc. Trans.,* 14, 1129, 1986.

347. **Schmitt, J. J., Zimmermann, U. and Gessner, P.,** Electrofusion of osmotically treated cells: high and reproducible yields of hybridoma cells, *Naturwissenschaften,* 76, 122, 1989.

348. **Schmitt, J. J. and Zimmermann, U.,** Enhanced hybridoma production by electrofusion in strongly hypo-osmolar solutions, *Biochim. Biophys. Acta,* 983, 42, 1989.

349. **Maseehur Rehman, S. M., Perkins, S., Zimmermann, U. and Foung, S. K. H.,** Human hybridoma formation by hypo-osmolar electrofusion, in *Guide to Electroporation and Electrofusion,* Chang, D. C., Chassy, B., Saunders, J. A., and Sowers, A. E., Eds., Academic Press, San Diego, 1992, 523.

350. **Sun, F. Z. and Moor, R. M.,** Factors controlling the electrofusion of murine embryonic cells, *Bioelectrochem. Bioenerg.,* 21, 149, 1989.

351. **Rols, M.-P. and Teissié, J.,** Modulation of electrically induced permeabilization and fusion of Chinese hamster ovary cells by osmotic pressure, *Biochemistry,* 29, 4561, 1990.

352. **Stenger, D. A., Kubiniec, R. T., Purucker, W. J., Liang, H. and Hui, S. W.,** Optimization of electrofusion parameters for efficient production of murine hybridomas, *Hybridoma,* 7, 505, 1988.

353. **Morikawa, H., Asada, M. and Yamada, Y.,** Effects of physical parameters upon protoplast electrofusion, *Plant Cell Physiol.,* 29, 659, 1988.

354. **Foung, S. K. H., Perkins, S., Kafadar, K., Gessner, P. and Zimmermann, U.,** Development of microfusion techniques to generate human hybridomas, *J. Immunol. Meth.,* 134, 35, 1990.

355. **Perkins, S., Zimmermann, U. and Foung, S. K. H.,** Parameters to enhance human hybridoma formation with hypoosmolar electrofusion, *Hum. Antib. Hybridomas,* 2, 155, 1991.

356. **Klöck, G., Wisnewski, A. V., El-Bassiouni, E. A., Ramadan, M. I., Gessner, P., Zimmermann, U. and Kresina, T. F.,** Human hybridoma generation by hypo-osmolar electrofusion: characterization of human monoclonal antibodies to *Schistosoma mansoni* parasite antigens, *Hybridoma,* 11, 469, 1992.

357. **Vollmers, H. P., v. Landenberg, P., Dämmrich, J., Stulle, K., Wozniak, E., Ringdörfer, C., Müller-Hermelink, H. K., Herrmann, B. and Zimmermann, U.,** Electro-immortalization of B lymphocytes isolated from stomach carcinoma biopsy material, *Hybridoma,* 12, 221, 1993.

358. **v. Landenberg, P., Dämmrich, J., Stulle, K., Wozniak, E., Ringdörfer, C., Herrmann, B., Müller-Hermelink, H. K., Zimmermann, U. and Vollmers, H. P.,** Immortalization of stomach carcinoma infiltrating B-lymphocytes from biopsy material by electrofusion, *Oncol. (Life Sci. Adv.),* 12, 35, 1993.

359. **Schmitt, J. J., Zimmermann, U. and Neil, G. A.,** Efficient generation of stable antibody forming hybridoma cells by electrofusion, *Hybridoma,* 8, 107, 1989.

360. **Kowalski, M., Hannig, K., Klöck, G., Gessner, P., Zimmermann, U., Neil, G. A. and Sammons, D. W.,** Electrofused mammalian cells analyzed by free-flow electrophoresis, *BioTechniques,* 9, 332, 1990.

361. **van der Steege, G. and Tempelaar, M. J.,** Comparison of electric field mediated DNA-uptake and fusion properties of protoplasts of *Solanum tuberosum, Plant Sci.,* 69, 103, 1990.

362. **Pelc, R.,** The electric field strength distribution in sample chambers commonly used in electrofusion of cells, *Bioelectrochem. Bioeng.,* 26, 205, 1991.

363. **Grosse, C. and Schwan, H. P.,** Cellular membrane potentials induced by alternating fields, *Biophys. J.,* 63, 1632, 1992.

364. **Mehrle, W., Naton, B. and Hampp, R.,** Determination of physical membrane properties of plant cell protoplasts via the electrofusion technique: prediction of optimal fusion yields and protoplast viability, *Plant Cell Rep.,* 8, 687, 1990.

365. **Zimmermann, U. and Scheurich, P.,** Fusion of *Avena sativa* mesophyll cell protoplasts by electrical breakdown, *Biochim. Biophys. Acta,* 641, 160, 1981.

366. **Morikawa, H., Hayashi, Y., Hirabayashi, Y., Asada, M. and Yamada, Y.,** Cellular and vacuolar fusion of protoplasts electrofused using platinum microelectrodes, *Plant Cell Physiol.,* 29, 189, 1988.

367. **Stewart-Savage, J. and Grey, R. D.,** The cell cycle governs the onset of spherulation of *Xenopus* eggs fused by an electric field, *Develop. Growth and Differ.,* 29, 229, 1987.

368. **Mitani, T., Utsumi, K. and Iritani, A.,** Developmental ability of enucleated bovine oocytes matured *in vitro* after fusion with single blastomeres of eight-cell embryos matured and fertilized *in vitro, Mol. Reprod. Dev.,* 34, 314, 1993.

369. **Herzog, R., Müller-Wellensiek, A. and Voelter, W.,** Fusion tierischer Tumorzellen im elektrischen Feld, *Chemiker-Zeitung,* 110, 11, 1986.

370. **Komari, T., Saito, Y., Nkaido, F. and Kumashiro, T.,** Efficient selection of somatic hybrids in *Nicotiana tabacum* L. using a combination of drug-resistance markers introduced by transformation, *Theor. Appl. Genet.,* 77, 547, 1989.

371. **Kishinami, I. and Widholm, J. M.,** Auxotrophic complementation in intergeneric hybrid cells obtained by electrical and dextran-induced protoplast fusion, *Plant Cell Physiol.,* 28, 211, 1987.

372. **Lucy, J. A. and Ahkong, Q. F.,** An osmotic model for fusion of biological membranes, *FEBS Lett.,* 199, 1, 1986.

373. **Lucy, J. A. and Ahkong, Q. F.,** Osmotic forces and the fusion of biomembranes, in *Molecular Mechanisms of Membrane Fusion,* Ohki, S., Doyle, D., Flanagan, T. D., Hui, S. W., and Mayhew, E., Eds., Plenum Press, New York, 1988, 163.

374. **Dimitrov, D. S. and Jain, R. K.,** Membrane stability, *Biochim. Biophys. Acta,* 779, 437, 1984.

375. **Kuzmin, P. I., Pastushenko, V. F., Abidor, I. G., Sukharev, S. I., Barbul, A. I. and Chizmadzhev, Yu. A.,** Theoretical analysis of a cell electrofusion mechanism, *Biologicheskie Membrany,* 5, 600, 1988.

376. **Zhelev, D. V., Dimitrov, D. S. and Doinov, P.,** Correlation between physical parameters in electrofusion and electroporation of protoplasts, *Bioelectrochem. Bioenerg.,* 20, 155, 1988.

377. **Abidor, I. G. and Sowers, A. E.,** Kinetics and mechanism of cell membrane electrofusion, *Biophys. J.,* 61, 1557, 1992.

378. **Chernomordik, L. V. and Sowers, A. E.,** Evidence that the spectrin network and a nonosmotic force control the fusion product morphology in electrofused erythrocyte ghosts, *Biophys. J.,* 60, 1026, 1991.

379. **Phansiri, S., Taniguchi, T. and Maeda, E.,** Morphological study of soybean protoplasts in early electrofusion process, *Jpn. J. Crop Sci.,* 60, 550, 1991.

380. **Baldwin, J. M., O'Reilly, R., Whitney, M. and Lucy, J. A.,** Surface exposure of phosphatidylserine is associated with the swelling and osmotically-induced fusion of human erythrocytes in the presence of Ca^{2+}, *Biochim. Biophys. Acta,* 1028, 14, 1990.

381. **Melikyan, G. B., Abidor, I. G., Chernomordik, L. V. and Chailakhyan, L. M.,** Electrostimulated fusion and fission of bilayer lipid membranes, *Biochim. Biophys. Acta,* 730, 395, 1983.

382. **Wilhelm, C., Winterhalter, M., Zimmermann, U. and Benz, R.,** Kinetics of pore size during irreversible electrical breakdown of lipid bilayer membranes, *Biophys. J.,* 64, 121, 1993.

383. **Dimitrov, D. S. and Sowers, A. E.,** A delay in membrane fusion: lag times observed by fluorescence microscopy of individual fusion events induced by an electric field pulse, *Biochemistry,* 29, 8337, 1990.

384. **Lucy, J. A.,** Salient features of artificially induced cell fusion, *Biochem. Soc. Trans.,* 14, 250, 1986.

385. **Cheng-Zhi, J.,** Dynamics of membrane skeleton in fused red cell membranes, *J. Cell Sci.,* 90, 93, 1988.

386. **Zheng, Q. and Chang, D. C.,** Reorganization of cytoplasmic structures during cell fusion, *J. Cell Sci.,* 100, 431, 1991.

387. **Wu, Y., Sjodin, R. A. and Sowers, A. E.,** Distinct mechanical relaxation components in pairs of erythrocyte ghosts undergoing fusion, *Biophys. J.,* 66, 114, 1994.

388. **Zheng, Q. and Chang, D. C.,** Dynamic changes of microtubule and actin structures in CV-1 cells during electrofusion, *Cell Motility and the Cytoskeleton,* 17, 345, 1990.

389. **Verhoek-Köhler, B., Hampp, R., Ziegler, H. and Zimmermann, U.,** Electro-fusion of mesophyll protoplasts of *Avena sativa, Planta,* 158, 199, 1983.

390. **Wu, Y., Montes, J. G. and Sjodin, R. A.,** Determination of electric field threshold for electrofusion of erythrocyte ghosts: comparison of pulse-first and contact-first protocols, *Biophys. J.,* 61, 810, 1992.

391. **Sowers, A. E.,** The long-lived fusogenic state induced in erythrocyte ghosts by electric pulses is not lateral mobile, *Biophys. J.,* 52, 1015, 1987.

392. **Day, R. N. and Hinkle, P. M.,** Transient dopaminergic inhibition of prolactin release from hybrid cells derived by fusion of normal rat pituitary and GH_4C_1 tumor cells, *Endocrinology,* 122, 2165, 1988.

393. **Podestá, E. J., Solano, A. R., Molina Y Vedia, L., Paladini, A., Sánchez, M. L. and Torres, H. N.,** Production of steroid hormone and cyclic AMP in hybrids of adrenal and Leydig cells generated by electrofusion, *Eur. J. Biochem.,* 145, 329, 1984.

394. **Osbourn, J. K., Plaskitt, K. A., Watts, J. W. and Wilson, T. M. A.,** Tobacco mosaic virus coat protein and reporter gene transcripts containing the TMV origin-of-assembly sequence do not interact in double-transgenic tobacco plants: Implications for coat protein-mediated protection, *Mol. Plant-Microbe Interact.,* 2, 340, 1989.

395. **Kwon, O. C., Chung, C. H., Sato, T., Taniguchi, T. and Maeda, E.,** Ultrastructure of electrofused products from pansy (*Viola tricolor*) mesophyll and wild viola (*V. patrinii*) petiole callus protoplasts, *Jpn. J. Crop Sci.,* 61, 469, 1992.

396. **Johnson, R. T. and Rao, P. N.,** Mammalian cell fusion: induction of premature chromosome condensation in interphase nuclei, *Nature,* 226, 717, 1970.

397. **Brüderlein, S. and Gebhart, E.,** Double minutes in prematurely condensed chromatin of human tumor cells, *Cancer Genet. Cytogenet.,* 16, 145, 1985.

398. **Cervenka, J. and Camargo M.,** Premature chromosome condensation induced by electrofusion, *Cytogenet. Cell Genet.,* 45, 169, 1987.

399. **Bertsche, U. and Zimmermann, U.,** Analysis of X-ray induced aberrations in mammalian chromosomes by electrofusion induced premature chromosome condensation, *Radiat. Environ. Biophys.,* 27, 201, 1988.

400. **Schnabl, H. and Zimmermann, U.,** Immobilization of plant protoplasts, in *Biotechnology in Agriculture and Forestry, Vol. 8: Plant Protoplast and Genetic Engineering I.,* Bajaj, Y. P. S., Ed., Springer-Verlag, Berlin, 1989, 63.

401. **Geisen, K., Deutschländer, H., Gorbach, S., Klenke, C. and Zimmermann, U.,** Function of barium alginate-microencapsulated xenogenic islets in different diabetic mouse models, in *Frontiers in Diabetes Research. Lessons from Animal Diabetes III.,* Shafrir, E., Ed., Smith-Gordon, London, 1990, 142.

402. **Zekorn, T., Horcher, A., Siebers, U., Schnettler, R., Klöck, G., Hering, B., Zimmermann, U., Bretzel, R. G. and Federlin, K.,** Barium-cross-linked alginate beads: a simple, one-step method for successful immunoisolated transplantation of islets of Langerhans, *Acta Diabetol.,* 29, 99, 1992.

403. **Zekorn, T., Siebers, U., Horcher, A., Schnettler, R., Zimmermann, U., Bretzel, R. G. and Federlin, K.,** Alginate coating of islets of Langerhans: *in vitro* studies on a new method for microencapsulation for immuno-isolated transplantation, *Acta Diabetol.,* 29, 41, 1992.

404. **Zimmermann, U., Klöck, G., Federlin, K., Hannig, K., Kowalski, M., Bretzel, R. G., Horcher, A., Entenmann, H., Siebers, U. and Zekorn, T.,** Production of mitogen-contamination free alginates with variable ratios of mannuronic acid to guluronic acid by free flow electrophoresis, *Electrophoresis,* 13, 269, 1992.

405. **Klöck, G., Siebers, U., Pfeffermann, A., Schmitt, J., Houben, R., Federlin, K. and Zimmermann, U.,** Immunoisolation transplantierter Langerhansscher Inseln: Mikrokapseln mit definierter molekularer Ausschlußgrenze, *Immun. Infekt.,* 21, 183, 1993.

406. **Zimmermann, U. and Steudle, E.,** Physical aspects of water relations of plant cells, *Adv. Bot. Res.,* 6, 45, 1978.

407. **Berggren, P.-O. and Sohtell, M.,** Microelectrode studies of D-glucose- and K⁺-induced changes in membrane potential of electrofused insulin-producing cells, *FEBS Lett.,* 202, 367, 1986.

408. **Salhani, N., Vienken, J., Zimmermann, U., Ward, M., Davey, M. R., Clothier, R. H., Balls, M., Cocking, E. C. and Lucy, J. A.,** Haemoglobin synthesis and cell wall regeneration by electric field-induced interkingdom heterokaryons, *Protoplasma,* 126, 30, 1985.

409. **Bopp-Buhler, M.-L., Schnettler, R., Bethmann, B., Zimmermann, U. and Gimmler, H.,** Intra- and interspecific electrofusion of *Dunaliella* cells and yeast protoplasts, *J. Plant Physiol.,* 144, 385, 1994.

410. **Gaertig, J., Kiersnowska, M. and Iftode, F.,** Induction of cybrid strains of *Tetrahymena thermophila* by electrofusion, *J. Cell Sci.,* 89, 253, 1988.

411. **Ruthe, H.-J. and Adler, J.,** Fusion of bacterial spheroplasts by electric fields, *Biochim. Biophys. Acta,* 819, 105, 1985.

412. **Rols, M.-P., Bandiera, P., Laneelle, G. and Teissié, J.,** Obtaining of viable hybrids between *Corynebacteria* by electrofusion, *Studia biophysica,* 119, 37, 1987.

413. **Fikus, M., Grzesiuk, E., Marszalek, P., Rózycki, S. and Zielinski, J.,** Electrofusion of *Neurospora crassa* slime cells, *FEMS Microbiol. Lett.,* 27, 123, 1985.

414. **Sonnenberg, A. S. M. and Wessels, J. G. H.,** Heterokaryon formation in the basidiomycete *Schizophyllum commune* by electrofusion of protoplasts, *Theor. Appl. Genet.,* 74, 654, 1987.

415. **Fikus, M., Marszalek, P., Rózycki, S. and Zielinski, J.,** Dielectrophoresis and electrofusion of *Neurospora crassa* slime, *Studia biophysica,* 119, 73, 1987.

416. **Peberdy, J. F.,** Developments in protoplast fusion in fungi, *Microbiol. Sci.,* 4, 108, 1987.

417. **Zhang, J.-M., Zheng, Y.-P. and Chen, M.-Y.,** Study on intergenetic electrofusion of protoplasts between *Lentinus* and *Pleurotus, Acta Bot. Yunnanica,* 14, 283, 1992.

418. **Halfmann, H. J., Emeis, C. C. and Zimmermann, U.,** Electro-fusion and genetic analysis of fusion products of haploid and polyploid *Saccharomyces* yeast cells, *FEMS Microbiol. Lett.,* 20, 13, 1983.

419. **Halfmann, H. J., Emeis, C. C. and Zimmermann, U.,** Electro-fusion of haploid *Saccharomyces* yeast cells of identical mating type, *Arch. Microbiol.,* 134, 1, 1983.

420. **Förster, E. and Emeis, C. C.,** Enhanced frequency of karyogamy in electrofusion of yeast protoplasts by means of preceding G1 arrest, *FEMS Microbiol. Lett.,* 34, 69, 1986.

421. **Emeis, C. C.,** Intergeneric hybridization of yeasts by electrofusion, *Studia biophysica,* 119, 31, 1987.

422. **Broda, H. G., Schnettler, R. and Zimmermann, U.,** Parameters controlling yeast hybrid yield in electrofusion: the relevance of pre-incubation and the skewness of the size distributions of both fusion partners, *Biochim. Biophys. Acta,* 899, 25, 1987.

423. **Tsoneva, I., Doinov, P. and Dimitrov, D. S.,** Electrofusion of fragile mutants of *Saccharomyces cerevisiae, FEMS Microbiol. Lett.,* 60, 61, 1989.

424. **Tsoneva, I. C.,** Electric field-induced fusion of yeast protoplasts. The role of different physical and chemical factors, *Acta Microbiol. Bulgarica,* 24, 53, 1989.

425. **Vondrejs, V., Pavlícek, I., Kothera, M. and Palková, Z.,** Electrofusion of oriented *Schizosaccharomyces pombe* cells through apical protoplast-protuberances, *Biochem. Biophys. Res. Commun.,* 166, 113, 1990.

426. **Chiplonkar, J. M.,** Electrofusion of mosquito cells, *In Vitro Cell. Dev. Biol.,* 28A, 230, 1992.

427. **Bates, G. W. and Hasenkampf, C. A.,** Culture of plant somatic hybrids following electrical fusion, *Theor. Appl. Genet.,* 70, 227, 1985.

428. **Bates, G. W.,** Electrical fusion for optimal formation of protoplast heterokaryons in *Nicotiana, Planta,* 165, 217, 1985.

429. **Watts, J. W., Doonan, J. H., Cove, D. J. and King, J. M.,** Production of somatic hybrids of moss by electrofusion, *Mol. Gen. Genet.,* 199, 349, 1985.

430. **Kirsten, U., Jacob, H.-E., Tesche, M. and Kluge, S.,** Isolierung und Elektrofusion von Koniferenprotoplasten, *Silvae Genet.,* 35, 186, 1986.

431. **Okamura, T., Nagata, S., Misono, H. and Nagasaki, S.,** Interspecific electrofusion between protoplasts of *Streptomyces antibioticus* and *Streptomyces fradiae, Agric. Biol. Chem.,* 52, 1433, 1988.

432. **Okamura, T., Nagata, S., Misono, H. and Nagasaki, S.,** New antibiotic-producing *Streptomyces* TT-strain, generated by electrical fusion of protoplasts, *J. Ferment. Bioeng.,* 67, 221, 1989.

433. **Yakovenko, K. N., Khendogii, N. V., Sukharev, S. I., Abidor, I. G. and Troitskii, N. A.,** Fusion of *Erwinia chrysanthemi* spheroplasts in an electric field, *Mol. Genet. Mikrobiol. Virusol.,* 5, 30, 1990.

434. **Yu, B.-S. and Yuan, L. R.,** Studies of recombination of *Streptomyces mycarofaciens* protoplast electrofusion, *Chin. J. Antibiot.,* 16, 323, 1991.

435. **Han, L.,** Interspecific hybridization of *Streptomyces* by electrofusion, *Chin. J. Biotechnol.,* 7, 285, 1991.

436. **Tsuzuki, K. and Murano, R.,** Studies on the electrofusion of the bacterial cells II, *Bull. Nippon Vet. Zootech. Coll.,* 39, 72, 1990.

437. **Zhang, J.-Y., Lu, W.-Y., Han, W.-H. and Li, H.-L.,** Protoplast electrofusion of *Streptomyces aureofaciens, Chin. J. Antibiot.,* 18, 163, 1993.

438. **Künkel, W., Groth, I., Jacob, H.-E., Risch, S., Harnisch, M., May, R., Berg, H. and Katenkamp, U.,** Electrofusion of protoplasts of *Penicillium chrysogenum, Studia biophysica,* 119, 35, 1987.

439. **Murkovic, M., Steiner, W. and Esterbauer, H.,** Electrofusion of *Trichoderma reesei* protoplasts, *Lett. Appl. Microbiol.,* 5, 107, 1987.

440. **Ushijima, S., Nakadai, T. and Uchida, K.,** Interspecific electrofusion of protoplasts between *Aspergillus oryzae* and *Aspergillus sojae, Agric. Biol. Chem.,* 55, 129, 1991.

441. **Korom, K., Franko, A., Vagvölgyi, C. S. and Ferenczy, L.,** Optimization of electrofusion for protoplasts of *Aspergillus nidulans, Acta Microbiol. Hung.,* 38, 224, 1991.

442. **Urano, N., Sahara, H. and Koshino, S.,** Conversion of a non-flocculent brewer's yeast to flocculent ones by electrofusion. 1. Identification and characterization of the fusants by pulsed field gel electrophoresis, *J. Biotechnol.,* 28, 237, 1993.

443. **Urano, N., Sato, M., Sahara, H. and Koshino, S.,** Conversion of a non-flocculent brewer's yeast to flocculent ones by electrofusion. 2. Small-scale brewing by fusants, *J. Biotechnol.,* 28, 249, 1993.

444. **Aarnio, T. H. and Suihko, M.-L.,** Electrofusion of an industrial baker's yeast strain with a sour dough yeast, *Appl. Biochem. Biotechnol.,* 27, 65, 1991.

445. **Nagodawithana, T. W. and Trivedi, N. B.,** Yeast selection for baking, in *Yeast Strain Selection,* Panchal, C. J., Ed., Marcel Dekker, New York, 1990, 139.

446. **Tempelaar, M. J., Drenth-Diephuis, L. J., Saat, T. A. W. M. and Jacobsen, E.,** Spontaneous and induced loss of chromosomes in slow-growing somatic hybrid calli of *Solanum tuberosum* and *Nicotiana plumbaginifolia, Euphytica,* 56, 287, 1991.

447. **Bates, G. W.,** Electrofusion of plant protoplasts and the production of somatic hybrids, in *Guide to Electroporation and Electrofusion,* Chang, D. C., Chassy, B. M., Saunders, J. A., and Sowers, A. E., Eds., Academic Press, San Diego, 1992, 249.

448. **Daunay, M. C., Chaput, M. H., Sihachakr, D., Allot, M., Vedel, F. and Ducreux, G.,** Production and characterization of fertile somatic hybrids of eggplant (*Solanum melongena* L.) with *Solanum aethiopicum* L., *Theor. Appl. Genet.,* 85, 841, 1993.

449. **Gilissen, L. J. W., van Staveren, M. J., Verhoeven, H. A. and Sree Ramulu, K.,** Somatic hybridization between potato and *Nicotiana plumbaginifolia.* 1. Spontaneous biparental chromosome elimination and production of asymmetric hybrids, *Theor. Appl. Genet.,* 84, 73, 1992.

450. **Yamaguchi, J. and Shiga, T.,** Characteristics of regenerated plants *via* protoplast electrofusion between melon (*Cucumis melo*) and pumpkin (interspecific hybrid, *Cucurbita maxima* × *C. moschata*), *Japan J. Breed.,* 43, 173, 1993.

451. **Montané, M. H., Alibert, G. and Teissié, J.,** Genetic investigations of somatic hybrids between tobacco albino strains obtained by electrofusion, *Studia biophysica,* 119, 89, 1987.

452. **Gilmour, D. M., Davey, M. R. and Cocking, E. C.,** Production of somatic hybrid tissues following chemical and electrical fusion of protoplasts from albino cell suspensions of *Medicago sativa* and *M. borealis, Plant Cell Rep.,* 8, 29, 1989.

453. **De Vries, S. E., Jacobsen, E., Jones, M. G. K., Loonen, A. E. H. M., Tempelaar, M. J., Wijbrandi, J. and Feenstra, W. J.,** Electrofusion of biochemically well characterized nitrate reductase deficient *Nicotiana plumbaginifolia* mutants. Studies on optimization and complementation, *Plant Sci.,* 51, 105, 1987.

454. **Negrutiu, I., De Brouwer, D., Watts, J. W., Sidorov, V. I., Dirks, R. and Jacobs, M.,** Fusion of plant protoplasts: a study using auxotrophic mutants of *Nicotiana plumbaginifolia,* Viviani, *Theor. Appl. Genet.,* 72, 279, 1986.

455. **Kohn, H., Schieder, R. and Schieder, O.,** Somatic hybrids in tobacco mediated by electrofusion, *Plant Sci.,* 38, 121, 1985.

456. **Toriyama, K. and Hinata, K.,** Diploid somatic-hybrid plants regenerated from rice cultivars, *Theor. Appl. Genet.,* 76, 665, 1988.

457. **Wijbrandi, J., van Capelle, W., Hanhart, C. J., van Loenen Martinet-Schuringa, E. P. and Koornneef, M.,** Selection and characterization of somatic hybrids between *Lycopersicon esculentum* and *Lycopersicon peruvianum, Plant Sci.,* 70, 197, 1990.

458. **Morikawa, H., Hayashi, Y., Ohnishi, N., Yamamoto, Y., Sato, F. and Yamada, Y.,** Culture of electrically induced heterokaryons of plant cells rich in secondary metabolites, *Plant Cell Physiol.,* 29, 1201, 1988.

459. **Puite, K. J., Roest, S. and Pijnacker, L. P.,** Somatic hybrid potato plants after electrofusion of diploid *Solanum tuberosum* and *Solanum phureja, Plant Cell Rep.,* 5, 262, 1986.

460. **Mattheij, W. M. and Puite, K. J.,** Tetraploid potato hybrids through protoplast fusions and analysis of their performance in the field, *Theor. Appl. Genet.,* 83, 807, 1992.

461. **Spangenberg, G., Osusky, M., Oliveira, M. M., Freydl, E., Nagel, J., Pais, M. S. and Potrykus, I.,** Somatic hybridization by microfusion of defined protoplast pairs in *Nicotiana:* morphological, genetic, and molecular characterization, *Theor. Appl. Genet.,* 80, 577, 1990.

462. **Finch, R. P., Slamet, I. H. and Cocking, E. C.,** Production of heterokaryons by the fusion of mesophyll protoplasts of *Porteresia coarctata* and cell suspension-derived protoplasts of *Oryza sativa:* a new approach to somatic hybridization in rice, *J. Plant Physiol.,* 136, 592, 1990.

463. **Puite, K., Ten Broeke, W. and Schaart, J.,** Inhibition of cell wall synthesis improves flow cytometric sorting of potato heterofusions resulting in hybrid plants, *Plant Sci.,* 56, 61, 1988.

464. **Chaput, M.-H., Sihachakr, D., Ducreux, G., Marie, D. and Barghi, N.,** Somatic hybrid plants produced by electrofusion between dihaploid potatoes: BF15 (H1), Aminca (H6) and Cardinal (H3), *Plant Cell Rep.,* 9, 411, 1990.

465. **Obi, I., Kakutani, T., Imaizumi, N., Ichikawa, Y. and Senda, M.,** Surface charge density of hetero-fused plant protoplasts: an electrophoretic study, *Plant Cell Physiol.,* 31, 1031, 1990.

466. **Schmitz, P. and Schnabl, H.,** Regeneration and evacuolation of protoplasts from mesophyll, hypocotyl and petioles from *Helianthus annuus* L., *J. Plant Physiol.,* 135, 223, 1989.

467. **Naton, B., Ecke, M. and Hampp, R.,** Production of fertile hybrids by electrofusion of vacuolated and evacuolated tobacco mesophyll protoplasts, *Plant Sci.,* 85, 197, 1992.

468. **Han San, L., Vedel, F., Sihachakr, D. and Rémy, R.,** Morphological and molecular characterization of fertile tetraploid somatic hybrids produced by protoplast electrofusion and PEG-induced fusion between *Lycopersicon esculentum* Mill. and *Lycopersicon peruvianum* Mill., *Mol. Gen. Genet.,* 221, 17, 1990.

469. **Suh, J.-W. and Lee, K.-W.,** Electrofusion of tobacco and pea protoplasts, *Korean J. Bot.,* 29, 1, 1986.

470. **Terada, R., Kyozuka, J., Nishibayashi, S. and Shimamoto, K.,** Plantlet regeneration from somatic hybrids of rice (*Oryza sativa* L.) and barnyard grass (*Echinochloa orizycola* Vasing), *Mol. Gen. Genet.,* 210, 39, 1987.

471. **Chand, P. K., Davey, M. R., Power, J. B. and Cocking, E. C.,** An improved procedure for protoplast fusion using polyethylene glycol, *J. Plant Physiol.,* 133, 480, 1988.

472. **Matsumuto, E.,** Interspecific somatic hybridization between lettuce (*Lactuca sativa*) and wild species *L. virosa, Plant Cell Rep.,* 9, 531, 1991.

473. **Morikawa, H., Kumashiro, T., Kusakari, K., Iida, A., Hirai, A. and Yamada, Y.,** Interspecific hybrid plant formation by electrofusion in *Nicotiana, Theor. Appl. Genet.,* 75, 1, 1987.

474. **Hayaschi, Y., Kyozuka, J. and Shimamoto, K.,** Hybrids of rice (*Oryza sativa* L.) and wild *Oryza* species obtained by cell fusion, *Mol. Gen. Genet.,* 214, 6, 1988.

475. **Bates, G. W.,** Asymmetric hybridization between *Nicotiana tabacum* and *N. repanda* recipient protoplast fusion: transfer of TMV resistance, *Theor. Appl. Genet.,* 80, 481, 1990.

476. **Bates, G. W.,** Transfer of tobacco mosaic virus resistance by asymmetric protoplast fusion, in *Progress in Plant Cellular and Molecular Biology,* Nijkamp, H. J. J., Van Der Plas, W., and Van Aartrijk, J., Eds., Kluwer Academic Press, Dordrecht, 1990, 293.

477. **Hagimori, M., Matsui, M., Matsuzaki, T., Shinozaki, Y., Shinoda, T. and Harada, H.,** Production of somatic hybrids between *Nicotiana benthamiana* and *N. tabacum* and their resistance to aphids, *Plant Sci.,* 91, 213, 1993.

478. **Takamizo, T., Spangenberg, G., Suginobu, K.-I. and Potrykus, I.,** Intergeneric somatic hybridization in Gramineae: somatic hybrid plants between tall fescue (*Festuca arundinacea* Schreb.) and Italian ryegrass (*Lolium multiflorum* Lam.), *Mol. Gen. Genet.,* 231, 1, 1991.

479. **Saito, W., Ohgawara, T., Shimizu, J., Ishii, S. and Kobayashi, S.,** *Citrus* cybrid regeneration following cell fusion between nucellar cells and mesophyll cells, *Plant Sci.,* 88, 195, 1993.

480. **Van Wert, S. L. and Saunders, J. A.,** Electrofusion and electroporation of plants, *Plant Physiol.,* 99, 365, 1992.

481. **Jadari, R., Sihachakr, D., Rossignol, L. and Ducreux, G.,** Transfer of resistance to *Verticillium dahliae* Kleb. from *Solanum torvum* S. W. into potato (*Solanum tuberosum* L.) by protoplast electrofusion, *Euphytica.,* 64, 39, 1992.

482. **Akagi, H., Sakamoto, M., Negishi, T. and Fujimura, T.,** Construction of rice cybrid plants, *Mol. Gen. Genet.,* 215, 501, 1989.

483. **Barth, S., Voeste, D. and Schnabl, H.,** Somatic hybrids of sunflower (*Helianthus annuus* L.) identified at the callus stage by isoenzyme analysis, *Bot. Acta,* 106, 100, 1993.

484. **Pupilli, F., Scarpa, G. M., Damiani, F. and Arcioni, S.,** Production of interspecific somatic hybrid plants in the genus *Medicago* through protoplast fusion, *Theor. Appl. Genet.,* 84, 792, 1992.

485. **De Vries, S. E. and Tempelaar, M. J.,** Electrofusion and analysis of potato somatic hybrids, in *Biotechnology in Agriculture and Forestry, Vol. 3: Potato,* Bajaj, Y. P. S., Ed., Springer-Verlag, Berlin, 1989, 211.

486. **Yang, Z.-Q., Shikanai, T. and Yamada, Y.,** Asymmetric hybridization between cytoplasmic male-steril (CMS) and fertile rice (*Oryza sativa* L.) protoplasts, *Theor. Appl. Genet.,* 76, 801, 1988.

487. **Hagimori, M., Nagaoka, M., Kato, N. and Yoshikawa, H.,** Production and characterization of somatic hybrids between the Japanese radish and cauliflower, *Theor. Appl. Genet.,* 84, 819, 1992.

488. **Gaff, D. F., Ziegler, H. and Zimmermann, U.,** Electrofusion of protoplasts from desiccation tolerant grass species with desiccation sensitive grass protoplasts, *J. Plant Physiol.,* 120, 375, 1985.

489. **Nugent, G. and Gaff, D. F.,** Electrofusion of protoplasts from desiccation tolerant species and desiccation sensitive species of grasses, *Biochem. Physiol. Pflanzen,* 185, 93, 1989.

490. **Spangenberg, G., Freydl, E., Osusky, M., Nagel, J. and Potrykus, I.,** Organelle transfer by microfusion of defined protoplast-cytoplast pairs, *Theor. Appl. Genet.,* 81, 477, 1991.

491. **Kranz, E., Bautor, J. and Lörz, H.,** *In vitro* fertilization of single, isolated gametes of maize mediated by electrofusion, *Sex. Plant Reprod.,* 4, 12, 1991.

492. **Faure, J.-E., Mogensen, H. L., Dumas, C., Lörz, H. and Kranz, E.,** Karyogamy after electrofusion of single egg and sperm cell protoplasts from maize: cytological evidence and time course, *The Plant Cell,* 5, 747, 1993.

493. **Kranz, E. and Lörz, H.,** *In vitro* fertilization with isolated, single gametes results in zygotic embryogenesis and fertile maize plants, *The Plant Cell,* 5, 739, 1993.

494. **Kranz, E., Bautor, J. and Lörz, H.,** Electrofusion-mediated transmission of cytoplasmic organelles through the *in vitro* fertilization process, fusion of sperm cells with synergids and central cells, and cell reconstitution in maize, *Sex. Plant Reprod.,* 4, 17, 1991.

495. **Robl, J. M. and Stice, S. L.,** Prospects for the commercial cloning of animals by nuclear transplantation, *Theriogenology,* 31, 75, 1989.

496. **Tzagoloff, A. and Myers, A. M.,** Genetics of mitochondrial biogenesis, *Annu. Rev. Biochem.,* 55, 249, 1986.

497. **Ferris, S. D., Sage, R. D. and Wilson, A. C.,** Evidence from mtDNA sequences that common laboratory strains of inbred mice are descended from a single female, *Nature,* 295, 163, 1982.

498. **Kanka, J., Fulka, J., Jr., Fulka, J. and Petr, J.,** Nuclear transplantation in bovine embryo: fine structural and autoradiographic studies, *Mol. Reprod. Dev.,* 29, 110, 1991.

499. **Cheong, H. T., Takahashi, Y. and Kanagawa, H.,** Developmental capacity of reconstituted mouse embryos: influences of nucleus and cytoplasm sources, *J. Vet. Med. Sci.,* 54, 1099, 1992.

500. **Modlinski, J. A. and Smorag, Z.,** Preimplantation development of rabbit embryos after transfer of embryonic nuclei into different cytoplasmic environment, *Mol. Reprod. Dev.,* 28, 361, 1991.

501. **Stice, S. L. and Robl, J. M.,** Nuclear reprogramming in nuclear transplant rabbit embryos, *Biol. Reprod.,* 39, 657, 1988.

502. **Prather, R. S., Sims, M. M. and First, N. L.,** Nuclear transplantation in early pig embryos, *Biol. Reprod.,* 41, 414, 1989.

503. **Kurischko, A. and Berg, H.,** Electrofusion of rat and mouse blastomeres, *Bioelectrochem. Bioenerg.,* 15, 513, 1986.

504. **Collas, P., Balise, J. J., Hofmann, G. A. and Robl, J. M.,** Electrical activation of mouse oocytes, *Theriogenology,* 32, 835, 1989.

505. **Ware, C. B., Barnes, F. L., Maiki-Laurila, M. and First, N. L.,** Age dependence of bovine oocyte activation, *Gamete Res.,* 22, 265, 1989.

506. **Kono, T., Iwasaki, S. and Nakahara, T.,** Parthenogenetic activation by electric stimulus of bovine oocyte matured *in vitro, Theriogenology,* 32, 569, 1989.

507. **Onodera, M. and Tsunoda, Y.,** Parthenogenetic activation of mouse and rabbit eggs by electric stimulation *in vitro, Gamete Res.,* 22, 277, 1989.
508. **Marcus, G. J.,** Activation of cumulus-free mouse oocytes, *Mol. Reprod. Dev.,* 26, 159, 1990.
509. **Ozil, J. P.,** The parthenogenetic development of rabbit oocytes after repetitive pulsatile electrical stimulation, *Development,* 109, 117, 1990.
510. **Collas, P. and Robl, J. M.,** Factors affecting the efficiency of nuclear transplantation in the rabbit embryo, *Biol. Reprod.,* 43, 877, 1990.
511. **Maruyama, Y., Kita, M., Imai, H., Tokunaga, T. and Tsunoda, Y.,** Examination of the suitable condition for the parthenogenetic activation and electrofusion of a porcine enucleated oocyte with a pseudo-blastomere, *Anim. Sci. Technol. (Jpn.),* 62, 757, 1991.
512. **Rickords, L. F. and White, K. L.,** Electrofusion-induced intracellular Ca^{2+} flux and its effect on murine oocyte activation, *Mol. Reprod. Dev.,* 31, 152, 1992.
513. **Ushijima, H. and Eto, T.,** Production of a calf from a nuclear transfer embryo using *in vitro* matured oocytes, *J. Reprod. Dev.,* 38, 61, 1992.
514. **Takano, H., Koyama, K., Kozai, C., Kato, Y. and Tsunoda, Y.,** Effect of aging of recipient oocytes on the development of bovine nuclear transfer embryos *in vitro, Theriogenology,* 39, 909, 1993.
515. **Westhusin, M. E., Pryor, J. H. and Bondioli, K. R.,** Nuclear transfer in the bovine embryo: a comparison of 5-day, 6-day, frozen-thawed, and nuclear transfer donor embryos, *Mol. Reprod. Dev.,* 28, 119, 1991.
516. **Cadoret, J.-P.,** Electric field-induced polyploidy in mollusc embryos, *Aquaculture,* 106, 127, 1992.
517. **Fan, Z., Shi, L. and Yin, H.,** Electrofusion of fish embryonic cells *in vivo, Aquaculture,* 111, 308, 1993.
518. **Gao, X., Li, G., Du, M., Fen, Y., Cao, M., Lu, D. and Li, S.,** Electric field mediated fusion of blastomeres of goldfish and its developmental potential, *Acta Zool. Sin.,* 36, 199, 1990.
519. **Yamaha, E. and Yamazaki, F.,** Electrically fused-egg induction and its development in the goldfish *Carassius auratus, Int. J. Dev. Biol.,* 37, 291, 1993.
520. **Köhler, G. and Milstein, C.,** Continuous cultures of fused cells secreting antibody of predefined specificity, *Nature,* 256, 495, 1975.
521. **Karsten, U., Papsdorf, G., Roloff, G., Stolley, P., Abel, H., Walther, I. and Weiss, H.,** Monoclonal anti-cytokeratin antibody from a hybridoma clone generated by electrofusion, *Cancer Clin. Oncol.,* 21, 733, 1985.
522. **Abel, H., Stolley, P., Dreyer, G., Walther, I. and Junge, C.,** Hybridoma cells produced by electrofusion in a homogeneous electric field, *Acta Biotechnol.,* 6, 287, 1986.
523. **Mangoldt, D., Schumann, I. and Stelzner, A.,** Hybridome production by electrofusion, *Allerg. Immunol.,* 33, 63, 1987.
524. **Ohnishi, K., Chiba, J., Goto, Y. and Tokunaga, T.,** Improvement in the basic technology of electrofusion for generation of antibody-producing hybridomas, *J. Immunol. Meth.,* 100, 181, 1987.
525. **Tsoneva, I., Panova, I., Doinov, P., Dimitrov, D. S. and Strahilov, D.,** Hybridoma production by electrofusion: monoclonal antibodies against the Hc antigene of *Salmonella, Studia biophysica,* 125, 31, 1988.
526. **Vuento, M., Nikkilä, L. and Rauvala, H.,** Cell fusion in electric field in the production of monoclonal antibodies against soluble and cellular polypeptide antigens, in *Protides of the Biological Fluids Proceedings,* Peeters, H., Ed., Vol. 35, Pergamon Press, Oxford, U.K., 1987, 407.
527. **Zimmermann, U., Schmitt, J. J. and Kleinhans, P.,** The potential of electrofusion for hybridoma production in *Clinical Applications of Monoclonal Antibodies,* Hubbard, R. and Marks, V., Eds., Plenum Press, New York, 1988, 3.
528. **Seidel, B. and Fiebig, H.,** Electrofusion as a tool in hybridoma technique — comparison with the PEG-technique, *Studia biophysica,* 130, 197, 1989.

529. **Tsoneva, I., Tomov, T., Panova, I. and Strahilov, D.,** Effective production by electrofusion of hybridomas secreting monoclonal antibodies against Hc-antigen of *Salmonella, Bioelectrochem. Bioenerg.,* 24, 41, 1990.

530. **Blancher, A., Calvas, P., Conte, P. and Teissié, J.,** Electric field-induced hybridomas: targeting by immunological and physical methods, *Bioelectrochem. Bioenerg.,* 25, 295, 1991.

531. **Kwekkeboom, J., de Groot, C. and Tager, J. M.,** Efficient electric-field-induced generation of hybridomas from human B lymphocytes without prior activation *in vitro, Hum. Antibod. Hybridomas,* 3, 48, 1992.

532. **Gravekamp, C., Bol, S. J. L., Hagemeijer, A. and Bolhuis, R. L. H.,** Production of human T-cell hybridomas by electrofusion, in *Human Hybridomas and Monoclonal Antibodies,* Engelman, E. G., Foung, S. K., and Larrick, J., Eds., 1985, 323.

533. **Gravekamp, C., Santoli, D., Vreugdenhil, R., Collard, J. G. and Bolhuis, R. L. H.,** Efforts to produce human cytotoxic T-cell hybridomas by electrofusion and PEG fusion, *Hybridoma,* 6, 121, 1987.

534. **Glassy, M. C.,** Production methods for generating human monoclonal antibodies, *Hum. Antibod. Hybridomas,* 4, 154, 1993.

535. **Neil, G. A. and Urnovitz, H. B.,** Recent improvements in the production of antibody-secreting hybridoma cells, *TIBTECH,* 6, 209, 1988.

536. **O'Hare, M. J. and Edwards, P. A. W.,** Human monoclonal antibodies and the LICR-LON-HMy2 system, in *Immunology Series, Human Hybridomas: Diagnostic and Therapeutic Applications,* Strelkauskas, A. J., Ed., Vol. 30, Marcel Dekker, New York, 1987, 47.

537. **Pratt, M., Mikhalev, A. and Glassy, M. C.,** The generation of Ig-secreting UC 729-6 derived human hybridomas by electrofusion, *Hybridoma,* 6, 469, 1987.

538. **Gathuru, J. K., Miyoshi, I. and Naiki, M.,** Electrically stimulated fusion of human UC 729-6 cell line potentially useful for generating antibody-secreting human-human hybridomas, *Jpn. J. Vet. Res.,* 35, 109, 1987.

539. **Meng, Y. G. and Trawinski, J.,** Identification of mouse J chain in human IGM produced by an electrofusion-derived mouse-human heterohybridoma, *J. Immunol.,* 141, 2684, 1988.

540. **Glaser, R. W., Jahn, S. and Grunow, R.,** Development of specific human mab's by a small scale electrofusion technique: the influence of some physical and chemical factors on hybridoma yield of human peripheral blood lymphocytes × CB-F7 fusions, *Allerg. Immunol.,* 35, 123, 1989.

541. **Foung, S. K. H. and Perkins, S.,** Electric field-induced cell fusion and human monoclonal antibodies, *J. Immunol. Meth.,* 116, 117, 1989.

542. **Wallace, E. F., Foung, S. K. H., Bradbury, K., Pask, S. L. and Grumet, F. C.,** Generation of a human hybridoma producing pure anti-HLA-A2 monoclonal antibody, *Hum. Immunol.,* 28, 65, 1990.

543. **Uobe, K.,** Analysis of oral carcinoma antigens by monoclonal antibodies, *J. Osaka Odontol. Soc.,* 56, 197, 1993.

544. **Chiba, J.,** A strategy for the production of human monoclonal antibodies, *Hum. Cell,* 1, 31, 1988.

545. **Banchereau, J., dePaoli, P., Valle, A., Garcia, E. and Rousset, F.,** Long-term human B cell lines dependent on interleukin-4 and antibody to CD40, *Science,* 251, 70, 1991.

546. **Steenbakkers, P. G. A., van Meel, F. C. M. and Olijve, W.,** A new approach to the generation of human or murine antibody producing hybridomas, *J. Immunol. Meth.,* 152, 69, 1992.

547. **Steenbakkers, P. G. A., van Wezenbeek, P. M. G. F. and Olijve, W.,** Immortalization of antigen selected B cells, *J. Immunol. Meth.,* 163, 33, 1993.

548. **Steenbakkers, P. G. A., van Wezenbeek, P. M. G. F., van Zanten, J. and The, T. H.,** Efficient generation of human anti-cytomegalovirus IgG monoclonal antibodies from preselected antigen-specific B cells, *Hum. Antibod. Hybridomas,* 4, 166, 1993.

549. **Zimmermann, U., Love-Homan, L., Gessner, P., Clark, D., Klöck, G., Johlin, F. C. and Neil, G. A.,** Generation of a human monoclonal antibody to Hepatitis C virus JRA1 by activation of peripheral blood lymphocytes and hypo-osmolar electrofusion, *Hepatology,* 18, 265A, 1993.

550. **Glassy, M. C. and Pratt, M.,** Generation of human hybridomas by electrofusion, in *Electroporation and Electrofusion in Cell Biology,* Neumann, E., Sowers, A. E., and Jordan, C. A., Eds., Plenum Press, New York, 1990, 271.

551. **Krenn, V., von Landenberg, P., Wozniak, E., Kißler, C., Müller-Hermelink, H.-K., Zimmermann, U. and Vollmers, H. P.,** Efficient immortalization of rheumatoid synovial tissue B lymphocytes. A comparison between the techniques of electric field-induced and PEG fusion, *Hum. Antibod. Hybridomas,* 6, 47, 1995.

552. **Zimmermann, U., Love-Homan, L., Gessner, P., Clark, D., Klöck, G., Johlin, F. C., and Neil, G. A.,** Generation of a human monoclonal antibody to hepatitis C virus, JRA1 by activation of peripheral blood lymphocytes and hypo-osmolar electrofusion, *Hum. Antibod. Hybridomas,* 6, 77, 1995.

553. **Hedrich, R. and Marten, I.,** Malate-induced feedback regulation of plasma membrane anion channels could provide a CO_2 sensor to guard cells, *EMBO J.,* 12, 897, 1993.

Chapter 5

CELL MOTION IN TIME-VARYING FIELDS: PRINCIPLES AND POTENTIAL

Günter Fuhr, Ulrich Zimmermann, and Stephen G. Shirley

CONTENTS

I. Introduction 260

II. "Ponderomotive" Forces 261
    A. Theoretical Background 261
    B. Cell Movement 266

III. Dielectrophoresis and Levitation 268
    A. Positive Dielectrophoresis 269
    B. Negative Dielectrophoresis 272
    C. Levitation, Traps, and Cages 275

IV. Electrorotation 283
    A. Historical Background 283
    B. Electrorotation of Single Cells 285
        1. Experimental 285
        2. The Interpretation of Rotation Spectra 288

V. Travelling-Waves Drives 300
    A. Electrohydrodynamic Fluid Pumping 300
    B. Travelling-Wave-Induced Particle and Cell Motion 303

VI. Conclusions 305

References 307

Appendices 324
    A. Short Description of the "Dipole Concept" 324
    B. Multishells in the Field of Multiple Electrodes 325

0-8493-4476-X/96/$0.00+$.50
© 1996 by CRC Press Inc.

I. INTRODUCTION

Scientists are very often confronted by unexpected findings that demand new approaches, and they discover that yesterday's "sophisticated tools" are today's "blunt instruments." This experience is clearly demonstrated in the preceding Chapters, 1, 2, and 4. The use of electric field pulses of high strength and short duration has resulted in a new, very powerful technology for control and manipulation of the genome and cytosol of living cells. Indisputably, intense field pulse techniques allow numerous cell manipulations which were impossible using previous chemical or viral methods.

More recently, microfabrication techniques pioneered for semiconductor manufacturing have also led to a renaissance of cell manipulation techniques that use electric fields much weaker than the intense pulses needed to cause fusion. These include translational movement and/or rotation, as well as the orientation of cells induced by direct current, alternating, rotating, and travelling fields. "Dielectrophoresis" is a term used to signify particle movement induced by a non-uniform field. Particle movement may also be caused by travelling electrical waves. The term "electrorotation" indicates the spinning of a particle caused by a rotating field.

In contrast to cell electrophoresis,[1,2] cell (and particle) movement in response to time-varying fields is dependent on field-induced polarization. The resultant forces increase with the volume of the particle, and are, therefore, often referred to as "body" or "ponderomotive" forces. The experimental and theoretical foundations for understanding polarization-induced particle movement were established about one century ago[3,4] and extended in the 1920s and 1930s.[5-8] Although experimental work with cells[9] started in 1939, only recently have the applications of these phenomena sparked widespread interest in biology and biotechnology. This is due largely to the increasing availability of microfabricated devices and nanotechnology for silicon-based integrated systems.

Single electrodes were miniaturized in the sixties by electrochemists seeking a way to optimize voltametric measurements (reviewed in Reference 10). The interest in the development of miniaturized biosensors contributed considerably to the development of microstructures.[11-13] This process, which started nearly 30 years ago, has accelerated enormously over the last 10 years due to the progress in semiconductor technology.

However, less was done with living cells in semiconductor-fabricated microstructures. Starting in the eighties, electrode systems on glass and silicon substrates were used for the cultivation of adherently growing cells, whose behavior was monitored electrically and optically.[14-22] Strong, high-frequency electric fields are tolerated in the short term and do not significantly inhibit growth, division, and cell function when applied for hours or days.[23,24] It was soon realized that microstructures are promising tools for the study of model "neuronal networks;" cells can be electrically stimulated from outside, allowing the investigation of brain function.[25-33]

Microstructures are also beginning to produce revolutionary applications of ponderomotive forces in biology, biomedicine, and biotechnology. There are: micromotors;[34-42] micropumps and micro-scale cell separators;[43,44] nearly unlimited possibilities for cell manipulation[23,40,45-49] (including dielectrophoresis, levitation, and the electrofusion of single heterologous cell pairs); possibilities for the collection of cells (and very small particles and macromolecules) from highly dilute solutions,[23,40,43,50] and for the high frequency field trapping of cells in small volumes which allows the formation of "artificial tissues."[23,40] The scaling down of electrodes allows useful fields to be imposed in physiological (highly conducting) media with modest heat production,[24] which was impossible before. It also enables fields to be structured on a scale comparable with the diameter of a cell (or smaller) and allows the exploitation of non-linear effects.

This chapter provides a detailed discussion of the phenomena of dielectrophoresis, electrorotation, levitation, field trapping, and the travelling-wave-induced transport of cells. The principles that underlie the theoretical and technical advances in this area, as well as the current state-of-the-art and the potential use of ponderomotive force techniques in biology, biomedicine, and biotechnology are reviewed. This chapter is written for biologists. However, the techniques described here, with their potential for handling small numbers of molecules, may also be of interest to chemists, physicists, and engineers; for them, the appendices and the comprehensive reference list of this chapter may serve as a useful guideline to special experimental conditions and detailed theoretical considerations.

II. "PONDEROMOTIVE" FORCES

A. THEORETICAL BACKGROUND

Two kinds of methods can induce polarization and cause net movement of a particle in an alternating field. Figure 1a illustrates one — the application of an inhomogeneous field in a fixed direction.[9,51-54] Alternatively, as shown in Figures 1b and 1c, rapid angular or linear movement of the electric field relative to the particle will cause motion.[3,6,55-58]

Dielectrophoresis, levitation, and the trapping and caging of particles and cells belong to the first group; electrorotation and travelling-wave drives to the second. While such a classification is convenient, a technique could employ both strategies as shown below.

A quantitative expression for the force developed on an ideal, homogeneous dielectric by a non-uniform field in air or vacuum[59] was given as early as 1904 and for a particle in a rotating field[3] in 1906. Since then, a large number of publications on the theory of ponderomotive forces have appeared.[60-87] The rapid increase of interest in the last thirty and, especially, in the last ten years has centered mainly on the fields of particle (dust, bubbles, etc.) precipitators,[54,88-106] cellular measurements,[51-54] and electrofusion (see Chapter 4).

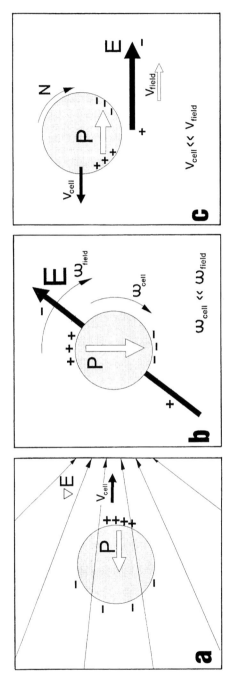

FIGURE 1.   The relationships of field and polarization for three electric-field-techniques of cell manipulation. In order to concentrate on the fundamental effects, a structureless model particle is shown. In all cases the polarization (P) is associated with the appearance of induced surface charges. (a) A highly divergent field exerts more force on the charges induced on the high-field side of the particle than those on the low-field side. The particle moves towards the higher field — positive dielectrophoresis. (b) The polarization mechanism is too slow to keep up with the rapidly rotating field (radian velocity ω_{field}), and the interaction between field and induced charges gives rise to a torque leading to slow cell spinning (ω_{cell}, here co-field rotation). (c) An electric-field vector sweeps rapidly past the cell, with velocity, v_{field}. The induced polarization persists for a short time after the field has passed, and the interaction of the induced charges with the field in its new position cause a linear force on the particle leading to cell motion (v_{cell}, here anti-field motion). This is a travelling-wave drive. Fields are usually inhomogeneous, so particle levitation and a torque are also generated (see also Figure 1 in Chapter 1 and Figure 4 in Chapter 3).

These effects of alternating and rotating electrical fields can be quantitatively understood if cells, their constituent parts, and the surrounding medium are treated as "lossy" dielectrics.[107-115] A dielectric is a polarizable material and "lossy" means that energy dissipation must be considered in addition to the dielectric properties (permittivity). Energy dissipation is due to ionic currents flowing through a resistive medium and (at higher frequencies) to the viscous damping of oscillating ions or molecules. The sum of the ion transport and viscous effects is described by the conductivity.

Permittivity and conductivity may be compared in magnitude by defining a complex conductivity $\tilde{\sigma}$:

$$\tilde{\sigma} = \sigma + i\omega\varepsilon\varepsilon_0 \tag{1}$$

where σ is the conductivity, ω is the radian frequency ($=2\pi f$), ε is the permittivity, ε_0 is the permittivity of free space $= 8.85 \cdot 10^{-14}$ F cm^{-1} and $i^2 = -1$. The physical basis of this mathematical description is that the displacement current has a phase shift of 90° with respect to the alternating field.

It is generally assumed that the conductivity, σ, and permittivity, ε, do not change appreciably over the frequency range of interest.* Equation 1 then shows that the complex conductivity changes with frequency and is dominated by the conductivity at low frequencies and by the permittivity at high ones. Such changes are most significant when the radian frequency is close to the reciprocal of one of the charging time constants ($\omega \approx 1/\tau \approx \sigma/\varepsilon_0\varepsilon$) and are termed dispersions[107-115] (see also Chapter 1). All dielectric materials show dispersive effects in some frequency region.[116]

When a lossy dielectric body in a medium is polarized by a field, there is an accumulation of electric charge at the interfaces (see Figure 1 in Chapter 4). The magnitude of this charge depends on the radian frequency (ω) and on the relationships between the permittivities and the conductivities. The interaction of induced charge and applied field gives rise to a driving force on the particle if the field is inhomogeneous or if the field vector is moving (e.g., in the case of electrorotation or travelling-wave drives, see below).

The situation becomes more complex when we consider structured particles such as cells, because shelled or multi-shelled particles[82,117-119] exhibit additional changes in polarization with frequency. Cells are surrounded by a plasma membrane and include compartments which are separated by lipoprotein membranes. Especially for cells in suspension, this multi-shelled structure must be acknowledged.

The membranes have very much lower conductivity than the aqueous compartments, and the aqueous compartments are not identical (the ionic composition of the cytosol can be very different to that of the external medium).

---

* However, it is important to remember that the conductivity and permittivity arise from physical processes which can themselves have characteristic frequency dependencies. Permittivity and conductivity do change, and the assumption of constancy is an approximation. Changes lead to the occurrence of more complex dispersions.

Essentially, there are (at least) three dielectrics in series (medium, membrane, cytosol). Thus, the polarization of each dielectric interface is frequency-dependent, and there are (at least) two dispersions in the kHz to MHz range. At kHz frequencies there is a large induced trans-membrane potential and only a small ohmic current flow through the membrane (see Chapter 6) and virtually no capacitive current. Due to the dominance of membrane permittivity in the MHz range, the transmembrane potential is very low (see also Chapter 1), and the capacitive current flowing through the membrane is very much larger than the ohmic one (i.e., from an electronics standpoint, the membrane is near short circuit at high frequencies).

For calculation of the frequency-dependent, field-mediated forces acting on the particles or cells, we have to distinguish between analytical and numerical solutions. Analytical solutions suffer from the disadvantage that they require (greatly) simplifying assumptions to be made. For instance, spherical, cylindrical, or ellipsoidal geometry is often assumed for the cell; dielectrics are assumed to be isotropic; and the field strength and gradient near the cell are assumed to be known. Much detail can be lost in the process. However, analytical solutions yield equations relating forces, torques, or rotation and travelling speeds to field frequency via a number of parameters (whose values we wish to know). These equations provide a good overview of the process, and it is relatively easy to fit real data to the curves to get values for the unknown parameters. The results, however, do depend on the assumptions which have been made. For instance, there have been electrorotation measurements made on erythrocytes.[120-122] These cells have the well known biconcave shape. An analytical solution for the field-mediated forces on bodies of this shape has not been found, and parameters were determined by fitting data to spherical or ellipsoidal model. For comparing values obtained under slightly different conditions this is acceptable. However, it does not yield absolute values.

Typical numerical procedures are finite-element analysis and the finite difference methods.[123] Numerical methods need fewer assumptions (though there is still a critical dependency on those which are made). Bodies of complex and asymmetric shape can be modelled as well as the individual electrode geometries. The "solution" is a two or three dimensional distribution of field and force over the whole electrode space from which particle trajectories can be obtained. From the interactions between the induced charges at dielectric boundaries and the charges on the electrodes, cell motion and rotation can be calculated[123] and frequency dependent spectra obtained. Another advantage of modelling is that it allows consideration of non-uniform energy dissipation, local heating, convective streaming, and other anisotropies and gradients.[24]

Interestingly, the numerical approach makes no distinction between dielectrophoresis, rotation, levitation, or travelling-wave-induced motion. All that is obtained is a force and a torque acting on a particle. The conceptual separation of the phenomena is, in a sense, the result of assumptions made for analytical purposes. With increasing miniaturization of the electrode arrays, this blurring of the phenomenological boundaries is likely to become more apparent (see below).

For all their advantages, numerical methods require much computational power. To calculate particle behavior in a three-dimensional electrode system and, simultaneously, fit measured data is (currently) practically impossible. Modelling is usually used for the design of complex electrode systems and for the understanding of cell motion under various field and boundary conditions.[123-125]

The fields produced by the relatively simple electrode arrangements used for the majority of biological work with dielectrophoresis and electrorotation prior to 1988 could be described sufficiently well by equations with analytical solutions. However, the more recently developed arrays of microelectrodes (microstructures), especially those required for travelling waves or field cages, are complex in form and have non-uniform charge distributions. In these cases, the only practical solutions are numerical, such as those described in Appendix B. The numerical approach can give the field distribution over the entire neighborhood of the array, and the paths which particles will follow can also be predicted.

There are several physical assumptions that can be made to allow (analytical or numeric) calculations to proceed. One is the dipole concept[3,117,119,126,127] (see Appendix A), which is based on the consideration of the forces that arise from the interaction of induced dipoles with the applied field. It has been criticized on the grounds that it is based on energy principles in electrostatics and energy loss in the medium is not taken into account.[127] However, most dielectrophoretic and electrorotation measurements are made in poorly conducting media.

Alternatively, it is possible to use modelling in terms of resistance/capacitance circuit systems,[110,128] or to proceed by integration of the Maxwell stress tensor over a surface enclosing the cell (see Chapter 3).[69,70,72-74,80,123,127] The latter approach is generally accepted.

All approaches lead to relationships between motion and frequency and can yield values for cell and membrane parameters. However, as noted above, the values obtained do depend on the model used. While, in most cases, this is not serious, it has been recently shown that integration of the stress tensor for non-spherical particles can lead to results significantly different from those of the dipole approach, even in loss-free media.[129]

A recurring theme of much theoretical work has been the problem of inhomogeneous objects such as cells. In most cases, it has proved possible to use the expression developed for a single shelled sphere (enclosing an essentially homogeneous medium)[107,117] in a repetitive manner for calculations on multi-shelled objects.[118,119,126] Ellipsoidal shells can be treated analogously.[60]

In addition to the above assumptions, there are other, implicit ones. For example, magnetic fields are neglected. While these have no influence on the cell, they do have an effect on the development of the electric field, especially at high frequencies (MHz range).

B. CELL MOVEMENT

The induced charge on the cell interfaces oscillates with the same frequency as the applied field. Since both charge and field change sign twice every cycle, the resulting force does not average to zero (unless charge and field are exactly 90° out of phase). The cell responds to this average force, there being enough inertia and viscous drag to eliminate any movement at the field frequency. The viscosity of aqueous media is such that cell movement is relatively slow compared to that of the field. For example, fields of 100 V cm^{-1} rotating in the kHz to MHz range elicit rotation speeds of perhaps a few Hz; travelling waves of about 100 V cm^{-1} r.m.s. (root-mean-square) with "phase speeds" (the speed of movement of a positive or negative crest) of the order of 100 m s^{-1} induce particle, and in some cases fluid, speeds of up to only about 100 μm s^{-1}. From a technical point of view this represents the principle of an asynchronous dielectric motor or linear stepper.[36,130,131]

The velocity of cell motion (translational and rotational) depends, under ideal conditions, on the square of the field strength.[85,127,132-136] The direction of cell motion is governed by the complex dielectric properties of the cells and the surrounding medium. The literature on dielectrophoresis and rotation of cells and other particles uses terms such as positive and negative dielectrophoresis[52,54] and co-field and anti-field rotation.[137] Travelling-waves can also induce cell motion in the same direction as, or opposite to the wave direction.[50,138] The reason for these variations in the direction of the "ponderomotive" forces is that the sign of the total surface charge on each hemisphere of a suspended particle depends on the polarization of the medium as well as that of the particle (Figure 2).

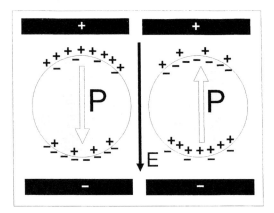

FIGURE 2. A body exposed to a steady or alternating homogeneous field in a polarizable medium exhibits polarization, either positive (right) or negative (left) depending on the dielectric properties of body and medium and on the frequency. There are two accumulations of charge at each interface. Often, for simplicity, only their sum is shown (Chapter 4, Figure 1), but this misses the important point that the net induced charge resides on the inside surface in the case of positive polarization and on the outside surface in the case of negative polarization. The object itself acquires no net charge and can be described by the resultant dipole ($\vec{P}$).

As shown on the right of Figure 2, the polarization of the particle can be greater than that of the medium (always the case in vacuum or air). In this case the net negative charge faces the positive electrode, and the net positive charge faces the negative electrode; the polarization vector $\vec{P}$ is anti-parallel to the field vector $\vec{E}$. In an inhomogeneous field the particle is attracted to regions of higher field strength (Figure 1a). This is positive dielectrophoresis (see also Chapter 4).

The reverse occurs (Figure 2, left) when the medium polarization contributes more charge. The net surface charges are the same as those on the facing electrodes, and $\vec{P}$ is parallel to $\vec{E}$. In an inhomogeneous field the particle is repelled from regions of high field strength. This is negative dielectrophoresis.

In the case of a rotating field, the angular displacement of the induced polarization of a cell must also be considered. The rotation can therefore be either in the same direction as the field (co-field) or opposite it (anti-field). The same terminology can be used for travelling-wave motion because the effects involved are very similar.

By measuring the speed of cell movement at different frequencies, it is possible to obtain electrokinetic force spectra. These reflect the changes in polarization with frequency as noted above. The calculated dielectrophoresis, electrorotation, and travelling-wave drive spectra for a plant protoplast (single-membrane model) are shown in Figure 3. It can be seen that all of these effects exhibit complex frequency dependencies.

Calculations such as those in Figure 3 depend upon a knowledge of the permittivity and conductivity of cell constituents. Conversely, calculations from experimentally obtained spectra can yield values for these important parameters. Until recently, this kind of data had to be obtained from impedance measurements on tissue or concentrated cell suspensions (reviewed in References 128, 139–141) with the disadvantage that the polarization of the "average cell" had to be deduced from a mixture equation (for more details, see the following sections and Appendix B). Electrokinetic spectra methods require much less material (one cell), and they work with real cells, not "average" ones. Further advantages are the avoidance of "small difference in large quantity" errors and the avoidance (if desired) of complications arising from cell-cell interactions.

The situation with living cells (but also with particles) becomes even more complicated when field-induced movement of microaggregates is considered. Chains of particles (which play an important role in electrofusion, see Chapter 4) may have larger dipole moments than might be expected from cell-number. Therefore, chains move considerably faster than single cells. It has been shown[77] that the dipole moment of a chain of two conducting spheres should be about 4.8 times that of a single sphere. This multi-body problem is difficult to solve analytically because of the existence of significant multipoles in adjacent particles. However, it was solved by using the method of images and selecting appropriate conditions.[77] For short chains of non-conducting spheres in non-conducting media, an approximation considering only dipole-dipole interactions gave reasonable results.[142] This approximation also appears usable for predicting turn-over frequencies of chains of particles in most cases.[84]

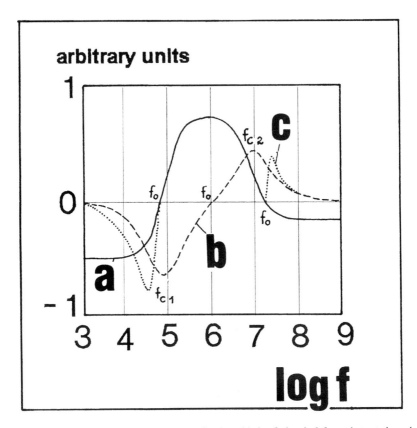

FIGURE 3. Theoretical frequency spectra for three kinds of electrical force that act through induced polarization on a model cell consisting of a conducting interior surrounded by a single, thin, relatively insulating spherical shell suspended in a low conductivity, aqueous liquid. In all cases the force, torque, and velocity units are arbitrary and the following (frequency independent) electrical properties are assumed: Conductivity (σ): Medium, 10^{-4}; Shell, 10^{-9}; Interior, 10^{-3} S cm^{-1}. Permittivity (ε): Medium, 80; Shell, 8; Interior, 50. (a) The dielectrophoretic force tends to push the cell away from high field regions at low and high frequencies (negative dielectrophoresis). Between about 0.1 and 10 MHz, the movement is towards the electrodes (positive dielectrophoresis); this frequency range is used for cell collection in electrofusion (Chapter 4). (b) The torque (or rotation speed of a cell) exerted by a rotating electrical field has two extrema, which occur at frequencies, f_{c1} and f_{c2}. (c) The velocity of a particle driven by an electrical travelling wave. At low and high frequencies, negative dielectrophoresis levitates the particle and velocity is proportional to the travelling-wave force. As the frequency changes to give positive dielectrophoresis, the particle may at first roll along the electrodes but it is eventually immobilized, so that the broad peaks become extremely sharp on one side.

III. DIELECTROPHORESIS AND LEVITATION

Dielectrophoresis is fundamental to the levitation, trapping, and travelling-wave-induced transport of cells. For reasons that will become apparent, negative (rather than positive) dielectrophoresis is of particular interest in this chapter.

However, the enormous impact of positive dielectrophoresis on the electrofusion of cells (see Chapter 4) demands a brief review of this subject area. Further exploitation of this phenomenon is certainly possible and should be encouraged. Particle, cell, and perhaps even molecular filters may be anticipated, and there is potential for particle concentrators and focusing devices.

A. POSITIVE DIELECTROPHORESIS

The first observations of dielectrophoresis in liquids date back to the 19th century.[143] In very strong, non-uniform fields quinine sulphate particles were seen to be attracted to high field intensities and also to each other. The latter effect caused particles to collect along the field lines (this was later termed pearl-chaining[144]). However, this very early work[143] used direct current fields, so that some (but not all) of the particle movement may have been electrophoretic in origin. A quantitative understanding of the effect seems to have been absent, and indeed the first theoretical description of the force exerted on a polarizable material in a non-uniform field came at the beginning of this century.[59]

High-frequency fields with no direct current component were first used for work on particle suspensions in 1927.[144] These fields cannot give rise to electrophoretic effects. The formation of pearl chains of fat particles which was observed in such high frequency fields[144] was therefore definitely due to dielectrophoresis alone. Pearl chaining of cells[9] was first reported in 1939, and later (in the fifties and sixties) it was carefully analyzed by many workers.[145-148] Around 1960, the term "dielectrophoresis" was introduced into the literature[149] and seems to have become firmly established in the scientific (especially the biological) language, despite criticisms.[150-152] Although the term "dielectrophoresis" is appealing (like the term "electroporation," see discussion about terminology in Chapter 1), it does not describe cell movement in response to a non-uniform alternating field properly. The fundamental effect operating in dielectrophoresis is displacement: either the displacement of medium from regions of high field by particles of higher polarizability (permittivity or dielectric constant) or the displacement of particles by medium of higher polarizability. The expression "dielectric migration," used to distinguish dielectrophoresis from other effects operating in electrostatic separation[89] is, therefore, a good description of the process. In fact, the concentration of high dipole-moment components in the region of the largest field inhomogeneity[153] was recognized in 1938 as one of the possible effects of high fields on a liquid mixture. However, we will continue to use "dielectrophoresis" as an operational term since the word has become so firmly entrenched.

Starting in the sixties, the most widespread application of dielectrophoresis was the collection of cells, chloroplasts, mitochondria, and other microparticles (by pearl-chaining).[51-54,88,154-159] The general belief at that time was that the minimum size of particle that could be collected by dielectrophoresis is limited by Brownian motion. Arguments based on the effective diffusional force[54] also lead to the expectation that, ultimately, random thermal motion will overwhelm electrically induced motion as particle sizes decrease.

At first glance, the above arguments may lead to the conclusion that particles smaller than several hundred nanometers cannot be collected by dielectrophoresis.[54] However, collection of very small particles, even of relatively large macromolecules, is possible provided that very high fields, large inhomogeneities, or both are used. The high degree of inhomogeneity required for dielectrophoresis, the field geometry, and the mode of operation can be controlled by the way the electrodes are connected. We have mentioned that chained particles may have larger dipoles than isolated ones. In addition, hydration shells, highly polarizable ionic atmospheres, and other effects apparently act to increase the range of particle sizes amenable to dielectrophoresis (and also for trapping and caging, see Section III.C). First reports on the concentration of macromolecules from their solution by means of dielectrophoresis were published about 40 years ago by some authors.[160,161] The difficulty in microscopic observation of this collection was avoided by using the capacitance change due to the formation of a macromolecule-enriched atmosphere around a fine wire electrode. More recently, DNA threads of several microns length were collected and oriented by positive dielectrophoresis.[162-165]

The separation of biological particles according to passive electrical properties and size was frequently claimed as a potential application for dielectrophoresis. Analogy with electrostatic dust precipitators suggests that apparatus based on matrix electrodes and positive dielectrophoresis should very effectively filter living cells out of a suspension and, therefore, also be capable of separating living cells from permeabilized, dead cells and debris (because these are less polarizable). Differences in dielectrophoretic yield or, in some cases, separations have been achieved[51-53,90,92,157,166-173] but, this approach has not attained popularity.

A third application of positive dielectrophoresis is in the study of the dielectric properties of particles. Since dielectrophoresis depends upon the polarizability of the particle in the suspension medium, the magnitude of the dielectrophoretic force is a measure for the dielectric properties of the cell. Measurements were initially made by holding a particle or cell against gravity[174-178] by means of positive dielectrophoresis. A further method of assessment of the dielectrophoretic force is to time the movement of a particle in a simple field geometry.[179-181] Dielectrophoretic velocity has also been assessed by light-scattering.[182]

As noted in Section II, dielectrophoretic spectra can also be used to estimate the dielectric properties of particles or cells, and spectra-based methods are much more informative than simple force measurements. Cellular data obtained in the kHz to low-MHz region can be interpreted in terms of membrane properties.[183-185] This application is very similar to that of electrorotation, as expected from the close theoretical link between these two effects (References 74, 80, 186–188 and Figure 4). Indeed, several studies and reviews have been concerned with comparison of data obtained from these two methods.[184,186,189-191] In those cases where analyses have been carried out together to give particle properties, it can be said that dielectrophoresis and electrorotation do agree broadly with one another.[186] To a large extent, the techniques are complimentary (see Section II) as the maxima of the dielectrophoresis spectra tend to occur near the zeros of the rotation spectra and, by measuring both, good precision can be obtained over a wide frequency range. The comparison can be extended to include impedance measurements on suspensions[187,188,192] (see also Section IV).

A further application of positive (and of negative) dielectrophoresis is to use the pearl chaining phenomenon to visualize electric-field distribution. Although the principle of this is more than a century old,[143] the simultaneous application of bacteria (for positive dielectrophoresis) and latex particles (for negative dielectrophoresis)[193] is novel (see Figures 7 and 8 in Chapter 4).

Almost all the work cited above relied on positive dielectrophoresis and conventional electrode systems (although the effect of negative dielectrophoresis on cells and particles was recognized[52,54]). Therefore, devices for cell separation or selection were plagued by problems caused by cells sticking to the electrodes and problems resulting from the strongest effects being confined (usually) to the very small, high-field volume close to a small electrode (see also discussion in Chapter 4). The proposal of one author[54,88,154,156] to increase the useful volume for particle separation (with possible application in a flow system), the "isomotive force" electrode, did not find widespread application. Such electrode arrangements are interesting but have been rendered unnecessary because most electrode problems can be avoided by use of negative rather than positive dielectrophoresis.

However, recent improvements in the production of microelectrodes by the use of photo- and electron beam-lithography have not only resulted in a rapid development of techniques based on negative dielectrophoresis, but they have also given rise to a resurgence of interest in the application of positive dielectrophoresis.[162,164,165,172] Photo-lithography was first applied to the production of microelectrodes for biological purposes[194] as early as 1960 and to a cellular handling device in the early eighties.[14,195-197] Resulting improvements to cellular dielectrophoretic techniques included the introduction of "meander" chambers[196,197] and devices consisting of interdigitated or castellated electrodes.[124,172,198-203] The main advantage of the interdigitated and other re-entrant electrode arrays over earlier systems is that they give a highly inhomogeneous field over a large proportion of the chamber volume. These properties have facilitated the use of positive dielectrophoresis for electrofusion of oocytes,[195] the diagnosis of cellular viability,[171] the separation of cells of different types (on a batch basis),[172,189] the selective handling of macromolecules,[162-165,204] and dielectrophoretic spectroscopy.[124,200-202,205-208]

The new electrodes also have brought benefits to the image analysis,[205] photometry,[124,203] and light-scattering[182] methods used to assess cell movement, collection, and dispersal on the electrodes. In fact, the highly inhomogeneous field makes it relatively easy to derive a large photometric or other signal, which is an average over a large number of uniformly-sized sub-chambers. This can lead to a rapid analytical result. However, care must be taken in interpreting the photometric results with some electrode geometries, because cell collection can result from negative as well as positive dielectrophoresis if there are significant field minima within the inter-electrode space;[124,172] also, cell size and density play important roles in the generation of photometric signals. The use of polarized light to detect differences in cell orientation[208] (which can be expected to be characteristic for the part of the chamber in which collection is occurring) may possibly be of help in interpreting this kind of data.

Some experimentally observed phenomena associated with positive dielectrophoresis still remain to be fully explained. One of these is the hopping of cells onto the top of planar electrodes in low-frequency (<500 Hz) fields (quoted in Reference 124), while the inter-electrode volumes characteristic of negative dielectrophoresis (see below) at higher frequencies remain unoccupied. A second such phenomenon is the formation of static and dynamic cord-like accumulations of cells lying approximately along the isopotentials between electrodes energized in the range 10 Hz to 10 kHz.[209] Negative dielectrophoresis (together with other effects such as particle-particle interactions and electrophoresis) is suggested as the probable cause of both effects.

B. NEGATIVE DIELECTROPHORESIS

Whereas positive dielectrophoresis is well-known to biologists and immunologists because of its use in cell electrofusion, negative dielectrophoresis (the exclusion of poorly polarizable objects from high-field regions such as the neighborhood of the electrodes) is relatively unknown. This is surprising because the effect can be easily demonstrated with synthetic particles (e.g., coated silica,[54] PVC,[210] latex,[193] and polystyrene[211]). The potential of negative dielectrophoresis was first realized for the stable levitation (trapping) of air bubbles (or of weakly polarizable micro-particles) at particular positions in a polarizable medium.[212-219] Levitation could be accomplished, independent of the buoyancy (negative or positive) of the particle. For bubbles in corn oil, silicone oil, etc., an inverted "V" formation of two plate electrodes was found to give confinement along the vertical axis. However, true trapping (see below) requires a cylindrical symmetry about the vertical axis, such as a ring electrode with disk electrodes above and/or below it.[214-219] Trapping occurs just above the ring (for a bubble) or just beneath it (for a negatively buoyant particle). The technique can be used to give precision measurements of the dipole moment of dielectric particles and is suitable for both single particles and field-aligned pearl-chains.

Despite early recognition that certain field frequency ranges gave zero collection of cells at the electrode[51-54] and that these probably corresponded to negative dielectrophoresis, the first quantitative measurements of the negative dielectrophoresis of cells over a range of frequencies[159] were published in 1981. It seems that such measurements were made possible by the introduction of the axial-flow arrangement a few years previously.[54,158] This is a laminar-flow system where the different migration speeds of differing cells passing between two long, parallel electrodes causes them to be collected at one or other of several outlets.[158,159] The system was used when collection of cells on the electrodes was not desired; it allowed the effects of both positive and negative dielectrophoresis on microorganisms to be exploited.[159] Later, other authors[183,220] used a similar geometry, but with sedimentation under gravity rather than flow, to carry out measurements of negative (and positive) dielectrophoresis.

One reason for the poor recognition and worse characterization of negative dielectrophoresis of cells by biologists is that effects were usually assessed as the "Dielectrophoretic Yield," that is the number or volume of particles that collected on the electrode(s). With the older pin-pin or cylindrical electrodes, this can reflect only positive dielectrophoresis.

The newer, edge-on planar electrodes and interdigitated arrangements may give collection near to the electrodes under negative dielectrophoresis.[124,125,172] This means that dielectrophoretic yield has become uninterpretable and obsolete. The term "Dielectrophoretic Response" (the rate of increase in length of the chains formed in the central region between parallel electrodes)[221] has been introduced and better characterizes negative (and positive) dielectrophoresis. (Particle chains still form under negative dielectrophoresis, but not on the electrode, see photographs in References 125, 193, 210.)

Another reason for negative dielectrophoresis of cells remaining relatively unknown for biologists and immunologists lies in the media which have, up to now, been used. Dilute, low-conductivity, electrofusion media give positive dielectrophoresis in the normal frequency range (say 0.1 to 10 MHz, Figure 4). This is because the effective conductivity of most cells is higher than that of the medium in this frequency range. Due to the frequency-dependent nature of cellular electrical properties and the effect of conductivity as well as permittivity, negative dielectrophoresis does actually occur at both high and (at least in some cases) low frequencies in such media.[40,125,186]

Negative dielectrophoresis can be elicited and strengthened by increasing the polarizability of the medium, either by increasing the permittivity[40] or the conductivity[23,24,186] (Figure 4). Whereas the permittivity of a medium can only be increased by a factor of about two to three (for example, by the addition of glycylglycine or related compounds[40]), the conductivity can, in principle, be varied over a large range (several decades). However, conventional electrode systems can only be used with poorly conducting media (conductivities up to about 1 mS cm^{-1}) because of gas bubble formation at the metal surfaces and heating effects. This is why only solutions of low salinity could be used in most previous electrofusion work.

Field manipulation of cells in saline solutions (e.g., mammalian cell culture media, Ringer's solution, and even sea water) has recently become possible. As was noted above, this was due to the reproducible fabrication of (sub-) microelectrode arrays (planar electrode arrays, stacked planar arrangements of azimuthally symmetrical electrodes, etc.) by means of the processes developed for semiconductor technology. Very small electrodes can be made by electron beam lithography[23,222-224] (Plate 1*). Electrode gaps of less than 100 nm are possible if electron scattering processes are corrected.

* Plate 1 follows page 276.

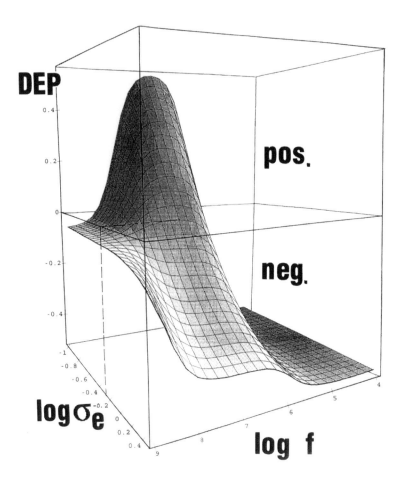

FIGURE 4. Calculation of the dielectrophoretic force (DEP) as a function of the field frequency (f) and the external conductivity (σ_e) for a single shell sphere (parameters as in Figure 3). In the perspective of the figure, high conductivities are at the front. Positive dielectrophoresis occurs over a fairly wide frequency range in poorly conducting solutions, but in highly conductive media (medium conductivity > cell conductivity) only negative dielectrophoresis is found. Note that conductivities are expressed in S m⁻¹.

Thermal problems are much reduced in these systems. As the device is scaled down, the surface-to-volume ratio increases, allowing better heat removal, and the internal distances over which heat must be transported decrease, resulting in smaller temperature rises. This allows higher electrical conductivities to be used. A second advantage of miniature systems is that edge effects become more important, perhaps even dominant. Edge effects, which may normally be neglected for conventional electrodes, can create strong local field gradients which are beneficial for (positive and negative) dielectrophoresis in more conductive solutions (Plate 1). The use of microstructures also makes it easier to avoid adverse electrode processes (e.g., electrolysis). This is because microelectrode structures have the significant characteristic that the very small spacings allow high field strengths and steep gradients to be produced from rather small applied voltages. Thus 10 volts applied across a gap of 5 μm results in a field strength of $2 \cdot 10^4$ V cm^{-1}. Such high field strengths are not necessarily associated with electrical breakdown of the membrane and other damaging effects (see Chapters 1, 2, and 4), if the applied frequency and the thermal design are appropriate.

Following the development of appropriate microstructures and the widespread realization that most cells show negative dielectrophoresis at some frequency, several groups[183-185,225-227] made measurements of the positive to negative crossover frequencies f_o (see Figure 3). These parameters are also of importance in particle trapping and caging[76,227] (see below). The f_o values (particularly that of the first positive to negative crossover in the low frequency range) are expected to increase with the conductivity of the medium for low particle conductivities.[23] There is an approximate analogy to the electro-rotational (and travelling-wave drive) f_c values (see below). Increasing the conductivity of the medium shifts the whole dielectrophoresis spectrum to higher frequencies, extending the low frequency region where negative dielectrophoresis is observed. If the conductivity of the medium exceeds that of the cytosol, cells exhibit only negative dielectrophoresis over the whole frequency range.[23,24]

C. LEVITATION, TRAPS, AND CAGES

As noted above, negative dielectrophoresis causes particles to be repelled from electrodes (high field regions) and to migrate to field minima. This effect can be used to levitate cells or particles. Levitation can be easily achieved above a planar array of microelectrodes[36,40,130,228] because the peri-electrode field is very intense and highly inhomogeneous. Cells above such electrodes experience a force with a component directed vertically upwards. Cell and particle levitation produced by negative dielectrophoresis in microstructures is always stable in the sense that a particle moving upwards is moving into a low-field, low-gradient region and so experiencing less upward force, and, therefore, no feed-back control is necessary. This is a decisive advantage over cell levitation using positive dielectrophoresis, which is inherently unstable and requires complex compensation strategies in any practical device.[177,178,226,229]

High frequency fields can be used to exclude cells from regions. Fibroblasts do not settle from suspension onto a surface protected by an array of energized electrodes; however, cells which grow adherently into such a region seem to suffer no serious adverse effects[23,24] (Figure 5).

Levitation can occur over almost any array (including travelling-wave drives, see below) but electrodes can be designed specifically to use the force to trap particles* (Plate 1) or cells (Figure 6, a and c). Trapping of cells was

---

* Particles are trapped at a field minimum as motion is damped by the viscous medium. This is quite different to the case of atomic particles trapped in vacuum.[230-232] These spin (undamped) on a saddle point in the electromagnetic field and an additional torque is required for stability.[232]

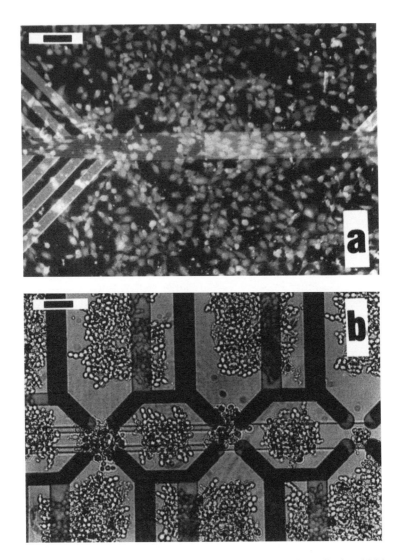

FIGURE 5. Cultivation of cells under prolonged high frequency field application. (a) Mouse fibroblasts after one day in culture medium (σ_e = 14 mS cm^{-1}) on a structure of 1 μm spaced interdigitated microelectrodes. The horizontal lines are 32 electrodes of 500 nm width. Fluoresceindiacetate labelling shows that nearly all the cells exhibit an intact membrane system. The adhered cells were experiencing a field strength of more than 1 kV cm^{-1} (5 MHz) for 24 hours. The bar is 30 μm. (b) Yeast cells after 5 hours of cultivation in growth medium (6 mS cm^{-1}). Some cells were levitated and trapped by negative dielectrophoresis. Normal growth was observed for cells trapped in the field cages and for those outside, simply between electrodes. The field strength in the cage was 500 V cm^{-1} (2 MHz). The bar is 25 μm. (Photograph by courtesy of Müller, T.)

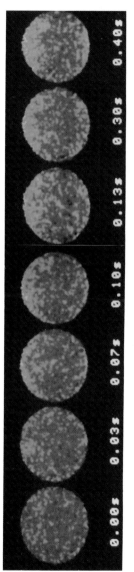
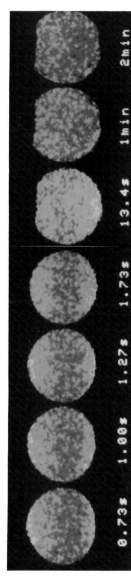

CHAPTER 1, PLATE 1. Time-resolved imaging of intracellular free Ca²⁺ in a sea urchin egg after electropermeabilization (field and medium conditions as in Chapter 1, Figure 8). Changes in Ca²⁺ were imaged by pre-loading the egg with the fluorescent probe indo-1 before field application and by using "dual-view color microscopy" (i.e., simultaneous imaging of the green and blue components of the fluorescence). The fluorescence of the dye is greenish at low Ca²⁺ and turns bluish at high Ca²⁺. Note that two transient Ca²⁺ influxes occur, an early influx through the hemisphere facing the positive electrode which lasted about 1 s and a slower, but more extensive influx from the negative side lasting tens of seconds. (From Kinosita, K., Itoh, H., Ishiwata, S., Hirano, K., Nishizaka, T., and Hayakawa, T., *J. Cell Biol.*, 115, 67, 1991. With permission.)

CHAPTER 5, PLATE 1. Examples of electron-beam-fabricated planar electrode structures with lines and spaces of 500 nm. These allow collection of micron and submicron particles and, in some cases, macromolecules. Excitation signals of several volts and MHz-frequencies are usually applied. This type of structure can be used for positive as well as for negative dielectrophoresis.

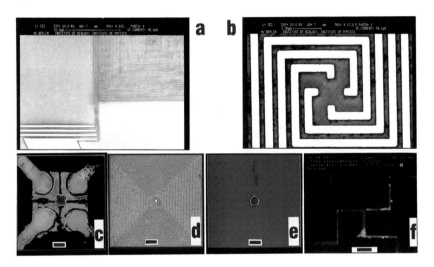

(a) and (b) Scanning electron micrograph of a meander-shaped electrode structure made on silicon by the electron beam fabrication process. The central part (magnified in b) is a tetrode system suitable for the application of alternating or rotating electric fields. The white lines are gold electrodes of 500 nm thickness. The central part is the smallest field trap which is used today. Field strengths of up to several tens of kV cm^{-1} can be applied to aqueous solutions. (c) Four, 2-μm spaced electrodes trapping 1 μm, fluorescently marked latex particles. The bar is 5 μm and false color imaging is by a confocal laser scanning microscope, CLSM-Leica. (d) A single 1 μm particle trapped in a weak electrolyte solution. The bar is 2 μm (fluorescence false color imaging [CLSM]). (e) 14 nm latex beads are collected between four electrodes fabricated by electron-beam lithography on a quartz glass wafer. Prior to field application, particle aggregates (if any exist) are smaller than 100 nm. With field, a large aggregate forms in a few seconds. Minimum manipulable particle size is less than sometimes assumed[54] (false color imaging as described above; bar, 2 μm). (f) Concentration of porin trimers in a detergent/water solution. The fluorescently marked porins concentrate in some regions of the electrode system (fluorescence image; bar, 50 μm). (Photographs courtesy of Frank, H., Gerardino, A., and Müller, T.)

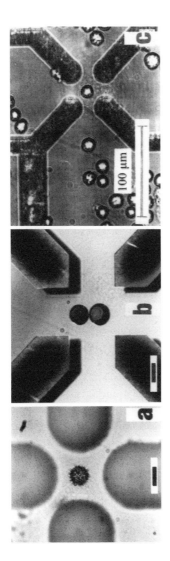

FIGURE 6. Particles levitated above a four-electrode micro-chamber (a, c) or trapped in an eight-electrode cage (b). (a) Pollen of *Helianthus annuus* L (bar is 50 μm). (b) Two Sephadex spheres (bar is 50 μm). (c) Hybridoma cell in culture medium.

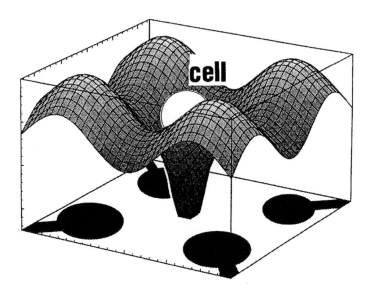

FIGURE 7. Illustration of the force distribution in a quadrupole field trap. Four planar electrodes (black circles) produce a time-average, electric field funnel. The vertical component of force is the same at all points on the curved surface. Negative dielectrophoresis levitates and stably traps a cell at a height where the electric force counterbalances the sedimentational force.

suggested early,[54] but practical devices require the use of negative dielectrophoresis and microstructures. The high degree of field inhomogeneity necessary for trapping means that the electrode dimensions must be comparable to those of the cell to be levitated.

The interplay of the negative dielectrophoretic force with the opposing gravitational force creates a surface on which a particle experiences no vertical force. For a trap,[23,40] this surface should be funnel-like, deep, and steep sided, and horizontal force components should be directed towards the center. A minimum of three (individually addressable) electrodes are required. Best trapping occurs when high frequency alternating voltages* are applied to a symmetric four electrode structure (e.g., arrays of four azimuthally symmetrical, co-planar electrodes as shown in Figure 7 to create a quadrupole;[40] see also Figure 6, a and c). This generates the required "field funnel." Thus, upon excitation with an alternating field, all particles initially within the funnel are trapped and will eventually come to rest at the bottom of the funnel assuming some viscous damping to be present. The boundaries of the trap depend on the field strength (and the cell properties) at a given frequency. If the field strength is increased, the lowest part of the funnel increases in height so that the particle is further levitated; at the same time the sides of the funnel become less steep. Thus, at sufficiently high field strengths the trapped particles can escape; the trap becomes a simple levitator.

---

* High frequency traps and cages can also be produced by rotating fields as described in Section IV. A rotating field vector leads to an additional torque and, in turn, to slow cell or particle spinning.[23,123] In a three-dimensional microelectrode system, rotation around any axis can be induced.

Negative dielectrophoresis can be used to manipulate small particles.[23,233] A single bead of 1 μm has been trapped, and even 14 nm particles have been collected between very finely spaced electrodes and observed by fluorescence microscopy (Plate 1). Even trapping of macromolecules can be envisaged because of the reasons outlined in Section III.A.

Not only open-topped traps but also closed cages, consisting of stacked arrays of three, four, six, or more azimuthally symmetrical electrodes, are possible.[23,123] In cages, such as the octopole field cage shown in Figures 6b and 8, particles are trapped in the minimum-field zone (i.e., at the center of the cage) solely by electric forces. Other forces, such as gravity and buoyancy lead only to a small displacement of the caged particles and are not obligatory for trapping. Thus, in contrast to open-topped traps, cages allow positioning of particles with densities comparable to those of the surrounding fluid.

Because the effects of buoyancy and gravity are insignificant, cages are useful both for single-particle (cell) work and for making droplet-, particle-, and cell-microaggregates (Figures 6b and 9). The production of such aggregates is known as field-casting. Field-casting of microaggregates may have significant advantages for bringing mechanically sensitive particles or cells together, because no external compression or contact is required. Particles or cells can be cross-linked with tannin, glutaraldehyde, or other agents to prevent dissolution when the field is removed (Figure 9). Thus, the field-casting process may have great potential in biotechnology and biomedicine. Many applications can be envisaged such as: the development of "artificial tissues;" the encapsulation of cell systems (such as Langerhans's islets[234-237]) in thin polymeric gel matrices; the cultivation of (adherent) cells on novel microcarrier systems;[238] electroinjection of DNA on a single-cell level (Chapters 1 and 2) and, particularly, electrofusion of cells remote from the electrodes (Chapter 4).*

The design of micro-cages requires the calculation of field-induced forces acting on the particles (or cells). Due to the complexicity of such systems (e.g., anisotropy of the cells, induction of multipoles in the particles, the geometry and arrangement of the multi-electrodes, thermal gradients and convection in fluid, etc.), the force distribution in the cages can only be calculated numerically as described in Section II and Appendix B. Such calculations lead to spherical, cushion-shaped or other surfaces on which there is constant radially directed force.[123] This means that small particles in such a field will form an aggregate of similar shape (see also Figure 9). Since the geometry of the electrodes determines the force distribution of the cage, ring- and meander-shaped, toroidal, cylindrical, or star-like "force fields" (and, in turn, correspondingly shaped microaggregates) can also be created.

---

* Electrofusion of pairs of heterologous cells in microstructures has been under investigation since 1990,[49,239,240] but without great success. The use of cages and more conductive media seems promising to us. However, with physiological media, the field line distribution differs considerably from that in poorly conducting media (Figure 10, see also Chapter 4, Figure 7). Due to the absence of field constriction in the contact zone of two aligned cells, breakdown should occur nearly everywhere not just in this zone.

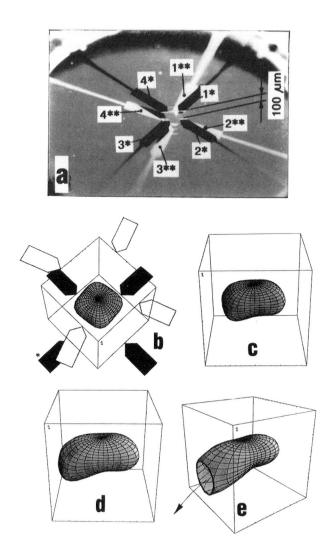

FIGURE 8.   (a) Stereoscopic image of an eight-electrode field cage. Each of two glass wafers
carries four planar electrodes (el. 1*–4*, el. 1**–4**) made using semiconductor technologies.
Only the tips of the electrodes are in direct contact to the solution to avoid unnecessary electric
losses. (b) The surface of constant centripetal force for a negative dielectrophoresis cage when all
electrodes are driven with equal amplitude. A particle would be trapped inside this (closed) surface.
An aggregate of similar geometry would form if a large number of small particles were trapped.
(c) to (e) Progressive decreases in the amplitude applied to the starred electrode in (b) would displace
a trapped cell from the center of the cage and, finally, release it through the opening illustrated in (e).

Deformations of the cage surface can be achieved by varying the amplitude,
frequency, or phase of the voltage signals at one or more electrodes. Thus,
under controlled conditions, the cage can be opened so that particles can be
inserted into or released from the cage (Figure 8). Furthermore, varying the

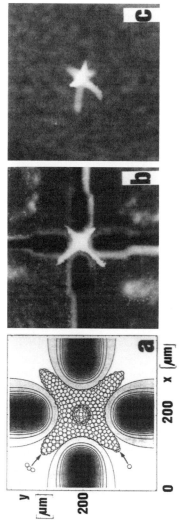

FIGURE 9. The formation of an irregular particle aggregate in an octopole field cage. Only the upper four of the eight electrodes are shown. (a) The lines show the surface of constant field strength squared at the mid plane of the cage. (b) An aggregate of 1 μm diameter amino-functional Latex beads, which have been treated with glycidylacrylate to enable photoinduced cross-linking. (c) The field-cast microaggregate removed from the cage. (From Fuhr, G., Müller, T., Schnelle, Th., Hagedorn, R., Voigt, A., Fiedler, S., Arnold, W. M., Zimmermann, U., Wagner, B., and Heuberger, A., *Naturwissenschaften*, 81, 528, 1994. With permission.)

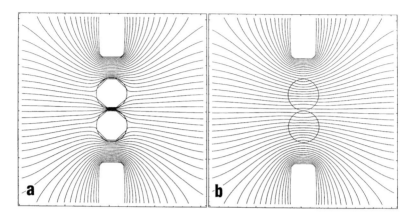

FIGURE 10. Equipotential lines around cells in close proximity. Close spacing of the lines indicates high field strengths. (a) The situation in low conductivity solutions commonly used for electrofusion. High fields develop in the cell regions directed toward the electrodes and also between the cells. This latter region is essential for cell fusion. (Chapter 4). (b) By contrast, in highly conductive solutions such as mammalian cell culture media, the contact area between cells is less influenced by the field which makes fusion of cells less probable. (Courtesy of Th. Schnelle.)

amplitude at two electrode pairs in an octopole cage (see above) can lead, under some circumstances, to the generation of two stable field minima in which particles can be trapped.

A discussion of field traps and cages must also include a brief review of positioning and trapping of particles or cells by electrophoretic and optical forces. Electrophoretic trapping of particles, cells, and macromolecules in quadrupole (or other multipole) fields has been successfully used, but requires feed-back control.[241] Furthermore, only a single permanently charged particle or molecule can be trapped, a disadvantage compared to trapping by negative dielectrophoresis. However, new applications in biology, biomedicine, and biotechnology may be envisaged when both techniques are combined. It is easy to provide a direct current feedback signal via one (or more) electrode(s) of the micro-cage.

Laser manipulation[242-249] of solid particles and liquid droplets* has been reported since 1970. The number of applications of laser traps increased

---

* Particles in a laser beam experience two types of force. The first is the scattering force; even a transparent particle will scatter light and, if the refractive index of the particle exceeds that of the medium, the momentum transfer from photons scattered away from the beam pushes the particle to regions of higher intensity as well as along the direction of propagation. The latter component of the force also acts on absorptive or reflective particles and is usually referred to as the radiation pressure. A second force is the gradient force, analogous to positive dielectrophoresis. For a particle considerably smaller than one wavelength, this can be considered as the interaction between the dipole induced in the particle and the field gradient. In a single beam trap, the gradient force must be made stronger than the scattering force, otherwise the particle will move away in response to the radiation pressure. In the case of small particles, it can be shown theoretically that trapping is not possible with the scattering force alone.[250]

dramatically with the introduction of the strongly focused single-beam technique.[246] The strong focusing requires a lens of relatively short focal-length, preferably of high numerical aperture. Such lenses are used in many microscopes, and it is therefore not surprising that the technique was soon applied to cell biology.[247-249] This sort of trap is a freely positionable, orientation-independent device termed "optical tweezers."[247]

Both optical and field traps have promising properties and applications. The considerable differences in device geometry and in frequency range between the field and optical techniques make it likely that certain cell systems or macromolecules can be handled optically, and others electrically.

IV. ELECTROROTATION

A. HISTORICAL BACKGROUND

A steady electric field induces a retarding torque in a rotating, conductive sphere (electromagnetic braking).[251] Starting in 1896, it was then discovered that an intense electrostatic field would cause a dielectric body immersed in a non-conducting liquid to rotate.[252-254] This effect is usually referred to as Quincke rotation. At the end of the last century, several workers demonstrated that a rotating field can induce a torque in a stationary dielectric body. From the theoretical point of view, this is the opposite effect to electromagnetic braking.[255] Voltmeters[55,56] and instruments for detecting dielectric losses[256-259] were constructed on this principle, and rotating field induction devices[56] formed one class of early electrostatic motors (reviewed in Reference 260).

Several laboratories used rotating fields for the measurement of the dielectric properties of macroscopic bodies at this time. Frequencies of a few Hz could be generated by mechanically rotating an electrode array,[257,258] and a single low frequency such as 40 Hz from the main supply was sometimes used.[56,256] Frequencies of some MHz were generated by spark excitation,[258,259] but very high frequencies (up to 75 MHz) only became available with the development of thermionic valves.[261]

Technical improvements, such as the production of relatively uniform rotating fields,[262] enabled the demonstration that the direction of rotation of an insulating dielectric cylinder could be either co-field or anti-field, depending on the conductivity of the suspending liquid. It was probably these experiments which prompted the development of a detailed theory allowing the interpretation of rotating field measurements.[3] The most important of these equations relates the torque (L) exerted on a homogeneous sphere (with radius a) by a rotating field (of strength, E) and radian frequency (ω) to the electrical properties of sphere and medium:

$$L = 12\pi\varepsilon_0\varepsilon_e a^3 E^2 \frac{(\varepsilon_e\sigma_i - \varepsilon_i\sigma_e)(\omega/\omega_c)}{(\varepsilon_i + 2\varepsilon_e)(\sigma_i + 2\sigma_e)(1 + (\omega/\omega_c)^2)} \tag{2}$$

where: σ_i and σ_e represent the conductivities of the sphere and liquid; ε_i and ε_e represent the dielectric constants (these properties are assumed to be independent of frequency); and ε_0 is the permittivity of free space. ω_c (the frequency at which maximum torque occurs) $= (\sigma_i + 2\sigma_e)/\varepsilon_0(\varepsilon_i + 2\varepsilon_e)$.

Equation 2 shows that the direction of the torque (and hence of any rotation) depends on the sign of the term $(\varepsilon_e\sigma_i - \varepsilon_i\sigma_e)$, i.e., it depends only on the electrical properties of sphere and medium. The size of the torque varies with frequency, and there is a maximum even if the electrical properties remain constant.

The effects of relaxation processes due to ionic mobilities and molecular events[5] were considered around 1920, and rotation methods were used to demonstrate their existence in liquids.[6,261] The observations of the effects of fields on the rotational behavior of suspended particles and objects[4,263,264] led to the formulation of a theory explaining how rotating fields could be generated from static fields near lossy objects.[265]

The subject of rotating fields then became rather neglected, although there were occasional publications about electrostatic motors (reviewed in Reference 131). Around 1960 there was a revival of interest stemming from the need for a contactless method of measurement of dielectric properties, especially conduction and energy loss due to molecular dispersion (Born-Lertes effect). A series of papers on rotating-field measurements in liquids[266-273] and one on photoconduction in cadmium sulphide[274] appeared. Rotating fields were used to produce the synchronous orientation of asbestos fibers. This technique rendered the fibers more detectable by light-scattering and it found industrial interest.[275] At this time, studies of the effect of rotating fields on microscopic particles[276] addressed how the average orientation effect changed with frequency.

There was also interest in the subject of cell rotation in non-rotating fields. In 1960 the first report that living cells may rotate when subjected to an alternating (non-rotating) field of 1 to 100 MHz appeared.[211] The rotation could be observed either when two cells had approached each other very closely (by positive dielectrophoresis) or when the field had detached a fragment from a cell. There was also a report of a rotation or realignment of erythrocytes by a mechanism which appeared to involve cell deformation.[277] However, it is likely that the voltage levels used were causing artifacts due to heating or membrane breakdown (see Chapter 1). As interest in the application of electric fields to living cells increased, it was found that suspended yeast cells could be made to rotate anywhere in the field between electrodes.[52,278,279]

It was observed that cell "spinning" was not reproducible over the course of an experiment and that rotation often had a sporadic nature. This random and unpredictable cell rotation led to the (now discarded) hypothesis that cell rotation originates in the coupling of the external field to intrinsic cellular radio-frequency oscillators. The effect was termed "spin resonance."[280-282] Only cells in certain short-lived physiological states (e.g., in certain parts of the life cycle) were supposed to come into resonance with the field. However, observations of relatively sharp frequency dependencies have not been confirmed later by other groups using advanced electrorotation techniques.

In 1981, the phenomenon of multi-cell rotation was discovered. Conditions were published for the reproducible and predictable rotation of cells in alternating fields of a few kHz to a few MHz.[283] Rotation could be induced in high suspension densities of yeast, erythrocytes, plant protoplasts, and mammalian cells. Rotation of practically all cells of any given type could be observed in a frequency range characteristic of that type. Careful investigation revealed that rotation occurred when at least two cells were in close proximity and the line through their centers was not exactly parallel or perpendicular to the field. This contrasted with some earlier reports[54] and led to the suggestion[284] that the rotation was due to an interaction between the dipoles induced in each cell. For rotation to occur, those dipoles must be appreciably phase-delayed with respect to the applied field. The resulting phase-shifted field in the immediate vicinity of each cell adds to the applied field to produce an (orientation-dependent) rotating component. Rotation speed is determined by the dielectric properties of the cells involved and the mechanical properties of their surfaces.

The idea that an intrinsic cellular oscillator was necessary for rotation was finally laid to rest by the demonstration that liposomes[285] and ion-exchange particles[286] also showed "multi-cell" rotation. However, it was several more years before the induced dipole explanation of "cellular spin resonance" was fully accepted.[287]

All the above cellular work made use of non-rotating fields (at least non-rotating applied fields). However, it did lead to the prediction and demonstration that a single cell should spin in a rotating electric field.[132,288] In fact it was the interpretation of the multi-cell phenomenon in dielectric terms, rather than the much earlier macroscopic work on rotating fields, which led to this.

B. ELECTROROTATION OF SINGLE CELLS
1. Experimental

It is difficult to gain detailed information about cell or particle properties from multi-cell rotation because of the electrical and mechanical particle-particle interactions. The use of a rotating field circumvents this problem and makes it possible to get meaningful data, even from single cells in heterogeneous populations.[132,288] The absence of interactions allows the method to be used as a non-invasive, single particle measurement technique. There are a considerable number of results, both practical[120-122,132,134,137,192,288-350] and theoretical.[126,127,129,133,351-372]

Unlike dielectrophoresis, the generation of rotating fields requires several (three or more, optimally four) periodically varying voltages with precise phase shifts relative to each other.[132,288,291-293] Fields which are essentially homogeneous can be produced by the use of four or more relatively large (several mm) electrodes (Figure 11). These may be flat plates[132,137,288,323] or

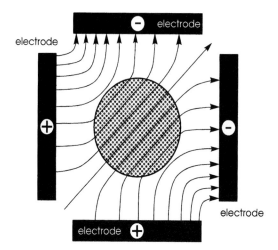

FIGURE 11. Four plate electrodes generate a near uniform rotating field of near constant strength in the shaded area if there are 90° phase shifts between the alternating voltages. In this area, rotation (see Figure 1b) can occur with little interference from dielectrophoresis and with minimal preferred-orientation effects.

specially shaped.[186,320,371] In most work, a rotating field is generated by applying four 90° phase-shifted voltages to four electrodes at right angles to each other.

The shape and volume of the electrorotation chamber are also important design considerations. Although the electrorotation of a single cell is completely independent of chamber volume, large volumes (relative to cell size) are recommended, as they allow rotation to be studied in regions remote from the electrodes, avoiding cell alignment complications. Cell suspension densities should also be low to avoid cell-cell interactions. Thus electrorotation has considerable advantages over impedance measurements which must be made on suspensions of high density. In addition, voltage/current signals must be recorded from the electrodes. Rotation data can be analyzed without consideration of the electrode polarization processes which interfere with impedance measurements, again because electrorotation can use larger volumes. A disadvantage of electrorotation when compared to impedance measurements is that relatively high fields (>10 V cm⁻¹) have to be used.

Electrorotation chambers must have a satisfactory field distribution, and they must be mechanically robust and optically transparent. A tried and tested chamber is described in Reference 323. The electrode spacing is about 1.4 mm; this is large enough to create a nearly uniform field — at least in the center of the chamber. Opposite electrodes may be connected to a conventional conductometer to allow solution conductivity to be monitored. The critical optical part is disposable.

To date, almost all electrorotation experiments have been performed in macroscopic electrode chambers using low-conductivity media to reduce heat production. The osmotic pressure of these nearly ion-free media is maintained by sugars or other non-conductive solutes. Most cells can survive only limited exposure to such media.

The waveform of the driving voltage is an important consideration. Ideally, fields with only a single frequency component are needed when induced cell movement is to be used for analytical purposes. Sinusoidal voltages should be applied to the electrodes. Such voltages were used in much of the early work on electrorotation.[126,132,288,320,322] However, square waves (of 1:1, or even 1:3, mark:space) may be used instead of sine waves with little loss of accuracy.[291,293]

For dielectric spectroscopy using electrorotation, it is not sufficient to generate sine-wave voltages with exact phase shifts. The voltages must remain constant in phase and, especially, in amplitude over a large frequency range. There are few if any analog circuits which do not give a phase change with frequency. Feedback loops can compensate for amplitude errors, but such loops usually fail (due to ringing or overshoot) when frequency changes and/or phase reversals must be carried out rapidly (a few hundred or thousand times per second).[137,304] It is best if signal generation is carried out digitally; exact phase shifts can be generated independent of frequency or its rate of change.

Advances in the field of electrorotation were made possible by the introduction of the "null frequency" method, which allows the frequencies producing rotation maxima to be found quickly and accurately[137,304] (Figure 12). If the rotation speed of a cell is plotted against the logarithm of the frequency, it is frequently found that the curve is symmetric about the frequency, f_c, of maximum rotation (see Equation 2 and upper curve of Figure 12). Also, regardless of whether the rotation is co-field or anti-field, a cell reverses its direction of spin when the direction of the field rotation is reversed. This means that if two, contra-rotating fields (of different frequencies) are applied, it is possible for their effects to cancel and for there to be no time-average resultant torque. For measurement purposes, the fields must be of equal amplitude and have a constant frequency ratio to each other. When the geometric mean ($f_g = \sqrt{f_{low} \cdot f_{high}}$) of the field frequencies matches the frequency of the rotation maximum, the induced torques cancel and the cell stops rotating (the zero crossing point in the lower part of Figure 12). Essentially, the electronics differentiates the rather broad maximum to produce an easily visible null point. The "null frequency" method requires exact phase shifts and amplitudes over the whole frequency range to work properly. Chopping between the two fields (applying them alternately in rapid succession) rather than applying them simultaneously reduces the cost and complexity. In this case, the timing also becomes critical and the changeover period must be less than the rotational time constant due to the moment of inertia of the cell.

An alternative type of generator,[319,334] which enables rotation spectra to be recorded with good precision, keeps one signal constant and varies the parameters of the other. A similar principle is used in feedback-controlled levitation (see above), where a frequency giving positive dielectrophoresis is used to compensate the force due to negative dielectrophoresis.[226,229]

rotation speed

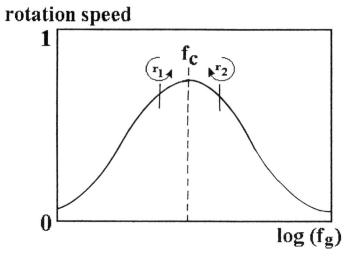

rotation speed

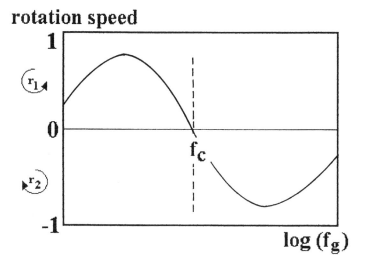

FIGURE 12. The upper curve shows rotation speed when a cell is driven by a single rotating field. The frequency of maximum rotation, f_c, is difficult to identify precisely under such a drive. In the "null-frequency" method two contra-rotating fields (with frequencies f and f/4) are applied alternately.[137] E.g., the lower frequency field (r_1) is rotating anti-clockwise and the higher frequency one (r_2) is rotating clockwise. Cell rotation under this regime is shown in the lower curve. The cell stops rotating when the frequencies reach $f_c/2$ and $2f_c$. This null point is easily detected.

2. The Interpretation of Rotation Spectra

The main features of electrorotation spectra for cells are fairly well understood. As noted above, at low frequencies, a cell has a very large apparent permittivity arising mainly from membrane charging. From Equation 2, such a cell should show strong anti-field rotation, and this is, in fact, found.[288]

Treatment of cells with membrane destructive agents will abolish or even reverse this rotation showing that plasma membrane-charging is the dominant mechanism of polarization in the kHz range in low conductivity media.[299,308] Identification of the membrane-charging time constant as the determining factor, allows the field frequency for optimum anti-field rotation (f_c) to be predicted in terms of membrane parameters (see below). Finding this frequency experimentally is best achieved by the "null frequency" technique.

At high frequencies, the reactance of the membrane becomes negligible and the particle conductivity essentially becomes equal to the cytoplasmic conductivity. Co-field rotation occurs. This demonstrates that a living cell cannot be modelled by assuming a homogeneous particle.[133,357]

Much more detailed interpretations are possible but cellular rotation spectra are complex for reasons discussed in Section II. This is not a "fault" of electro-rotation methods but a reflection of the fact that as the resolution of techniques improves, the amount of information gatherable increases. Many different ways for the calculation of rotational behavior can be found in the literature.[3,5,74,126,127,319,346,356,361,369,372-374] Bearing in mind the assumptions and considerations in Section II and in Appendix A, we develop expressions for the torques acting on different, structured spheres, starting with a homogeneous dielectric sphere and building up to multi-shelled bodies. We consider only dipole moments; this simplifies the equations. Equation 2 can be written as follows if small terms are neglected:[126,127,351]

$$L = K \frac{\omega\tau}{1+(\omega\tau)^2} \tag{3}$$

where K is a combination of the passive electric parameters of the body and the surrounding solution combined with constants, the field strength, E, and the radius, a. τ is the relaxation time of the relevant charge induction:

$$\tau = \varepsilon_0(\varepsilon_i + 2\varepsilon_e)/(\sigma_i + 2\sigma_e) \tag{4}$$

The frequency-dependent torque exhibits a linear increase at low frequencies, shows a maximum at $\omega\tau = 1$, and decreases with further increasing frequency proportional to $1/\omega$. Therefore, in a semi-logarithmic plot there is a symmetrical peak (Figure 12, upper curve).

As noted above, this model cannot explain the rotational behavior of a living cell since cells exhibit two peaks, an anti-field and a co-field peak. We therefore introduce a very thin shell corresponding to the membrane. The torque acting on the cell can now be written in two terms* as each new dielectric leads to an additional dispersion.

---

* This is an approximation, valid if the two peaks are well separated.

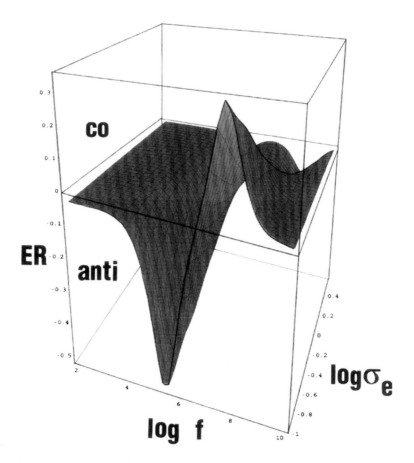

FIGURE 13. Three-dimensional plot of the rotation (ER) of a single shelled sphere as a function of the frequency (f) and the external conductivity (σ_e). Both frames show the same surface. (a) Low external conductivity is in the foreground, and frequency increases left to right. Both, anti- and co-field peaks are clearly seen. (b) High external conductivity is in the foreground, and frequency increases right to left. In highly conductive solutions such as most physiological media, both peaks are now anti-field and shifted toward higher frequencies. The calculations were based on a cell of radius 6 μm with conductivity 5 mS cm^{-1} and permittivity 50, bounded by a membrane of 6 nm thickness (specific conductivity 10^{-10} S cm^{-1}, permittivity 6) in a medium of permittivity 80. Note, conductivities in the Figure are expressed in S m^{-1}.

$$L = K_1 \frac{\omega\tau_1}{1+(\omega\tau_1)^2} + K_2 \frac{\omega\tau_2}{1+(\omega\tau_2)^2} \tag{5}$$

The constants K_1 and K_2 combine all conductivities, permittivities, the field strength, the thickness of the shell, and some other constants. The new shell leads to an additional peak. The two peaks are clearly separated when physiological values for the electrical properties of living cells are substituted into the equations. Depending on the passive electric properties of the cell and the

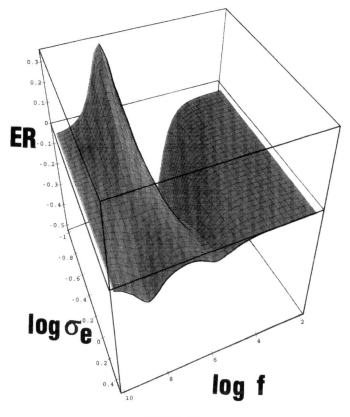

FIGURE 13b.

exterior, K_1 can be negative, whereas K_2 remains positive. This is exactly the situation we observe in all experiments with cells in low conductivity media: anti-field-rotation at low frequencies and co-field-rotation in the MHz-range (Figure 13a). In highly conductive media, K_2 is also negative and two anti-field peaks occur (Figure 13b). The extrema are at $\omega\tau_1 = 1$ and $\omega\tau_2 = 1$, corresponding to the characteristic frequencies $f_{c1} = \omega_{c1}/2\pi$ and $f_{c2} = \omega_{c2}/2\pi$. From these characteristic values typical cell parameters can be obtained (see below).

In many cases, it can be shown that rotation spectra cannot be fitted accurately on the basis of a single shell model. The reason is that a cell contains liquid, viscous and solid components, and membrane bound compartments, etc. Different types of charge accumulate at the dielectric interfaces at differing angles to the rotating field vector, and this can lead to complicated force patterns within the cell.[126,359,363,375] In some circumstances, rotation of (moveable) cellular components[332] can be observed or there can be mechanical stress produced which is used in electrofusion[307] (see Chapter 4).

In the case of plant protoplasts, there is a range of medium conductivities where the rotation maximum is much broader than expected from the single

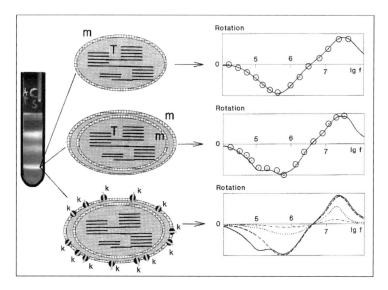

FIGURE 14. Measured and calculated rotation spectra of chloroplasts fractionated on a Percoll step gradient. Two types could be isolated from the lowest band: (i) with only one intact outer membrane, (ii) with intact double envelope membrane system. There are differences in the rotation spectra (two peaks or three peaks, respectively). Double envelope chloroplasts treated with the ionophore, gramicidin (k, lower part of figure), give spectra consistent with a single-shell model. (Osmotic shrinking increases membrane separation and gives similar results.[186]) Rotation spectra reflect the properties of enveloping membranes (m) but not those of the thylacoids (T).

shell model.[297,353] This is due to the fact that the relative magnitude and phase of the voltages induced in the plasma membrane and in the vacuolar membrane (tonoplast) change appreciably as the field frequency is scanned across the rotation maximum. At lower frequencies (which are used when measuring in low conductivity media), the plasma membrane dominates the rotational behavior. At higher frequencies the inner membranes also appear to be influencing the rotational behavior considerably.[297]

In mammalian cells, the large nucleus (which can occupy up to 85% of the cell volume) also contributes to the rotation spectrum,[303] whereas "imprints" of smaller compartments such as mitochondria are usually not seen as they are shielded against the field by the highly conductive cytosol.

However, once organelles (e.g., chloroplasts from higher plant cells) have been isolated, the effect of their double membrane system is clearly seen in the rotation spectrum[186] (Figure 14). Structures such as the cell walls of bacteria, yeast, and plant cells,[133,298,308,310,324,338] the zona pellucida of oocytes,[305,311,325,332,375] the membrane vitellina of insect oocytes,[321] and the exine and intine of pollen[346] also show "signatures" in rotation spectra.

In principle, such rotation spectra can be described by the introduction of a second shell. In this case, the torque can be written as follows (with the proviso that the three peaks are non-overlapping):

$$L = K_1 \frac{\omega\tau_1}{1+(\omega\tau_1)^2} + K_2 \frac{\omega\tau_2}{1+(\omega\tau_2)^2} + K_3 \frac{\omega\tau_3}{1+(\omega\tau_3)^2} \qquad (6)$$

However, the half-width of each peak is approximately one decade of frequency. Due to the similarity of the charge relaxation processes of cellular components three separate peaks are very seldom found.[346] Then Equation 6 becomes more complex.

With each additional shell, at least three more unknown quantities are introduced. Multi-shell models can explain the complexity of the spectrum in terms of the conductivities and permittivities of cellular components and compartments and allow these parameters to be deduced by curve fitting. Several variables have to be determined by the fitting of experimental data but the process often leads to satisfactory results, particularly if the two-shell model is used.[133,311,332,346] Interestingly, the values deduced by such analyses are not average values for the compartment. They represent the dielectric properties of the equatorial area as the field in the polar area does not contribute significantly to rotation. In this respect, single cell electrorotation can be considered as a technique with spatial resolution.

However, multi-shell models are not always superior, particularly if markedly non-spherical cells are studied. The biconcave shape of erythrocytes is difficult to idealize. The best agreement between experiment and theory (as also in the case of impedance measurements) is obtained by assuming a single-shell ellipsoid.[368,369,372,376,377]

A general problem with curve fitting is that the number of unknowns can be too large to allow unambiguous values to be assigned to any of them (this also applies to impedance measurements and dielectrophoresis). In theory, the situation can be improved by taking spectra at several different conductivities. However, the simultaneous fitting of several spectra is a difficult computation, not yet carried out.

Another strategy is to measure both the rotation and dielectrophoresis spectra (at different conductivities) on the same cell or organelle.[186,191] As noted in Section II, these are theoretically related and related in practice if the correct model is being used. Very often, however, a single-shell model gives quite satisfactory results for fitting the rotation curve but not the dielectrophoretic curve (or vice versa). Possible reasons are an oversimplified model or non-linear effects. This becomes particularly evident if data from both spectra are used in a Cole-Cole plot (a control which should always be carried out) since flat and deformed curves appear rather than semicircles.[83,126,186] Regardless of the above, determination of both spectra is always recommended because the resolution and accuracy of the two techniques varies differently with frequency as mentioned in Section III.

With the current state-of-the-art, the best approach for extraction of membrane and cellular data from single-cell electrorotation is to use analytical solutions despite their limitations, especially when comparative values are needed. Analytical solutions can be found for any multi-shelled sphere, cylinder, or ellipsoid provided appropriate experimental conditions are used. For example, if rotation studies are performed in low-conductivity media and the rotation spectrum can be approximated by a single-shelled model, the measurement of the variation of f_c (the frequency of maximum anti-field rotation) with external conductivity yields valuable membrane data. For this case, we can assume that the conductivity of interior of the cell is high compared to the external medium (see Chapter 1). If the specific membrane conductivity per unit area, (G_m), is relatively low, then (with the addition of a surface conductance term) the following equation can be derived:

$$f_c \cdot a = \frac{\sigma_e}{\pi \cdot C_m} + \frac{a \cdot G_m}{2\pi \cdot C_m} + \frac{K_S}{\pi \cdot a \cdot C_m} \tag{7}$$

where: f_c is the frequency giving maximum rotation, a is the radius of the cell, σ_e is the external conductivity, C_m is the specific membrane capacity, G_m is the specific membrane conductance per unit membrane area, and K_s is the surface conductance.

In Equation 7 the frequency is normalized against radius to take into account the variations in size found in typical cell populations. This equation is very useful for experimental work on single cells and provides pressing additional reasons for the use of low conductivities. Equation 7 states that a plot of the normalized frequency, (f_c), of the anti-field rotation peak versus the external conductivity yields a straight line. Experiments using thin-walled glass or agar spheres filled with electrolyte solution as macroscopic models of cells and, also, measurements on isolated plant vacuoles confirmed this prediction.[354,357]

For various types of cells, a linear relationship between f_c and σ_e was also found in the frequency range of 10 to 1000 kHz[136,137,288,345,348,350] (Figure 15, a and c). From the slope of the straight line, the specific membrane capacity, C_m (μF cm^{-2}), and the whole cell capacitance, $C_c = 4\pi a^2 C_m$ (pF) can be obtained according to Equation 7. The C_m values of cell systems with smooth surface and ideal geometry[136,288,294,309,345,348,350] were in the range of 0.5 to 1 μF cm^{-2}. These values agree well with C_m values derived for similar cell types at physiological conductivity from impedance measurements as well as from current-voltage and from charge pulse measurements (see Chapters 1 and 3). The membrane changes seen by rotation in hypo-osmotically swollen cells[336,345] were also in good agreement with impedance data on the same phenomenon,[378,379] when allowance was made for the difference in the types of cells

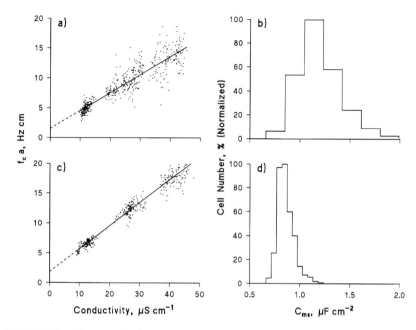

FIGURE 15. Electrorotation data (a,c) on mouse myeloma cells and the corresponding distributions of single cell specific membrane capacity, C_{ms} (b, d) in iso-osmolar (a, b, 280 mOsm) and hypo-osmolar (c, d, 150 mOsm) inositol media. The field frequency causing maximum cell rotation (f_c), the cell radius (a), and the medium conductivity (σ_e) were recorded for each cell. The slopes and intercepts of the least squares straight lines were used to calculate the mean ($<C_{ms}>$), standard deviation, and coefficient of variation for the two samples. The change in C_{ms} and the decreasing heterogeneity that accompany the flattening of the membrane in hypo-osmolar conditions are clear from the histograms.

used. In many cases, the ability to work with very small numbers of cells, even with individual cells, makes the rotation method more informative. Thus impedance measurements[380] on mouse splenocytes (a mixture of B and T lymphocytes) gave an average C_m value of 0.86 µF cm^{-2}. Using rotation, unstimulated B and T lymphocytes could be measured separately, yielding C_m values of 0.93 µF cm^{-2} and 0.76 µF cm^{-2}, respectively.[331] When the ratio of T:B in splenocyte sample was taken into account, the rotational data led to an expected mixed-population value in the range 0.845 to 0.88 µF cm^{-2}, in excellent agreement with the impedance data. The agreement between the rotation and impedance data shows — at least for the membrane capacity — that the normal state of the cell is apparently not affected in the short-term by the unphysiological environment during rotation.

Electrorotation of single cells has also yielded useful data for the evaluation of changes in membrane surface induced by various treatments. Insight into alterations of the membrane surface can be obtained by comparing the changes in the total capacitance (C_c) with those in the apparent specific membrane capacity (C_m).

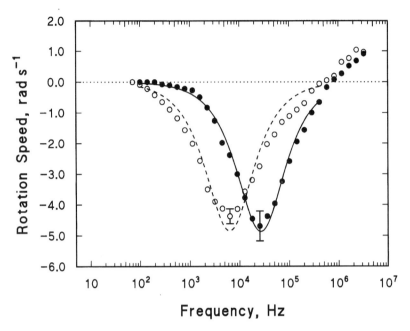

FIGURE 16.   Rotation spectra of mouse myeloma cells in the presence of 5 μg ml⁻¹ [W(CO)₅CN]⁻ (the open circles represent the mean of 5 cells) and untreated (the solid circles represent the mean of 3 cells). The mobile charges introduced by the tungsten complex into the membrane result in a four- to fivefold increase in the apparent membrane capacity over the control data. f_c changes from 27 kHz (control) to 6.3 kHz, and asymmetry enters the spectrum (there are deviations from the symmetric theoretical curve for treated cells). No significant differences in maximum rotation speed exist. Medium conductivity was 42 μS cm⁻¹ in both cases. (Unpublished data of Schenk, W. A., Sukhorukov, V. L., and Zimmermann, U.)

This is because C_c reflects the capacity of the total membrane area, whereas C_m is calculated for a smooth membrane surface assuming an average radius for the cell under investigation (determined from measurements of the radii of many cells under the microscope). C_m values, which are sometimes considerably larger than 1 μF cm⁻², are, therefore, found when microvilli are present or when the plasma membrane is folded. For example, giant murine myeloma cells obtained by electrofusion of many individual cells exhibited C_m values of about 3 μF cm⁻² because of pronounced membrane folding as seen under the light microscope.[336] However, unusually high C_m values are also expected (unpublished data) if the membranes contain mobile charges[381,382] (Figure 16). Such effects on the rotation spectrum, which were observed by myeloma cells after dotting the membranes with large metal-organic compounds (Schenk, W.A., Sukhorukov, V. L., and Zimmermann, U., manuscript in preparation), must be carefully distinguished from surface structure effects. Electrogenic transport systems which contain mobile charges should be visible in spectra.

Electropermeabilization of aphidicolin-synchronized L-mouse cells led to a reduction of both C_c and of C_m suggesting a dramatic loss and smoothing of plasma membrane,[348] presumably involving endocytosis-like vesiculation (see Chapter 1).

In another system, the hypo-osmolar swelling of cultured myeloma cells, rotation measurements were able to give excellent data regarding the reduction of the apparent C_m that occurs due to flattening of microvilli.[345] The concomitant, over-proportional increase in C_c suggested an additional incorporation of internal membrane material.

Equation 7 shows that $f_c \cdot a$ values from single cells, measured at various external conductivities can give an approximation* to the specific membrane capacity values of single cells, C_{ms}. From the data shown in Figure 15 (b and d), the mean and standard deviation of C_{ms} can be extracted. In the case of mouse myeloma cells in isotonic media, the coefficient of variation of C_{ms} was 25–30% implying considerable variations in the degree of villiation among cells.[350] With hypo-osmolar swelling, the microvilli disappear, cell surfaces become smoother, and the heterogeneity is reduced (coefficient of variation of C_{ms} < 10%).

Determination of the specific membrane conductivity, G_m (S cm^{-2}), from the intercept of $f_c \cdot a$ versus σ_e is much more difficult than that of C_m values. The first problem is that the values usually exhibit large statistical uncertainties, because the effect of physiological values of G_m on the rotation spectrum or on the f_c value, is rather small (except in the case of very large cells). The only way to increase the accuracy of the measurement on a given type of cell, assuming that apparatus of the highest available resolution is already in use, is to decrease the medium conductivity still further. Another complication is that the tangential conductivity of extra-cellular structures (e.g., the glycocalix) in Equation 7 cannot be neglected for most cell systems. This term has a very similar effect on electrorotation (or dielectrophoresis or impedance) measurements to that of membrane conductivity, particularly for small cells. It seems that the significant contribution of this surface conductance to the overall conductivity of the cell is often responsible for the comparatively high values of G_m that have been reported from electrorotation experiments. This conclusion is corroborated by findings on insulating particles,[306] isolated cell walls,[324] and isolated zonae pellucidae[305] (see also Reference 137). For these systems f_c is also an accurately linear function of σ_e. The analysis of the data have demonstrated that insulating particles, as well as isolated yeast cell walls and isolated zonae pellucidae, show a measurable conductance. Apparently the surface conductance increases the effective particle conductance significantly. This is particularly the case for small particles.

However, despite the large contribution of surface conductance to membrane conductance, reliable G_m values do seem to have been obtained in some systems. As may be expected from the above, the systems in question were large vesicles or cells, in all cases without appreciable surface structures and mobile charges. For example, the conductivity produced by gramicidin in osmotically swollen thylakoid vesicles[294] showed good agreement with literature values for gramicidin in artificial bilayers. In other reports,[311,325,332] the

---

* The reason that individual C_{ms} values are approximations is that the Y intercept in Figure 15 (a and c) is an average over the cell population. However, the values are accurate enough to be useful since this is usually small compared to the $f_c \cdot a$ values of individual cells.

membrane conductivity of rabbit and mouse oocytes (after proteolytic removal of the zona pellucida) showed a large fertilization-mediated increase in G_m. Furthermore, in a number of systems, large increases in G_m have been seen after challenge of cells with destructive agents.[120,308,313,314]

As noted above, the difficulties associated with the measurement of membrane conductivity are not unique to rotation; the causes and the effects are the same for all techniques that use external fields to induce membrane potentials. It is, therefore, not surprising that there are few (if any) reliable measurements (other than upper limit values) of plasma membrane conductivity using the other non-invasive techniques — dielectrophoresis or impedance spectroscopy. The fact that some good data has been obtained with electrorotation is due to the possibility of using very low conductivity media, combined with the excellent resolution available on single cells (the quantity f_c can be measured with a resolution of about 2% when using the "null frequency" method).

A third cellular parameter that is, in principle, available from rotation measurements is the internal conductivity. But, at least until recently, only a small number of groups could make meaningful measurements of this parameter. There were two reasons for this. The first is that physiological values of internal conductivity exert a dominant role on rotation only at frequencies in the VHF range (30 MHz and higher), where precision measurements are difficult and precision generators are required (these have now been developed and are commercially available). Secondly, whereas yeast, bacteria, plant protoplasts, and pollen tolerate low salt media, certain mammalian cells (see Chapter 1) lose their ionic content quite rapidly. For erythrocytes it was shown[326] that the loss is accelerated by application of significant alternating fields. This means that values of internal conductivity obtained after more than a few minutes in low-conductivity media have little to do with the normal state — at least for mammalian cells. Other methods, such as size measurements below and above the membrane breakdown voltage using a hydrodynamically focusing particle analyzer (see Chapter 1) are superior because physiological solutions can be used.

The ability to measure internal conductivity may soon improve following the introduction of microstructured electrode arrangements.[23,24] As noted in Section III.B, the absence of thermal problems allows the use of high-conductivity media. In the microelectrode systems described above, there are essentially no conductivity limitations. The electrode voltages required, although further into the VHF range, are considerably smaller than those needed for large chambers and are, therefore, easier to generate. This opens up completely new experimental possibilities for characterizing single cells in physiological solutions such as culture media or sea water[23,24,383] (Figure 17).

As expected from the theoretical considerations in Sections II to IV and in Appendix A, two anti-field peaks occur if the conductivity of the medium

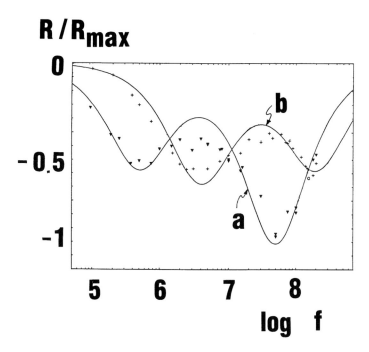

FIGURE 17. Typical rotation-spectrum (R/Rmax) of a mouse 3T3 fibroblast (curve a, σ_e = 13.8 mS cm^{-1}) and a plant protoplast (curve b, σ_e = 26 mS cm^{-1}) measured in highly conductive media. (From Fuhr, G., Glasser, H., Müller, T., and Schnelle, Th., *Biochim. Biophys. Acta,* 1201, 353, 1994. With permission.)

exceeds that of the cytosol (Figure 13b).* Above an external conductivity of 100 mS cm^{-1} (that is above sea water conductivity: 10 to 90 mS cm^{-1}), the anti-field peak related to membrane charging disappears and the rotation spectrum is determined only by the dielectric constants and the external conductivity.

Cells can be positioned by negative dielectrophoresis and rotation studied in strongly inhomogeneous field under physiological conditions. The electrode arrays described for traps and other planar microelectrode arrangements may be used to produce cell rotation.[23,40,123,384] Although many applications may arise from these new possibilities, the complimentary rotation measurements in low-conductivity media are still useful and even necessary. In high-conductivity solutions, rotation is slower (by a factor of 5 to 10); this demands more accurate electronic equipment.

---

* This effect is expected to occur when the ionic concentration of the medium equals that of the cell interior because the ionic mobility is damped in the viscous cytosol.

V. TRAVELLING-WAVE DRIVES

Travelling-waves can be used both to pump liquids and to induce particle movement. Experimentally, a travelling-wave is produced if appropriate voltages (comparable to those used in electrorotation) are applied in repeating succession to many parallel (or near parallel) electrodes arranged stripwise. Phase-shifted signals are necessary to induce field propagation.[57,58,385] A field vector of the travelling-wave propagates at right angles to the electrode strips. The rotating fields described above are, essentially, "wrapped round" travelling-wave drives.

A. ELECTROHYDRODYNAMIC FLUID PUMPING

Theoretical and experimental reports on the pumping of liquids by travelling-waves have appeared from the mid-sixties onward.[57,58,374,385-390] Motion of less conductive fluids like corn oil was studied by using macroscopic ring-shaped fluid pumps (with diameters of several tens of centimeters). The aim of this work was to pump oil through pipelines, but this was not successful.

Two phenomena can be used to pump fluids electrohydrodynamically. Both involve polarization and charge relaxation, but the relaxation can occur at either dielectric interfaces[385] or in the bulk solution[57] (Figure 18). The simplest pump utilizing charge relaxation at interfaces consists of segmented electrodes on the top of a channel. A gas (low permittivity and highly insulating) lies between the electrodes and the liquid to be pumped (Figure 18a). There is a discontinuity in conductivity and permittivity at the interface between the liquid and gas. The travelling-wave-induced charges at this interface lag the wave of image surface charge on the electrodes. Therefore, a time-average electrical surface traction is produced which moves the fluid with the travelling-wave. The same effect can be achieved by layering two or more non- or poorly mixable liquids of different passive electrical properties. (Note — for each electrode arrangement, there is an optimum distance between the electrodes and the interface.)

A pump based on charge relaxation in the bulk solution must create (or utilize) an anisotropy in the liquid. This is because Maxwell's equations dictate that no space charge can be induced in an isotropic medium. Usually, pure liquids and electrolyte solutions do not have the necessary monotonic, time-stable inhomogeneities. But the imposition of a temperature gradient can create the required anisotropy as the electrical properties of most liquids, especially electrolyte solutions, depend on the temperature. A temperature gradient is, therefore, accompanied by gradients of conductivity and permittivity.

A schematic representation of the physical situation is shown in Figure 18b. If a fluid filled channel is cooled at the bottom and warmed at the top, the electrical conductivity increases linearly from the bottom to the top (for most liquids).[57] A travelling-wave will induce charges in the bulk of such a liquid. Because these charges lag behind the field, there is a driving force which acts in the liquid, not at an interface. The pumping behavior is determined only by the dielectric properties of the liquid. The direction of pumping depends on the direction of the travelling-wave and on frequency; it can easily be reversed by changing the direction of the wave.

Microstructures offer great advantages, in particular — the ability to pump water and electrolyte solutions. The first microstructures to use the principle were semiconductor-based pump-devices[43,391,392] (Figure 18c). Such electrode arrangements for travelling-wave drives have typical dimensions of several hundred nanometers up to some tens of micrometers and need signal amplitudes of 0.5 V up to 50 V.[43,391,393] However, pump operation becomes more and more effective with decreasing dimensions and, in many cases, the necessary heating is provided by the excitatory current itself. As noted above (Section III.B) microsystems fabricated in silicon technology can have electrodes with pitches down to 100 nm and can pump aqueous solutions with excitation voltages ranging from less than 1 V up to several volts.[23]

There is a characteristic frequency, f_c, at which maximum fluid streaming occurs (similar to electrorotation). In the case of fluid motion due to charge relaxation in bulk solution, the streaming behavior is more complex than when pumping is induced by charge relaxation at interfaces.* The reason is that no single relaxation time is present; rather there is a superposition of different relaxing charges along the temperature gradient.[43,390] The optimal frequency to pump an aqueous solution with a conductivity of 10 μS cm^{-1} is approximately 200 kHz (phase-shifted by 90° in four signals); for 100 μS cm^{-1} it is 2 MHz; and so forth.

In this frequency range, most liquids (water, alcohol, butanol, and their mixtures) stream against the direction of field propagation. However, forward streaming is observed in the Hz range (see also Reference 388), a phenomenon whose detailed interpretation is missing. Detailed calculations[43] show that all polarizable liquids which exhibit temperature-dependent conductivity and permittivity can be pumped. Conducting or insulating liquids can be pumped, at different excitation frequencies, without modification to the electrode device. This makes the pumping technique astonishingly universal.

Travelling-wave-drive micropumps have important advantages. They operate without moving elements and can be dried and refilled without noticeable changes in the pumping behavior. The flow direction can be controlled in ways which would not be possible by mechanical impellers.

Direct contact between the fluid and the electrodes can be avoided by covering the metallic surface with thin dielectric layers (lacquer, SiO_2, Si_3N_4). At sufficiently high frequencies (MHz range), such surface coatings do not alter the qualitative behavior of the pump since the field can bridge an isolating layer of less than 0.5 μm thickness.[23]

Devices consisting of microchannels (with heights of 20 to 100 μm and widths between 10 and several hundred μm) individually driven by groups of

---

* Mathematically, solutions to the Poisson equation rather than the Laplace are required as there are space charges.

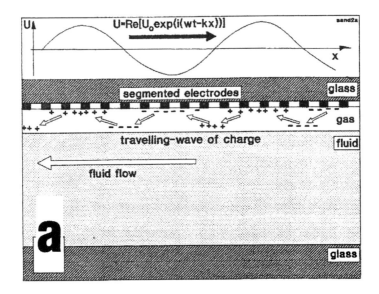

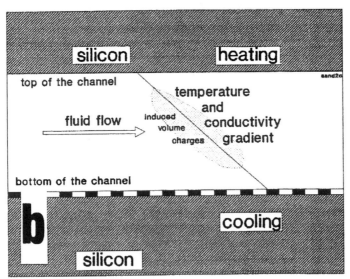

FIGURE 18. Two methods of pumping fluids by means of travelling-electric waves are: (a) Force development at fluid-gas or fluid-fluid interfaces. (b) Pumping by induced space charges in an externally generated temperature gradient. (c) A micropump for aqueous liquids based on the thermopump principle. The array of parallel electrodes is covered by a transparent glass plate with a 100 μm-width channel (the electrodes are not interrupted by the channel). The travelling-wave and the thermal gradient produced by the electrodes combine to create a linear force in the liquid, which may be conductive if the frequency of the travelling-wave is made sufficiently high. (From Fuhr, G., Schnelle, Th., and Wagner, B., *J. Micromech. Microeng.*, 4, 217, 1994. With permission.)

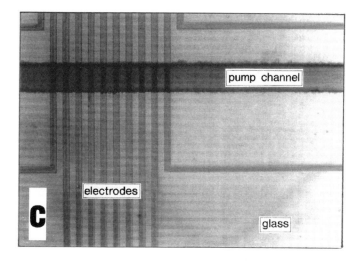

FIGURE 18c.

travelling-wave pumps open up new perspectives for technical, chemical, biological, and medical devices. It seems possible to use such pumps for electronically controlled mixing systems. Another application could be the micro-scale separation and collection of different components (e.g., high performance liquid chromatography effluent) as they pass through a branch point at the input to several micropumps. The ability to produce a counter-flow,[43] even in capillary systems, may be of particular interest, suggesting applications in micro-scale apparatus for counter current separations or for mixing adjacent microdrips. Particle-trapping behavior (see above) could be exploited to implement an electrically controlled filter.[43,393]

B. TRAVELLING-WAVE-INDUCED
PARTICLE AND CELL MOTION

Travelling-waves can also induce cell or microparticle motion. Investigations started only recently.[50,394] In these studies, the propagation velocity of the field and the particle were similar; this is termed "synchronous operation." Synchronous operation requires long relaxation times, and the trajectories of particle motion show complicated loops.[50,394] Linear motion can be better achieved in a strongly asynchronous operating mode in microchannels.[23,43,395,396]

As noted above, it is now possible[23,50,392-395] to fabricate planar strip electrodes on wafer substrates with an accuracy in the (sub-) micron-range using complimentary metal oxide silicon techniques (Figure 19) and electron beam lithography.

The physical description of cell motion near planar strip electrode arrays is similar to that explained above for electrorotation. A small particle exposed to

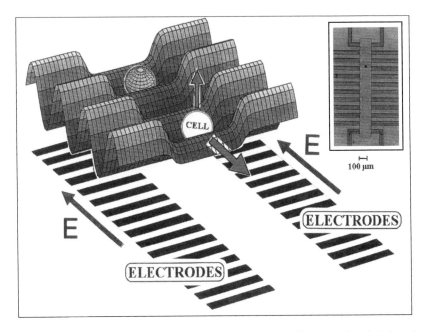

FIGURE 19. Arrangement of gapped travelling-wave electrodes illustrating the principle and the direction of field propagation E, cell levitation, and motion. A microstructure fabricated in silicon technology with pollen grain (black point in the channel) is shown as photograph. Due to negative dielectrophoresis cells are focused toward the gap.

a travelling-wave becomes polarized and, if the field is applied from one side (say the bottom of a liquid-filled channel), the polarization is more or less asymmetric.[395-397] Depending upon whether the delay between the imposed field and the induced dipole moment is greater or less than half of the travelling-wave period, the dielectric particle or cell moves in the direction of field propagation or in the opposite direction, respectively (Figure 3 and, for more details, Reference 395). Each suspended particle is driven individually, and its motion occurs at right angles to the electrode strips.

However, the behavior of particles is more complex than in the case of electrorotation in low-conductivity media. This is due to the superposition of dielectrophoretic effects. With weak positive dielectrophoresis, cells and particles are attracted toward the electrodes and roll over the surface.[395] With increasing positive dielectrophoresis they are pressed onto the electrodes more strongly and no motion can occur. Therefore, forward and reverse motion (in analogy to co-field and anti-field rotation) exist, but the velocity-frequency curves are cut off sharply when positive dielectrophoresis occurs (see Figure 3).

With negative dielectrophoresis, by contrast, the particles and cells are levitated and move freely suspended in the liquid, retarded only by Stoke's friction. Thus, travelling-wave excitation of closely spaced electrode arrays as shown in Figure 19 induces directed, contactless transport of particles.[50,138,395,396] An interesting observation is that excitation will tend to concentrate particles (which show negative dielectrophoresis in the gap) at the exit from the electrode region.[43] Despite continuous pumping of the liquid, the particles become trapped in this region and exhibit a remarkable rolling motion. This effect still requires a reasonable physical explanation but is used to concentrate particles.

Generation of cell motion by a travelling-wave can best be achieved at frequencies near f_c. For living cells in low conductivity media, negative dielectrophoresis occurs in the upper MHz range (see Section III.B). In highly conductive solutions, negative dielectrophoresis occurs exclusively and linear motion can be induced quite easily, even for fragile cells. However, it has to be noted that increasing the conductivity reduces the particle speed[23] by a factor of 5 to 10.

As a first application, there is a two-dimensional micromanipulator device consisting of two crossing channels, each covered by electrode strips.[44] Latex, Sephadex, and other polymer spheres can be moved quickly and accurately in all four directions. The propagation velocities of particles with diameters of 10 to 100 μm are up to 500 μm s^{-1}.

There is no doubt travelling-wave-induced linear particle (and cell) motion will open up new avenues in biology, biomedicine, biotechnology, and environmental research. These include:

1. the separation of cells or of inert particles coated with virus, macromolecules, etc.;
2. the rapid transfer of cells from unphysiological into nutrition media (as needed in electrofusion, see Chapter 4);
3. the concentration of particles and macromolecules by the formation of rolling aggregates;
4. studies of short-term exposure of cells to toxins, effectors, inhibitors, or other biological compounds (including possible uses in the study of binding kinetics);
5. many multi-stage processes made possible by electrode arrangements combining traps and travelling-wave drives.

With the recent advances in extreme miniaturization of electrode assemblies, a rapid development of this field can now be foreseen.[23,393,395]

VI. CONCLUSIONS

We are at the dawn of a new epoch of analytical and bio-processing devices based on the technology of microfabrication. Already, there is an array of devices which can achieve effects which were previously impossible. There are traps, cages, and "conveyor-belts" for cells. It is possible to create bio-repellant surfaces by means of "force fields." This has far-reaching application in biotechnology and medicine where reactor, sensor, and implant surfaces must be kept free of growth. Contactless casting of homo- and heterogeneous particle aggregates into complex shapes is possible. Tiny pumps with no moving parts are a reality. Steep, moving, or stationary pH[23] and temperature[24,43,391] gradients can be produced on a micrometer scale.

A major strength of these techniques is the ease with which they can be combined with one another. Cell sorters based on cages and travelling-wave drives can be imagined. Micro mixers and fraction collectors based on micropumps have been mentioned. Arrays of field cages in combination with gradients lead to compartmentalized chemical reaction systems in bulk solutions. Other reaction systems might have a liquid pumped through a series of cages each containing its own (replaceable) cell (or enzyme immobilized on a bead). Very sensitive sensors could be based on tens or hundreds of cages built into the side wall of a small channel through which extremely small amounts of fluid could stream.

Techniques such as dielectrical field flow fractionation,[398-400] thermophoresis, and directed thermo diffusion,[401,402] which, until now, have been unworkable, may become practical propositions due to the steep gradients available from microstructures. It is very likely that microparticle (and even macromolecule) separators and concentrators will be developed.

Microelectrode arrays also combine well with other methodologies — optical, ultrasonic, magnetic, laser manipulation, etc. It should also be remembered that electromanipulation does not require the use of aqueous media at moderate temperatures. Organic solvents and/or low temperatures can be used; this may be helpful in some chemical applications.

Dielectric spectroscopy will be increasingly used in cellular biology as techniques advance and theoretical understanding deepens. Spectroscopic and other microstructure-based methods may begin to answer some other questions posed in the next chapter. We can also expect dielectric spectroscopy to have a significant impact at the organelle and even molecular aggregate levels.

Questions remain, of course, about the effect of electric fields on biological systems; some of these are discussed in the next chapter. Microstructures allow the prolonged application of fields to cells. The first results[23,24] suggest that cells tolerate MHz fields quite well, at least as far the criteria of morphology, growth, motility, division, and anchorage time are concerned. Of course, there may be more subtle effects, and this field is deserving of study.

As stated at the beginning of this chapter, technical advances in one field can often lead to predictable and unpredictable developments in other fields. Electric micro-tools are now emerging from the laboratory, and we can expect spectacular developments. There is need for both basic research and interdisciplinary work.

REFERENCES

1. **Schütt, W. and Klinkmann, H.,** *Cell Electrophoresis,* W. de Gruyter, Berlin, 1985.
2. **Hannig, K. and Heidrich, H.-G.,** *Free-Flow Electrophoresis,* GIT Verlag, Darmstadt, 1990.
3. **Lampa, A.,** Über Rotationen im elektrostatischen Drehfelde, *Wien. Ber. 2a,* 115, 1659, 1906.
4. **v. Lang, V.,** Versuche im elektrostatischen Wechselfelde, *Sitzungsberichte der Akademie der Wissenschaften in Wien 2a,* 116, 976, 1907.
5. **Born, M.,** Über die Beweglichkeit der elektrolytischen Ionen, *Z. Phys.,* 1, 221, 1920.
6. **Lertes, P.,** Untersuchungen über Rotationen von dielektrischen Flüssigkeiten im electrostatischen Drehfeld, *Z. Phys.,* 4, 315, 1921.
7. **Fürth, R.,** Die absolute Bestimmung von Dielektrizitätskonstanten mit der Ellipsoidmethode, *Z. Phys.,* 44, 256, 1927.
8. **Krasny-Ergen, W.,** Nicht-thermische Wirkungen elektrischer Schwingungen auf Kolloide, *Hochfrequenztechn. Elektroakust.,* 40, 126, 1936.
9. **Liebesny, P.,** Athermic short wave therapy, *Arch. Phys. Ther.,* 19, 736, 1939.
10. **Heinze, J.,** Ultramicroelectrodes in electrochemistry, *Angew. Chem., Int. Ed. Engl.,* 32, 1268, 1993.
11. **Göpel, W.,** Chemical sensing, molecular electronics and nanotechnology: interface technologies down to the molecular scale, *Sensors and Actuators,* B4, 7, 1991.
12. **Wise, K. D. and Najafi, K.,** Microfabrication techniques for integrated sensors and microsystems, *Science,* 254, 1335, 1991.
13. **Sethi, R. S.,** Transducer aspects of biosensors, *Biosensors & Bioelectronics,* 9, 243, 1994.
14. **Giaever, I. and Keese, C. R.,** Monitoring fibroblast behavior in tissue culture with an applied electric field, *Proc. Natl. Acad. Sci. U.S.A.,* 81, 3761, 1984.
15. **Dow, J. A. T., Clark, P., Connolly, P., Curtis, A. S. G. and Wilkinson, C. D. W.,** Novel methods for the guidance and monitoring of single cells and simple networks in culture, *J. Cell. Sci. Suppl.,* 8, 55, 1987.
16. **Giaever, I. and Keese, C. R.,** Fractal motion of mammalian cells, *Physica D,* 38, 128, 1989.
17. **Kowolenko, M., Keese, C. R., Lawrence, D. A. and Giaever, I.,** Measurement of macrophage adherence and spreading with weak electric fields, *J. Immun. Meth.,* 127, 71, 1990.
18. **O'Neill, C., Jordan, P., Riddle, P. and Ireland, G.,** Narrow linear strips of adhesive substratum are powerful inducers of both growth and total focal contact area, *J. Cell Sci.,* 95, 577, 1990.
19. **Giaever, I. and Keese, C. R.,** Micromotion of mammalian cells measured electrically, *Proc. Natl. Acad. Sci. U.S.A.,* 88, 7896, 1991.
20. **Cooper, J. M., Barker, J. R., Magill, J. V., Monaghan, W., Robertson, M., Wilkinson, C. D. W., Curtis, A. S. G. and Moores, G. R.,** A review of research in bioelectronics at Glasgow University, *Biosensors & Bioelectronics,* 8, 22, 1993.
21. **Weibezahn, K. F., Knedlitschek, G., Dertinger, H., Bier, W., Schaller, T. and Schubert, K.,** An *in vitro* tissue model in mechanically processed microstructures, *BIOforum,* 17, 49, 1994.
22. **Singhvi, R., Kumar, A., Lopez, G. P., Stephanopoulos, G. N., Wang, D. I. C., Whitesides, G. M. and Ingber, D. E.,** Engineering cell shape and function, *Science,* 264, 696, 1994.
23. **Fuhr, G., Müller, T., Schnelle, Th., Hagedorn, R., Voigt, A., Fiedler, S., Arnold, W. M., Zimmermann, U., Wagner, B. and Heuberger, A.,** Radio-frequency microtools for particle and live cell manipulation, *Naturwissenschaften,* 81, 528, 1994.
24. **Fuhr, G., Glasser, H., Müller, T. and Schnelle, Th.,** Cell manipulation and cultivation under A.C. electric field influence in highly conductive culture media, *Biochim. Biophys. Acta,* 1201, 353, 1994.

25. **Connolly, P., Clark, P., Curtis, A. S. G., Dow, J. A. T. and Wilkinson, C. D. W.,** An extracellular microelectrode array for monitoring electrogenic cells in culture, *Biosensors & Bioelectronics,* 5, 223, 1990.

26. **Fromherz, P., Offenhäusser, A., Vetter, T. and Weis, J.,** A neuron-silicon junction: a Retzius cell of the leech on an insulated-gate field-effect transistors, *Science,* 252, 1290, 1991.

27. **Fromherz, P., Dambacher, K. H., Ephardt, H., Lambacher, A., Müller, C. O., Neigl, R., Schaden, H., Schenk, O. and Vetter, T.,** Fluorescent dyes as probes of voltage transients in neuron membranes, *Ber. Bunsenges. Phys. Chem.,* 95, 1333, 1991.

28. **Bradley, R. M., Smoke, R. H., Akin, T. and Najafi, K.,** Functional regeneration of glossopharyngeal nerve through micromachined sieve electrode arrays, *Brain Res.,* 594, 84, 1992.

29. **Jimbo, Y. and Kawana, A.,** Electrical stimulation and recording from cultured neurons using a planar electrode array, *Bioelectrochem. Bioenerg.,* 29, 193, 1992.

30. **Jimbo, Y., Robinson, H. P. C. and Kawana, A.,** Simultaneous measurement of intracellular calcium and electrical activity from patterned neural networks in culture, *IEEE Trans. Biomed. Eng.,* 40, 804, 1993.

31. **Clark, P., Britland, S. and Connolly, P.,** Growth cone guidance and neuron morphology on micropatterned laminin surfaces, *J. Cell Sci.,* 105, 203, 1993.

32. **Tatic-Lucic, S., Tai, Y.-C. and Wright, J. A.,** Silicon-micromachined neurochips for *in vitro* studies of cultured neural networks, *The 7th International Conference on Solid-State Sensors & Actuators,* 1993, 943.

33. **Kovacs, G. T. A., Storment, C. W., Halks-Miller, M., Belczynski, C. R., Della Santina, C. C., Jr., Lewis, E. R. and Maluf, N. I.,** Silicon-substrate microelectrode arrays for parallel recording of neural activity in peripheral and cranial nerves, *IEEE Trans. Biomed. Eng.,* 41, 567, 1994.

34. **Trimmer, W. S. N. and Gabriel, K. J.,** Design considerations for a practical electrostatic micro-motor, *Sensors and Actuators,* 11, 189, 1987.

35. **Bart, S. F., Lober, T. A., Howe, R. T., Lang, J. H. and Schlecht, M. F.,** Design considerations for micromachined electric actuators, *Sensors and Actuators,* 14, 269, 1988.

36. **Fuhr, G., Hagedorn, R., Glaser, R. and Gimsa, J.,** Dielektrische Motoren, *Elektrie,* 2, 45, 1989.

37. **Bart, S. F. and Lang, J. H.,** An analysis of electroquasistatic induction micromotors, *Sensors and Actuators,* 20, 97, 1989.

38. **Tai, Y.-C., Fan, L.-S. and Muller, R. S.,** IC-processed micro-motors: design, technology and testing, *Proc. IEEE Int. Micro Electro Mechanical Systems Workshop,* Salt Lake City, UT, 20–22 Feb. 1989, 1.

39. **Tavrow, L. S., Bart, S. F., Lang, J. H. and Schlecht, M. F.,** A LOCOS process for an electrostatic microfabricated motor, *Sensors and Actuators,* A21–A23, 893, 1990.

40. **Fuhr, G., Arnold, W. M., Hagedorn, R., Müller, T., Benecke, W., Wagner, B. and Zimmermann, U.,** Levitation, holding, and rotation of cells within traps made by high-frequency fields, *Biochim. Biophys. Acta,* 1108, 215, 1992.

41. **Dhuler, V. R., Mehregany, M. and Phillips, S. M.,** Micromotor operation in a liquid environment, *IEEE Solid State Sensor & Actuator Workshop* (Hilton Head, SC), 1992, 10.

42. **Hagedorn, R., Fuhr, G., Müller, Th., Schnelle, Th., Schnakenberg, U. and Wagner, B.,** Design of asynchronous dielectric micromotors, *J. Electrostat.,* 33, 159, 1994.

43. **Müller, T., Arnold, W. M., Schnelle, Th., Hagedorn, R., Fuhr, G. and Zimmermann, U.,** A traveling-wave micropump for aqueous solutions: comparison of 1 g and μg results, *Electrophoresis,* 14, 764, 1993.

44. **Fuhr, G., Fiedler, S., Müller, Th., Schnelle, Th., Glasser, H., Lisec, T. and Wagner, B.,** Particle micromanipulator consisting of two orthogonal channels with travelling-wave electrode structures, *Sensors and Actuators,* A41–42, 230, 1994.

45. **Masuda, S., Washizu, M. and Nanba, T.,** Novel method of cell fusion in field constriction area in fluid integrated circuit, *IEEE Trans. Ind. Appl.,* 25, 732, 1989.

46. **Washizu, M., Nanba, T. and Masuda, S.,** Handling biological cells using a fluid integrated circuit, *IEEE Trans. Ind. Appl.,* 26, 352, 1990.
47. **Washizu, M., Jones, T. B. and Kaler, K. V. I. S.,** Higher-order dielectrophoretic effects: levitation at a field null, *Biochim. Biophys. Acta,* 1158, 40, 1993.
48. **Washizu, M.,** Electrostatic manipulation of biological objects, *J. Electrostat.,* 25, 109, 1990.
49. **Urano, N., Kamimura, M., Nanba, T., Okada, M., Fujimoto, M. and Washizu, M.,** Construction of a yeast cell fusion system using a fluid integrated circuit, *J. Biotech.,* 20, 109, 1991.
50. **Masuda, S., Washizu, M. and Iwadare, M.,** Separation of small particles suspended in liquid by nonuniform traveling field, *IEEE Trans. Ind. Appl.,* IA–23, 474, 1987.
51. **Crane, J. S. and Pohl, H. A.,** A study of living and dead cells using dielectrophoresis, *J. Electrochem. Soc.,* 115, 584, 1968.
52. **Pohl, H. A. and Crane, J. S.,** Dielectrophoresis of cells, *Biophys. J.,* 11, 711, 1971.
53. **Pohl, H. A.,** Dielectrophoresis, applications to the characterization and separation of cells, in *Methods of Cell Separation,* Catsimpoolas, N., Ed., Plenum Press, New York, Vol. 1, 1977, 67.
54. **Pohl, H. A.,** *Dielectrophoresis,* Cambridge University Press, Cambridge, 1978.
55. **Arnò, R.,** Campo elettrico rotante per mezzo di differenze di potenziali alternative, *Atti Accad. Naz. Lincei Rend.,* 1, 284, 1901.
56. **Arno, R.,** Ueber ein rotierendes elektrisches Feld und durch elektrostatische Hysteresis bewirkte Rotationen, *Elektrotechnische Z.,* 2, 17, 1893.
57. **Melcher, J. R. and Firebaugh, M. S.,** Traveling-wave bulk electroconvection induced across a temperature gradient, *Phys. Fluids,* 10, 1178, 1967.
58. **Melcher, J. R. and Taylor, G. I.,** Electrohydrodynamics: a review of the role of interfacial shear stresses, *Ann. Rev. Fluid Mech.,* 1, 111, 1969.
59. **Abraham, M.,** *Einführung in die Maxwellsche Theorie der Elektrizität,* Teubner Verlag, Leipzig, 1904.
60. **Stepin, L. D.,** Dielectric permeability of a medium with non-uniform ellipsoidal inclusions, *Soviet Physics — Technical Physics,* 10, 768, 1965 (Engl. Trans *Zhurnal Tekhnicheskoi Fiziki,* 35, 996, 1965).
61. **Sher, L. D.,** Dielectrophoresis in lossy dielectric media, *Nature,* 220, 695, 1968.
62. **Pohl, H. A.,** Theoretical aspects of dielectrophoretic deposition and separation of particles, in *Dielectrophoretic and Electrophoretic Deposition,* Pohl, H. A. and Pickard, W. F., Eds., The Electrochemical Society, New York, 1969, 12–18.
63. **Crane, J. S. and Pohl, H. A.,** Theoretical models of cellular dielectrophoresis, *J. Theor. Biol.,* 37, 15, 1972.
64. **Masuda, S. and Kamimura, T.,** Approximate methods for calculating a non-uniform travelling field, *J. Electrostat.,* 1, 351, 1975.
65. **Jones, T. B.,** Dielectrophoretic force calculation, *J. Electrostat.,* 6, 69, 1979.
66. **Benguigui, L. and Lin, I. J.,** More about the dielectrophoretic force, *J. Appl. Phys.,* 53, 1141, 1982.
67. **Benguigui, L. and Lin, I. J.,** The dielectrophoresis force, *Am. J. Phys.,* 54, 447, 1986.
68. **Martinez, G. and Sancho, M.,** Integral equation methods in electrostatics, *Am. J. Phys.,* 51, 170, 1983.
69. **Mognaschi, E. R. and Savini, A.,** The action of a non-uniform electric field upon lossy dielectric systems — ponderomotive force on a dielectric sphere in the field of a point charge, *J. Phys. D: Appl. Phys.,* 16, 1533, 1983.
70. **Sauer, F. A.,** Forces on suspended particles in the electromagnetic field, in *Coherent Excitations in Biological Systems,* Fröhlich, H. and Kremer, F., Eds., Springer-Verlag, Berlin, 1983, 134.
71. **Lin, I. J. and Jones, T. B.,** General conditions for dielectrophoretic and magnetohydrostatic levitation, *J. Electrostat.,* 15, 53, 1984.

72. **Sauer, F. A.,** Interaction-forces between microscopic particles in an external electromagnetic field, in *Interactions Between Electromagnetic Fields and Cells,* Chiabrera, A., Nicolini, C. and Schwan, H. P., Eds., Plenum Press, New York, 1985, 181.

73. **Chizmadzhev, Yu. A., Kuzmin, P. I. and Pastushenko, V. Ph.,** Theory of the dielectrophoresis of vesicles and cells, *Biologicheskie Membrany,* 2, 1147, 1985.

74. **Pastushenko, V. Ph., Kuzmin, P. I. and Chizmadzhev, Yu. A.,** Dielectrophoresis and electrorotation: a unified theory of spherically symmetrical cells, *Studia biophysica,* 110, 51, 1985.

75. **Schwan, H. P.,** EM-field induced force effects, in *Interactions Between Electromagnetic Fields and Cells,* Chiabrera, A., Nicolini, C. and Schwan, H. P., Eds., Plenum Press, New York, 1985, 371.

76. **Shalom, A. L. and Lin, I. J.,** Principles of high-gradient dielectrophoresis with a rod matrix and determination of particle trajectories, *J. Electrostat.,* 18, 39, 1986.

77. **Jones, T. B.,** Dipole moments of conducting particle chains, *J. Appl. Phys.,* 60, 2226, 1986.

78. **Jones, T. B.,** Dielectrophoretic force in axisymmetric fields, *J. Electrostat.,* 18, 55, 1986.

79. **Jones, T. B. and Rubin, B.,** Forces and torques on conducting particle chains, *J. Electrostat.,* 21, 121, 1988.

80. **Pastushenko, V. Ph., Kuzmin, P. I. and Chizmadzhev, Yu. A.,** Dielectrophoresis and electrorotation of cells: unified theory for spherically symmetric cells with arbitrary structure of membrane, *Biologicheskie Membrany,* 5, 65, 1988.

81. **Sancho, M., Martinez, G. and Llamas, M.,** Multipole interaction between dielectric particles, *J. Electrostat.,* 21, 135, 1988.

82. **Turcu, I. and Lucaciu, C. M.,** Dielectrophoresis: a spherical shell model, *J. Phys. A: Math. Gen.,* 22, 985, 1989.

83. **Jones, T. B. and Kaler, K. V. I. S.,** Relationship of rotational and dielectrophoretic cell spectra, *Proc. Annual EMBS EE Conf.,* Nov. 1990.

84. **Jones, T. B.,** Frequency-dependent orientation of isolated particle chains, *J. Electrostat.,* 25, 231, 1990.

85. **Foster, K. R., Sauer, F. A. and Schwan, H. P.,** Electrorotation and levitation of cells and colloidal particles, *Biophys. J.,* 63, 180, 1992.

86. **Grosse, C. and Schwan, H. P.,** Cellular membrane potentials induced by alternating fields, *Biophys. J.,* 63, 1632, 1992.

87. **Paul, R., Kaler, K. V. I. S. and Jones, T. B.,** A nonequilibrium statistical mechanical calculation of the surface conductance of the electrical double layer of biological cells and its application to dielectrophoresis, *J. Phys. Chem.,* 97, 4745, 1993.

88. **Pohl, H. A. and Plymale, C. E.,** Continuous separations of suspensions by non-uniform electric fields in liquid dielectrics, *J. Electrochem. Soc.,* 107, 390, 1960.

89. **Ralston, O. C.,** *Electrostatic Separation of Mixed Granular Solids,* Elsevier, Amsterdam, 1961.

90. **Black, B. C. and Hammond, E. G.,** Separation by dielectric distribution: theory, *J. Amer. Oil Chem. Soc.,* 42, 931, 1965.

91. **Zebel, G.,** Deposition of aerosol flowing past a cylindrical fiber in a uniform electric field, *J. Colloid. Sci.,* 20, 522, 1965.

92. **Black, B. C. and Hammond, E. G.,** Separation by dielectric distribution: application to the isolation and purification of soybean phosphatides and bacterial spores, *J. Amer. Oil Chem. Soc.,* 42, 936, 1965.

93. **Hall, H. J. and Brown, R. F.,** A new electrostatic liquid cleaner, *Lubric. Eng.,* 488, 1966.

94. **Verschure, R. H. and Ijlst, L.,** Apparatus for continuous dielectric-medium separation of mineral grains, *Nature,* 211, 619, 1966.

95. **Robinson, R.,** The early history of the electrodeposition and separation of particles, in *Dielectrophoretic and Electrophoretic Deposition,* Pohl, H. A. and Pickard, W. F., Eds., The Electrochemical Society, New York, 1969, 1.

96. **Masuda, S., Fujibayashi, K. and Ishida, K.,** Elektrodynamisches Verhalten aufgeladener Aerosolteilchen im inhomogenen Wechselfeld und seine Anwendungsmöglichkeiten in der Staubtechnik, *Staub — Reinhalt. Luft,* 30, 449, 1970.

97. **Masuda, S. and Fujibayashi, K.,** Electrodynamics of charged dust particles within an AC quadrupole electric field. Theoretical treatment, *Electric. Engineer. in Japan,* 90, 1, 1970, translation of *Denki Gakkai Zasshi,* 90, 861, 1970 (in Japanese).

98. **Molinari, G. and Viviani, A.,** Analytical evaluation of the electro-dielectrophoretic forces acting on spherical impurity particles in dielectric fluids, *J. Electrostat.,* 5, 343, 1978.

99. **Molinari, G. and Viviani, A.,** Experimental results and computer simulation of impurity particles motion in N-hexane under D.C. and A.C. conditions, *J. Electrostat.,* 5, 355, 1978.

100. **Benguigui, L. and Lin, I. J.,** Dielectrophoretic filtration of nonconductive liquids, *Sep. Sci. Technol.,* 17, 1003, 1982.

101. **Lin, I. J. and Benguigui, L.,** Dielectrophoretic filtration of liquids. II. Conducting liquids, *Sep. Sci. Technol.,* 17, 645, 1982.

102. **Lin, I. J. and Benguigui, L.,** High-intensity, high-gradient electric separation and dielectric filtration of particulate and granular materials, *J. Electrostat.,* 13, 257, 1982.

103. **Benguigui, L. and Lin, I. J.,** Phenomenological aspect of particle trapping by dielectrophoresis, *J. Appl. Phys.,* 56, 3294, 1984.

104. **Lin, I. J. and Benguigui, L.,** Dielectrophoretic filtration in time-dependent fields, *Sep. Sci. Technol.,* 20, 359, 1985.

105. **Lin, I. J., Shalom, A. L. and Benguigui, L.,** Dewatering of organic liquids by high-gradient dielectrophoretic separation, *Filtration & Separation,* 24, 196, 1987.

106. **Benguigui, L., Shalom, A. L. and Lin, I. J.,** Influence of the sinusoidal field frequency on dielectrophoretic capture of a particle on a rod, *J. Phys. D: Appl. Phys.,* 19, 1853, 1986.

107. **Fricke, H.,** A mathematical treatment of the electric conductivity and capacity of disperse systems, *Phys. Rev.,* 24, 575, 1924.

108. **Dänzer, H.,** Theorie des Verhaltens biologischer Körper im Hochfrequenzfeld, in *Ultrakurzwellen in ihren medizinisch-biologischen Anwendungen,* Dänzer, H., Hollman, H. E., Rajewsky, B., Schaefer, H. and Schliephake, E., Eds., Georg Thieme Verlag, Leipzig, 1938, 191.

109. **Schwan, H. P.,** Electrical properties of tissue and cell suspensions, in *Advances in Biological and Medical Physics,* Lawrence, J. H. and Tobias, C. E., Eds., Academic Press, New York, 5, 1957, 147.

110. **Cole, K. S.,** *Membranes, Ions and Impulses,* University of California Press, Berkeley and Los Angeles, 1968.

111. **Cole, K. S.,** Dielectric properties of living membranes, in *Physical Principles of Biological Membranes,* Snell, F., Wolken, J., Iverson, G. and Lam, J., Eds., Gordon and Breach, New York, 1970, 1.

112. **Grant, E. H., Sheppard, R. J. and South, G. P.,** *Dielectric Behaviour of Biological Molecules in Solution,* Clarendon Press, Oxford, 1978.

113. **Pethig, R.,** *Dielectric and Electronic Properties of Biological Materials,* Wiley & Sons, Chichester, 1979.

114. **Stoy, R. D., Foster, K. R. and Schwan, H. P.,** Dielectric properties of mammalian tissues from 0.1 to 100 MHz: a summary of recent data, *Phys. Med. Biol.,* 27, 501, 1982.

115. **Pethig, R. and Kell, D. B.,** The passive electrical properties of biological systems: their significance in physiology, biophysics and biotechnology, *Phys. Med. Biol.,* 32, 933, 1987.

116. **Arnold, W. M., Gessner, A. G. and Zimmermann, U.,** Dielectric measurements on electro-manipulation media, *Biochim. Biophys. Acta,* 1157, 32, 1993.

117. **Pauly, H. and Schwan, H. P.,** Über die Impedanz einer Suspension von kugelförmigen Teilchen mit einer Schale. Ein Modell für das dielektrische Verhalten von Zellsuspensionen und von Proteinlösungen, *Z. Naturforsch.,* 14b, 125, 1959.

118. **Stepin, L. D.,** Permittivity of a medium with inhomogeneous spherical inclusions, *Soviet Physics — Technical Physics,* 9, 1348, 1965 (Engl. Trans. *Zhurnal Tekhnicheskoi Fiziki,* 34, 1742, 1964).

119. **Irimajiri, A., Hanai, T. and Inouye, A.,** A dielectric theory of "multi-stratified shell" model with its application to a lymphoma cell, *J. Theor. Biol.,* 78, 251, 1979.

120. **Glaser, R. and Fuhr, G.,** Electrorotation of single cells — a new method for assessment of membrane properties, in *Electric Double Layers in Biology,* Blank, M., Ed., Plenum Press, New York, 1986, 227.

121. **Gimsa, J., Pritzen, C. and Donath, E.,** Characterization of virus-red-cell interaction by electrorotation, *Studia biophysica,* 130, 123, 1989.

122. **Donath, E., Egger, M. and Pastushenko, V. Ph.,** Dielectric behavior of the anion exchange protein of human red blood cells. Theoretical analysis and comparison to electrorotation data, *Bioelectrochem. Bioenerg.,* 23, 337, 1990.

123. **Schnelle, Th., Hagedorn, R., Fuhr, G., Fielder, S. and Müller, T.,** Three-dimensional electric field traps for manipulation of cells — calculation and experimental verification, *Biochim. Biophys. Acta,* 1157, 127, 1993.

124. **Pethig, R., Huang, Y., Wang, X.-B. and Burt, J. P. H.,** Positive and negative dielectrophoretic collection of colloidal particles using interdigitated castellated microelectrodes, *J. Phys. D: Appl. Phys.,* 24, 881, 1992.

125. **Wang, X.-B., Huang, Y., Burt, J. P. H., Markx, G. H. and Pethig, R.,** Selective dielectrophoretic confinement of bioparticles in potential energy wells, *J. Phys. D: Appl. Phys.,* 26, 1278, 1993.

126. **Fuhr, G.,** Über die Rotation dielektrischer Körper in rotierenden Feldern, *Habilitation/Dissertation,* Humboldt-Universität zu Berlin, 1985.

127. **Sauer, F. A. and Schlögl, R. W.,** Torques exerted on cylinders and spheres by external electromagnetic fields: a contribution to the theory of field induced cell rotation, in *Interactions Between Electromagnetic Fields and Cells,* Chiabrera, A., Nicolini, C. and Schwan, H. P., Eds., New York, Plenum Press, 1985, 203.

128. **Schwan, H. P.,** Determination of biological impedances, in *Physical Techniques in Biological Research,* Vol. 6, Nastuk, W. L., Ed., Academic Press, New York, 1963, 323.

129. **Paul, R. and Kaler, K. V. I. S.,** Effects of particle shape on electromagnetic torques: a comparison of the effective-dipole-moment method with the Maxwell-stress-tensor method, *Phys. Rev. E.,* 48, 1491, 1993.

130. **Fuhr, G., Hagedorn, R., Müller, T., Benecke, W., Schnakenberg, U. and Wagner, B.,** Dielectric induction micromotors: field levitation and torque-frequency characteristics, *Sensors and Actuators,* 32, 525, 1992.

131. **Bollée, B.,** Elektrostatische Motoren, *Philips technische Rundschau,* 6/7, 175, 1970.

132. **Arnold, W. M. and Zimmermann, U.,** Rotation of an isolated cell in a rotating electrical field, *Naturwissenschaften,* 69, 297, 1982.

133. **Fuhr, G. and Kuzmin, P.,** Behavior of cells in rotating electric fields with account to surface charges and cell structures, *Biophys. J.,* 50, 789, 1986.

134. **Fuhr, G., Müller, T. and Hagedorn, R.,** Reversible and irreversible rotating field-induced membrane modifications, *Biochim. Biophys. Acta,* 980, 1, 1989.

135. **Mahaworasilpa, T. C., Coster, H. G. L. and George, E. P.,** Forces on biological cells due to applied alternating (AC) electric fields. I. Dielectrophoresis, *Biochim. Biophys. Acta,* 1193, 118, 1994.

136. **Zimmermann, U. and Arnold, W. M.,** The interpretation and use of the rotation of biological cells, in *Coherent Excitations in Biological Systems,* Fröhlich, H. and Kremer, F., Eds., Springer-Verlag, Berlin, 1983, 211.

137. **Arnold, W. M. and Zimmermann, U.,** Electro-rotation: development of a technique for dielectric measurements on individual cells and particles, *J. Electrostat.,* 21, 151, 1988.

138. **Fuhr, G., Hagedorn, R., Müller, T., Wagner, B. and Benecke, W.,** Linear motion of dielectric particles and living cells in microfabricated structures induced by traveling electric fields, *Proc. IEEE - MEMS,* 91, Japan, Nara, 1991, 259.

139. **Ackmann, J. J. and Seitz, M. A.,** Methods of complex impedance measurements in biologic tissue, *CRC Critical Reviews in Biomedical Engineering,* 11(4), 281, 1984.

140. **Kell, D. B.,** The principles and potential of electrical admittance spectroscopy: an introduction, in *Biosensors,* Turner, A. P. F., Karube, I. and Wilson, G. S., Eds., Oxford University Press, Oxford, 1987, 427.

141. **Schmukler, R.,** Impedance spectroscopy: the measurement of electrical impedance of biologic materials, in *Electrical Trauma: The Pathophysiology, Manifestations and Clinical Management,* Lee, R. C., Cravalho, E. G. and Burke, J. F., Eds., Cambridge University Press, Cambridge, 1992, Chap. 13.

142. **Jones, T. B., Miller, R. D., Robinson, K. S. and Fowlkes, W. Y.,** Multipolar interactions of dielectric spheres, *J. Electrostat.,* 22, 231, 1989.

143. **Weiler, W.,** Zur Darstellung elektrischer Kraftlinien, *Z. phys. chem. Unterricht,* 4, 194, 1893.

144. **Muth, E.,** Ueber die Erscheinung der Perlschnurkettenbildung von Emulsionspartikelchen unter Einwirkung eines Wechselfelds, *Kolloid-Z.,* 41, 97, 1927.

145. **Wildervanck, A., Wakim, K. G., Herrick, J. F. and Krusen, F. H.,** Certain experimental observations on a pulsed diathermy machine, *Arch. Phys. Med.,* 40, 45, 1959.

146. **Saito, M. and Schwan, H. P.,** The time constants of pearl-chain formation, *Proc. 4th. Ann. Tri-service Conf. Biol. Eff. Microwave Radiation,* 1, 1960, 85.

147. **Schwan, H. P. and Sher, L. D.,** Alternating-current field-induced forces and their biological implications, *J. Electrochem. Soc.,* 116, 22C, 1969.

148. **Schwan, H. P. and Sher, L. D.,** Electrostatic field-induced forces and their biological implications, in *Dielectrophoretic and Electrophoretic Deposition,* Pohl, H. A. and Pickard, W. F., Eds., The Electrochemical Society, New York, 1969, 107.

149. **Pohl, H. A.,** The motion and precipitation of suspensoids in divergent electric fields, *J. Appl. Phys.,* 22, 869, 1951.

150. **Zimmermann, U.,** Electric field-mediated fusion and related electrical phenomena, *Biochim. Biophys. Acta,* 694, 227, 1982.

151. **Barnaby, E., Bryant, G., George, E. P. and Wolfe, J.,** Deformation and motion of cells in electric fields, *Studia biophysica,* 127, 45, 1988.

152. **Barnaby, E., Bryant, G. and Wolfe, J.,** What is "dielectrophoresis"?, *Bioelectrochem. Bioenerg.,* 19, 347, 1988.

153. **Müller, F. H.,** Dielektrische Polarisation von Flüssigkeiten in ungleichförmigem Felde, *Wiss. Veröffentlichungen der Siemens-Werke,* 17, 20, 1938.

154. **Pollock, J. K. and Pohl, H. A.,** Dielectrophoretic cell sorting, in *Modern Bioelectrochemistry,* Gutmann, F. and Keyzer, H., Eds., Plenum Press, New York, 1986, 445.

155. **Ting, I. P., Jolley, K., Beasley, C. A. and Pohl, H. A.,** Dielectrophoresis of chloroplasts, *Biochim. Biophys. Acta,* 234, 324, 1971.

156. **Pohl, H. A. and Pollock, K.,** Electrode geometries for various dielectrophoretic force laws, *J. Electrostat.,* 5, 337, 1978.

157. **Pohl, H. A. and Hawk, I.,** Separation of living and dead cells by dielectrophoresis, *Science,* 152, 647, 1966.

158. **Pohl, H. A. and Kaler, K.,** Continuous dielectrophoretic separation of cell mixtures, *Cell Biophysics,* 1, 15, 1979.

159. **Pohl, H. A., Kaler, K. and Pollock, K.,** The continuous positive and negative dielectrophoresis of microorganisms, *J. Biol. Phys.,* 9, 67, 1981.

160. **Debye, P., Debye, P. P. and Eckstein, B. H.,** Dielectric high-frequency method for molecular weight determinations, *Phys. Rev.,* 94, 1412, 1954.

161. **Debye, P., Debye, P. P., Eckstein, B. H., Barber, W. A. and Arquette, G. J.,** Experiments on polymer solution in inhomogeneous electrical fields, *J. Chem. Phys.,* 22, 152, 1954.

162. **Washizu, M. and Kurosawa, O.,** Electrostatic manipulation of DNA in microfabricated structures, *IEEE Trans. Ind. Appl.,* 26, 1165, 1990.

163. **Shimamoto, N., Kabata, H., Kurosawa, O. and Washizu, M.,** Visualisation of single molecule of RNA polymerase and its application for detecting sliding motion on T7 DNA dielectrophoretically manipulated, in *Structural Tools for the Analysis of Protein-Nucleic Acid Complexes,* Lilley, D. M. J., Heumann, H. and Suck, D., Eds., Birkhäuser Verlag, Basel, 1992, 241.

164. **Washizu, M., Suzuki, S., Kurosawa, O., Nishizaka, T. and Shinohara, T.,** Molecular dielectrophoresis of bio-polymers, *Proc. IEEE/IAS Annual Meeting,* 1992, 1446.

165. **Washizu, M., Kurosawa, O., Arai, I., Suzuki, S. and Shimamoto, N.,** Applications of electrostatic stretch-and-positioning of DNA, *Conference Record IEEE/IAS Annual Meeting,* Toronto, Oct. 1993, 1629.

166. **Mason, B. D. and Townsley, P. M.,** Dielectrophoretic separation of living cells, *Can. J. Microbiol.,* 17, 879, 1971.

167. **Glaser, R., Pescheck, Ch., Krause, G., Schmidt, K. P. and Teuscher, L.,** Dielektrophorese als Grundlage für ein neues Verfahren zur präparativen Zelltrennung, *Z. Allg. Mikrobiol.,* 19, 601, 1979.

168. **Krause, G., Schade, W., Glaser, R. and Gräger, B.,** Anwendung der Dielektrophorese für die präparative Trennung thermotoleranter Hefezellen, *Z. Allg. Mikrobiol.,* 22, 175, 1982.

169. **Lopez, M. C., Iglesias, F. J., Santamaria, C. and Dominguez, A.,** Dielectrophoretic behaviour of *Saccharomyces cerevisiae:* effect of cations and detergents, *FEMS Microbiol. Lett.,* 24, 149, 1984.

170. **Lopez, M. C., Iglesias, F. J., Santamaria, C. and Dominguez, A.,** Dielectrophoretic behavior of yeast cells: effect of growth sources and cell wall and a comparison with fungal spores, *J. Bacteriol.,* 162, 790, 1985.

171. **Archer, G. P., Betts, W. B. and Haigh, T.,** Rapid differentiation of untreated, autoclaved and ozone-treated *Cryptosporidium parvum* oocysts using dielectrophoresis, *Microbios,* 73, 165, 1993.

172. **Markx, G. H., Talary, M. S. and Pethig, R.,** Separation of viable and non-viable yeast using dielectrophoresis, *J. Biotech.,* 32, 29, 1994.

173. **Büschl, R., Ringsdorf, H. and Zimmermann, U.,** Electric field-induced fusion of large liposomes from natural and polymerizable lipids, *FEBS Lett.,* 150, 38, 1982.

174. **Chen, C. S. and Pohl, H. A.,** Biological dielectrophoresis: the behavior of lone cells in a nonuniform electric field, *Ann. N.Y. Acad. Sci.,* 238, 176, 1974.

175. **Pohl, H. A. and Pethig, R.,** A new method of dielectric constant determination using non-uniform electric field effects, *Quantum Theoretical Research Group,* 57, 1, 1975.

176. **Pohl, H. A. and Pethig, R.,** Dielectric measurements using non-uniform electric field (dielectrophoretic) effects, *J. Phys. E,* 10, 190, 1977.

177. **Jones, T. B. and Kraybill, J. P.,** Active feedback-controlled dielectrophoretic levitation, *J. Appl. Phys.,* 60, 1247, 1986.

178. **Kaler, K. V. I. S. and Jones, T. B.,** Dielectrophoretic spectra of single cells determined by feedback-controlled levitation, *Biophys. J.,* 57, 173, 1990.

179. **Stoicheva, N., Tsoneva, I. and Dimitrov, D. S.,** Protoplast dielectrophoresis in axisymmetric fields, *Z. Naturforsch.,* 40c, 735, 1985.

180. **Stoicheva, N. and Dimitrov, D. S.,** Dielectrophoresis of single protoplasts: effects of increased conductivity caused by calcium ions, *Studia biophysica,* 111, 17, 1986.

181. **Tsoneva, I. C., Zhelev, D. V. and Dimitrov, D. S.,** Red blood cell dielectrophoresis in axisymmetric fields, *Cell Biophysics,* 8, 89, 1986.

182. **Kaler, K. V. I. S., Fritz, O. G. and Adamson, R. J.,** Dielectrophoretic velocity measurements using quasi-elastic light scattering, *J. Electrostat.,* 21, 193, 1988.

183. **Marszalek, P., Zielinski, J. J., Fikus, M. and Tsong, T. Y.,** Determination of electric parameters of cell membranes by a dielectrophoresis method, *Biophys. J.,* 59, 982, 1991.

184. **Gimsa, J. and Donath, E.,** Are dispersions in dielectrophoretic and electrorotational cell spectra in the 10 kHz region related to protein properties?, in *Electricity and Magnetism in Biology and Medicine,* Blank, M., Ed., San Francisco Press, San Francisco, 1993, 159.

185. **Gascoyne, P. R. C., Pethig, R., Burt, J. P. H. and Becker, F. F.,** Membrane changes accompanying the induced differentiation of Friend murine erythroleukemia cells studied by dielectrophoresis, *Biochim. Biophys. Acta,* 1149, 119, 1993.

186. **Fuhr, G., Rösch, P., Müller, T., Dressler, V. and Göring, H.,** Dielectric spectroscopy of chloroplasts isolated from higher plants — characterization of the double-membrane system, *Plant Cell Physiol.,* 31, 975, 1990.

187. **Wang, X.-B., Huang, Y., Hölzel, R., Burt, J. P. H. and Pethig, R.,** Theoretical and experimental investigations of the interdependence of the dielectric, dielectrophoretic and electrorotational behaviour of colloidal particles, *J. Phys. D: Appl. Phys.,* 26, 312, 1993.

188. **Wang, X.-B., Pethig, R. and Jones, T. B.,** Relationship of dielectrophoretic and electrorotational behaviour exhibited by polarized particles, *J. Phys. D: Appl. Phys.,* 25, 905, 1992.

189. **Huang, Y., Hölzel, R., Pethig, R. and Wang, X.-B.,** Differences in the AC electrodynamics of viable and non-viable yeast cells determined through combined dielectrophoresis and electrorotation studies, *Phys. Med. Biol.,* 37, 1499, 1992.

190. **Pethig, R.,** Application of A.C. electrical fields to the manipulation and characterisation of cells, in *Automation in Biotechnology, Proc. of the 4th Toyota Conference,* 21–24 October 1990, Karube, I., Ed., Elsevier, Amsterdam, 1991, 159.

191. **Hölzel, R., Lamprecht, I. and Mischel, M.,** Characterization of cells by dielectrophoretic techniques, in *Physical Characterization of Biological Cells,* Schütt, W., Klinkmann, H., Lamprecht, I. and Wilson, T., Eds., Verlag Gesundheit, Berlin, 1991, 273.

192. **Fuhr, G. and Glaser, R.,** Electrorotation — a new method of dielectric spectroscopy, *Studia biophysica,* 127, 11, 1988.

193. **Mehrle, W., Hampp, R., Zimmermann, U. and Schwan, H. P.,** Mapping of the field distribution around dielectrophoretically aligned cells by means of small particles as field probes, *Biochim. Biophys. Acta,* 939, 561, 1988.

194. **Wirtschafter, J. D.,** The photomechanical preparation of multiple lead electrodes, *Rev. Sci. Instr.,* 31, 63, 1960.

195. **Richter, H.-P., Scheurich, P. and Zimmermann, U.,** Electric field-induced fusion of sea urchin eggs, *Develop., Growth and Differ.,* 23, 479, 1981.

196. **Zimmermann, U., Vienken, J. and Greyson, J.,** Electrofusion: a novel hybridization technique, in *Biotech 84 Europe,* 1, 231, 1984.

197. **Förster, E. and Emeis, C. C.,** Quantitative studies on the viability of yeast protoplasts following dielectrophoresis, *FEMS Microbiol. Lett.,* 26, 65, 1985.

198. **Benguigui, L. and Lin, I. J.,** Dielectrophoresis in two dimensions, *J. Electrostat.,* 21, 205, 1988.

199. **Inoue, T., Pethig, R., Al-Ameen, T. A. K., Burt, J. P. H. and Price, J. A. R.,** Dielectrophoretic behaviour of *Micrococcus lysodeikticus* and its protoplast, *J. Electrostat.,* 21, 215, 1988.

200. **Price, J. A. R., Burt, J. P. H. and Pethig, R.,** Applications of a new optical technique for measuring the dielectrophoretic behaviour of micro-organisms, *Biochim. Biophys. Acta,* 964, 221, 1988.

201. **Burt, J. P. H., Al-Ameen, T. A. K. and Pethig, R.,** An optical dielectrophoresis spectrometer for low frequency measurements on colloidal suspensions, *J. Phys. E: Sci. Instrum.,* 22, 952, 1989.

202. **Burt, J. P. H., Pethig, R., Gascoyne, P. R. C. and Becker, F. F.,** Dielectrophoretic characterisation of Friend murine erythroleukaemic cells as a measure of induced differentiation, *Biochim. Biophys. Acta,* 1034, 93, 1990.

203. **Pethig, R.,** Dielectric-based biosensors, *Biochem. Soc. Trans.,* 19, 21, 1991.

204. **Washizu, M.,** Electrostatic manipulation of biological objects, *J. Electrostat.,* 25, 109, 1990.

205. **Gascoyne, P. R. C., Huang, Y., Pethig, R., Vykoukal, J. and Becker, F. F.,** Dielectrophoretic separation of mammalian cells studied by computerized image analysis, *Meas. Sci. Technol.,* 3, 439, 1992.

206. **Talary, M. S. and Pethig, R.,** Optical technique for measuring the positive and negative dielectrophoretic behaviour of cells and colloidal suspensions, *Proc. IEEE Meas. Technol.,* 141, 395, 1994.

207. **Markx, G. H., Huang, Y., Zhou, X.-F. and Pethig, R.,** Dielectrophoretic characterization and separation of micro-organisms, *Microbiol.,* 140, 585, 1994.

208. **Fomchenkov, V. M. and Gavrilyuk, B. K.,** The study of dielectrophoresis of cells using the optical technique of measuring, *J. Biol. Phys.,* 6, 29, 1979.

209. **Mischel, M., Pohl, H. A., Lamprecht, I., Blode, H. and Strube, K. H.,** Dynamic and static cord-structure: a new dielectrophoretic phenomenon, *J. Biol. Phys.,* 11, 81, 1983.

210. **Santamaria, C., Iglesias, F. J. and Dominguez, A.,** Dielectrophoretic deposition in suspensions of macromolecules: polyvinylchloride and Sephadex G-50, *J. Colloid Interface Sci.,* 103, 508, 1985.

211. **Teixeira-Pinto, A. A., Nejelski, L. L., Cutler, J. L. and Heller, J. H.,** The behaviour of unicellular organisms in an electromagnetic field, *Exp. Cell Res.,* 20, 548, 1960.

212. **Parmar, D. S. and Jalaluddin, A. K.,** Dielectrophoretic forces in liquids, *Jap. J. Appl. Phys.,* 13, 793, 1974.

213. **Jones, T. B. and Bliss, G. W.,** Bubble dielectrophoresis, *J. Appl. Phys.,* 48, 1412, 1977.

214. **Jones, T. B. and Kallio, G. A.,** Dielectrophoretic levitation of spheres and shells, *J. Electrostat.,* 6, 207, 1979.

215. **Kallio, G. A. and Jones, T. B.,** Dielectric constant measurements using dielectrophoretic levitation, *IEEE Trans. Indust. Appl.,* IA–16, 69, 1980.

216. **Jones, T. B. and McCarthy, M. J.,** Electrode geometries for dielectrophoretic levitation, *J. Electrostat.,* 11, 71, 1981.

217. **Jones, T. B.,** Cusped electrostatic fields for dielectrophoretic levitation, *J. Electrostat.,* 11, 85, 1981.

218. **Jones, T. B. and Loomans, L. W.,** Size effect in dielectrophoretic levitation, *J. Electrostat.,* 14, 269, 1983.

219. **Tombs, T. N. and Jones, T. B.,** Effect of moisture on the dielectrophoretic spectra of glass spheres, *IEEE Trans. Ind. Appl.,* 29, 281, 1993.

220. **Marszalek, P., Zielinski, J. J. and Fikus, M.,** Experimental verification of a theoretical treatment of the mechanism of dielectrophoresis, *Bioelectrochem. Bioenerg.,* 22, 289, 1989.

221. **Wharton, S. A. and Riley, P. A.,** The effect of agents influencing cytoskeletal elements upon the dielectrophoretic behavior of cultured epidermal cells, *J. Biol. Phys.,* 9, 87, 1981.

222. **Heuberger, A.,** *Mikromechanik: Mikrofertigung mit Methoden der Halbleitertechnologie,* Springer-Verlag, Berlin, 1989.

223. **Menz, W. and Bley, P.,** *Mikrosystemtechnik für Ingenieure,* VCH Verlag, Weinheim, 1993.

224. **Büttgenbach, S.,** *Mikromechanik: Einführung in Technologie und Anwendungen,* 2nd ed., Teubner Verlag, Stuttgart, 1994.

225. **Gimsa, J. and Glaser, R.,** New aspects for theory and application of dielectrophoresis and electrorotation of biological cells, in *Proc. of Eleventh School on Biophysics of Membrane Transport,* Poland, 1992, 156.

226. **Kaler, K. V. I. S., Xie, J.-P., Jones, T. B. and Paul, R.,** Dual-frequency dielectrophoretic levitation of *Canola* protoplasts, *Biophys. J.,* 63, 58, 1992.

227. **Huang, Y. and Pethig, R.,** Electrode design for negative dielectrophoresis, *Meas. Sci. Technol.,* 2, 1142, 1991.

228. **Washizu, M.,** Precise calculation of dielectrophoretic force in arbitrary field, *J. Electrostat.,* 29, 177, 1992.

229. **Kaler, K. V. I. S., Xie, J.-P. and Jones, T. B.,** Double frequency levitation of bioparticles, *Proc. IEEE Trans.,* 13, 1031, 1991.

230. **Paul, W. and Steinwedel, H.,** Ein neues Massenspektrometer ohne Magnetfeld, *Z. Naturforsch.,* 8a, 448, 1953.

231. **Paul, W., Osberghaus, O. and Fischer, E.,** Ein Ionenkäfig, *Forschungsberichte des Wirtschafts- und Verkehrsministeriums Nordrhein-Westfalen,* Report 415, Westdeutscher Verlag, Köln, 1958.

232. **Paul, W.,** Nobel Lecture, 1989; Elektromagnetische Käfige für geladene und neutrale Teilchen, *Phys. Bl.,* 46, 227, 1990.

233. **Bordoni, F., Carelli, P., De Gasperis, G. and Leoni, R.,** Submicron electrodes for negative dielectrophoresis, *Micromechanical Technologies 94 Berlin,* 1994, 465.

234. **Schnabl, H. and Zimmermann, U.,** Immobilization of plant protoplasts, in *Biotechnology in Agriculture and Forestry, Vol. 8: Plant Protoplasts and Genetic Engineering I,* Bajaj, Y. P. S., Ed., Springer-Verlag, Berlin, 1989, 63.

235. **Geisen, K., Deutschländer, H., Gorbach, S., Klenke, C. and Zimmermann, U.,** Function of barium alginate-microencapsulated xenogenic islets in different diabetic mouse models, in *Frontiers in Diabetes Research. Lessons from Animal Diabetes III.,* Shafrir, E., Ed., Smith-Gordon, London, 1990, 142.

236. **Klöck, G., Frank, H., Houben, R., Zekorn, T., Horcher, A., Siebers, U., Wöhrle, M., Federlin, K. and Zimmermann, U.,** Production of purified alginates suitable for use in immunoisolated transplantation, *Appl. Microbiol. Biotech.,* 40, 638, 1994.

237. **Gröhn, P., Klöck, G., Schmitt, J., Zimmermann, U., Horcher, A., Bretzel, R. G., Hering, B. J., Brandhorst, D., Brandhorst, H., Zekorn, T. and Ferderlin, K.,** Large-scale production of Ba^{2+}-alginate-coated islets of Langerhans for immunoisolation, *Exp. Clin. Endocrinol.,* 102, 380, 1994.

238. **Hirtenstein, M. and Clark, J.,** Microcarrier-bound mammalian cells, in *Immobilized cells and organelles, Vol. 1,* Mattiasson, B., Ed., CRC Press, Boca Raton, FL, 1983, 57.

239. **Sato, K., Kawamura, Y., Tanaka, S., Uchida, K. and Kohida, H.,** Individual and mass operation of biological cells using micromechanical silicon devices, *Sensors and Actuators,* A23, 948, 1990.

240. **Matsumoto, N., Matsue, T. and Uchida, I.,** Paired cell alignment using a jagged microarray electrode, *Bioelectrochem. Bioenerg.,* 34, 199, 1994.

241. **Eigen, M. and Rigler, R.,** Sorting single molecules: application to diagnostics and evolutionary biotechnology, *Proc. Natl. Acad. Sci. U.S.A.,* 91, 5740, 1994.

242. **Ashkin, A.,** Acceleration and trapping of particles by radiation pressure, *Phys. Rev. Lett.,* 24, 156, 1970.

243. **Roosen, G. and Imbert, C.,** Optical levitation by means of two horizontal laser beams: a theoretical and experimental study, *Phys. Lett.,* 59A, 6, 1976.

244. **Roosen, G.,** A theoretical and experimental study of the stable equilibrium positions of spheres levitated by two horizontal laser beams, *Opt. Commun.,* 21, 189, 1977.

245. **Ashkin, A. and Dziedzic, J. M.,** Observation of radiation-pressure trapping of particles by alternating light beams, *Phys. Rev. Lett.,* 54, 1245, 1985.

246. **Ashkin, A., Dziedzic, J. M., Bjorkholm, J. E. and Chu, S.,** Observation of a single-beam gradient force optical trap for dielectric particles, *Optics Lett.,* 11, 288, 1986.

247. **Block, S. M.,** Optical tweezers: a new tool for biophysics, *Noninvasive Techniques in Cell Biology,* Wiley-Liss, New York, 1990, 375.

248. **Weber, G. and Greulich, K. O.,** Manipulation of cells, organelles and genomes by laser microbeam and optical trap, *Int. Rev. Cytol.,* 133, 1, 1992.

249. **Visscher, K., Brakenhoff, G. J. and Krol, J. J.,** Micromanipulation by "multiple" optical traps created by a single fast scanning trap integrated with the bilateral confocal scanning laser microscope, *Cytometry,* 14, 105, 1993.

250. **Ashkin, A. and Gordon, J. P.,** Stability of radiation-pressure particle traps: an optical Earnshaw theorem, *Optics Lett.,* 8, 10, 511, 1983.

251. **Hertz, H. R.,** Ueber die Vertheilung der Electricität auf der Oberfläche bewegter Leiter, *Wied. Ann.,* 13, 266, 1881.

252. **Quincke, G.,** Ueber Rotationen im constanten electrischen Felde, *Ann. Phys. Chem.,* 59, 417, 1896.

253. **Heydweiller, A.,** Ueber Rotationen im constanten electrischen Felde, *Verhandlungen der physikalischen Gesellschaft (Braunschweig),* 16, 32, 1897.

254. **v. Schweidler, E. R.,** Über Rotationen im homogenen elektrischen Felde, *Sitzungsberichte der Akademie der Wissenschaften in Wien 2a,* 106, 526, 1897.

255. **Jones, T. B.,** Quincke rotation of spheres, *IEEE Trans. Ind. Appl.,* 20, 845, 1984.

256. **Arnò, R.,** Sulla isteresa dielettrica viscosa, *Atti Accad. Naz. Lincei Rend.,* 5, 262, 1896.

257. **Threlfall, R.,** On the conversion of electric energy in dielectrics. I, *Phys. Rev.,* 4, 457, 1897.

258. **Threlfall, R.,** On the conversion of electric energy in dielectrics. II, *Phys. Rev.,* 5, 21, 1897.
259. **Threlfall, R.,** On the conversion of electric energy in dielectrics. III, *Phys. Rev.,* 5, 65, 1897.
260. **Jefimenko, O. D.,** *Electrostatic Motors: Their History, Types and Principles of Operation,* Electret Scientific Company, Star City, West Virginia, 1973.
261. **Lertes, P.,** Der Dipolrotationseffekt bei dielektrischen Flüssigkeiten, *Z. Phys.,* 6, 56, 1921.
262. **v. Lang, V.,** Versuche im elektrostatischen Drehfelde, *Wien. Ber. 2a.,* 115, 211, 1906.
263. **Lampa, A.,** Über eine einfache Anordnung zur Herstellung eines elektrostatischen Drehfeldes, *Sitzungsberichte der Akademie der Wissenschaften in Wien 2a,* 116, 987, 1907.
264. **Lampa, A.,** Theorie der Drehfelderscheinung im einfachen elektrostatischen Wechselfeld, *Sitzungsberichte der Akademie der Wissenschaften in Wien 2a,* 120, 1007, 1911.
265. **Krasny-Ergen, W.,** Der Feldverlauf im Bereiche sehr kurzer Wellen; spontane Drehfelder, *Hochfrequenztechn. Elektroakust.,* 49, 195, 1937.
266. **Grossetti, E.,** Born-Lertes effect on the dipolar rotation of liquids, *Nuovo Cimento,* 10, 193, 1958.
267. **Carrelli, A. and Marinaro, M.,** On the mechanical moment of rotation of mixtures of liquids in rotating electric fields, *Nuovo Cimento,* 11, 262, 1959.
268. **Grossetti, E.,** Dipolar rotation effect in liquids, *Nuovo Cimento,* 13, 350, 1959.
269. **Grossetti, E.,** Effect of the dipolar rotation of liquids III, *Nuovo Cimento,* 18, 21, 1960.
270. **Grossetti, E.,** Mechanical momentum of solutions of liquids in a rotating electric field, *Nuovo Cimento,* 19, 1, 1961.
271. **Grossetti, E.,** Rotational mechanical moments of electrolyte solutions in a rotating high-frequency electric field, *Nuovo Cimento,* 21, 395, 1961.
272. **Pickard, W. F.,** On the Born-Lertes rotational effect, *Nuovo Cimento,* 21, 316, 1961.
273. **Pickard, W. F.,** Electrical force effects in dielectric liquids, *Progr. Dielec.,* 6, 1, 1965.
274. **Ogawa, T.,** Measurement of the electrical conductivity and dielectric constant without contacting electrodes, *J. Appl. Phys.,* 32, 583, 1961.
275. **Lilienfeld, P., Elterman, P. B. and Baron, P.,** Development of a prototype fibrous aerosol monitor, *Am. Ind. Hyg. Ass. J.,* 40, 276, 1979.
276. **Funakoshi, H.,** Theory of the birefringence caused by the orientation of particles in the rotating electric field, *Ann. Inst. of Stat. Math., Suppl. V,* 45, 1968.
277. **Füredi, A. A. and Ohad, I.,** Effects of high-frequency electric fields on the living cell. I. Behaviour of human erythrocytes in high-frequency electric fields and its relation to their age, *Biochim. Biophys. Acta,* 79, 1, 1964.
278. **Crane, J. S. and Pohl, H. A.,** A study of living and dead cells using dielectrophoresis, in *Dielectrophoretic and Electrophoretic Deposition,* Pohl, H. A. and Pickard, W. F., Eds., The Electrochemical Society, New York, 1969, 127.
279. **Mischel, M. and Lamprecht, I.,** Dielectrophoretic rotation in budding yeast cells, *Z. Naturforsch.,* 35 c, 1111, 1980.
280. **Pohl, H. A.,** Natural oscillating fields of cells, in *Coherent Excitations in Biological Systems,* Fröhlich, H. and Kremer, F., Eds., Springer-Verlag, Berlin, 1983, 199.
281. **Pohl, H. A. and Lamprecht, I.,** Wechselfelder umgeben wachsende Zellen, *Umschau,* 6, 366, 1985.
282. **Rivera, H. and Pohl, H. A.,** Cellular spin resonance (CSR), in *Modern Bioelectrochemistry,* Gutmann, F. and Keyzer, H., Eds., Plenum Press, New York, 1986, 431.
283. **Zimmermann, U., Vienken, J. and Pilwat, G.,** Rotation of cells in an alternating electric field: the occurrence of a resonance frequency, *Z. Naturforsch.,* 36 c, 173, 1981.
284. **Holzapfel, C., Vienken, J. and Zimmermann, U.,** Rotation of cells in an alternating electric field: theory and experimental proof, *J. Membrane Biol.,* 67, 13, 1982.
285. **Hub, H.-H., Ringsdorf, H. and Zimmermann, U.,** Rotation of polymerized vesicles in an alternating electric field, *Angew. Chem. Int. Ed. Engl.,* 21, 134, 1982.
286. **Küppers, G., Wendt, B. and Zimmermann, U.,** Rotation of cells and ion exchange beads in the MHz-frequency range, *Z. Naturforsch.,* 38c, 505, 1983.
287. **Pollock, J. K., Pohl, D. G. and Pohl, H. A.,** Cellular spin resonance inversion by ion-induced dipoles, *Int. J. Quantum Chem.,* 32, 19, 1987.

288. **Arnold, W. M. and Zimmermann, U.,** Rotating-field-induced rotation and measurement of the membrane capacitance of single mesophyll cells of *Avena sativa, Z. Naturforsch.,* 37c, 908, 1982.

289. **Glaser, R., Fuhr, G. and Gimsa, J.,** Rotation of erythrocytes, plant cells and protoplasts in an outside rotating electric field, *Studia biophysica,* 96, 11, 1983.

290. **Mischel, M. and Pohl, H. A.,** Cellular spin resonance: theory and experiment, *J. Biol. Phys.,* 11, 98, 1983.

291. **Pilwat, G. and Zimmermann, U.,** Rotation of a single cell in a discontinuous rotating electric field, *Bioelectrochem. Bioenerg.,* 10, 155, 1983.

292. **Arnold, W. M. and Zimmermann, U.,** Electric field-induced fusion and rotation of cells, in *Biological Membranes,* Vol. 5, Chapman, D., Ed., Academic Press, London, 1984, 389.

293. **Fuhr, G., Hagedorn, R. and Göring, H.,** Cell rotation in a discontinuous field of a 4-electrode chamber, *Studia biophysica,* 102, 221, 1984.

294. **Arnold, W. M., Wendt, B., Zimmermann, U. and Korenstein, R.,** Rotation of a single swollen thylakoid vesicle in a rotating electric field. Electrical properties of the photosynthetic membrane and their modification by ionophores, lipophilic ions and pH, *Biochim. Biophys. Acta,* 813, 117, 1985.

295. **Fuhr, G., Hagedorn, R. and Göring, H.,** Separation of different cell types by rotating electric fields, *Plant Cell Physiol.,* 26, 1527, 1985.

296. **Fuhr, G., Hagedorn, R. and Müller, T.,** Cell separation by using rotating electric fields, *Studia biophysica,* 107, 23, 1985.

297. **Gimsa, J., Fuhr, G. and Glaser, R.,** Interpretation of electrorotation of protoplasts, II. Interpretation of experiments, *Studia biophysica,* 109, 5, 1985.

298. **Kaler, K. V. I. S. and Johnston, R. H.,** Spinning response of yeast cells to rotating electric fields, *J. Biol. Phys.,* 13, 69, 1985.

299. **Arnold, W. M., Geier, B. M., Wendt, B. and Zimmermann, U.,** The change in the electro-rotation of yeast cells effected by silver ions, *Biochim. Biophys. Acta,* 889, 35, 1986.

300. **Egger, M., Donath, E., Ziemer, S. and Glaser, R.,** Electrorotation — a new method for investigating membrane events during thrombocyte activation. Influence of drugs and osmotic pressure, *Biochim. Biophys. Acta,* 861, 122, 1986.

301. **Müller, T., Fuhr, G., Hagedorn, R. and Göring, H.,** Influence of dielectric breakdown on electrorotation, *Studia biophysica,* 113, 203, 1986.

302. **Wicher, D., Gündel, J. and Matthies, H.,** Measuring chamber with extended applications of the electro-rotation. Alpha- and beta-dispersion of liposomes, *Studia biophysica,* 115, 51, 1986.

303. **Ziervogel, H., Glaser, R., Schadow, D. and Heymann, S.,** Electrorotation of lymphocytes — the influence of membrane events and nucleus, *Biosci. Rep.,* 6, 973, 1986.

304. **Arnold, W. M. and Zimmermann, U.,** Electrorotation: experimental methods and results, in *Proc. of the 13th Ann. NE Bioengineering Conference,* Foster, K. R., Ed., IEEE, New York, 1987, 514.

305. **Arnold, W. M., Schmutzler, R. K., Schmutzler, A. G., van der Ven, H., Al-Hasani, S., Krebs, D. and Zimmermann, U.,** Electro-rotation of mouse oocytes: single-cell measurements of zona-intact and zona-free cells and of the isolated zona pellucida, *Biochim. Biophys. Acta,* 905, 454, 1987.

306. **Arnold, W. M., Schwan, H. P. and Zimmermann, U.,** Surface conductance and other properties of latex particles measured by electrorotation, *J. Phys. Chem.,* 91, 5093, 1987.

307. **Fuhr, G., Müller, T., Wagner, A. and Donath, E.,** Electrorotation of oat protoplasts before and after fusion, *Plant Cell Physiol.,* 28, 549, 1987.

308. **Geier, B. M., Wendt, B., Arnold, W. M. and Zimmermann, U.,** The effect of mercuric salts on the electro-rotation of yeast cells and comparison with a theoretical model, *Biochim. Biophys. Acta,* 900, 45, 1987.

309. **Glaser, R. and Fuhr, G.,** Electrorotation — the spin of cells in rotating high frequency electric fields, in *Mechanistic Approaches to Interactions of Electric and Electromagnetic Fields with Living Systems,* Blank, M. and Findl, E., Eds., Plenum Press, New York, 1987, 271.

310. **Hölzel, R. and Lamprecht, I.,** Cellular spin resonance of yeast in a frequency range up to 140 MHz, *Z. Naturforsch.,* 42c, 1367, 1987.

311. **Fuhr, G., Geissler, F., Müller, T., Hagedorn, R. and Torner, H.,** Differences in the rotation spectra of mouse oocytes and zygotes, *Biochim. Biophys. Acta,* 930, 65, 1987.

312. **Arnold, W. M.,** Analysis of optimum electro-rotation technique, *Ferroelectrics,* 86, 225, 1988.

313. **Arnold, W. M., Zimmermann, U., Pauli, W., Benzing, M., Niehrs, C. and Ahlers, J.,** The comparative influence of substituted phenols (especially chlorophenols) on yeast cells assayed by electro-rotation and other methods, *Biochim. Biophys. Acta,* 942, 83, 1988.

314. **Arnold, W. M., Zimmermann, U., Heiden, W. and Ahlers, J.,** The influence of tetraphenylborates (hydrophobic anions) on yeast cell electro-rotation, *Biochim. Biophys. Acta,* 942, 96, 1988.

315. **Egger, M., Donath, E., Spangenberg, P., Bimmler, M., Glaser, R. and Till, U.,** Human platelet electrorotation change induced by activation: inducer specificity and correlation to serotonin release, *Biochim. Biophys. Acta,* 972, 265, 1988.

316. **Engel, J., Donath, E. and Gimsa, J.,** Electrorotation of red cells after electroporation, *Studia biophysica,* 125, 53, 1988.

317. **Freitag, R., Arnold, W. M., Zimmermann, U. and Schügerl, K.,** Electrorotation; a new method for characterizing cells, in *DECHEMA Biotechnology Conferences 3,* VCH Verlagsgesellschaft, 1989, 479.

318. **Georgieva, R. and Glaser, R.,** Electrorotation of lidocaine-treated human erythrocytes, in *Electromagnetic Fields and Biomembranes,* Markov, M. and Blank, M., Eds., Plenum Press, New York, 1988, 263.

319. **Gimsa, J., Donath, E. and Glaser, R.,** Evaluation of the data of simple cells by electrorotation using square-topped fields, *Bioelectrochem. Bioenerg.,* 19, 389, 1988.

320. **Gimsa, J., Glaser, R. and Fuhr, G.,** Remarks on the field distribution in four electrode chambers for electrorotational measurements, *Studia biophysica,* 125, 71, 1988.

321. **Freitag, R., Schügerl, K., Arnold, W. M. and Zimmermann, U.,** The effect of osmotic and mechanical stresses and enzymatic digestion on the electro-rotation of insect cells (*Spodoptera frugiperda*), *J. Biotech.,* 11, 325, 1989.

322. **Hoelzel, R.,** Sine-quadrature oscillator for cellular spin resonance up to 120 MHz, *Med. Biol. Eng. Comp.,* 26, 102, 1988.

323. **Arnold, W. M. and Zimmermann, U.,** Measurement of dielectric properties of single cells or other particles using direct observation of electro-rotation, *Proceeding of the First International Conference on Low Cost Experiments in Biophysics (Cairo 18–20 Dec. 1989),* 1989, 1.

324. **Arnold, W. M., Jäger, A. H. and Zimmermann, U.,** The influence of yeast strain and of growth medium composition on the electro-rotation of yeast cells and of isolated walls, *DECHEMA Biotechnology Conferences,* 3, VCH Verlagsgesellschaft, 653, 1989.

325. **Arnold, W. M., Schmutzler, R. K., Al-Hasani, S., Krebs, D. and Zimmermann, U.,** Differences in membrane properties between unfertilised and fertilised rabbit oocytes demonstrated by electro-rotation. Comparison with cells from early embryos, *Biochim. Biophys. Acta,* 979, 142, 1989.

326. **Georgiewa, R., Donath, E., Gimsa, J., Löwe, U. and Glaser, R.,** Ac-field-induced KCl leakage from human red cells at low ionic strengths. Implications for electrorotation measurements, *Bioelectrochem. Bioenerg.,* 22, 255, 1989.

327. **Georgiewa, R., Donath, E. and Glaser, R.,** On the determination of human erythrocyte intracellular conductivity by means of electrorotation: influence of osmotic pressure, *Studia biophysica,* 133, 185, 1989.

328. **Gündel, J., Wicher, D. and Matthies, H.,** Electrorotation as a viability test for isolated single animal cells, *Studia biophysica,* 133, 5, 1989.

329. **Wicher, D. and Gündel, J.,** Electrorotation of multi- and oligolamellar liposomes, *Bioelectrochem. Bioenerg.,* 21, 279, 1989.

330. **Wicher, D. and Gündel, J.,** Existence of a low-frequency limit for electrorotation experiments, *Studia biophysica,* 134, 223, 1989.
331. **Hu, X., Arnold, W. M. and Zimmermann, U.,** Alterations in the electrical properties of T and B lymphocyte membranes induced by mitogenic stimulation. Activation followed by rotation of single cells, *Biochim. Biophys. Acta,* 1021, 191, 1990.
332. **Müller, T., Fuhr, G., Geissler, F. and Hagedorn, R.,** Rotation spectra of mouse eggs up to 35 MHz: experiments and theoretical interpretation, *Studia biophysica,* 139, 77, 1990.
333. **Alder, G. M., Arnold, W. M., Bashford, C. L., Drake, A. F., Pasternak, C. A. and Zimmermann, U.,** Divalent cation-sensitive pores formed by natural and synthetic melittin and by Triton X-100, *Biochim. Biophys. Acta,* 1061, 111, 1991.
334. **Gimsa, J., Glaser, R. and Fuhr, G.,** Theory and application of the rotation of biological cells in rotating electric fields (electrorotation), in *Physical Characterization of Biological Cells,* Schütt, W., Klinkmann, H., Lamprecht, I. and Wilson, T., Eds., Verlag Gesundheit, Berlin, 1991, 295.
335. **Gimsa, J., Marszalek, P., Loewe, U. and Tsong, T. Y.,** Dielectrophoresis and electrorotation of neurospora slime and murine myeloma cells, *Biophys. J.,* 60, 749, 1991.
336. **Arnold, W. M., Klarmann, B. G., Sukhorukov, V. L. and Zimmermann, U.,** Membrane accommodation in hypo-osmotically-treated, and in giant electrofused cells, *Biochem. Soc. Trans.,* 20, 120S, 1992.
337. **Gimsa, J. and Glaser, R.,** New aspects for theory and application of dielectrophoresis and electrorotation of biological cells, in *Proc. Eleventh School on Biophysics of Membrane Transport,* Poland, May 1992, 156.
338. **Hölzel, R. and Lamprecht, I.,** Dielectric properties of yeast cells as determined by electrorotation, *Biochim. Biophys. Acta,* 1104, 195, 1992.
339. **Gimsa, J., Marszalek, P., Loewe, U. and Tsong, T. Y.,** Electroporation in rotating electric fields, *Bioelectrochem. Bioenerg.,* 29, 81, 1992.
340. **Coghlan, A.,** Sticky beads put bacteria in a spin, *New Scientist,* 138, 1873, 21, 1993.
341. **Beardsley, T.,** Putting a spin on parasites, *Sci. Am.,* 87, 1993.
342. **Gimsa, J., Löwe, U., Marszalek, P. and Tsong, T. Y.,** Electroporation of cell membranes by rotating electric fields, in *Electricity and Magnetism in Biology and Medicine,* Blank, M., Ed., San Francisco Press, San Francisco, 1993, 44.
343. **Kakutani, T., Shibatani, S. and Senda, M.,** Electrorotation of barley mesophyll protoplasts, *Bioelectrochem. Bioenerg.,* 31, 85, 1993.
344. **Shibatani, S., Minami, K., Senda, M. and Kakutani, T.,** Electrorotation of vacuoles isolated from barley mesophyll cells, *Bioelectrochem. Bioenerg.,* 29, 327, 1993.
345. **Sukhorukov, V. L., Arnold, W. M. and Zimmermann, U.,** Hypotonically induced changes in the plasma membrane of cultured mammalian cells, *J. Membrane Biol.,* 132, 27, 1993.
346. **Müller, T., Küchler, L., Fuhr, G., Schnelle, Th. and Sokirku, A.,** Dielektrische Einzelzellspektroskopie an Pollen verschiedener Waldbaumarten — Charakterisierung der Pollenvitalität, *Silvae Genetica,* 42, 311, 1993.
347. **Djuzenova, C. S., Sukhorukov, V. L., Klöck, G., Arnold, W. M. and Zimmermann, U.,** Effect of electric field pulses on the membrane-bound immunoglobulins of LPS-activated murine B-lymphocytes: correlation with the cell cycle, *Cytometry,* 15, 35, 1994.
348. **Sukhorukov, V. L., Djuzenova, C. S., Arnold, W. M. and Zimmermann, U.,** DNA, protein and plasma-membrane incorporation by arrested mammalian cells, *J. Membrane Biol.,* 142, 77, 1994.
349. **Wang, X.-B., Huang, Y., Gascoyne, P. R. C., Becker, F. F., Hölzel, R. and Pethig, R.,** Changes in Friend murine erythroleukaemia cell membranes during induced differentiation determined by electrorotation, *Biochim. Biophys. Acta,* 1193, 330, 1994.
350. **Sukhorukov, V. L., Djuzenova, C. S., Frank, H., Arnold, W. M. and Zimmermann, U.,** Electropermeabilization and fluorescent tracer exchange: the role of whole cell capacitance, *Cytometry,* 21, 230, 1995.
351. **Hagedorn, R. and Fuhr, G.,** Calculation of rotation of biological objects in the electric rotation field, *Studia biophysica,* 102, 229, 1984.

352. **Lovelace, R. V. E., Stout, D. G. and Steponkus, P. L.,** Protoplast rotation in a rotating electric field: the influence of cold acclimation, *J. Membrane Biol.,* 82, 157, 1984.

353. **Fuhr, G., Gimsa, J. and Glaser, R.,** Interpretation of electrorotation of protoplasts, I. Theoretical considerations, *Studia biophysica,* 108, 149, 1985.

354. **Fuhr, G., Hagedorn, R. and Müller, T.,** Simulation of the rotational behaviour of single cells by macroscopic spheres, *Studia biophysica,* 107, 109, 1985.

355. **Schwan, H. P.,** Dielectric properties of the cell surface and electric field effects on cells, *Studia biophysica,* 110, 13, 1985.

356. **Wicher, D. and Gündel, J.,** Berechnungen zur Rotation biologischer Einzelzellen im rotierenden elektrischen Feld unter Zugrundelegung eines ein- bzw. zweischaligen Modells, *Forschungsergebnisse der Friedrich-Schiller-University Jena,* 26, 1, 1985.

357. **Fuhr, G., Glaser, R. and Hagedorn, R.,** Rotation of dielectrics in a rotating electric high-frequency field. Model experiments and theoretical explanation of the rotation effect of living cells, *Biophys. J.,* 49, 395, 1986.

358. **Fuhr, G. and Hagedorn, R.,** Rotating-field-induced membrane potentials and practical applications, *Studia biophysica,* 119, 97, 1987.

359. **Fuhr, G., Hagedorn, R., Glaser, R., Gimsa, J. and Müller, T.,** Membrane potentials induced by external rotating electrical fields, *J. Bioelectr.,* 6, 49, 1987.

360. **Fuhr, G. and Hagedorn, R.,** Dielectric rotation and oscillation — a principle in biological systems?, *Studia biophysica,* 121, 25, 1987.

361. **Schwan, H. P.,** Electro-rotation: the context with respect to other ponderomotive effects, *Proc. 13th Ann. N.E. Bioengineering Conference, Philadelphia,* March 12–13, 1987, 511.

362. **Turcu, I.,** Electric field induced rotation of spheres, *J. Phys. A: Math. Gen.,* 20, 3301, 1987.

363. **Fuhr, G. and Hagedorn, R.,** Grundlagen der Elektrorotation, in *Colloquia Pflanzenphysiologie,* Göring, H. and Hoffmann, P., Eds., Humboldt-Universität, Berlin, 1988.

364. **Schwan, H. P.,** Dielectric spectroscopy and electro-rotation of biological cells, *Ferroelectrics,* 86, 205, 1988.

365. **Turcu, I. and Lucaciu, C. M.,** Electrorotation: a spherical shell model, *J. Phys. A: Math. Gen.,* 22, 995, 1989.

366. **Jones, T. B. and Kaler, K. V. I. S.,** Relationship of rotational and dielectrophoretic cell spectra, *Proc. Ann. EMBS EE Conf.,* Nov., 1990.

367. **Radu, M.,** Cellular electrorotation — a general theoretical model, *Studia biophysica,* 137, 117, 1990.

368. **Egger, M., Donath, E., Kuzmin, P. I. and Pastushenko, V. Ph.,** Electrorotation of dumb-bell shaped particles, *Bioelectrochem. Bioenerg.,* 26, 383, 1991.

369. **Paul, R. and Otwinowski, M.,** The theory of the frequency response of ellipsoidal biological cells in rotating electrical fields, *J. Theor. Biol.,* 148, 495, 1991.

370. **Donath, E., Egger, M. and Pastushenko, V. Ph.,** Electrorotation behavior of aggregated particles. A theoretical study, *Bioelectrochem. Bioenerg.,* 31, 115, 1993.

371. **Hölzel, R.,** Electric field calculation for electrorotation electrodes, *J. Phys. D: Appl. Phys.,* 26, 2112, 1993.

372. **Kakutani, T., Shibatani, S. and Sugai, M.,** Electrorotation of non-spherical cells: theory for ellipsoidal cells with an arbitrary number of shells, *Bioelectrochem. Bioenerg.,* 31, 131, 1993.

373. **Jones, T. B. and Washizu, M.,** Equilibria and dynamics of DEP-levitated particles: multipolar theory, *J. Electrostat.,* 33, 199, 1994.

374. **Melcher, J. R.,** *Continuum Electromechanics,* MIT Press, Cambridge, Massachusetts, 1981.

375. **Fuhr, G., Geisler, F., Müller, T., Hagedorn, R. and Torner, H.,** Differences in rotation spectra of mouse oocytes and zygotes, *Biochim. Biophys. Acta,* 930, 65, 1987.

376. **Sokirku, A. V.,** The electrorotation of axisymmetrical cell, *Biologicheskie Membrany,* 6, 587, 1992.

377. **Gimsa, J., Schnelle, Th., Zechel, G. and Glaser, R.,** Dielectric spectroscopy of human erythrocytes: investigation under the influence of nystatin, *Biophys. J.,* 66, 1244, 1994.

378. **Davey, C. L., Kell, D. B., Kemp, R. B. and Meredith, R. W. J.,** On the audio- and radio-frequency dielectric behaviour of anchorage-independent, mouse L929-derived LS fibroblasts, *Bioelectrochem. Bioenerg.,* 20, 83, 1988.

379. **Irimajiri, A., Asami, K., Ichinowatari, T. and Kinoshita, Y.,** Passive electrical properties of the membrane and cytoplasm of cultured rat basophil leukemia cells. II. Effects of osmotic perturbation, *Biochim. Biophys. Acta,* 896, 214, 1987.

380. **Asami, K., Takahashi, Y. and Takashima, S.,** Dielectric properties of mouse lymphocytes and erythrocytes, *Biochim. Biophys. Acta,* 1010, 49, 1989.

381. **Zimmermann, U., Büchner, K.-H. and Benz, R.,** Transport properties of mobile charges in algal membranes: influence of pH and turgor pressure, *J. Membrane Biol.,* 67, 183, 1982.

382. **Wang, J., Zimmermann, U. and Benz, R.,** Contribution of electrogenic ion transport to impedance of the algae *Valonia utricularis* and artificial membranes, *Biophys. J.,* 67, 1582, 1994.

383. **Fuhr, G. and Shirley, S. G.,** Cell handling and characterisation using micron and submicron electrode arrays — state of the art and perspectives of semiconductor microtools, *Micro Mechanics Europe, Workshop Digest Pisa '94,* 1994, 164.

384. **Iwazawa, J., Imae, Y. and Kobayasi, S.,** Study of the torque of the bacterial flagellar motor using a rotating electric field, *Biophys. J.,* 64, 925, 1993.

385. **Melcher, J. R.,** Traveling-wave induced electroconvection, *Phys. Fluids,* 9, 1548, 1966.

386. **Jolly, D. C. and Melcher, J. R.,** Electroconvective instability in a fluid layer, *Proc. R. Soc. London,* A, 314, 269, 1970.

387. **Crowley, J. M.,** The efficiency of electrohydrodynamic pumps in the attraction mode, *J. Electrostat.,* 8, 171, 1980.

388. **Ehrlich, R. M. and Melcher, J. R.,** Bipolar model for traveling-wave induced nonequilibrium double-layer streaming in insulating liquids, *Phys. Fluids,* 25, 1785, 1982.

389. **Melcher, J. R.,** Electrodynamic surface waves, in *Waves on Fluid Interfaces,* Academic Press, New York, 1983, 167.

390. **Krein, P. T.,** A theoretical study of the thermal electrohydrodynamic induction pump with harmonics and inhomogeneous fluids, *Phys. Fluids,* 27, 315, 1984.

391. **Fuhr, G., Hagedorn, R., Müller, T., Benecke, W. and Wagner, B.,** Microfabricated electrohydrodynamic (EHD) pumps for liquids of higher conductivity, *J. Microelectromech. Systems,* 1, 141, 1992.

392. **Bart, S. F., Tavrow, L. S., Mehregany, M. and Lang, J. H.,** Microfabricated electrohydrodynamic pumps, *Sensors and Actuators,* A21, 193, 1990.

393. **Fuhr, G., Schnelle, Th. and Wagner, B.,** Travelling wave-driven microfabricated electrohydrodynamic pumps for liquids, *J. Micromech. Microeng.,* 4, 217, 1994.

394. **Masuda, S., Washizu, M. and Kawabata, I.,** Movement of blood cells in liquid by nonuniform traveling field, *IEEE Trans. Ind. Appl.,* 24, 217, 1988.

395. **Hagedorn, R., Fuhr, G., Müller, T. and Gimsa, J.,** Traveling-wave dielectrophoresis of microparticles, *Electrophoresis,* 13, 49, 1992.

396. **Fuhr, G., Hagedorn, R., Müller, T., Benecke, W., Wagner, B. and Gimsa, J.,** Asynchronous traveling-wave induced linear motion of living cells, *Studia biophysica,* 140, 79, 1991.

397. **Wang, X.-B., Huang, Y., Becker, F. F. and Gascoyne, P. R. C.,** A unified theory of dielectrophoresis and travelling wave dielectrophoresis, *J. Phys. D: Appl. Phys.,* 27, 1571, 1994.

398. **Giddings, J. C.,** Field-flow fractionation, *Sep. Sci. Technol.,* 19, 831, 1984.

399. **Davis, J. M. and Giddings, J. C.,** Feasibility study of dielectrical field-flow fractionation, *Sep. Sci. Technol.,* 21, 969, 1986.

400. **Giddings, J. C.,** Field-flow fractionation: analysis of macromolecular, colloidal, and particulate materials, *Science,* 260, 456, 1993.

401. **Rousselet, J., Salome, L., Ajdari, A. and Prost, J.,** Directional motion of brownian particles induced by a periodic asymmetric potential, *Nature,* 370, 446, 1994.

402. **Derjaguin, B. V., Churaev, N. V. and Muller, V. M.,** *Poverkhnostnye Sily,* Moskva, Nauka, 1985, 322.

APPENDICES

Examples of analytical (A) and numeric (B) calculations of high frequency electric field-induced forces on microparticles and cells (dielectrophoresis, electrorotation)

APPENDIX A. SHORT DESCRIPTION
OF THE "DIPOLE CONCEPT"

The calculation uses the following steps:

1. A well defined cell geometry is assumed, e.g., a sphere or a shelled sphere consisting of different isotropic dielectrics.
2. The actual field strength without cell is assumed to be known. It is calculated from the geometry and arrangement of the electrodes.
3. The potential inside each of the dielectrics is taken to consist of a constant field and dipole field components but neglecting higher orders of multipoles, therefore the name "dipole concept."
4. From the boundary conditions at each interface a system of equations is developed which can be solved for multi-shell cases.
5. Time dependent fields are represented by complex (real and imaginary) quantities ($\vec{E} = \mathrm{Re}(\vec{E} * \exp(i\omega t))$).
6. Now a resulting dipole moment ($\vec{P}*$) with real and imaginary part can be obtained ($\vec{P}* \approx \vec{E}*$).
7. It can easily be shown that the time averaged torque (L) can be expressed by the cross product:

$$\vec{L} = \langle \vec{E} \times \vec{P} \rangle \qquad (A1)$$

and corresponds to:

$$\vec{L} \approx \mathrm{Im}(\vec{P}*) \qquad (A2)$$

The time averaged dielectrophoretic force ($\vec{F}$) requires the knowledge of the field gradient across the cell:

$$\vec{F} = \langle (\vec{P} \circ \vec{\nabla}) \vec{E} \rangle \qquad (A3)$$

and is proportional to the real part of ($\vec{P}*$):

$$\vec{F} \approx \mathrm{Re}(\vec{P}*) \qquad (A4)$$

A more general approach is to sum the force or torque over the surface S of the particle, by evaluating the Maxwell stress tensor $\overleftrightarrow{T}$:

$$\overleftrightarrow{T} = \tilde{\varepsilon}_e \{ \vec{E} \, (\vec{n} \, \circ \, \vec{E}) - 0.5 \vec{E}^2 \, \circ \, \vec{n} \} \qquad (A5)$$

where $\vec{n}$ is the unit vector directed perpendicularly outward from the surface. It is assumed that the cell is suspended in a liquid (in most cases an aqueous solution) of permittivity $\tilde{\varepsilon}_e$ (relative permittivity multiplied with that of the vacuum).

The dielectrophoretic force and rotational torque are given by integrals similar to the dipole expressions:

$$\vec{F} = \int_S (\overleftrightarrow{T} \, \circ \, \vec{n}) dS \qquad (A6)$$

$$\vec{L} = \int_S (\overleftrightarrow{T} \, \times \, \vec{r}) dS \qquad (A7)$$

This approach can in principle allow for any degree of electrode complexity or inhomogeneity of the field. However, determination of $\vec{E}$ developed by a highly non-uniform field over the surface of a particle with a complex inner structure may be more than a trivial problem.

APPENDIX B. MULTISHELLS IN THE FIELD OF MULTIPLE ELECTRODES

Of particular interest for microstructures are numeric methods for calculating the effects of a large number of electrodes and where the field is highly non-uniform and the cell is not small compared with the gap between the electrodes. The complexity of this problem excludes analytical solutions.

To determine the potential Φ in such a space we have, apart from the boundary conditions, to satisfy:

$$\text{Poisson's equation}: \quad \vec{\nabla}(\tilde{\varepsilon}(\vec{r})\vec{\nabla}\Phi(\vec{r})) = -\rho(\vec{r}) \qquad (A8)$$

$$\text{and conservation of charge}: \quad \vec{\nabla}\vec{j}(\vec{r},t) + \frac{\partial}{\partial t}\rho(\vec{r},t) = 0 \qquad (A9)$$

where $\vec{\nabla}$ denotes the del vector operator and ρ is the space charge density. In order to neglect magnetic fields the electric currents are assumed to be small. Provided Ohm's law:

$$\vec{j} = -\sigma\vec{\nabla}\Phi \qquad (A10)$$

is valid (usually the case, except at very high field strengths) one obtains:

$$\vec{\nabla} \left[\left(\sigma(\vec{r}) + \tilde{\varepsilon}(\vec{r}) \frac{\partial}{\partial t} \right) \vec{\nabla} \Phi(\vec{r}, t) \right] = 0 \qquad \text{(A11)}$$

For potentials which vary as sinusoidal functions of time, this yields:

$$\vec{\nabla} \left[\left(\sigma(\vec{r}) + i\omega\tilde{\varepsilon}(\vec{r}) \right) \vec{\nabla} \varphi(\vec{r}) \right] = 0 \qquad \text{(A12)}$$

where we have introduced the complex potential $\varphi(\vec{r})$ described by:

$$\Phi(\vec{r}, t) = \mathrm{Re}\,(\varphi(\vec{r}) \exp(i\omega t)) \qquad \text{(A13)}$$

Each electrode is replaced by an appropriate number of point charges (from five to several hundred, depending upon the complexity of the structure, the required accuracy, and the available computing power). For the sake of simplicity we will assume that a particle can be described sufficiently well by a homogeneous sphere model with radius a (in the case of a cell, a single shell model has to be introduced).

The first step is to determine these surface charges for given values of the potential at the electrodes Φ^k, in the absence of particles. This may be done using standard techniques such as those of finite difference or integral equations. The distribution of surface charges will certainly change if particles approach the electrodes. However, these effects are comparatively small and will therefore be neglected. If the concentration of the particles becomes appreciable (as may happen in a trap) this is only an approximation, and in any case the dipole-dipole interaction of the cells must also be taken into account. The potential inside and outside the sphere satisfies the Laplace equation if the inhomogeneities due to temperature gradients, etc. are neglected. Now the potential can be written in the form:

$$\varphi^e(\vec{r}) = \frac{1}{4\pi\varepsilon_e} \sum_{k=1}^{n} \frac{q_k}{|\vec{r} - \vec{r}_k|} + \sum_{l=0}^{\infty} \sum_{m=-l}^{l} b_{lm} \left(\frac{r}{a} \right)^{-l-1} Y_{lm}(\vartheta, \varphi) \quad \text{for } r = |\vec{r}| \geq a \qquad \text{(A14)}$$

$$\varphi^i(\vec{r}) = \sum_{l=0}^{\infty} \sum_{m=-l}^{l} a_{lm} \left(\frac{r}{a} \right)^{l} Y_{lm}(\vartheta, \varphi) \qquad \text{for } r \leq a \qquad \text{(A15)}$$

where $Y_{lm}(\vartheta, \varphi)$ stands for the spherical functions satisfying the angular part of the Laplace equation. The complex constants a_{lm}, b_{lm} can be determined from the boundary conditions:

$$\varphi^i(a) = \varphi^e(a) \tag{A16}$$

$$(\sigma_i + i\omega\tilde{\varepsilon}_i)\frac{\partial}{\partial r}\varphi^i\bigg|_a = (\sigma_e + i\omega\tilde{\varepsilon}_e)\frac{\partial}{\partial r}\varphi^e\bigg|_a \tag{A17}$$

resulting in:

$$a_{lm} = \frac{g_{lm}(2l+1)(\sigma_e + i\omega\tilde{\varepsilon}_e)}{\tilde{\varepsilon}_e\left((l+1)\sigma_e + l\sigma_i + i\omega((l+1)\tilde{\varepsilon}_e + l\tilde{\varepsilon}_i)\right)} \tag{A18}$$

$$b_{lm} = \frac{g_{lm}l(\sigma_e - \sigma_i + i\omega(\tilde{\varepsilon}_e - \tilde{\varepsilon}_i))}{\tilde{\varepsilon}_e\left((l+1)\sigma_e + l\sigma_i + i\omega((l+1)\tilde{\varepsilon}_e + l\tilde{\varepsilon}_i)\right)} \tag{A19}$$

with the restriction:

$$b_{lm} = 0 \quad \text{for} \quad l = 0 \text{ (no induced monopoles)} \tag{A20}$$

and where:

$$g_{lm} = \frac{(-1)^m}{(2l+1)}\sum_{k=1}^n \frac{q_k}{r_k}\left(\frac{a}{r_k}\right)^l Y_{l-m}(\vartheta_k,\varphi_k) \tag{A21}$$

from which the forces between the induced multipoles in the cell and the surface charges of the electrodes can easily be derived. After performing the time average over one period we get for the dielectrophoretic force omitting higher multipoles:

$$\langle\vec{F}\rangle = \left|\frac{z_1}{z_2}\right|a^3\sum_{i,j}^n q_iq_j\frac{\vec{r}_i\vec{r}_j^2 - 3\vec{r}_j[\vec{r}_j\vec{r}_i]}{8\pi\tilde{\varepsilon}_e|\vec{r}_j|^5|\vec{r}_i|^3}\cos[(n_i - n_j)\beta - \alpha(\omega)] \tag{A22}$$

and for the torque:

$$\langle\vec{L}\rangle = \left|\frac{z_1}{z_2}\right|a^3\sum_{i,j}^n q_iq_j\frac{\vec{r}_i\times\vec{r}_j}{8\pi\tilde{\varepsilon}_e|\vec{r}_j|^3|\vec{r}_i|^3}\cos[(n_i - n_j)\beta - \alpha(\omega)] \tag{A23}$$

with:

$$z_1 = \frac{\sigma_e - \sigma_i}{2\tilde{\varepsilon}_l + \tilde{\varepsilon}_i} + i\omega \frac{\varepsilon_e - \varepsilon_i}{2\varepsilon_e - \varepsilon_i}, \quad z_2 = \frac{2\sigma_e + \sigma_i}{2\tilde{\varepsilon}_e + \tilde{\varepsilon}_i} + i\omega \tag{A24}$$

and:

$$\alpha(\omega) = \arg(z_1) - \arg(z_2) \tag{A25}$$

The term $(n_i - n_j)\beta$ corresponds to the phase shift of the surface charges q_i and q_j. If we use, e.g., four electrodes (0...3) with phases of 0, $^\pi/_2$, π, $^3/_2\pi$, then n_i simply stands for the number of electrodes and $\beta = {}^\pi/_2$. The restriction to induced dipoles is justified by the small amplitude of forces developed by induced quadrupoles, hextupoles, and higher multipoles. However, for particles on the axis of radially symmetrical electrode arrays, the induced dipole may be cancelled out, and in this case the higher multipoles should be considered.

Whereas the analytical expressions developed above provide a description of the frequency-dependence of the force in idealized fields, the above expressions enable numerical calculation of the forces produced by complicated arrays of realistically shaped electrodes. Apart from the field and force distribution, particle trajectories can also be calculated.

Chapter 6

ELECTRIC PROPERTIES OF THE MEMBRANE AND THE CELL SURFACE

Roland Glaser

CONTENTS

I. Introduction 330

II. The Structure of the Cell Membrane Electric Field 332
   A. The Potential Profile $\psi(x)$ Perpendicular
      to The Membrane Surface 332
     1. The Transmembrane Potential $\Delta\psi$ 332
       a. The Transmembrane Potential
         in Non-Equilibrium Steady State 333
       b. The Transmembrane Potential at
         Donnan-Equilibrium 336
     2. The Surface Potential ψ_o 339
       a. The Molecular Basis of Membrane Surface Charges 339
       b. The Gouy-Chapman and Stern Theory
         for Double Layers on Planar Surfaces 340
       c. Peculiarities of the Electrostatics
         of the Cell Membrane 341
     3. Current-Induced Concentration Gradients and Fields 343
   B. The Potential Profile $\psi(y,z)$ in the Plane of the Membrane 346
     1. The Situation Outside the Membrane 346
     2. The Lateral Electric Field Inside the Membrane 347

III. The Functional Consequences of the Membrane Electric Field 347
   A. Experimental Evidence for a Control Function
      by the Membrane Electric Field 348
   B. Molecular Mechanisms 349
     1. Surface Potential Effects 349
     2. Field Effects 350
       a. Field Effects in the Lipid Bilayer 350
       b. Field Effects in Intrinsic Membrane Proteins 351

IV. Conclusions and Future Directions 353

References 354

0-8493-4476-X/96/$0.00+$.50
© 1996 by CRC Press Inc.

329

I. INTRODUCTION

The term electromanipulation of cells, at present, refers to the application of relatively strong and supraphysiological intensity electric fields to cells in order to induce movement (e.g., dielectrophoresis, electrorotation, travelling-wave drive) or to alter their membrane and/or internal structures (e.g., electropermeabilization, electrofusion). The majority of this book is devoted to these powerful procedures, their underlying mechanisms, and their application to basic research, biotechnology, biomedical research, and genetic engineering. It must be kept in mind, however, that the application of low intensity electric (or magnetic and electromagnetic fields) to cells, or to intact organisms *in vivo,* may also exert profound biological influences that we are only now beginning to understand.

The further development of existing electromanipulation technology and the discovery of many exciting new applications will depend, in part, upon a more complete knowledge of electric field effects on the cell membrane and other cellular constituents. Interpretation of dielectrophoresis and electrorotation data and the restoration of the cell's electric system after electromanipulation require detailed knowledge of the effects of electric fields. The acquisition of such knowledge will presuppose a detailed consideration of the interactions of low level electric field effects on biological systems and of the "electric structure" of the cell. Such information will be useful beyond the applications to biotechnology in the understanding of fundamental processes in cell physiology and cell-cell communication.

In this chapter I will first describe the electric structure of the cell surface per se, and subsequently discuss some functional consequences of electric field modifications at the cellular and subcellular level. Possible mechanisms underlying these processes will also be discussed. The electric structure of the cell expresses a number of defined properties which themselves result from the molecular structure and functions of membrane and intracellular components. These electric properties can be classified as follows:

1. Passive electric properties, i.e., the capacity and conductivity of the plasmamembrane and of the cytoplasm. As described by the "RC-circle," or the equivalent Cole-Cole plot, these determine the impedance of the cell (see Chapter 5).
2. Electrostatic equilibrium properties, based upon the molecular structure and distribution of fixed charges. These consist of rapidly equilibrating electric double layers and Debye-Hückel ionic clouds. These properties also include the distribution of ions described by the Donnan-equilibrium, which passively adjusts to the ionic redistribution driven by ATP-requiring pumps.
3. Properties based on an electrodynamic system and originating directly as a result of the activity of rheogenic (i.e., electric current inducing) pumps

or indirectly by triggering an electrochemical gradient (built up by electrically neutral ionic pumps), as results of membrane excitation.*

These three properties form a functional network and control a number (perhaps the majority) of cellular physiological processes. A classical example for such an electrophysiological control system is the excitation of nerves and muscles. Today we know that these are processes in which electric field effects (modification of transport proteins) are combined with chemical shifts (e.g., changes in Ca^{2+}-concentrations). This process is controlled by delicately balanced feedback-mechanisms (for example, ion permeability $\rightarrow$ membrane potential $\rightarrow$ field strength $\rightarrow$ modification of transport proteins $\rightarrow$ ion permeability), but the entire cascade of events is initially triggered by chemical as well as by electric or other physical stimuli.

The electrochemical trigger mechanisms occurring in nerve and muscle cells are one of the best known examples of a widespread mechanism that is common to nearly all cells. This is evidenced by the discovery and characterization of many other field-sensitive transporters and channels that influence a vast number of cellular mechanisms.[1-4] Many other influences (hormone signaling, the effects of toxins, antibiotics, immunological reactions, viral infection, and so on) can modify the membrane permeability and trigger the release of electrochemical energy stored in the cell. A number of correlations have been found between an *in vivo* change of the transmembrane potential and various cell functions (including proliferation, fertilization, secretion, vesiculation, shape transformation, cell movement, and others), even though precise electrochemical mechanisms have not yet been clarified for many of these activities.[5]

Electrically mediated influences on cell function are not restricted to changes in the transmembrane potential. Changes in the surface potential on both the inner and outer surfaces of the membrane can also influence various cellular functions. Such changes can exert direct effects on membrane bound macromolecules electrostatically. Alternatively, they may build up electric "microcompartments" on the membrane surface, exerting specific influences on ionic strength, pH-value, conductivity, and osmotic pressure. In this respect they can indirectly influence membrane bound proteins as well as interact with a variety of inter- and intra-cellular processes indirectly.

---

* The term "rheogenic" characterizes non-electroneutral transport processes, directly transferring charge and therefore inducing an electric current through the membrane. The more popular term "electrogenic" is more specific and refers to carriers which change the transmembrane potential of the cell. This is not the same. The Na^+- K^+-ATPase of human erythrocytes, for example, is "rheogenic" (see Figure 2) but not "electrogenic," because of the high electric conductivity of the band-3 carrier, which short-circuits the transported charges. The word "electrogenic" is more common, but in this paper in the interest of physical exactness we will differentiate between these two terms.

II. THE STRUCTURE OF THE
CELL MEMBRANE ELECTRIC FIELD

The cell membrane is structurally a "mosaic" comprised of a lipid bilayer interspersed with a variety of integral and associated proteins, carbohydrates, and other more complex structures. Its electric field and its corresponding electric potential is thus a function $\psi(x,y,z)$ of the coordinates x (perpendicular to the membrane) and y,z (in the plane parallel to the membrane surface). In living cells this potential is generally in a non-equilibrium state, sustained by entropy-producing ion pumps. The double-layers around the fixed charges, however, are in thermodynamic equilibrium. In this system, a broad "hierarchy" of time constants exists. The double layers are established much faster than the functionally realized transfer of charges. They are also established faster than the ions can be moved by the energy-consuming pumps or than changes in local concentrations effected by transmembrane fluxes and electrochemical gradients.[6] Even in the case of quickly depolarizing nerves ($\tau \approx 1$ ms), the ion distributions around the corresponding channels are never thrown out of local electrochemical equilibrium.

A. THE POTENTIAL PROFILE $\psi(x)$
PERPENDICULAR TO THE MEMBRANE SURFACE

As indicated in Figure 1, the function $\psi(x)$ is determined by an electric phase potential difference, or transmembrane potential ($\Delta\psi$) between the cytoplasm and extracellular fluid, and two surface potentials at the external (ψ_{oe}) and internal (ψ_{oi}) interface of the membrane. The phase potential difference ($\Delta\psi$) corresponds with the so-called "membrane potential," measured electrophysiologically as the "resting potential" or "action potential" by microelectrodes. The external surface potential (ψ_{oe}) can be calculated from the ζ-potential, which is available from electrokinetic methods such as cell electrophoresis.

1. The Transmembrane Potential $\Delta\psi$

From the electrophysiological point of view, the living cell is a multicompartmental system, consisting of multiple microphases (organelles including mitochondria, nucleus, Golgi vesicles, and so on). All of these compartments exhibit differences in the electrochemical and electric potential in relation to the cell plasma, which itself shows a potential difference with respect to the extracellular medium. In thermodynamic equilibrium, this is a heterogeneous Donnan-system where the electric and osmotic differences are governed by the charges of the corresponding macromolecular compounds. In living cells this Donnan-equilibrium is thrown out of balance by metabolically driven pumps. This process results in a non-equilibrium steady state with properties (including membrane potential, ionic concentrations, volume, and so on) very different from those of the Donnan-equilibrium. The Donnan-state, however, can be quickly re-established by electric breakdown of the membrane (see Chapters 1 and 2), or by increase of the membrane permeability with

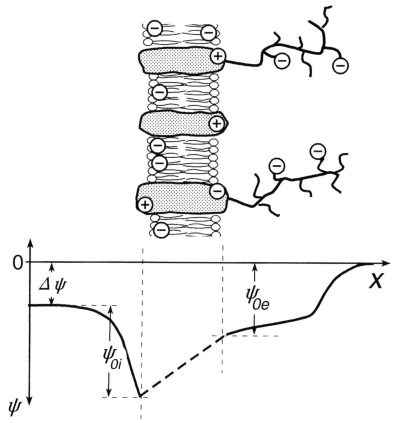

FIGURE 1. A schematic potential profile $\psi(x)$ through a cell membrane, indicating the definitions of the transmembrane potential ($\Delta\psi$) as well as the inner (ψ_{oi}) and outer (ψ_{oe}) surface potentials.

certain toxins, antibiotics, or viruses. Therefore, in all experiments using electromanipulation the possibility of transforming the cell into a Donnan-equilibrium must be taken into consideration. A brief discussion of each of these states follows.

a. The Transmembrane Potential in Non-Equilibrium Steady State

The classical view on the electrophysiology of the cell is centered upon an electrochemical gradient resulting from the actions of Na^+-K^+-ATPases, pumping potassium into, and sodium out of, the cell. The resulting large internal potassium concentration coupled with an equally large external sodium concentration produces a diffusion potential via the selective permeabilities of these ions. The permeabilities to Na^+ and K^+ are tightly regulated by specialized membrane proteins. This model was established some years ago by experiments on giant axons: in these particular cells it reflects the actual situation very well. However, subsequent investigation of the transmembrane potential of other cell types including lymphocytes, tumor cells, and others indicated that a more complex situation generally occurs.*

The dramatic change in the theory of the origin and maintenance of the membrane potential in the majority of cells was also a result of increasing knowledge of the molecular mechanisms of ion transporters (Figure 2). It is now known that the simple, passive ionic leaks proposed by the classical theory are uncommon in most cells. Such pores do occur in highly specialized nerve cells and can also be produced using artificial ion transporters, such as valinomycin.

In general, each cell membrane contains a multitude of very specific transporters that are obligatorily electroneutral or have a very definite electric conductance. One must distinguish between two distinct types of ionic pumps. The first of these may be termed electroneutral pumps, exchanging for example H^+ for Na^+ in 1:1 stoichiometry. These pumps create an electrochemical gradient which can, however, produce a permeability controlled membrane potential. The second broad category of pumps are rheogenic. These pumps generate an electric current directly by transporting a net charge in one direction. This is the case for Ca^{2+}- or H^+-pumps, as well as for Na^+-K^+-ATPases. The latter transport three sodium ions, but only two potassium ions in the opposite direction, resulting in a net flow of cations.

It is impossible to differentiate between the two types of potentials, namely those generated directly by electrogenic pumps and those based upon diffusion potentials, simply on the basis of measurements of $\Delta\psi$. In cell-manipulation applications, however, this difference is evident owing to its effects on quite different kinds of cellular reactions. These would include the inhibition of ionic pumps by special drugs or short-circuiting them by electric breakdown of the membrane (see Chapter 1). Under these circumstances the effect of *rheogenic pumps* immediately vanishes. The pump-induced current in this case is short-circuited, and the membrane potential changes over a few seconds. In contrast to this, the *diffusion potential*, as the result of the membrane electrochemical gradient which was built up by electroneutral or even electrogenic pumps, will change with a time constant, governed by the process of the induced ion leakage. If only the pump is inhibited, or, if the induced membrane leak is moderate, the membrane potential falls slowly and frequently initiates a sequence of various processes.[9-11] A nerve cell can, for example, generate action potentials for hours, even if its Na^+-K^+-ATPase is fully blocked by the drug ouabain.

Another important peculiarity of rheogenic pumps is that they drive a circulating current through the membrane, into the cellular environment, and then back into the cell. The consequent external field profile can be measured by the vibrating electrode technique.[12-15] The structure and functional role of these fields is discussed in detail below.

---

*   The evidence for widespread deviation from the classical model was not readily accepted at first.[7,8]

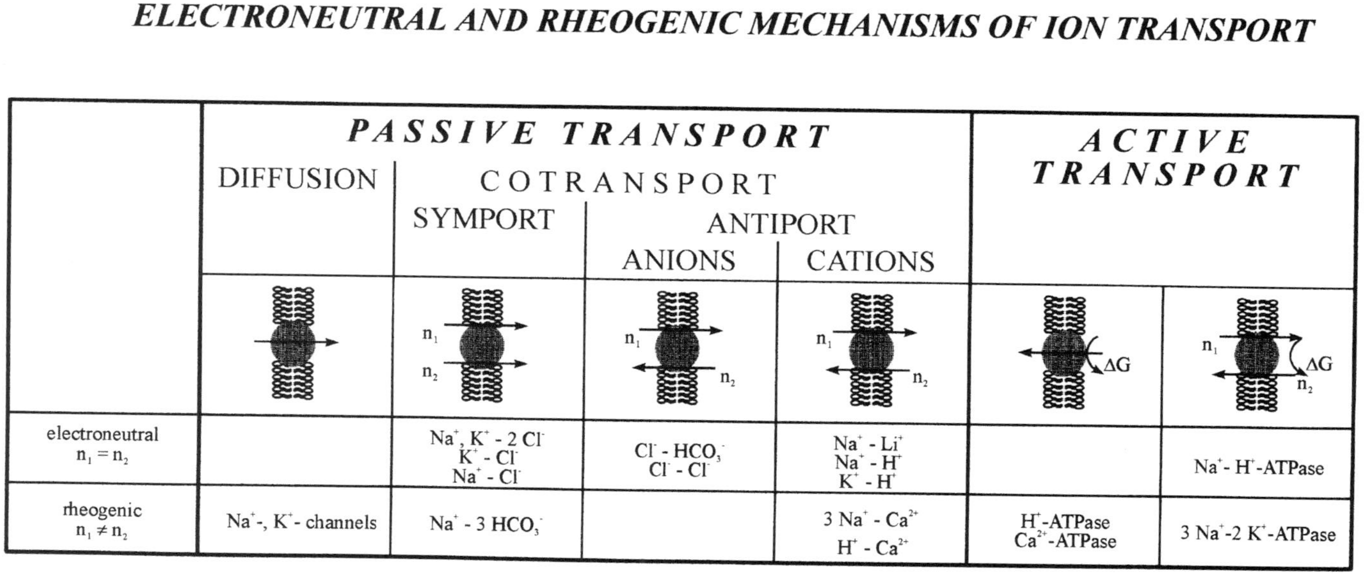

ELECTRONEUTRAL AND RHEOGENIC MECHANISMS OF ION TRANSPORT

	PASSIVE TRANSPORT				ACTIVE TRANSPORT	
	DIFFUSION	COTRANSPORT				
		SYMPORT	ANTIPORT			
			ANIONS	CATIONS		
electroneutral $n_1 = n_2$						
rheogenic $n_1 \neq n_2$	Na^+, K^+ - channels	$Na^+ - 3 HCO_3^-$		$3 Na^+ - Ca^{2+}$ $H^+ - Ca^{2+}$	H^+-ATPase Ca^{2+}-ATPase	$3 Na^+ - 2 K^+$-ATPase

Examples (electroneutral row): SYMPORT — $Na^+, K^+ - 2 Cl^-$; $K^+ - Cl^-$; $Na^+ - Cl^-$. ANTIPORT ANIONS — $Cl^- - HCO_3^-$; $Cl^- - Cl^-$. ANTIPORT CATIONS — $Na^+ - Li^+$; $Na^+ - H^+$; $K^+ - H^+$. ACTIVE TRANSPORT — $Na^+ - H^+$-ATPase.

FIGURE 2. Schematic presentation of electroneutral and rheogenic mechanisms of ion transport, and examples of corresponding transport systems found in various cells.

b. The Transmembrane Potential at Donnan-Equilibrium

In contrast to the non-equilibrium compounds of the transmembrane potential as generated by electrolyte diffusion or by rheogenic pumps, the Donnan-potential is an electric potential difference at equilibrium state and therefore unable to induce a current.[16] On the other hand, its *gradient* generates an electric field which is responsible for lipid interactions and for the particular distribution of the ions inducing the generation of local osmotic pressure as well as pH differences. The Donnan-osmotic state of the cell thus determines its volume and is therefore most responsible for the fate of cells after electric breakdown (see Chapter 1).

The Donnan-equilibrium in human erythrocytes has been studied in experimental and theoretical detail,[17-21] including the investigation of the Donnan-osmotic hemolysis of erythrocytes[22] caused by melittin (a component of bee venom). Many other reactions of cells to a variety of factors are the result of a dramatic increase in ionic permeability and subsequent transition to the Donnan-state.

If the membrane permeability of a living cell is increased, or if the cell is transferred to a medium with altered composition (for example, into low ionic-strength solutions for the purpose of electrorotation, dielectrophoresis, etc.), or alternatively if its ion pumps are inhibited, a series of quasi-equilibrium states will be reached[18] (Figure 3). Each of these states is associated with a particular cell volume and membrane potential. Thus in the case of erythrocytes, about 10 minutes after transferring them into a low-ionic strength solution, the internal Cl^--concentration and the pH will be in equilibrium with the external medium. This is a sort of a Donnan-equilibrium state, in which, however, the K^+- and Na^+-contents of the cell are *not* altered, because the equilibration of these ions requires some hours. In this situation (the so-called "C-STATE"), the fixed charges include not only the pH-dependent hemoglobin charges, but also (and principally) the charges of K^+ as well as other intracellular ions. As indicated in Figure 4, the transmembrane potential in this situation is pH-dependent, but to a much lower degree than in the D-STATE.

If the membrane permeability of the cells under these conditions is significantly increased (for example by electric breakdown, ionophores, or toxins), Na^+ and K^+ can also exchange quickly, the only fixed charges remaining in the cell being due to hemoglobin (the so-called "D-STATE"). Given the buffer-equation for the hemoglobin charge density (ρ_{Hb}), an isoelectric point of 6.9, and an internal pH-value determined by $\Delta\psi$ through the Nernst equation, a feedback loop is established as follows: $\Delta\psi \rightarrow pH_i \rightarrow \rho_{Hb} \rightarrow \Delta\psi$. In this situation, the transmembrane potential becomes strongly dependent on the external pH (Figure 4).

Such simple calculations are inappropriate for cells other than erythrocytes. The large number of proteins and reticular structures present in most cells all participate in the eventual Donnan-equilibrium, which is therefore complex. Nevertheless, a number of results obtained using erythrocytes can be transferred to other cells, at least qualitatively. One such finding is that all effects

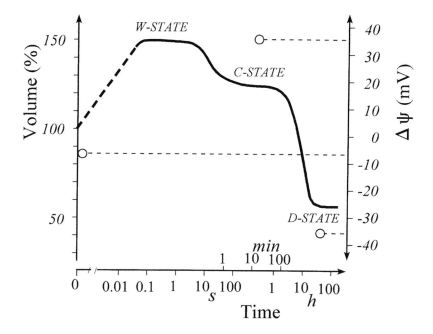

FIGURE 3. Volume changes of human erythrocytes transferred at $t = 0$ from a physiological solution into a hypo-osmolar solution with low ionic strength (10 mM NaCl, 120 mM sucrose, pH 7.4, 37°C). Note that the time axis is logarithmic. The data were calculated using the corresponding equation sets for stationary states[18,19] and time constants of 91 s^{-1}, 5 · 10^{-2} s^{-1}, 3.2 · 10^{-5} s^{-1} for the establishment of the W-STATE, the C-STATE, and the D-STATE, respectively. The W-STATE is a quasi-stationary state where only water is fully equilibrated. The transmembrane potential $\Delta\psi$ for the normal state (–8 mV), for the C-STATE (+33.5 mV), and for the D-STATE (–36.2 mV) are listed as points. The time course of the transmembrane potential at the transition states is not predictable, because it depends strongly on the corresponding values of the ionic permeability. Note that in this case, cells attain three different quasi-stationary volumes in succession, although remaining in the same solution.

upon the transmembrane potential can induce Donnan-osmotic volume alterations. The volume of the cell, of course, is involved in the feedback loop above. Volume modification leads to a change of charge *densities* even when the *number* of charges per cell remains constant. This phenomenon explains why some iso-osmolar solutions can cause cells to shrink (Figure 4). Modification of the integrity of the membrane barrier, by electric breakdown for example, also influences the process of volume regulation (see Chapter 1). If cells are suspended in pure electrolyte solutions, breakdown of the membrane barrier will lead to cytolysis, because the volume of cells in D-STATE increases in iso-osmolar solutions as sucrose is progressively replaced by NaCl. The equations of Donnan-equilibrium give rapid increases in cell volume (as V → ∞; Figure 4) when the proportion of electrolyte (assumed freely membrane-permeable) in the surrounding medium approaches 100%.

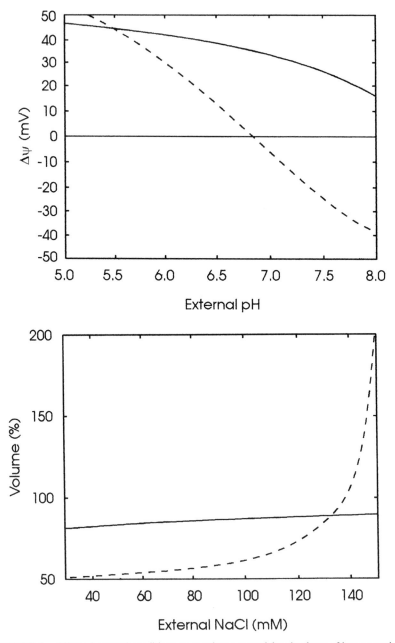

FIGURE 4.   Effect of external conditions on membrane potential and volume of human erythrocytes in iso-osmolar NaCl-sucrose solutions (290 mosmol kg⁻¹, T = 25°C) in C-STATE (solid line) and D-STATE (dashed line) (see text). Top: The transmembrane-potential as a function of external pH (c_{NaCl} = 30 mM). Bottom: The volume as a function of external NaCl-concentration (pH = 7.4).

2. The Surface Potential ψ_o

a. The Molecular Basis of Membrane Surface Charges

As noted above, the membrane consists of a mosaic of proteins, embedded in a bilayer of phospholipids and anchored to the internal cytoskeleton. The lipids have a specific distribution between the two layers and a lateral allocation which is strongly influenced by the protein mosaic. This structure exhibits a multitude of fixed charges of different origin, different polarity, and different isoelectric points.

The surface potential and the surface charge of cell membranes were first investigated by measurement of the ζ-potential by cell electrophoresis.[23-31] More recently, fluorescent probes and spin labels were introduced to investigate the surface potential, as well as the potential at definite distances from the real membrane surface.[32-35] These methods allow verification of the calculations of surface potential determined by ζ-potential measurements. Furthermore they provide experimental evidence for the applicability of the double-layer theories. The lateral distribution of surface charges are measured by electron microscopical analysis of the distribution of charged colloids of high electron density.[36-38]

Most cells are negatively charged under physiological conditions. Only some of these charges, however, are fixed in the membrane plane. They are located predominantly in a surface coat, or glycocalyx, formed by glycoprotein filaments (see Figure 1).

The surface charges of the membrane proper include various charged groups in the hydrophilic portions of proteins and the ionizable groups of phospholipids. The latter are of special interest and consist principally of phosphatidylserine (PS), which carries one negative charge at physiological pH. In human erythrocytes, PS comprises about 15–20% of the total lipids and is located exclusively in the inner membrane leaflet. In other cells, PS is also located predominantly on the inner side of the membrane. The asymmetrical organization of phospholipids in the membranes, in the steady state, is maintained by an ATP-driven pump, the aminophospholipid translocase. This pump selectively pumps PS and phosphatidylethanolamine from the outer to the inner monolayer.[39-41] The $8.4 \cdot 10^7$ PS-molecules per human erythrocyte result in a surface charge density of $\sigma = -0.09$ C cm^{-2} and obviously induce a considerable inner surface potential.[21]

The glycocalyx (Figure 1) consists of a loose layer of glycoprotein filaments, which extend from the intrinsic membrane protein glycophorin, as well as from glycolipids. The distal (outer) portions of these molecules bear neuraminic acids (= sialic acids). At physiological pH these acids are dissociated and are the principal determinant of the negative surface potential of cells: a single human erythrocyte carries $2 \cdot 10^7$ sialic acid charges. These charges can be removed selectively by neuraminidase (this enzyme plays a role in the process of capacitation of sperm in the female genital tract and is released by viruses during cell infection).

The structure of the glycocalyx is determined by the electrostatic interaction of its own intrinsic charges. Depending upon the ionic strength and therefore upon the Debye-Hückel parameter, κ, (see Equation 2 below), these molecules are either closely attached to the membrane or in a more loose arrangement. The latter can result from the effects of electrostatic repulsion. Measurements with spin labels, however, indicate that the sialic acid groups in the glycocalyx protrude into the aqueous phase and do not significantly influence the surface potential at the external bilayer interface of human erythrocytes.[35]

b. The Gouy-Chapman and Stern Theory
for Double Layers on Planar Surfaces

The simplest approach to the calculation of the electrostatic situation near the cell surface is based upon the double-layer models proposed in 1879 by Helmholtz and developed further in the beginning of the 20th century by Gouy and by Chapman. A number of investigators have demonstrated that the Gouy-Chapman theory correlates well with experiments on artificial lipid membranes.[25,42-46]

The Gouy-Chapman theory assumes that the fixed charges are "smeared" uniformly over a plane and that the aqueous phase is a structureless medium with a uniform dielectric constant, containing ions as point charges. The function $\psi(x)$ near a surface with potential ψ_0 (for $x \rightarrow 0$) is determined by:

$$\psi = \psi_0 e^{(-\kappa x)} \tag{1}$$

where κ is the Debye-Hückel constant defined as follows:

$$\kappa = \sqrt{\frac{2e^2 z_v^2 Nc}{\varepsilon \varepsilon_0 kT}} \tag{2}$$

where N is the Avogadro number, ε is the dielectric constant, ε_0 is the permittivity of vacuum, e is the electronic charge, z_v is the valency of the symmetrical electrolyte, c is electrolyte concentration, k is the Boltzmann constant, and T is the temperature. $1/\kappa$ is, therefore, a distance (the Debye-Hückel length) indicating where the potential is reduced to $1/2.718$ of the surface potential ψ_0.

The surface potential ψ_0 depends on the surface charge σ in the following way:

$$\sigma = \kappa \frac{2\varepsilon \varepsilon_0 kT}{e z_v} \sinh\left(\frac{z_v e \psi_0}{2kT}\right) \tag{3}$$

For $z_v = 1$ and $T = 300$ K, the expression $z_v e/2kT = 19.34$ V^{-1}. For small potentials, i.e., $\psi \leq 0.01$ V, we can approximate: $\sinh(\alpha) \approx \alpha$. This linearization reduces Equation 3 to:

$$\sigma = \varepsilon\varepsilon_0 \kappa \psi_0 \tag{4}$$

This sort of linearization of exponential terms is very common, however one must not forget the strict limitation to low values of the potential. In some cases the potentials are higher, and therefore these approximations are not applicable.

The potential profile $\psi(x)$ near the surface of the membrane depends, therefore, not only on the surface charge density σ but also on the salt concentration c and, in a quadratic way, on the valence z_v of the ions.

An important extension of the Gouy-Chapman theory, introduced by Stern, takes into account adsorption processes. In this analysis, the Langmuir adsorption isotherm was used to modify the effective surface charge as follows:

$$\sigma = \frac{1}{K}(\sigma_{max} - \sigma)\left[A^-\right]_o \tag{5}$$

where σ is the number of singly charged molecules absorbed to the membrane per unit area, σ_{max} is its maximum, K is the dissociation constant, and $[A^-]_o$ is the concentration of the absorbing univalent species near the membrane surface. The value $[A^-]_o$ can be calculated in the simplest way from ψ_o using the Boltzmann equation:

$$\left[A^-\right]_o = \left[A^-\right]e^{\left(\frac{e\psi_0}{kT}\right)} \tag{6}$$

This approach is useful in a number of situations, but does not consider hydrophobic interactions, especially the hydrophobic adsorption of ions to uncharged membranes. The reader is cautioned that some serious contradictions occur if one attempts to apply this simple theory to highly charged membranes.[47]

c. Peculiarities of the Electrostatics of the Cell Membrane

The real molecular structure of the cell membrane as described before indicates clearly that the Gouy-Chapman-Stern approach can be considered only as a first approximation, even though it fits the experimental results surprisingly well.[44] In general, however, the correct description of the surface potential of the cell must take into consideration the following peculiarities:

1. The surface charges of the cell membrane are not "smeared," but are located at definite points in the mosaic-structure. In conditions of physiological ionic strength, their distance sometimes exceeds the Debye-Hückel length $1/\kappa$ (≈ 1 nm). This circumstance has been considered by a number of authors.[42,48-52] It seems that differences in the theoretical approach occur, but they will not be found by experimental measurements indicating only an over-all net effect, i.e., resulting from the effect over the whole surface. These calculations, however, are quite important

in explaining local effects of surface charges near reactive centers of surface proteins (see below).

2. Most charges at the outer surface of the cell membrane are localized in the glycocalyx. Therefore one must consider a space charge density (ρ in C cm^{-3}) rather than a surface charge density (σ in C cm^{-2}), i.e., a function: $\rho(x)$. Such calculations are possible by integration of the Poisson equation:

$$\nabla^2 \psi = -\frac{\rho(x)}{\varepsilon \varepsilon_0} \tag{7}$$

and by use of the Boltzmann equation for the distribution of the ions from the electrolyte as a function of ψ. This has been done using a number of assumptions for the function $\rho(x)$.[21,35,52-56] As a result, the function $\psi(x)$ is not a simple exponential according to Equation 1, but remains of some higher order depending on the thickness of the glycocalyx (Figure 1). In fact, the function $\rho(x)$ itself is dependent upon the electric field $\psi(x)$, because of the electrostatic interactions of the filaments in the glycocalyx. Therefore a solution to this problem is possible only if the control-loop $\psi(x) \leftrightarrow \rho(x)$ is taken into account. Such calculations are still more complicated when the overlapping double layers of two opposed cell surfaces are taken into consideration, which occurs in the case of cell-cell interaction and electrofusion (see Chapter 4).[57-59]

3. The surface potential of one side of the membrane depends to some extent on the charge density of the opposite side. This is of particular interest in erythrocytes because the charged PS molecules are located preferentially in the internal leaflet. One author[45] estimated that, for the case of identical electrolytes on both sides of the membrane (and therefore $\kappa_1 = \kappa_2 = \kappa$), Equation 4 can be modified to give the dependence of the surface potential of side 1 (ψ_{01}) upon the surface charge density (σ_2) of side 2:

$$\psi_{01} = \frac{\sigma_2}{\varepsilon \varepsilon_0} b \tag{8}$$

Here:

$$b = \frac{1}{2 + \kappa d \varepsilon / \varepsilon_m} \tag{9}$$

where ε_m is the dielectric constant of the membrane, and d is the membrane thickness.

4. It is not strictly correct to consider the cell membrane as planar, because it possesses curvature, especially in the presence of topological features

such as microvilli, spikes, or invaginations (see Chapter 1). Calculations have been performed to investigate the effect of these features on the structure of the double layer, as well as to draw conclusions about the shape-forming-force generated by the surface electric field.[60-65]

3. Current-Induced Concentration Gradients and Fields

The fluxes and ionic currents induced by external fields or by intracellular electrodes can influence the distribution of ions inside a charged membrane, in ion-conducting pores, or in the double layers at membrane surfaces. The function $\psi(x)$ is thereby also influenced, and unstirred aqueous layers on both membrane surfaces enforce this process. Such layers hinder the convection-mediated exchange of ions, so that the transmembrane ionic flux can only be carried into the bulk solution by diffusion. As a result, diffusion-limited layers may occur near the membrane, where the local ionic concentrations are determined by the direction of the corresponding flux.[66] Under artificial conditions (ion-exchange membranes), these layers can reach a thickness of hundreds of micrometers. They can be detected by laser diffraction[67] or by microelectrodes.[68-70] In living cells, such layers are much thinner, depending on ion transport rates.[71] Similarly, ionic currents shift ion concentrations in membrane channels or in membranes with ionic exchange properties.

Diffusion layers are, in general, non-equilibrium structures which generate electrochemical gradients with significant consequences for *in vivo* processes[72] as well as for cell electromanipulation.

Phenomenologically, these layers influence the voltage-current characteristics of the membrane,[73] including their effective capacity as well as their resistance.[74] In a system consisting of a membrane with an attached unstirred layer, application of an alternating field or a sinusoidal field pulse results in a phase shift:

$$\tan \varphi_t = \frac{\Delta V_m^0 \sin \varphi_m + \Delta V_u^0 \sin \varphi_u + \Delta V_b^0 \sin \varphi_b}{\Delta V_m^0 \cos \varphi_m + \Delta V_u^0 \cos \varphi_u + \Delta V_b^0 \cos \varphi_b} \qquad (10)$$

where ΔV^0 is the electric potential difference, and subscripts *t*, *m*, *u*, and *b* refer to total, membrane, unstirred layer, and bulk phases. Considering the external solution as purely resistive ($\varphi_b = 0$) and small values of φ_m and φ_u, one can rewrite this equation as follows:

$$\tan \varphi_t = \frac{\Delta V_m^0 \sin \varphi_m + \Delta V_u^0 \sin \varphi_u}{\Delta V_t^0} \qquad (11)$$

Using an RC-equivalent circuit, where R_t and C_t are the measured resistance and capacitance, and assuming for example a membrane with potassium and proton (or hydroxyl-) conductivity,[74] then:

$$C_t = C_m - \frac{I_h^0 B_h + I_k^0 B_k}{\omega I_t R_m^2 \sigma} \tag{12}$$

where:

$$B_h = \frac{1 - (D_0 / D_h)}{2K_h} \left[1 - \sqrt{2} \sin\left(\frac{K_h L + 4\pi}{4} \right) e^{-K_h L} \right] \tag{13}$$

and:

$$B_k = \frac{1 - (D_0 / D_k)}{2K_k} \left[1 - \sqrt{2} \sin\left(\frac{K_k L + 4\pi}{4} \right) e^{-K_k L} \right] \tag{14}$$

where: I is the current through the membrane (subscripts: t = total, h = proton, k = potassium, $K_h = (\omega/2D_h)^{1/2}$, $K_k = (\omega/2D_k)^{1/2}$, L is the unstirred layer thickness, D are the corresponding diffusion coefficients in the unstirred layer, ω is the radian frequency of the alternating field, and σ is the total ionic electric conductivity in the diffusion layer. This model indicates that even negative effective capacities can occur in membrane systems, if the diffusion layers are sufficiently strong.

A special voltage-current relationship results if two regions of opposite fixed charges are tightly connected.[73] In such a case a complicated potential profile occurs (see Figure 5). This system is an electrochemical analog of the P-N junction diode with a "depletion" layer between these two regions. In the region with positive fixed charges, the predominant mobile ions are negative, and correspondingly most of the mobile ions in the region with negative fixed-charges are cations. Remote from the depletion layer, the mobile-ion concentrations correspond to the Gouy-Chapman (or to the Donnan-) equilibrium. The fixed charges in the neighborhood of the junction, however, will be almost completely uncompensated ("depletion layer"). When a bias voltage is applied to those membranes, the concentration of the mobile ions will shift, leading to a change of the conductivity. If the applied voltage is sufficiently high, the conductivity of the membrane suddenly increases very rapidly. This is similar to a "punch-through" effect which occurs in P-N diodes. In this case the capacitance is vested in the depletion layer and is expected to vary with the membrane electric field, i.e., with the transmembrane potential ($\Delta\psi$).

Such "punch-through" effects (see also Chapter 1) were observed in membranes of *Nitella* and *Chara* even at membrane potentials far below the electric breakdown voltage (150–300 mV).[73,75,76] What could be the molecular basis of such a model? It seems unrealistic to propose extended ion-exchange areas of opposite charge in biological membranes. On the other hand, this behavior could be the result of a specific charge-distribution inside an ion conducting membrane protein. Further investigation applying the newly described

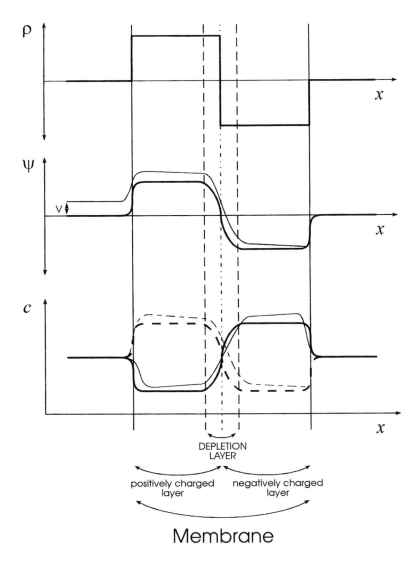

FIGURE 5. Schematic representation of the fixed charge density (ρ) of a system which consists of a double lattice charge membrane with a sharp contact region. The resulting potential profile (ψ) and the concentration of the mobile anions (c, dashed lines) and cations (c, solid lines). The thick lines indicate schematically the function at equilibrium, without phase potential difference ($\Delta\psi = 0$). If a voltage V is applied, a shift of the potential ψ, as well as of the concentrations c of mobile ions will occur, as indicated by the fine lines.

techniques discussed in this chapter and elsewhere in this book may provide answers to this question.

With respect to the mechanisms of the electric breakdown phenomenon discussed in Chapters 1 and 4 it is important to emphasize that the "punch-through" effect is the result of a voltage pulse lasting more than 10 ms. The time constant of this process is predicted by the diffusion of ions in the membrane. In contrast, electric breakdown occurs in the nanosecond range at higher values of the membrane potential.[75,76]

B. THE POTENTIAL PROFILE $\psi(y,z)$ IN THE PLANE OF THE MEMBRANE

1. The Situation Outside the Membrane

As pointed out in the Introduction, one must differentiate between two components, which together determine the electric field $\psi(y,z)$:

1. the membrane mosaic formed by proteins and lateral domains of charged or polarized lipids which represents an electrostatic equilibrium component;
2. a non-equilibrium or electrodynamic component, generated by ionic fluxes which penetrate the membrane at definite points.

The electrostatic mosaic is difficult to investigate, because methods for the measurement of charges (e.g., charged colloid electron microscopy) destroy the mosaic. A number of studies demonstrate the movement of fluorescence-labeled proteins under the influence of a laterally applied electric field (5–20 V cm^{-1}).[65,77-84] This movement, however, results not only from electrophoresis of charged membrane proteins, but is at least partly due to electro-osmotic streaming. The direction of the field-induced movement of membrane proteins, therefore, depends not only on net membrane charge, but also upon the relationship between its own ζ-potential to the mean ζ-potential of the whole cell surface.[81]

The most evident indication of extracellular electric fields is the electrocardiogram. It measures the sum, at the surface of the body, of potential differences which result from the electric fields induced by the propagation of cardiac muscular action potentials. Local excitation of each individual cell is accompanied by depolarization of the transmembrane potential, resulting in a lateral electric field.

By using the "vibrating electrode" it is possible to measure external fields, (Figure 6) generated by permanent ionic currents, driven by rheogenic pumps.[12,15,85-88] In this case a circuit exists where ions leave the cell at one point and re-enter elsewhere. This leads to measured external field strengths[15] up to the order of 0.1 V m^{-1}. It has been proposed that such fields are partly responsible for the regulation of morphogenesis.[13,87,89-91]

One author[92-94] indicated that this non-equilibrium system, consisting of both a rheogenic pump (driving the ions in one direction through the membrane) and a transporter (which acts as the return path for the current), results in the formation of dissipative structures that can thereby influence the distribution of both of these functional proteins. With respect to the lateral field of the membrane, it should be noted that owing to the larger concentration of mobile ions in the electric double layer, a surface conductivity exists which is higher than the conductivity of the surrounding medium (see also Chapter 5).[95,96]

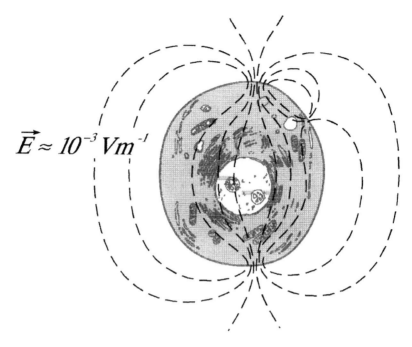

$$\vec{E} \approx 10^{-3}\, Vm^{-1}$$

FIGURE 6. The external and internal electric field of a cell as generated by a pump-leak system, corresponding to measurements obtained with the vibrating electrode.

2. The Lateral Electric Field Inside the Membrane

In contrast to the aqueous medium on both sides of the membrane, the hydrophobic core of the membrane phospholipid bilayer has a very low dielectric constant. Very large electric fields are therefore induced around charges inside the membrane. It is therefore likely that the lateral distribution of such molecules is influenced by mutual electrostatic interactions.[97,98] It is possible that such interaction between membrane proteins may also influence their functional state.[99,100]

Recently some investigators have examined whether "field-producing proteins" can influence "field-sensitive proteins" by means of an electric field in the membrane.[101] This field is significant when compared to the transmembrane field at distances closer than 5 nm, but falls off very rapidly at larger distances. An influence on a field-sensitive protein is only possible within this distance, unless association exists between the two proteins.

III. THE FUNCTIONAL CONSEQUENCES OF THE MEMBRANE ELECTRIC FIELD

It is evident that the electrostatic and electrodynamic structure of the living cell described above is not merely a passive consequence of the molecular design of the membrane and of its ionic fluxes. It must also be viewed as a functional structure crucial for numerous cellular physiological processes.[5] As shown below, this idea is based on an increasing body of experimental evidence correlating biological functions and changes in the electric structure of the membranes. The existence of such control functions is also supported by considerable theoretical evidence and by the results of experiments on model systems.

A. EXPERIMENTAL EVIDENCE FOR A CONTROL FUNCTION BY THE MEMBRANE ELECTRIC FIELD

The list of cellular functions that correlate with modifications of the transmembrane potential or of the surface charge is long and contains reactions with various degrees of complexity. A number of papers show that proliferating cells exhibit a lower transmembrane potential ($\Delta\psi < 37$ mV) than do non-dividing cells.[102-107] Cells with alternating mitotically active and inactive periods, such as cancer cells, appear to decrease their transmembrane potential during their mitogenic phase.

The aggregation of platelets becomes larger when their transmembrane potential ($\Delta\psi = -48 \pm 13$ mV) is depolarized ($\Delta\psi = -13 \pm 5$ mV) by increasing external K^+- concentration or by inhibiting ion pumps with ouabain.[108] This is probably related to calcium metabolism.[109] Many types of secretory cells show changes in membrane potential and resistance upon activation and initiation of secretion.[110-113] For example, L-cells exhibit spontaneous oscillations of membrane potential correlating with fluctuations in the cytoplasmic Ca^{2+}-concentrations. These changes are associated with pinocytosis and the formation of ring-like, rigid structures on the cell surface.[114,115]

Changes in the electric potential also accompany, in combination with changes in Ca^{2+}-levels, the process of egg fertilization.[116] The membrane potential of sea urchin eggs decreases after fertilization from a resting value of -70 mV down to -10 mV.[117] The electric field across the egg plasma membrane may regulate fertilization by affecting charged components in the sperm membrane. This "voltage dependence" of fertilization is a consequence of positively charged elements on the sperm membrane which must be inserted into the egg membrane, then move into the egg membrane via membrane fusion to effect fertilization. The voltage level necessary to block fertilization is characteristic of the sperm species.[118]

These results underscore the influence of changes of the membrane electric field on transport mechanisms. Besides the large number of directly voltage-dependent transporters, effects of the membrane and surface potentials on the kinetics of various transporters have been found.[1-4,119-125] Such effects are caused by modification of the kinetics of the transporting protein, rather than by a direct influence of the electric field.

The membrane potential can also function as a driving force for flagellar movement. In this case, the potential correlates directly with the mechanical properties of the membrane.[126-132] A related phenomenon is the transformation of washed human erythrocytes from echinocytic to stomatocytic morphology. This correlates with a change in the transmembrane potential from negative ($\psi_i - \psi_e$) to positive values.[19,133-135] The mechanical properties of erythrocyte membranes can also be investigated using external electric fields. In this case the area expansivity modulus of the erythrocyte membrane is affected by potential differences (from −1 V to +1 V) applied across the tip of a pipette into which the cells have been aspirated.[136,137]

B. MOLECULAR MECHANISMS

Some of the effects mentioned in the foregoing paragraph are rather loosely correlated with the membrane potential and are probably not causally related. However, a number of such effects are predicted by the laws of physics and physicochemistry. In addition, the existence of a number of causal connections can be proved experimentally by the use of several methods of modification of potential.

1. Surface Potential Effects

Considering the surface as a micro-phase with a surface potential ψ_o, a combination of the Nernst and Boltzmann equations predicts a local concentration c_{oi} of an ion i with a valence z_{vi} in relation to its bulk concentration c_i as follows:

$$c_{oi} = c_i e^{\frac{z_{vi} F \psi_0}{RT}} \tag{15}$$

where F is Faraday's constant, R is the gas constant, and T is the temperature. For example, at $T = 310$ K, a tenfold increase in the local concentration of monovalent cations ($z_{vi} = +1$) occurs at $\psi_o = -61$ mV. For bivalent cations ($z_{vi} = +2$) a tenfold concentration increase is realized at $\psi_o = -30.8$ mV. The increased effect of multivalent ions ($|z_{vi}| > 1$) is determined not only by Equation 15, but also by Equation 4 which describes the relation between surface charge (σ) and surface potential (ψ_o), including the z_v-dependence of the Debye-Hückel parameter (κ). The potential therefore has a much more complex dependence upon z_v than is indicated in Equation 15.

The above considerations explain why local pH values occur. The surface pH (pH_o) depends on the bulk pH in the following way:

$$pH_0 = pH + \frac{\psi_0 F}{RT \ln 10} \tag{16}$$

Therefore a shift of one pH-unit occurs at points with a local surface potential of only 61 mV.

These local concentrations and pH-values near surface charges, or even near fixed charges in the near vicinity of a functional group, are the direct cause of a number of observed phenomena. For example, the influence of surface charges on the activity of membrane bound enzymes, receptors, and transport proteins is well known.[56,138-155]

In this context the influence of surface charges and surface potential on electrodiffusional processes through the membrane down an electrochemical potential gradient should be mentioned. Local concentrations of ions in the double layers on each side of the membrane cannot influence the driving force of the ions between the two bulk phases, because each double layer is in equilibrium with its corresponding bulk phase. The difference in the electrochemical potentials of the bulk phases, therefore, is the same as that between the two surfaces. However, integration of the Nernst-Planck equation requires knowledge of the functions $\psi(x)$ and $c(x)$. The constant-field ($E = \mathrm{grad}\psi = $ constant) assumption of Goldman cannot, therefore, be applied in this setting.[156,157] Experimental confirmation of this came from the voltage-current curves taken on charged membranes.[158] One approximation is the use of ψ_o- and c_o-values in the Goldman equation, in order to apply the constant-field conditions exclusively to the membrane interior and to calculate the fluxes from surface to surface.[159]

2. Field Effects

The only real surface potential effect observed to date is the aforementioned generation of local micro-environments, resulting from surface potentials, near functional proteins. In all other cases, the function $\psi(x,y,z)$ or at least the potential profile $\psi(x)$ must be considered, including surface potentials as well as the transmembrane potential itself.

The field strength in the double layer as well as in the membrane itself is large, with values up to 10^5 V cm^{-1} in the x-direction. This is large enough to influence dipole orientation in the proteins as well as in the polar head groups of the lipids.

a. Field Effects in the Lipid Bilayer

Experiments with artificial lipid membranes (see also Chapter 3) containing lipids with ionizable groups have shown that the transition temperature between ordered and fluid state can be varied by changing the ionic environment.[160-163] The Gouy-Chapman theory predicts the observed decrease of the transition temperature with increasing *charge density* of the lipid. This effect has been measured for Li$^+$, Na$^+$, and K$^+$ ions.[164] Conversely, Ca^{2+} and Mg^{2+} ions increase the transition temperature of the lipid bilayer by charge neutralization.[165] Triggering of the phase transition by cations at constant temperature has also been observed.[166,167] Similar effects can be generated in multi-layers of dimyristoyl phosphatidylcholine and sodium oleate[168] by application of external fields of up to 10^5 V cm^{-1}.

This effect is produced by the electrostatic interaction of the lipid headgroups, changing their distance and orientation. X-ray results show a decrease in bilayer thickness, whereas the hydrocarbon chain packing stays essentially constant. The electrostatic repulsion of the head groups causes them to increase their separation, generating an electrostatic surface pressure.[165] The alteration of the isotropic tension in the lipid area of the membrane is probably also capable of generating an intracellular pressure.[169]

The difference of charge, dipole momentum, and pK of various lipids give rise to varying responses according to their ionic environment. One result of this is lateral phase-separation phenomena, which have been experimentally observed in monolayer experiments.[45] The lateral organization of mixed PS/PC membranes has been shown to be dependent upon the pH and on the concentrations of Ca^{2+} and of monovalent ions.[170] If the cell membrane is also partly separated into lateral domains of different lipids, the effect of the ionic composition of the surrounding medium will change between various points on the cell surface.

Modification of the biophysical parameters of biological membranes by electrostatic influences have also been measured. One example is the finding of a linear relationship between the changes in the pyrene excimer formation efficiency (which is a function of diffusion rates within the membrane) in intestinal brush-border membrane and charge shielding at the membrane surface.[171] The microviscosity of mitochondrial membrane lipids can also be changed by modifying the transmembrane potential with valinomycin.[161,162]

It is possible that the membrane electric field can also effect the distribution of the lipids between the two leaflets of the bilayer. The distribution of molecules (including cholesterol, fatty acids, etc.) between the two leaflets is considered to be a steady-state system, shifted from equilibrium by the phospholipid translocase.[39-41,172] Any disturbance of this system by, for example, a field-induced translocation of one molecular species, will subsequently lead to a new steady state, and thus affect all other molecules. In this context it should be mentioned that other charge-induced transformations in the membrane are also possible.[173-175] The dependence of the critical temperature of the echinocyte-stomatocyte transformation of human erythrocytes on the transmembrane potential, and the multi-exponential kinetics of this process, come under this heading.[135,176] The systematic variation of cell membrane lipid composition with the membrane potential corroborates the findings in some model lipid systems.[177]

b.   Field Effects in Intrinsic Membrane Proteins

One can consider membrane proteins as the genuine supporters of membrane functions. The influence of the membrane electric field on their structure and behavior has therefore been the subject of much investigation. The scope of this chapter allows only a summary of some general aspects of these mechanisms: the interested reader is referred to a number of recent reviews addressing these questions.[178-181]

In general, it can be expected that the properties of membrane proteins will be controlled by membrane electrostatics. This may be either a direct effect of the electric field or an indirect one acting via modification of the properties of

lipids as discussed above. The direct influences of the field on protein structures can be classified as follows:

1. charge displacement,
2. interfacial charge transport,
3. dipole re-orientation,
4. dipole induction.

A phenomenological description of such reactions is possible using
LeChatelier's principle of electrochemical reactions. In any cases, the fraction
of molecules (ξ), being affected by the field in the sense of an "on - off" switch,
is given by:

$$\xi = \frac{1}{1 + e^{(\Delta W + \Delta G_E)}} \tag{17}$$

where ΔW is the difference in the free energy between the two states in the
absence of the electric field, and ΔG_E is the corresponding energy difference
induced by the electric field.[178] For a two-state switch induced by a charge
displacement,

$$\Delta G_E = \frac{z_v F \Delta \psi}{RT} \tag{18}$$

where $\Delta \psi$ is the difference in electric potential between the two location of the
charges, F is the Faraday constant, z_v the valency of the displaced charge, and
RT is the thermal noise energy.

In the case of dipole reorientation:

$$\Delta G_E = \frac{\Delta \mu E}{RT} \tag{19}$$

where $\Delta \mu$ is the change in the component of the dipole moment along the
direction of the electric field (E). In the case of an induced dipole, G_E will
depend on the square of the field strength, because the induced dipole moment
is itself proportional to the electric field. As mentioned in the Introduction, the
best-investigated field-sensitive proteins are the voltage-gated channels, originally discovered in nerve and muscle cells (but since found in many other cells
as well). In membranes containing such proteins, "gating currents" which
directly reflect charge displacements associated with channel opening and
closing, can be measured. In cases where a large number of proteins can be
oriented, these electric signatures of their function can be measured directly.[182]

Dipole reactions in proteins are mostly coupled with formation of α-helices,
and therefore the electric field in the membrane must be considered as one of
the important driving forces for membrane protein folding and/or arrangement

of subunits within certain oligomeric proteins. Theoretical and experimental evidence suggests that α-helices have significant dipole moments, which can be represented approximately by the field generated by two opposite charges of magnitude $e/2$ placed at either end of the helix.[178] The ability of a peptide to assume an α-helical conformation renders it susceptible to a membrane potential; in this case, the membrane potential may act on the α-helical dipole moment rather than on the peptide's charge.[183]

The influence of an applied electric field on bacteriorhodopsin reconstituted into lipid vesicles has been investigated. Circular dichroism indicated that the field, irrespective of its direction, decreased the α-helical fraction of the protein. These results are interpreted as indicating the unfolding of edges of the helices due to submergence in the polar environment. This occurs when the lipid bilayer is electroconstricted and/or the helices are stretched by the electric field across the membrane.[184]

Some workers in the field have used the term "electroconformational coupling" to describe the field-induced energy and signal transduction in membranes.[180,185-192] This idea is engendered by the influence of strong alternating fields on membranes and implies a frequency-dependence of these effects (see also Chapter 1). Experimental evidence does, in fact, indicate a frequency-specific influence on transport processes, mostly Na^+-K^+-ATPases, as well as on other energy transduction processes.

IV. CONCLUSIONS AND FUTURE DIRECTIONS

The biological membrane possesses not only a complicated molecular structure but also, as a consequence of this, an extremely complex electric field. The field is generated electrostatically by fixed charges and electrodynamically by the ionic fluxes of living cells. This electric field influences all membrane constituents, as well as the external and internal environs of the membrane. As such, the electric field associated with the membrane is well positioned to function as an integral part of a "servo" control system, with feedback loops within the cell. As we have discussed, such systems are indispensable in regulating a large number of cellular and intercellular processes. The influence of electric or electromagnetic fields on the living cell and therefore, all types of electromanipulation, must consequently be considered as an artificial intrusion in this exquisitely balanced, functional structure.

For the most part, electromanipulation requires the use of relatively strong electric fields, i.e., the application of energy significantly exceeding the thermal noise energy. Under these conditions, the *in vivo* homeostasis of the membrane will change substantially, resulting in effects that are energetically understandable (at least qualitatively). A challenge and priority for future research lies in attaining a much better understanding of these processes at the molecular level.

Recently, an increasing body of reliable data has been published concerning the influence of extremely weak electric and electromagnetic fields on biological systems.[193,194] Low frequency electromagnetic fields, for example, influence Ca^{2+}-transport or the processes of gene transcription even at magnetic flux densities down to 40 μT.[193-200] The magnetic field probably influences the cells not directly but via the induced electric fields. In experimental set-ups used, these fields correspond to electric field strengths near the cell down to about 0.1 mV cm^{-1}. As already pointed out in Section II.B.1 of this chapter, this field strength is nevertheless of the order of the lateral membrane field which is generated by the cell itself. The answer to the question of whether the energy introduced by the weak externally applied fields is below the thermal noise level, depends on the proposed mechanism. Assuming some reasonable electric membrane parameters, one author came to the conclusion that the Nyquist-noise of a transmembrane resistor could be overcome by an external field of about 8.6 V cm^{-1}, whereas the Brownian movement of ions in the double layer can be influenced significantly[201] even by 0.041 mV cm^{-1}. These calculations assume the following parameters: bandwidth $\Delta f = 100$ Hz, membrane resistance $= 108 \ \Omega$, cell radius $= 10 \ \mu$m, and cell surface conductivity $= 3.2 \cdot 10^{-10}$ S.

The investigation of these mechanisms will present an additional and challenging direction for future research. This may provide further insights into the electric control system of the living cell and help us to understand other *in vivo* reactions. Equally important applications to practical research and medical electromanipulation may also result.

As in other disciplines, the application of first principles to electromanipulation often leads to a more rapid advance in knowledge than does empiricism. A thorough understanding of the consequences of applying electric fields of high and low strength to biological membranes and systems will also greatly enhance our ability to more cogently manipulate cells for the wide variety of new applications discussed elsewhere in this book. Despite a long history of scientific inquiry, the study of field influences on biological systems remains in its infancy. Based on developments to date it seems assured of a bright and eventful future.

REFERENCES

1. **Robinson, K. R.,** The responses of cells to electrical fields: a review, *J. Cell Biol.,* 101, 2023, 1985.
2. **Reynolds, J. A., Johnson, E. A. and Tanford, Ch.,** Incorporation of membrane potential into theoretical analysis of electrogenic ion pumps, *Proc. Natl. Acad. Sci. U.S.A.,* 82, 6869, 1985.
3. **Catteral, W. A.,** Structure and function of voltage sensitive ion channels, *Science,* 242, 50, 1988.
4. **Jan, L. Y. and Jan, Y. N.,** Voltage sensitive ion channels, *Cell,* 56, 13, 1989.

5. **Glaser, R.,** The influence of membrane electric field on cellular functions, in *Biophysics of the Cell Surface,* Glaser, R. and Gingell, D., Eds., Springer Series in Biophysics, Springer Verlag, Berlin, 1989, 5, 175.

6. **Plonsey, R.,** The nature of sources of bioelectric and biomagnetic fields, *Biophys. J.,* 39, 309, 1982.

7. **Bashford, C. L. and Pasternak, C. A.,** Plasma membrane potential of Lettré cells does not depend on cation gradients but on pumps, *J. Membrane Biol.,* 79, 275, 1984.

8. **Bashford, C. L. and Pasternak, C. A.,** Plasma membrane potential of some animal cells is generated by ion pumping, not by ion gradients, *TIBS,* 11, 113, 1986.

9. **Bashford, C. L., Alder, G. M., Gray, M. A., Micklem, K. J., Taylor, C. C., Turek, P. J. and Pasternak, C. A.,** Oxonol dyes as monitors of membrane potential: the effect of viruses and toxins on the plasma membrane potential of animal cells in monolayer culture and in suspension, *J. Cell Physiol.,* 123, 326, 1985.

10. **Bashford, C. L., Micklem, K. J. and Pasternak, C. A.,** Sequential onset of permeability changes in mouse ascites cells induced by Sendai virus, *Biochim. Biophys. Acta,* 814, 247, 1985.

11. **Bashford, C. L., Alder, G. M., Graham, J. M., Menestrina, G. and Pasternak, C. A.,** Ion modulation of membrane permeability: effect of cations on intact cells and on cells and phospholipid bilayers treated with pore forming agents, *J. Membrane Biol.,* 103, 79, 1988.

12. **Betz, W. J., Caldwell, J. H., Ribchester, R. R., Robinson, K. R. and Stump, R. F.,** Endogenous electric field around muscle fibres depends on the Na^+-K^+-pump, *Nature,* 287, 235, 1980.

13. **Jaffe, L. F.,** Control of development by steady ionic currents, *Fed. Proc.,* 40, 125, 1981.

14. **Jaffe, L. F., Robinson, K. R. and Nuccitelli, R.,** Local cation entry and self electrophoresis as an intracellular localization mechanism, *Ann. N.Y. Acad. Sci.,* 238, 372, 1974.

15. **Nuccitelli, R.,** Ionic currents in morphogenesis, *Experientia,* 44, 657, 1988.

16. **Overbeek, J. T. G.,** The Donnan equilibrium, *Prog. Biophys. Biophys. Chem.,* 6, 58, 1956.

17. **Freedman, J. C. and Hoffman, J. F.,** Ionic and osmotic equilibria of human red blood cells treated with nystatin, *J. Gen. Physiol.,* 74, 157, 1979.

18. **Brumen, M., Glaser, R. and Svetina, S.,** Osmotic states of the red blood cell, *Bioelectrochem. Bioenerg.,* 104, 227, 1979.

19. **Glaser, R.,** The shape of red blood cells as a function of membrane potential and temperature, *J. Membrane Biol.,* 51, 217, 1979.

20. **Glaser, R. and Donath, J.,** Stationary ionic states in human red blood cells, *Bioelectrochem. Bioenerg.,* 13, 71, 1984.

21. **Heinrich, R., Gaestel, M. and Glaser, R.,** The electric potential profile across the erythrocyte membrane, *J. Theoret. Biol.,* 96, 211, 1982.

22. **Wilbrandt, W.,** Osmotische Natur der sogenannten nicht osmotischen Hämolyse (Kolloidosmotische Hämolyse), *Pflügers Arch. Ges. Physiol.,* 245, 22, 1942.

23. **Eylar, E. H., Madoff, M. A., Brody, O. V. and Oncley, J. L.,** The contribution of sialic acid to the surface charge of the erythrocyte, *J. Biol. Chem.,* 237, 1992, 1962.

24. **Haydon, D. A. and Seaman, G. V. F.,** Electrokinetic studies on the ultrastructure of the human erythrocyte. I. Electrophoresis at high ionic strengths — the cell as a polyanion, *Arch. Biochem. Biophys.,* 122, 126, 1967.

25. **Haydon, D. A.,** The electrical double layer and electrokinetic phenomena, *Recent Progress in Surface Science,* 1, 94, 1969.

26. **Mehrishi, J. N. and Thomson, A. E. R.,** Relationship between pH and electrophoretic mobility for lymphocytes circulating in chronic lymphocytic leukaemia, *Nature,* 219, 1080, 1968.

27. **Mehrishi, J. N.,** Positively charged amino groups on the surface of normal and cancer cells, *Europ. J. Cancer,* 6, 127, 1970.

28. **Levine, S., Levine, M., Sharp, K. A. and Brooks, D. E.,** Theory of the electrokinetic behaviour of the human erythrocytes, *Biophys. J.*, 42, 127, 1983.
29. **Donath, E. and Lerche, D.,** Electrostatic and structural properties of the surface of human erythrocytes. I. Cell electrophoretic studies following neuraminidase treatment, *Bioelectrochem. Bioenerg.*, 7, 41, 1980.
30. **Donath, E. and Pastushenko, V.,** Electrophoretic study of cell surface properties. Theory and experimental applicability, *Bioelectrochem. Bioenerg.*, 7, 31, 1980.
31. **Donath, E. and Voigt, A.,** Electrophoretic mobility of human erythrocytes. On the applicability of the charged layer model, *Biophys. J.*, 49, 493, 1986.
32. **Fortes, P. A. G. and Hoffman, J. F.,** The interaction of fluorescent probes with anion permeability pathways of human red cells, *J. Membrane Biol.*, 16, 79, 1974.
33. **Fromherz, P. and Masters, B.,** Interfacial pH at electrically charged lipid monolayers investigated by the lipoid pH indicator method, *Biochim. Biophys. Acta*, 356, 270, 1974.
34. **Chiu, V. C. K., Mouring, D., Watson, B. D. and Haynes, D. H.,** Measurement of surface potential and surface charge densities of sarcoplasmic reticulum membranes, *J. Membrane Biol.*, 56, 121, 1980.
35. **Lin, G. S. B., Macey, R. I. and Mehlhorn, R. J.,** Determination of the electric potential at the external and internal bilayer aqueous interfaces of the human erythrocyte membrane using spin probes, *Biochim. Biophys. Acta*, 732, 683, 1983.
36. **Skutelsky, E. and Dannon, D.,** Electron microscopical analysis of surface charge labelling density at various stages of the erythroid line, *J. Membrane Biol.*, 2, 173, 1970.
37. **Marikovsky, Y. and Weinstein, R. S.,** Lateral mobility of negative charge sites at the surface of sheep erythrocytes, *Exp. Cell Res.*, 136, 169, 1981.
38. **Marikovsky, Y., Shlomai, Z., Asher, O., Lotan, R. and Ben Bassat, H.,** Distribution and modulation of surface charges of cells from human leukemia lymphoma lines at various stages of differentiation, *Cancer*, 58, 2218, 1986.
39. **Devaux, P. F.,** The aminophospholipid translocase: a transmembrane lipid pump — physiological significance, *NIPS*, 5, 53, 1990.
40. **Devaux, P. F.,** Static and dynamic lipid asymmetry in cell membranes, *Biochemistry,* 30, 1163, 1991.
41. **Devaux, P. F.,** Protein involvement in transmembrane lipid asymmetry, *Ann. Rev. Biophys. Biomol. Struct.*, 21, 417, 1992.
42. **Brown, R. H.,** Membrane surface charge: discrete and uniform modelling, *Progr. Biophys. Mol. Biol.*, 28, 343, 1974.
43. **McLaughlin, St.,** Electrostatic potentials at membrane solution interfaces, in *Current Topics in Membranes and Transport*, Bronner, F. and Kleinzeller, A., Eds., Academic Press, New York, 1977, 71.
44. **McLaughlin, St.,** The electrostatic properties of membranes, *Ann. Rev. Biophys. Biophys. Chem.*, 18, 113, 1989.
45. **Träuble, H.,** Membrane electrostatics, in *Structure of Biological Membranes*, Abrahamsson, S. and Pascher, J., Eds., Plenum Press, New York, 1977, 509.
46. **Shin, Y. K. and Hubbell, W. L.,** Determination of electrostatic potentials at biological interfaces using electron-electron double resonance, *Biophys. J.*, 61, 1443, 1992.
47. **Carnie, St. and McLaughlin, St.,** Large divalent cations and electrostatic potentials adjacent to membranes. A theoretical calculation, *Biophys. J.*, 44, 325, 1983.
48. **Cole, K. and Zeta, S.,** Potential and discrete versus uniform surface charges, *Biophys. J.*, 9, 465, 1969.
49. **Attwell, D. and Eisner, D.,** Discrete membrane surface charge distributions. Effect of fluctuations near individual channels, *Biophys. J.*, 24, 869, 1978.
50. **Levine, P. L.,** The solution of a modified Poisson-Boltzmann equation for colloidal particles in electrolyte solutions, *J. Coll. Interf. Sci.*, 51, 72, 1975.

51. **Winiski, A. P., McLaughlin, A. C., McDaniel, R. V., Eisenberg, M. and McLaughlin, St.,** An experimental test of the discreteness-of-charge effect in positive and negative lipid bilayers, *Biochemistry*, 25, 8206, 1986.

52. **Voigt, A. and Donath, E.,** Cell surface electrostatics and electrokinetics, in *Biophysics of the Cell Surface*, Glaser, R. and Gingell, D., Eds., Springer Series in Biophysics, Springer Verlag, Berlin, 5, 75, 1989.

53. **Donath, E. and Pastushenko, V.,** Electrophoretical study of cell surface properties. The influence of the surface coat on the electric potential distribution and on general electrokinetic properties of animal cells, *Bioelectrochem. Bioenerg.*, 6, 543, 1979.

54. **Ohki, S. and Oshima, H.,** Donnan potential and surface potential of a charged membrane and effect of ion binding on the potential profile, in *Electrical Double Layers in Biology*, Blank, M., Ed., Plenum Press, New York, 1986, 1.

55. **Ohshima, H. and Kondo, T.,** Electrostatic repulsion of ion penetrable charged membranes: Role of Donnan potential, *J. Theor. Biol.*, 128, 187, 1987.

56. **Ohshima, H. and Kondo, T.,** Potential and pH distribution across a membrane with a surface layer of ionizable groups, *J. Theor. Biol.*, 124, 191, 1987.

57. **Donath, E. and Voigt, A.,** Charge distribution within cell surface coats of single and interacting surfaces — a minimum free electrostatic energy approach. Conclusions for electrophoretic mobility measurements, *J. Theor. Biol.*, 101, 569, 1983.

58. **Lerche, D.,** The adhesiveness of the cell surface with regard to its electric and steric structure, *Bioelectrochem. Bioenerg.*, 8, 293, 1981.

59. **Voigt, A., Donath, E. and Heinrich, R.,** Free energy of the electrostatic interaction of cells with adjacent charged glycoprotein layer — a theoretical approach, *J. Theor. Biol.*, 98, 269, 1982.

60. **MacGillivray, A. D.,** Asymptotic solution of the Poisson-Boltzmann equation for a charged sphere, and implication, *J. Theor. Biol.*, 23, 205, 1969.

61. **Brenner, St. L. and Roberts, R. E.,** A variational solution of the Poisson-Boltzmann equation for a spherical colloidal particle, *J. Phys. Chem.*, 77, 2367, 1973.

62. **Grebe, R., Peterhänsel, G. and Schmid-Schönbein, H.,** Change of local charge density by change of local mean curvature in biological bilayer membranes, *Mol. Cryst. Liq. Cryst.*, 152, 205, 1987.

63. **Doulah, F. A., Coakley, W. T. and Tilley, D.,** Intrinsic electric field and membrane bending, *J. Biol. Phys.*, 12, 44, 1989.

64. **Winterhalter, M. and Helfrich, W.,** Effect of surface charge on the curvature elasticity of membranes, *J. Phys. Chem.*, 92, 6865, 1988.

65. **Ryan, T. A., Myers, J., Holowka, D., Bird, B., Webb, W. W.,** Molecular crowding on the cell surface, *Science*, 239, 61, 1988.

66. **Stroeve, P.,** Diffusion with chemical reaction in biological systems, *Adv. Exp. Med. Biol.*, 191, 45, 1985.

67. **Lerche, D.,** Temporal and local concentration changes in diffusion layers at cellulose membranes due to concentration differences between solutions on both sides of the membrane, *J. Membrane Biol.*, 27, 193, 1976.

68. **Antonenko, Y. N. and Yaguzhinsky, L. S.,** The peculiarities of reactions catalyzed by alcohol deoxyhydrogenase in unstirred layers adjacent to the bilayer lipid membrane, *Biochim. Biophys. Acta*, 861, 337, 1986.

69. **Antonenko, Y. N. and Bulychev, A. A.,** Effect of phloretin on the carrier-mediated electrically silent ion fluxes through the bilayer lipid membrane: measurements of pH shifts near the membrane by pH microelectrode, *Biochim. Biophys. Acta*, 1070, 474, 1991.

70. **Remis, D., Bulychev, A. A. and Rubin, A. B.,** Electro-induced disturbance of barrier properties of envelope membranes in isolated chloroplasts, *Biologicheskie Membrany*, 7, 382, 1990.

71. **Markin, V. S., Portnov, V. I., Simonova, M. V., Sokolov, V. S. and Cherny, V. V.,** Theory of penetration of remantadine and its analogues through membranes: intracellular pH change, unstirred layers and membrane potentials, *Biologicheskie Membrany*, 4, 502, 1987.

72. **Rasmussen, H. H., Mogul, D. J. and TenEick, R. E.,** On the effect of unstirred layers on K$^+$-activated electrogenic Na$^+$-pumping in cardiac Purkinje strands, *Biophys. J.*, 50, 827, 1986.

73. **Coster, H. G. L.,** A quantitative analysis of the voltage-current relationships of fixed charge membranes and the associated property of "punch-through", *Biophys. J.*, 5, 669, 1965.

74. **Homble, F. and Ferrier, J. M.,** Analysis of the diffusion-theory of negative capacitance — the role of K$^+$ and the unstirred layer thickness, *J. Theor. Biol.*, 131, 183, 1988.

75. **Coster, H. G. L. and Zimmermann, U.,** Dielectric breakdown in the membrane of *Valonia utricularis*. The role energy dissipation, *Biochim. Biophys. Acta*, 382, 410, 1975.

76. **Coster, H. G. L., Zimmermann, U.,** The mechanism of electrical breakdown in the membranes of *Valonia utricularis*, *J. Membrane Biol.*, 22, 73, 1975.

77. **Poo, M.-m.,** Molecular movements of receptors on cell surface, *Biorheology*, 16, 309, 1979.

78. **Poo, M.-m.,** Rapid lateral diffusion of functional ACh receptors in embryonic muscle cell membrane, *Nature*, 295, 332, 1982.

79. **Poo, M.-m., Lam, J. W., Orida, N. and Chao, A. W.,** Electrophoresis and diffusion in the plane of the cell membrane, *Biophys. J.*, 26, 1, 1979.

80. **Poo, M.-m. and Robinson, K. R.,** Electrophoresis of concanavalin A receptors along embryonic muscle cell membrane, *Nature*, 265, 602, 1977.

81. **McLaughlin, St. and Poo, M.-m.,** The role of electro osmosis in the electric-field-induced movement of charged macromolecules on the surfaces of cells, *Biophys. J.*, 34, 85, 1981.

82. **Brumfeld, V., Miller, I. R. and Korenstein, R.,** Electric field-induced lateral mobility of photosystem I in the photosynthetic membrane, *Biophys. J.*, 56, 607, 1989.

83. **Lin Liu, S., Adey, W. R. and Poo, M.-m.,** Migration of cell surface concanavalin A receptor in pulsed electric fields, *Biophys. J.*, 45, 1211, 1984.

84. **Orida, N. and Poo, M.-m.,** Electrophoretic movement and localization of acetylcholine receptors in the embryonic muscle cell membrane, *Nature*, 275, 31, 1978.

85. **Robinson, K. R. and Cone, R.,** Polarization of fucoid eggs by a calcium ionophore gradient, *Science*, 207, 77, 1980.

86. **Robinson, K. R. and McCaig, C.,** Electrical fields calcium gradients, and cell growth, *Ann. N.Y. Acad. Sci.*, 339, 132, 1980.

87. **Jaffe, L. F.,** Control of development by ionic currents, in *Membrane Transduction Mechanisms*, Cone, R. A. and Dowling, J. E., Eds., Raven Press, New York, 1979, 199.

88. **Sugahara, K., Caldwell, J. H. and Mason, R. J.,** Electrical currents flow out of domes formed by cultured epithelial cells, *J. Cell Biol.*, 99, 1541, 1984.

89. **Larter, R., Schmidt, S. and Ortoleva, P.,** Electrical effects in nonlinear physico-chemical systems: Field chemical wave interaction and bio-self-electrophoresis, in *Pattern Formation by Dynamic Systems and Pattern Recognition*, Haken, H., Ed., Springer Verlag, Berlin, 1979, 144.

90. **Larter, R. and Ortoleva, P.,** A theoretical basis for self electrophoresis, *J. Theor. Biol.*, 88, 599, 1981.

91. **Marx, J. L.,** Electric currents may guide development, *Science*, 211, 1147, 1981.

92. **Fromherz, P.,** Dissipative structures of ion channels in the fluid mosaic model of a membrane cable, *Ber. Bunsenges. Phys. Chem.*, 92, 1010, 1988.

93. **Fromherz, P.,** Self-organization of the fluid mosaic of charged channel proteins in membranes, *Proc. Natl. Acad. Sci. U.S.A.*, 85, 6353, 1988.

94. **Fromherz, P.,** Spatio-temporal patterns in the fluid mosaic model of membranes, *Biochim. Biophys. Acta*, 944, 108, 1988.

95. **McLaughlin, S. G. A., Szabo, G., Eisenman, G. and Ciani, S. M.,** Surface charge and the conductance of phospholipid membranes, *Proc. Natl. Acad. Sci. U.S.A.*, 67, 1268, 1970.

96. **Sakurai, I. and Kawamura, Y.,** Effect of a magnetic field on charge flow along the interface between phosphatidylcholine monolayer and water, *Biochim. Biophys. Acta*, 985, 347, 1989.

97. **Tsien, R. Y. and Hladky, S. B.,** Ion repulsion within membranes, *Biophys. J.*, 39, 49, 1982.

98. **Weinstein, J. N., Blumenthal, R., van Renswoude, J., Kempf, C. and Klausner, R. D.,** Charge clusters and the orientation of membrane proteins, *J. Membrane Biol.*, 66, 203, 1982.

99. **Konev, S. V. and Kaler, G. V.,** Transmembrane electric potential as a regulator of functional activities of biomembrane, *Biofizika*, 33, 1018, 1988.

100. **Fröhlich, H.,** The biological effects of microwaves and related questions, *Adv. Electronics Electron Phys.*, 53, 85, 1980.

101. **Brown, G. C.,** Electrostatic coupling between membrane proteins, *FEBS Lett.*, 260, 1, 1990.

102. **Bingelli, R. and Weinstein, R. C.,** Membrane potentials and sodium channels: hypothesis for growth regulation and cancer formation based on changes in sodium channels and gap junctions, *J. Theor. Biol.*, 123, 377, 1986.

103. **Cone, C. D., Jr.,** Electroosmotic interactions accompanying mitosis initiation in sarcoma cells *in vitro, Trans. N.Y. Acad. Sci.*, 31, 404, 1969.

104. **Cone, C. D., Jr.,** Variation of the transmembrane potential level as a basic mechanism of mitosis control, *Oncology*, 24, 438, 1970.

105. **Cone, C. D., Jr.,** Unified theory on the basic mechanism of normal mitotic control and oncogenesis, *J. Theor. Biol.*, 30, 151, 1971.

106. **Kiefer, H., Blume, A. J. and Kaback, H. R.,** Membrane potential changes during mitogenic stimulation of mouse spleen lymphocytes, *Proc. Nat. Acad. Sci. U.S.A.*, 77, 2200, 1980.

107. **Sachs, H. G., Stambrook, P. J. and Ebert, J. D.,** Changes in membrane potential during the cell cycle, *Exp. Cell Res.*, 83, 362, 1974.

108. **Friedhoff, L. T., Kim, E., Priddle, M. and Sonenberg, M.,** The effect of altered transmembrane ion gradients on membrane potential and aggregation of human platelets in blood plasma, *Biochem. Biophys. Res. Comm.*, 102, 832, 1981.

109. **Palés, J., López, A. and Gual, A.,** Platelet membrane potential as a modulator of aggregating mechanisms, *Biochim. Biophys. Acta*, 944, 85, 1988.

110. **Williams, J. A.,** Electrical correlates of secretion in endocrine and exocrine cells, *Fed. Proc.*, 40, 128, 1981.

111. **Mukai, M., Kyogoku, I. and Kuno, M.,** Calcium-dependent inactivation of inwardly rectifying K^+-channel in a tumor mast cell line, *Am. J. Physiol.*, 262, C84, 1992.

112. **Rorsman, P., Ashcroft, F. M. and Berggren, P.-O.,** Regulation of glucagon release from pancreatic A-cells, *Biochem. Pharmacol.*, 41, 1783, 1991.

113. **Rorsman, P. and Berggren, P.-O.,** Electrical bursting in islet beta cells — reply, *Nature*, 357, 28, 1992.

114. **Tsuchiya, W., Okada, Y., Yano, J., Murai, A., Miyahara, T. and Tanaka, T.,** Membrane potential changes associated with pinocytosis of serum lipoproteins in L cells, *Exp. Cell Res.*, 136, 271, 1981.

115. **Okada, Y., Tsuchiya, W., Yada, T., Yano, J. and Yawo, H.,** Phagocytic activity and hyperpolarizing responses in L-strain mouse fibroblasts, *J. Physiol.*, 313, 101, 1981.

116. **Taglietti, V.,** Early electrical responses to fertilization in sea urchin eggs, *Exp. Cell Res.*, 120, 448, 1979.

117. **Chambers, E. L. and de Armendi, J.,** Membrane potential, action potential and activation potential of eggs of the sea urchin, *Lytechinus variegatus, Exp. Cell Res.*, 122, 203, 1979.

118. **Jaffe, L. A., Gould Somero, M. and Holland, L. Z.,** Studies of the mechanism of the electrical polyspermy block using voltage clamp during cross species fertilization, *J. Cell Biol.*, 92, 616, 1982.

119. **Fitz, J. G., Lidofsky, S. D., Xie, M. H. and Scharschmidt, B. F.,** Transmembrane electrical potential difference regulates Na^+/HCO_3^- -cotransport and intracellular pH in hepatocytes, *Proc. Natl. Acad. Sci. U.S.A.,* 89, 4197, 1992.

120. **Barts, P. W. J. A. and Borst-Pauwels, G. W. F. H.,** Effects of membrane potential and surface potential on the kinetics of solute transport, *Biochim. Biophys. Acta,* 813, 51, 1985.

121. **Tabb, J. S., Kish, P. E., Vandyke, R. and Ueda, T.,** Glutamate transport into synaptic vesicles — roles of membrane potential, pH gradient, and intravesicular pH, *J. Biol. Chem.,* 267, 15412, 1992.

122. **Carter-Su, C. and Kimmich, G. A.,** Effect of membrane potential on Na^+-dependent sugar transport by ATP-depleted infertinal cells, *Am. J. Physiol.,* 238, C73, 1980.

123. **Fischer, H., Illek, B., Negulescu, P. A., Clauss, W. and Machen, T. E.,** Carbachol-activated calcium entry into HT-29 cells is regulated by both membrane potential and cell volume, *Proc. Natl. Acad. Sci. U.S.A.,* 89, 1438, 1992.

124. **Halperin, J. A., Brugnara, C., Tosteson, M. T., van Ha, T. and Tosteson, D. C.,** Voltage-activated cation transport in human erythrocytes, *J. Physiol.,* 257, C986, 1989.

125. **Mertz, L. M., Baum, B. J. and Ambudkar, I. S.,** Membrane potential modulates divalent cation entry in rat parotid acini, *J. Membrane Biol.,* 126, 183, 1992.

126. **Blum, J. J. and Hines, M.,** Biophysics of flagellar motility, *Quart. Rev. Biophys.,* 12, 103, 1979.

127. **Khan, S., Dapice, M. and Humayun, I.,** Energy transduction in the bacterial flagellar motor. Effects of load and pH, *Biophys. J.,* 57, 779, 1990.

128. **Meister, M. and Berg, H. C.,** The stall torque of the bacterial flagellar motor, *Biophys. J.,* 52, 413, 1987.

129. **Meister, M., Caplan, S. R. and Berg, H. C.,** Dynamics of a tightly coupled mechanism for flagellar rotation. Bacterial motility, chemiosmotic coupling, protonmotive force, *Biophys. J.,* 55, 905, 1989.

130. **Läuger, P.,** Torque and rotation rate of the bacterial flagellar motor, *Biophys. J.,* 53, 53, 1988.

131. **Jou, D., Perez-Garcia, C. and Clebot, J. E.,** Bacterial flagellar rotation as a nonequilibrium phase transition, *J. Theor. Biol.,* 122, 453, 1986.

132. **Fuhr, G. and Hagedorn, R.,** Dielectric rotation and oscillation — a principle in biological systems? *Studia biophysica,* 121, 25–36, 1987.

133. **Glaser, R.,** Echinocyte formation induced by potential changes of human red blood cells, *J. Membrane Biol.,* 66, 79, 1982.

134. **Glaser, R.,** Mechanisms of electromechanical coupling in membranes demonstrated by transmembrane potential-dependent shape transformations of human erythrocytes, *Bioelectrochem. Bioenerg.,* 30, 103, 1992.

135. **Hartmann, J. and Glaser, R.,** The influence of chlorpromazine on the potential-induced shape change of human erythrocyte, *Biosci. Rep.,* 11, 213, 1991.

136. **Katnik, Ch. and Waugh, R.,** Electric fields induce reversible changes in the surface to volume ratio of micropipette-aspirated erythrocytes, *Biophys. J.,* 57, 865, 1990.

137. **Katnik, Ch. and Waugh, R.,** Alterations of the apparent area expansivity modulus of red blood cell membrane by electric fields, *Biophys. J.,* 57, 877, 1990.

138. **Engasser, J. M. and Horvath, C.,** Electrostatic effects on the kinetics of bound enzymes, *Biochem. J.,* 145, 431, 1975.

139. **Wojtczak, L. and Nalecz, M. J.,** Surface charge of biological membranes as a possible regulator of membrane-bound enzymes, *Eur. J. Biochem.,* 94, 99, 1979.

140. **Kamo, N., Miyake, M., Kurihara, K. and Kobatake, Y.,** Physicochemical studies of taste reception. II. Possible mechanism of generation of taste receptor potential induced by salt stimuli, *Biochim. Biophys. Acta,* 367, 11, 1974.

141. **Ravindran, A. and Moczydlowski, E.,** Influence of negative surface charge on toxin binding to canine heart Na channels in planar bilayers, *Biophys. J.,* 55, 359, 1989.

142. **Gage, R. A., Van Wijngaarden, W., Theuvenet, A. P. R., Borst-Pauwels, G. W. F. H. and Verkleij, A. J.,** Inhibition of Rb$^+$-uptake in yeast by Ca$^+$ is caused by a reduction in the surface potential and not in the Donnan potential of cell wall, *Biochim. Biophys. Acta,* 812, 1, 1985.

143. **Fermini, B. and Nathan, R. D.,** Sialic acid and the surface charge associated with hyperpolarization-activated, inward rectifying channels, *J. Membrane Biol.,* 114, 61, 1990.

144. **Dörrscheidt-Käfer, M.,** Excitation-contraction coupling in frog sartorius and the role of the surface charge due to the carboxyl group of sialic acid, *Pflügers Archiv,* 380, 171, 1979.

145. **Dörrscheidt-Käfer, M.,** The interaction of ruthenium red with surface charges controlling excitation-contraction coupling in frog sartorius, *Pflügers Archiv,* 380, 181, 1979.

146. **Cukierman, S.,** Asymmetric electrostatic effects on the gating of rat brain sodium channels in planar lipid membranes, *Biophys. J.,* 60, 845, 1991.

147. **Woolley, P. and Teubner, M.,** Electrostatic interactions at charged lipid membranes. Calcium binding, *Biophys. Chem.,* 10, 335, 1979.

148. **Riddell, F. G. and Arumugam, S.,** Surface charge effects upon membrane transport processes: the effects of surface charge on the monensin-mediated transport of lithium ions through phospholipid bilayer studies by ^{7}Li-NMR spectroscopy, *Biochim. Biophys. Acta,* 945, 65, 1988.

149. **Blank, M.,** The surface compartment model: a theory of ion transport focused on ionic processes in the electric double layers at membrane protein surfaces, *Biochim. Biophys. Acta,* 906, 277, 1987.

150. **Blank, M.,** The surface compartment model (SCM): role of surface charge in membrane permeability changes, *Bioelectrochem. Bioenerg.,* 9, 615, 1982.

151. **Lakshminarayanaiah, N.,** Surface charges on membranes, *J. Membrane Biol.,* 29, 243, 1976.

152. **Britten, J. S. and Blank, M.,** The effect of surface charge on interfacial ion transport, *Bioelectrochem. Bioenerg.,* 4, 209, 1977.

153. **Lehotsky, J., Raeymaekers, L., Missiaen, L., Wuytack, F., Desmedt, H. and Casteels, R.,** Stimulation of the catalytic cycle of the Ca^{2+} pump of porcine plasma-membranes by negatively charged phospholipids, *Biochim. Biophys. Acta,* 1105, 118, 1992.

154. **Arcangeli, A., Del Bene, M. R., Becchetti, A., Wanke, E. and Olivotto, M.,** Effects of inhibitors of ion-motive ATPases on the plasma membrane potential of murine erythroleukemia cells, *J. Membrane Biol.,* 126, 123, 1992.

155. **Becchetti, A., Arcangeli, A., Del Bene, M. R., Olivotto, M. and Wanke, E.,** Intracellular and extracellular surface charges near Ca^{2+} channels in neurons and neuroblastoma cells, *Biophys. J.,* 63, 954, 1992.

156. **Goldman, D. E.,** Potential, impedance and rectification in membranes, *J. Gen. Physiol.,* 27, 37, 1943.

157. **Goldman, D. E.,** Membranes and ionic double layers, in *Perspectives in Membrane Biophysics,* Agin, D. P., Ed., Gordon and Breach, New York, 1972, 205.

158. **Hainsworth, A. H. and Hladky, S. B.,** Effects of double layer polarization on ion transport, *Biophys. J.,* 51, 27, 1987.

159. **Glaser, R., Bernhardt, I., Donath, E., Heinrich, R. and Gaestel, M.,** Ion transport in human red blood cells as a function of electric potential profile, in *Physical Chemistry of Transmembrane Ion Motion,* Spach, G., Ed., Elsevier, Amsterdam, 1983, 635.

160. **Corda, D., Pasternak, C. and Shinitzky, M.,** Increase in lipid microviscosity of unilamellar vesicles upon the creation of transmembrane potential, *J. Membrane Biol.,* 65, 235, 1982.

161. **O'Shea, P. S., Thelen, S., Petrone, G. and Azzi, A.,** Proton mobility in biological membranes: the relationship between membrane lipid state and proton conductivity, *FEBS Lett.,* 172, 103, 1984.

162. **O'Shea, P. S., Feuerstein-Thelen, S. and Azzi, A.,** Membrane-potential dependent changes of the lipid microviscosity of mitochondria and phospholipid vesicles, *Biochem. J.,* 220, 795, 1984.

163. **Cevc, G., Watts, A. and Marsh, D.,** Titration of the phase transition of phosphatidylserine bilayer membranes. Effects of pH, surface electrostatic, ion binding and head group, *Biochemistry*, 20, 4955, 1981.

164. **Träuble, H. and Eibl, H.,** Electrostatic effects on lipid phase transitions: membrane structure and ionic environment, *Proc. Natl. Acad. Sci. U.S.A.*, 71, 214, 1974.

165. **Strehlow, U. and Jähnig, F.,** Electrostatic interactions at charged lipid membranes. Kinetics of the electrostatically triggered phase transition, *Biochim. Biophys. Acta*, 641, 301, 1981.

166. **Jähnig, F.,** Electrostatic free energy and shift of the phase transition for charged lipid membranes, *Biophys. Chem.*, 4, 309, 1976.

167. **Jähnig, F., Harlos, K., Vogel, H. and Eibl, H.,** Electrostatic interactions at charged lipid membranes. Electrostatically induced tilt, *Biochemistry*, 18, 1459, 1979.

168. **Stulen, G.,** Electric field effects on lipid membrane structure, *Biochim. Biophys. Acta,* 640, 621, 1981.

169. **Malev, V. V. and Rusanov, A. I.,** Thermodynamics, mechanical and electrical properties of biomembranes, *J. Theor. Biol.*, 136, 295, 1989.

170. **Tokutomi, S., Ohki, K. and Ohnishi, S. I.,** Proton-induced phase separation in phosphatidylserine/phosphatidylcholine membranes, *Biochim. Biophys. Acta*, 596, 192, 1980.

171. **Ohyashiki, T. and Mohri, T.,** Effect of ionic strength on the membrane fluidity of rabbit intestinal brush-border membranes. A fluorescence probe study, *Biochim. Biophys. Acta*, 731, 312, 1983.

172. **Herrmann, A. and Müller, P.,** A model for the asymmetric lipid distribution in the human erythrocyte membrane, *Biosci. Rep.*, 6, 185, 1986.

173. **Benz, R. and Zimmermann, U.,** Evidence for the presence of mobile charges in the cell membrane of *Valonia utricularis*, *Biophys. J.*, 43, 13, 1983.

174. **Wang, J., Wehner, G., Benz, R. and Zimmermann, U.,** Influence of external chloride concentration on the kinetics of mobile charges in the cell membrane of *Valonia utricularis*: evidence for the existence of a chloride carrier, *Biophys. J.*, 59, 235, 1991.

175. **Wang, J., Zimmermann, U. and Benz, R.,** The voltage-dependent step of the chloride transporter of *Valonia utricularis* encounters a Nernst-Planck and not an Eyring type potential energy barrier, *Biophys. J.*, 64, 1004, 1993.

176. **Glaser, R. and Donath, J.,** Temperature and transmembrane potential dependence of shape transformations of human erythrocytes, *Bioelectrochem. Bioenerg.*, 27, 429, 1992.

177. **Clementz, T. and Christiansson, A.,** Transmembrane electrical potential affects the lipid composition of *Acholeplasma laidlawii, Biochemistry*, 25, 823, 1986.

178. **Honig, B. H., Hubbell, W. L. and Flewelling, R. F.,** Electrostatic interactions in membranes and proteins, *Ann. Rev. Biophys. Biophys. Chem.*, 15, 163, 1986.

179. **Neumann, E.,** Chemical electric field effects in biological macromolecules, *Prog. Biophys. Molec. Biol.*, 47, 197, 1986.

180. **Tsong, T. Y. and Astumian, R. D.,** Electroconformational coupling and membrane protein function, *Prog. Biophys. Molec. Biol.*, 50, 1, 1987.

181. **Tsong, T. Y.,** Electrical modulation of membrane proteins: enforced conformational oscillations and biological energy and signal transduction, *Ann. Rev. Biophys. Biophys. Chem.*, 19, 83, 1990.

182. **Keszthelyi, L. and Ormos, P.,** Protein electric response signals from dielectrically polarized systems, *J. Membrane Biol.*, 109, 193, 1989.

183. **de Kroon, A. I. P. M., de Gier, J. and de Kruijff, B.,** The effect of a membrane potential on the interaction of mastoparan X, a mitochondrial presequence, and several regulatory peptides with phospholipid vesicles, *Biochim. Biophys. Acta*, 1068, 111, 1991.

184. **Brumfeld, V. and Miller, J. R.,** Effect of membrane potential on the conformation of bacteriorhodopsin reconstituted in lipid vesicles, *Biophys. J.*, 54, 747, 1988.

185. **Tsong, T. Y., Liu, D. S., Chauvin, F., Gaigalas, A. and Astumian, R. D.,** Electroconformational coupling (ECC): an electric field induced enzyme oscillation for cellular energy and signal transductions, *Bioelectrochem. Bioenerg.*, 21, 319, 1989.

186. **Tsong, T. Y., Liu, D.-S., Chauvin, F. and Astumian, R. D.,** Resonance electro-conformational coupling: a proposed mechanism for energy and signal transductions by membrane proteins, *Biosci. Rep.*, 9, 13, 1989.

187. **Tsong, T. Y. and Astumian, R. D.,** Absorption and conversion of electric field energy by membrane bound ATPases, *Bioelectrochem. Bioenerg.*, 15, 457, 1986.

188. **Westerhoff, H. V., Tsong, T. Y., Chock, P. B., Chen, Y.-D. and Astumian, R. D.,** How enzymes can capture and transmit free energy from an oscillating electric field, *Proc. Natl. Acad. Sci. U.S.A.*, 83, 4734, 1986.

189. **Markin, V. S., Tsong, T. Y., Astumian, R. D. and Robertson, B.,** Energy transduction between a concentration gradient and an alternating electric field, *J. Chem. Phys.*, 93, 5062, 1990.

190. **Markin, V. S. and Tsong, T. Y.,** Electroconformational coupling for ion transport in an oscillating electric field — rectification versus active pumping, *Bioelectrochem. Bioenerg.*, 26, 251, 1991.

191. **Markin, V. S. and Tsong, T. Y.,** Frequency and concentration windows for the electric activation of membrane active transport system, *Biophys. J.*, 59, 1308, 1991.

192. **Markin, V. S. and Tsong, T. Y.,** Reversible mechano-sensitive ion pumping as a part of mechanoelectrical transduction, *Biophys. J.*, 59, 1317, 1991.

193. **Glaser, R.,** Current concepts of the interaction of weak electromagnetic fields with cells, *Bioelectrochem. Bioenerg.*, 27, 255, 1992.

194. **Cadossi, R., Bersani, F., Cossarizza, A., Zucchini, P., Emilia, G., Torelli, G. and Franceschi, C.,** Lymphocytes and low frequency electromagnetic fields, *FASEB J.*, 6, 2667, 1992.

195. **Goodman, R., Weisbrot, D., Uluc, A. and Henderson, A.,** Transcription in *Drosophila melanogaster* salivary gland cells is altered following exposure to low-frequency electromagnetic fields — analysis of chromosome 3R, *Bioelectromagnetics,* 13, 111, 1992.

196. **Greene, J. J., Skowronski, W. J., Mullins, J. M., Nardone, R. M., Penafield, L. M. and Meister, R.,** Delineation of electric and magnetic field effects of extremely low frequency electromagnetic radiation on transcription, *Biochem. Biophys. Res. Comm.*, 174, 742, 1991.

197. **Walleczek, J.,** Electromagnetic field effects on cells of the immune system — the role of calcium signaling, *FASEB J.*, 6, 3177, 1992.

198. **Walleczek, J. and Liburdy, R. P.,** Nonthermal 60 Hz sinusoidal magnetic-field exposure enhances $^{45}Ca^{2+}$ uptake in rat thymocytes: dependence on mitogen activation, *FEBS Lett.*, 271, 157, 1990.

199. **Liburdy, R.P.,** Calcium signaling in lymphocytes and ELF fields. Evidence for an electric field metric and a site of interaction involving the calcium ion channel, *FEBS Lett.*, 301, 53, 1992.

200. **Lindström, E., Lindström, P., Berglund A., Mild, A. B., and Lundgren, E.,** Intercellular calcium oscillations induced in T-cell line by a weak 50 Hz magnetic field, *J. Cell. Physiol.*, 156, 395, 1993.

201. **Polk, Ch.,** Counter-ion polarization and low frequency, low electric field intensity biological effects, *Bioelectrochem. Bioenerget.*, 28, 279, 1992.

Appendix A

PROTOCOLS FOR HIGH-EFFICIENCY ELECTROINJECTION OF "XENOMOLECULES" AND ELECTROFUSION OF CELLS

Ulrich Zimmermann

CONTENTS

I. Introduction 365

II. Electroinjection of "Xenomolecules" 366
   A. Bacteria 366
   B. Yeast Protoplasts.. 367
   C. Plant Protoplasts 368
   D. Mammalian Cells 368

III. Electrofusion of Cells.. 369
   A. Bacteria 369
   B. Yeast Protoplasts.. 369
   C. Plant Protoplasts 370
   D. Mammalian Cells 371

References 372

I. INTRODUCTION

The principles and the potential fields of application of electroinjection and electrofusion have been extensively reviewed in Chapters 1, 2, and 4. This chapter presents experimental conditions that, if carefully followed, will produce a high yield of hybrids, transformants, and manipulated cells, respectively.*

These protocols provide high efficiencies and high reproducibility. A minimum of manipulations are required when the "Electroinjection or Electrofusion Biojets" manufactured by Biomed (97531 Theres, Germany) as well as the pulse and fusion chambers described in Chapters 1 and 4 are used. These pieces of apparatus and the chambers were developed by the Department of Biotechnology, University of Würzburg; they have been tested for more than five years. Although the efficiency depends in part upon the power supply and the

---

* For cloning of mammalian embryos by electric field-mediated nuclear transplantation, see Reference 1.

pulse chambers, the protocols described in this chapter can be used as a guide for investigators to develop protocols specific for their unique conditions and reagents (e.g., cell lines, fusion partners, etc.; for technical details, see References 2-6). In my experience, direct application of any method is difficult, and it should be used as a framework for other investigators in the field.

II. ELECTROINJECTION OF "XENOMOLECULES"

A. BACTERIA

This section gives a protocol for electro-transformation of *Escherichia coli*. The procedure outlined below is based on the experience of A. Taketo (Chapter 2) and works sufficiently well with most (gram-negative) strains. It results in very high frequencies and efficiencies (10^6–10^{10} transformants per μg DNA).

1. Inoculate a fresh single colony (small size) into L broth (1% Bacto tryptone, 0.5% Bacto yeast extract, 0.5% NaCl) and shake the suspension over night at 37°C.
2. Dilute the seed culture 40–50 fold into L broth, and grow the cells under vigorous shaking until the cell density reaches about $3.5 \cdot 10^8$ cells per ml. The cell density can be determined by measurement of the turbidity.
3. Collect the bacteria by centrifugation and wash 3–4 times with a sufficient volume of chilled, distilled water.
4. Suspend the bacteria in a small volume of distilled water at a density of about $2 \cdot 10^{11}$ cells per ml and keep the suspension at 0°C.
5. Pipette 50 μl of the cell suspension into a small tube pre-cooled to 0°C and then add 10 μl of the DNA solution (dissolved in 10 mM Tris-HCl, pH 7.5; for DNA concentrations, see Chapter 2). Mix the suspension gently. Note that the presence of EDTA should be avoided (see Chapter 1).
6. Add 50 μl of a chilled solution of PEG 6000 (20% dissolved in distilled water) and mix the suspension immediately.
7. Transfer 90 μl of this mixture into a chilled (sterilized) pulse cuvette (electrode distance 0.1–1 cm). Note that electrodes made of aluminium may have toxic effects on the cells (see Chapter 1). Avoid gas bubbles in the solution.
8. Apply more than three exponentially decaying field pulses of high intensity (up to 25 kV cm⁻¹) and 100–400 μs duration. The time interval between consecutive pulses should be adjusted to at least 1 min. If the available pulser has only a low output voltage (e.g., Biorad gene pulser), apply a single, exponentially decaying field pulse of about 18 kV cm⁻¹ with a time constant of about 10 ms (electrode gap of the pulse cuvette 0.1 cm; pulser setting 1.75 kV, 25 μF, 400–600 Ω).
9. After pulsing, immediately dilute the suspension 4-fold with L broth and keep the suspension for 30 min at 37°C.

10. After further dilution as required, spread aliquots of the field-treated bacteria (with melted top agar) on selective plates and incubate them at 37°C until colonies are formed.

B. YEAST PROTOPLASTS

1. Grow the yeast cells in culture medium (e.g., YEP medium which contains 1% yeast extract, 2% peptone, and 2% glucose).[7] For further references, see Chapter 1.
2. Harvest cells from the logarithmic phase by centrifugation (3000 g).
3. After washing the cells with distilled water, incubate the cells in protoplast pre-incubation medium (50 mM EDTA pH 9, 35 mM β-mercaptoethanol) for 20–30 min at 37°C. Concentrate the cells with several centrifugations, resuspending the successive pellets as follows:

> 1st pellet: washing solution I (1.2 M sorbitol, 50 mM EDTA pH 7.5).
> 2nd pellet: washing solution II (1.2 M sorbitol, 50 mM Tris pH 7.5).
> 3rd pellet: enzyme solution [0.1 mM Ca^{2+}-acetate, 0.5 mM Mg^{2+}-acetate, 50 mM Tris, 1.2 M sorbitol, 0.6 mg ml^{-1} zymolyase (100,000 U/g; ICN, Costa Mesa, California 92626, USA) pH 7.5].

Keep the cells in this solution at a density of $5–7 \cdot 10^8$ cells per ml for 1.5–2 h at 37°C (for complete removal of the cell walls).

4. Wash the protoplasts with pulse medium (30 mM KCl, 1 mM $CaCl_2$, 0.3 mM KH_2PO_4, 0.85 mM K_2HPO_4, 1.2 M sorbitol); pellet the cells by centrifugation and resuspend them in pulse medium at a final density of 10^9 cells per ml.
5. Add the "xenomolecules" to be injected (plasmid DNA, RNA, protein, dyes, etc.). Mix the suspension by gently shaking the tube and transfer an aliquot into a pre-cooled (4°C) injection chamber (electrode gap 0.5–1 cm). Concentrations of 5–10 µg ml^{-1} are recommended for electroinjection of nucleic acids, whereas for electroincorporation of proteins higher concentrations (10–100 µg ml^{-1}) are optimal.
6. Apply three consecutive exponentially decaying field pulses of about 15–18 kV cm^{-1} strength and 10 µs duration (2 min pulse interval). After pulsing, keep the field-treated cells at 4°C for a few minutes before removing the chamber. Then transfer the cells (at room temperature) into 3 ml growth medium of appropriate composition (e.g., 1.2 M sorbitol, 0.67% yeast nitrogen base without amino acids) and keep them for at least 30 min at 22°C. The cells can then be used for analysis (e.g., determination of cell viability and of the efficiency of incorporation of dyes, proteins, etc. by optical and other means, see Chapter 1).

7. Selection for electrotransformed yeast protoplasts requires the following steps after field treatment: divide the suspension into 50 µl fractions, mix each fraction with 8 ml of a melted (49°C) protoplast regeneration agar medium, and pour the resulting mixture into a petri dish. Use regeneration agar media which are optimal for regeneration of the cell walls and which select for successfully electrotransformed cells. For example, products from parental cells which exhibit a defect in the synthesis of leucine should be plated (after electrotransfection with a vector carrying the complementary intact leucine gene) on medium consisting of 0.67% yeast nitrogen base without amino acids, 2% glucose, 30 µg ml⁻¹ histidine, 1.2 M sorbitol, and 3% agar. The first colonies are seen after 3–14 days depending on the species and the composition of the regeneration agar media.

C. PLANT PROTOPLASTS

1. Prepare and purify protoplasts according to standard protocols.[8] Ensure high purity and high viability of the protoplasts.
2. Wash the protoplasts in pulse medium (consisting e.g., of 450 mM mannitol, 0.1% bovine serum albumin, and 0.1 mM $CaCl_2$). Resuspend the cell pellet after centrifugation (about 200 g) at a final density of about 10^5 cells per ml.
3. Add the "xenomolecules" to be injected (linearized DNA, RNA, or protein) to the pre-cooled solution at concentrations described above for electroinjection of xenomolecules into yeast protoplasts.
4. Add the cell suspension to the pre-cooled injection chamber (electrode gap 0.5–1 cm).
5. Apply three consecutive exponentially decaying field pulses of either 10 kV cm⁻¹ strength and 15 µs duration or of 2–4 kV cm⁻¹ strength and 40 µs duration (1 min pulse interval). After pulse application, allow the chamber to warm to room temperature. Keep the field-treated protoplasts undisturbed for about 1 h.
6. Gently flush the cells from the chamber by using 3 ml of an appropriate growth medium.
7. Resuspend the cells in culture vessels containing an appropriate selection medium which selects for transformants and facilitates regeneration of the cell wall.

D. MAMMALIAN CELLS

1. Harvest the cells from the logarithmic phase by centrifugation. Adherent cells should be treated with 0.1 mg ml⁻¹ dispase (6 U ml⁻¹, grade 1, Boehringer, 68298 Mannheim, Germany) in complete growth medium for 30 min at 37°C.

2. Wash the cells twice with RPMI 1640 medium and then with pulse medium (30 mM KCl, 0.3 mM KH_2PO_4, 0.85 mM K_2HPO_4, and appropriate amounts of inositol to adjust the osmolality to 75–100 mosmol kg^{-1}).

3. Resuspend the cells in pre-cooled pulse medium at a final density of $2–10 \cdot 10^5$ cells per ml.

4. Add the "xenomolecules" to be injected at concentrations described in the electroinjection protocol for yeast protoplasts. Note that the presence of EDTA must be avoided (see Chapter 1).

5. Add the cell suspension to the pre-cooled injection chamber (electrode gap 0.5–1 cm) and apply a single exponentially decaying field pulse of about 4–8 kV cm^{-1} and 5 µs duration.

6. 2 min after pulse application, gently transfer the field-treated cells from the chamber into a pre-warmed (37°C) tube filled with 3 ml of RMPI 1640 medium without phenol red. Incubate the cells for at least 30 min at 37°C.

7. Determine the number of viable cells by the trypan blue assay.

8. Grow the cells in complete growth medium for 48 h at 37°C.

9. Transfer the cell into an appropriate selection medium. Colonies of adhering cells should be visible after about 1 week of cultivation.

Note that some cell lines may not tolerate strongly hypo-osmolar conditions. In this case, perform the electropermeabilization process of the cell membrane under (more) iso-osmolar conditions (about 280 mOsmol) and apply three (or more) consecutive breakdown pulses of about 5–10 kV cm^{-1} strength and 5–40 µs duration (1 min pulse interval). For further details, see Chapter 1.

III. ELECTROFUSION OF CELLS

A. BACTERIA

Protocols for fusion of bacterial protoplasts which can be used routinely for various applications are not available. The reader is referred to Reference 9.

B. YEAST PROTOPLASTS

1. Prepare yeast protoplasts according to the method described above (see also Reference 7 and for other References, Chapter 4).

2. Wash the protoplasts twice in fusion medium (1.2 M sorbitol, 0.1 mM Ca^{2+}-acetate, 0.5 mM Mg^{2+}-acetate) and resuspend them at a density of $2.5 \cdot 10^9$ cells per ml.

3. Mix the protoplasts of the two parental strains in a ratio of 1:1. Dilute 120 µl of this suspension with 280 µl fusion medium and then pipette gently (at least) 330 µl into a (sterilized) helical fusion chamber (Chapter 4) at room temperature (22°C).

4. Apply a constant alternating field of 275 V cm⁻¹ strength (550 V cm⁻¹ peak-to-peak value) and 2 MHz frequency for 120 s.
5. Inject two rectangular consecutive breakdown (fusion) pulses of 10–12 kV cm⁻¹ strength and 10–15 μs duration (1 s pulse interval). Adjust the alignment off time (AOT) of the alternating field during each pulse to 10 ms.
6. After pulsing, apply the alternating field for a further 60 s to keep the fusing cells in close contact (see Chapter 4).
7. Rinse the helical chamber with 670 μl of an appropriate growth medium (e.g., 1.2 M sorbitol, 0.67% yeast nitrogen base without amino acids). Divide the field-treated suspension into 50 μl fractions. Mix each fraction with 8 ml of melted agar (49°C) containing an appropriate selective medium. Finally, pour the melted agar into petri dishes and incubate the cells in a moist chamber (for further details, see protocol for electrotransfection of yeast protoplasts).

C. PLANT PROTOPLASTS

Plant protoplasts can usually be fused quite easily (for exceptions, see Chapter 4). The protocol described below allows fusion of protoplasts of different sizes and density and can also be applied to the fusion of vacuolated protoplasts with evacuolated or embryogenic protoplasts.[10]

1. Prepare plant protoplasts according to standard procedures.[8]
2. Wash the protoplasts in fusion medium consisting of 0.4–0.5 M mannitol, 0.1 mM Ca²⁺-acetate, 1–2 mg ml⁻¹ bovine serum albumin. Note that the osmolality depends critically on the tolerance of the protoplasts under investigation to hypo-osmolar conditions (for further details, see Chapter 4).
3. Mix the parental protoplasts at a ratio of about 1:1 and adjust the density to $5 \cdot 10^5$ cells per ml. Fill 200–300 μl of the cell suspension into the helical chamber or 5–10 μl into the adjustable plate microchamber (see Chapter 4).
4. Apply an amplitude-modulated, alternating field of 1.5 MHz frequency:[10] a field of 250–300 V cm⁻¹ strength (500–600 V cm⁻¹ peak-to-peak value) for about 3–5 s, followed automatically by a field of 45 V cm⁻¹ (90 V cm⁻¹ peak-to-peak value) for 20 s. When the protoplasts of the two fusion partners have similar sizes, an alternating field of constant amplitude (about 50–150 V cm⁻¹, depending on the size) can be used.
5. Apply a rectangular breakdown (fusion) pulse of 0.8–2 kV cm⁻¹ strength and 30 μs duration (22°C).
6. After pulsing, apply an alternating field of 250–300 V cm⁻¹ strength and 1.5 MHz frequency for an additional 30 s.

7. Keep the field-treated protoplasts for 10–30 min at 22°C without disturbance.

8. Rinse the chambers with 1–2 ml of an appropriate growth medium and cultivate the protoplasts according to standard plant tissue culture techniques.[8] Cell wall regeneration can be improved when the protoplasts are immobilized in an agarose or Ca^{2+}-cross-linked alginate matrix.

D. MAMMALIAN CELLS

The fusion protocol described below is optimal for the production of human monoclonal antibodies-secreting hybridoma cells by electrofusion of activated B cells with mouse-human hetero-myeloma cells.[11-16] However, with small modifications the protocol can also be applied successfully to homologous or heterologous fusions of other mammalian cells.[3]

Because the fusions of mammalian cells are performed very often with a very small number of cells, attention to detail is required to minimize cell loss. Timing is critical, so all supplies, equipment, and reagents must be prepared ahead of time and be available for immediate use. Minimize the length of time the cells spend in suboptimal conditions, such as in cell pellets (at the end of a wash) or while suspended in strongly hypo-osmolar solutions. All steps are performed (as described above for other cell species) at room temperature using sterile technique (for further important details, see Chapter 4).

1. B cells obtained from different sources (peripheral blood, biopsy material, spleen, etc., see Chapter 4) are isolated as described elsewhere.[11-16] The method of *in vitro* activation depends on the antigen system and the source of the B cells. Activation performed with 20 µg ml^{-1} lipopolysaccharide (Sigma, 82039 Deisenhofen, Germany) or 2.4 µg ml^{-1} phytohaemagglutinin (Biochrom KG, 12247 Berlin, Germany) is recommended.[11-15]

2. Cells of the mouse-human heteromyeloma fusion partner should be in the log-phase growth with optimal viability. (Dead cells interfere with the fusion process.) Best hybrid yields are obtained with the mouse-human heteromyeloma cell lines HAB-1 and H73C11.[11-15]

3. Harvest and mix the activated B cells and the myeloma fusion partner in a ratio of 1:5.

4. Concentrate the cells with several centrifugations (150–200 g for 10 min), resuspend and wash the successive pellets with fusion medium (0.1 mM Ca^{2+}-acetate, 0.5 mM Mg^{2+}-acetate, 1–2 mg ml^{-1} bovine serum albumin, and appropriate amounts of sorbitol to adjust the osmolality of the medium to 75–100 mOsmol). Note that the albumin is critical. Different batches of albumin can have pronounced effects on the fusion and hybrid yields. In

our laboratory, the bovine serum albumin Fraction V from Serva (No. 11930, 69115 Heidelberg, Germany) gave best results.

5. After washing, resuspend the cell mixture in fusion medium at a density of 1–$5 \cdot 10^6$ cells per ml. Fill 200–300 μl of the cell suspension into the helical chamber or 5–10 μl into the adjustable plate microchamber (see Chapter 4).

6. Apply a constant alignment field of 250–300 V cm^{-1} strength (500–600 V cm^{-1} peak-to-peak value) and of 1.5–2 MHz frequency for 30 s.

7. Inject a single rectangular breakdown (fusion) pulse of 1.25–2 kV cm^{-1} strength and 15 μs duration. (Note that under iso-osmolar conditions three consecutive pulses of about 2–3 kV cm^{-1} strength and 15 μs duration are required; pulse interval 1 s; AOT 10 ms.)

8. After pulsing, apply the alternating field again for 30 s.

9. Keep the cells for 5–10 min without disturbance. Then, gently flush the cells from the chamber with complete growth medium (without phenol red). When fusion is performed in the helical chamber, rinse the chamber with 2–3 ml of complete growth medium and then plate the cells into 4–6 wells of a 24-well plate. In the case of the adjustable plate microchamber, rinse with 100–700 μl of complete growth medium and then plate the cells in 1–7 wells of a 96-well plate. Cultivate the cells at 37°C in a 5% CO_2 enriched, water-saturated atmosphere.

10. After 24 hours, add 1 volume of appropriate selection medium (e.g., HAT medium). Human hybridoma colonies should be visible to the naked eye after 2–5 weeks.

11. Depending on the source of the B cells, the yield of human hybridomas can be enhanced considerably by co-culturing them with feeder cells (10^5 irradiated murine splenocytes per ml, prepared from BALB/c or NMRI mice; irradiation conditions: 3000 rad [30 Gy] for 10 min). Feeder layers are particularly recommended when the B cells were obtained from biopsy material.[14,15]

REFERENCES

1. **Robl, J. M., Collas, P., Fissore, R. and Dobrinsky, J.,** Electrically induced fusion and activation in nuclear transplant embryos, in *Guide to Electroporation and Electrofusion*, Chang, D. C., Chassy, B., Saunders, J. A. and Sowers, A. E., Eds., Academic Press, San Diego, 1992, 535.

2. **Zimmermann, U. and Urnovitz, H. B.,** Principles of electrofusion and electropermeabilization, *Meth. Enzym.*, 151, 194, 1987.

3. **Zimmermann, U., Gessner, P., Wander, M. and Foung, S. K. H.,** Electroinjection and electrofusion in hypo-osmolar solution, in *Electromanipulation in Hybridoma Technology*, Borrebaeck, C. A. K. and Hagen, I., Eds., Stockton Press, New York, 1990, 1.

4. **Neil, G. A. and Zimmermann, U.,** Electrofusion, *Meth. Enzym.*, 220, 174, 1993.
5. **Neil, G. A. and Zimmermann, U.,** Electroinjection, *Meth. Enzym.*, 221, 339, 1993.
6. **Maseehur Rehman, S. M., Perkins, S., Zimmermann, U. and Foung, S. K. H.,** Human hybridoma formation by hypo-osmolar electrofusion, in *Guide to Electroporation and Electrofusion*, Chang, D. C., Chassy, B., Saunders, J. A. and Sowers, A. E., Eds., Academic Press, San Diego, 1992, 523.
7. **Panchal, C. J.,** *Yeast Strain Selection*, Marcel Dekker, New York, 1990.
8. **Bajaj, Y. P. S.,** *Plant Protoplasts and Genetic Engineering I (Biotechnology in Agriculture and Forestry 8)*, Springer-Verlag, Berlin, 1989.
9. **Ruthe, H.-J. and Adler, J.,** Fusion of bacterial spheroplasts by electric fields, *Biochim. Biophys. Acta*, 819, 105, 1985.
10. **Klöck, G. and Zimmermann, U.,** Facilitated electrofusion of vacuolated x evacuolated oat mesophyll protoplasts in hypo-osmolar media after alignment with an alternating field of modulated strength, *Biochim. Biophys. Acta*, 1025, 87, 1990.
11. **Foung, S. K. H., Perkins, S., Kafadar, K., Gessner, P. and Zimmermann, U.,** Development of microfusion techniques to generate human hybridomas, *J. Immunol. Meth.*, 134, 35, 1990.
12. **Perkins, S., Zimmermann, U. and Foung, S. K. H.,** Parameters to enhance human hybridoma formation with hypoosmolar electrofusion, *Hum. Antib. Hybrid.*, 2, 155, 1991.
13. **Klöck, G., Wisnewski, A. V., El-Bassiouni, E. A., Ramadan, M. I., Gessner, P., Zimmermann, U. and Kresina, T. F.,** Human hybridoma generation by hypo-osmolar electrofusion: characterization of human monoclonal antibodies to *Schistosoma mansoni* parasite antigens, *Hybridoma*, 11, 469, 1992.
14. **Vollmers, H. P., v. Landenberg, P., Dämmrich, J., Stulle, K., Wozniak, E., Ringdörfer, C., Müller-Hermelink, H. K., Herrmann, B. and Zimmermann, U.,** Electro-immortalization of B lymphocytes isolated from stomach carcinoma biopsy material, *Hybridoma*, 12, 221, 1993.
15. **v. Landenberg, P., Dämmrich, J., Stulle, K., Wozniak, E., Ringdörfer, C., Herrmann, B., Müller-Hermelink, H. K., Zimmermann, U. and Vollmers, H. P.,** Immortalization of stomach carcinoma infiltrating B-lymphocytes from biopsy material by electrofusion, *Oncol. (Life Sci. Adv.)*, 12, 35, 1993.
16. **Perkins, S., Zimmermann, U., Gessner, P. and Foung, S. K. H.,** Formation of hybridomas secreting human monoclonal antibodies with mouse-human fusion partners, in *Electromanipulation in Hybridoma Technology*, Borrebaeck, C. A. K. and Hagen, I., Eds., Stockton Press, New York, 1990, 47.

INDEX

A

Acetate, 35, 200, 202, 367, 369-371
Acidic phospholipids, 28, 200
Acidiphilium sp., 110
Acoustic cell alignment, 182, 193, 201
Acrosome reaction, 57
Actin, 42, 46, 48, 49, 199
Action potential, 332, 334, 346
Activation of,
    B-cells, 228, 371
    oocytes, 54, 224-226
Active transport. *See* ATPases
Adenylate cyclase, 42
Adherent cells, 5, 39, 51, 177, 178, 193,
    260, 276, 279, 368. *See also*
    Membrane breakdown, pulsing in
    monolayers
Adhesion. *See* Cell, adhesion
Adipocytes, electroinjection, 43
Adjustable plate microchamber, 195, 196,
    370, 372
Adrenal chromaffine cells,
    electropermeabilization, 44
Adrenal medullary cells,
    electropermeabilization, 44
Adverse side effects. *See* Membrane
    breakdown, secondary processes
Agar, 109, 114, 115, 118, 367, 368
Agarose, 371
Agglutinating additives, 175, 179
Albinism, 220
Albumin, 30, 179, 195, 200-202, 368, 371,
    372
    electroinjection of, 47
Algae
    electrofusion, 197, 219
    electroinjection, 42
    electrotransfection, 58
Alginate, 217, 371
Alignment. *See* Cell apposition;
    Dielectrophoresis
Al^{3+}-ions, detrimental effects of, 35, 120,
    366
Alkane, 140
Alnus incana, electrotransfection, 58
α-helix, 352, 353
Alternating current fields. *See* Cell
    apposition, electric;
    Dielectrophoresis; Electrorotation

Aluminum ions, detrimental effects of, 35,
    120, 366
Aminophospholipid translocase, 28, 339,
    351
Amnion cells, electroinjection, 44
Ampicillin, 110
Anesthetics, 10, 197
Aneuploidy, 220
Anti-asparagine synthetase, electroinjection
    of, 44, 45
Antibiotics, 110, 331, 333
    resistance to, 221
Antibody, 28, 180. *See also* Monoclonal
    antibodies
    injection of, 40, 42, 44, 46, 52, 54
    production, 180, 202, 203, 227-230, 371,
    372
Anti-field rotation. *See* Electrorotation, anti-
    field
Antigen presentation, 49-51
Anti-vimentine, electroinjection of, 44, 45,
    48
Aphidicolin-synchronized cells, 38, 296. *See
    also* Cell, synchronization
Apposition. *See* Cell apposition
Area stretching, 138-141, 147-149
Artificial lipid bilayer membrane. *See* Lipid
    bilayer membrane
Artificial tissues, 261, 279, 281
Asbestos fibers, electrorotation, 284
Aspergillus sp., electrotransfection, 65
Aspiration pressure. *See* Micropipettes
Asymmetric membrane breakdown, 10, 12,
    13, 17-24, 34, 42, 43, 46, 52, 69,
    206, 207
Atomic particles, trapping of, 275
ATPases, 51, 331, 333-335, 353
ATP-level, 34, 35, 198, 214
Autolytic activity, 110
Autophagy, 50
Avena sativa, 10, 19, 205, 206
    electrofusion, 190
    electroinjection, 42
    electrotransfection, 58
Avidin-biotin complex, 179, 180
Avogadro number, 340
Axon. *See* Squid axon
Azolectin membranes, UO^{2+}-modified, 161,
    164

375

B

Bacillus cereus, electrotransfection, 108
Bacillus subtilis, electrotransfection, 116
Bacteria
 aerotactic, 7
 chemotactic, 4, 7, 10, 18, 22, 27
 gram-negative, 25, 366
  aerobic, 121
  anaerobic, 123
  facultatively anaerobic rods, 122
 gram-positive, 25
  cocci, 124
  rods, 125
 temperature-sensitive mutants, 120, 220
Bacteria motion, field-induced, 272, 298
Bacteria probes, visualization of field
  distribution, 204, 205, 271
Bacteria protoplasts
 electrofusion, 196, 197, 219, 369
 electrotransfection, 108
Bacteria transformation (transfection). *See
  also* Electrotransfection; High-
  intensity field pulses; Membrane
  breakdown; Plasmid
 colony-forming units, 115
 culture conditions for, 109, 110
 electric field conditions, 4, 6, 15, 16, 25,
  26, 108, 111-114, 127
 pitfalls, 120
 protocol, 108, 366, 367
 recipient cells, 109-111
 recipient density in, 111, 118, 119, 178,
  366
 resealing conditions, 114, 115
 transformation efficiency, 25, 112, 114,
  116, 117, 127
 transforming DNA, 116-120
Bacteria wall
 electrorotation, 292, 297
 electrotransfection through, 108-110, 114,
  127, 197
Bacteriorhodopsin, 353
Bacto tryptone, 366
Bacto yeast extract, 366
Ba^{2+} ions, 127, 200, 217
Baker's yeast, 219
Barley. *See Hordeum vulgare*
Band III protein, 198, 331
B-cells. *See also* Lymphocytes
 electrorotation, 295
 hybridomas, 179, 180, 195, 202, 203,
  227-230, 371, 372
Beads, 185, 204. *See also* Particle
Beam trap, 282

Bending deformation, 142, 143
Bending energy, 147
Beta vulgaris
 electroinjection, 42
 electrotransfection, 58
Biconcave shape, 264, 293
Bilayer. *See* Lipid bilayer membrane
Bilayer bridging electrofusion model, 209-
  213
Bio-affinity chromatography, 34
Biodegradation, 77
Biological competence, 220
Biopsy material, 203, 227, 229, 230, 371,
  372
Bio-repellant surfaces, 305
Biosensor, 261
Biotinylation, 180
Blastocyst stage, 224, 225
Blastomere, electrofusion, 178, 179, 181,
  197, 207, 216, 224-226, 365. *See
  also* Eggs, electrofusion; Oocytes,
  electrofusion
Bleb formation, 39, 127, 192. *See also*
  Field-induced, vesicle formation
Bleomycin, electroinjection of, 51, 78, 79
Blood circulation, 77, 78
B-lymphoblasts, 28
B-lymphocytes. *See* B-cells
Boltzmann constant, 340
Boltzmann distribution, 150, 151, 341, 342,
  349
Bone marrow cells, electrotransfection, 68
Born charging energy, 36, 148
Born-Lertes effect, 284
Bovine serum albumin. *See* Albumin
Bovine sphingomyelin, 141
Brassica sp.
 electrofusion, 222
 electrotransfection, 58-59
Breakdown. *See* Membrane breakdown
Breakdown voltage. *See* Membrane
  breakdown, voltage
Brevibacterium sp., electrotransfection, 109,
  116, 126
Brewing process, 219
Brownian motion, 146, 269, 354
Bubble levitation, 261, 272
Budding, 36, 39. *See also* Field-induced,
  vesicle formation
Buffers
 in electrofusion, 201
 in electropermeabilization, 18, 34
Buoyancy, 222, 272, 279. *See also* Cell,
  density

Butanol, 201
t-Butylhydroperoxide, 40
Butyrivibrio fibrisolvens, electrotransfection, 126

C

Ca^{2+}-activated channels, 19. *See also* Ion channels
Ca^{2+} ions
  and bilayer organization, 350, 351
  as contaminants in enzymes, 198, 199
  and electrofusion, 199, 200, 202, 211, 214, 369-371
  and electropermeabilization, 10, 34, 35, 367, 368
  and exocytosis, 30
  immobilization, 217, 371
  injection of, 43, 44
  in membrane resealing, 199, 209, 211
  metabolism, 348
  in oocyte activation, 226
  release, 44
  signalling, 51, 211
Ca^{2+}-phosphate transfection method. *See* Transfection, other methods
Ca^{2+}-pump, 334, 335
Cadmium sulphide, 284
Caging of cells, 261, 275-281, 306
Calculations, *See* Theories
Callus, 59, 63-65, 220, 222
Calmodulin, 199
Calves, cloning of, 226
Candida maltosa, electrotransfection, 65
Capacitance. *See* Membrane, capacitance; Whole-cell capacitance
Capacitation, 339
Capacitive current. *See* Current, capacitive
Capacitor discharge, 6, 32, 112, 113, 115, 118, 152. *See also* Discharge chamber
Carboxyfluorescein, 8, 9
Carboxyfluorescein diacetate, 8
Carcinoma cells
  electrochemotherapy, 78
  electroinjection, 44
  electrotransfection, 68, 69
Carrasius auratus, electrotransfection, 67
Carrier DNA, 68, 72, 73, 76, 117
Carriers. *See* Ion carriers
Carrot. *See Daucus carota*
CAT activity, 58-61, 63, 64, 67, 69-71, 73-75
Cattle, cloning of, 223-227

Caulobacter crescentus, electrotransfection, 126
CD4 receptor, electroinsertion of, 78
CD40 receptor, stimulation of, 228
Cell
  adhesion, 44, 230
  apposition. *See* Cell apposition
  ATP-level, 34, 35, 198, 214
  caging, 261, 275-281, 306
  Ca^{2+}-level, 34, 199
  colloid-osmotic pressure, 37, 38, 179, 192, 214
  confluence, 177, 178
  contact, 175, 177-179, 186-188, 192, 193, 196-198. *See also* Cell apposition
  culture, 109, 110
  cycle, 10, 24, 54, 73, 74, 109, 110, 114, 216, 284, 285, 348, 367, 368, 371
  cytoskeleton. *See* Cytoskeleton
  cytosol, access to, 6, 7-9, 34, 41, 199
  death. *See* Cell, lysis
  deformation, 144, 147, 187, 192, 205, 284
  density, 182, 191, 220-222, 272, 279
  dielectric properties. *See* Dielectric properties; Permittivity
  differentiation, 224
  division, 34, 260, 331, 348
  fusion. *See* Electrofusion
  growth, 34, 109, 260, 275, 276, 306, 331
  hopping, 272
  hydrostatic pressure. *See* Turgor pressure
  internal conductivity, 5, 14-17, 24, 26, 33, 275, 289, 292, 294, 328, 331
    measurement of 6, 298
  internal osmolality, 11, 331
  internal pH, 34, 331, 336
  levitation. *See* Levitation
  lysis, 7, 8, 14, 17, 24, 35, 120, 191, 201, 204, 208, 339
  manipulation. *See* Electrofusion; Electroinjection; Electrotransfection; Bacteria transformation
  membrane. *See* Membrane
  metabolism, 44-46, 49
  microvilli. *See* Microvilli
  multinucleated, 175, 178, 179, 188, 190, 208, 216, 217
  non-spherical, 111, 192, 264, 265, 293, 294
  orientation, 10, 111, 146, 192, 207, 272
  proliferation. *See* Cell, division
  rotation. *See* Electrorotation
  segregation, 191. *See also* Sedimentation

separation, 17, 187, 261, 270, 271, 305, 306
size, 4, 5, 17, 37, 182, 191, 204, 205, 207, 271, 294
sorting, 7, 207. *See also* Flow cytometry; Fluorescence activated flow cytometry
spinning, 262, 278, 284. *See also* Electrorotation
surface potential. *See* Surface, potential
surface-to-volume ratio, 6
synchronization, 38, 45, 296
trapping, 261, 268, 272, 275-283, 305
viability, 7-9, 34, 36, 49-51, 114, 191, 192, 197, 198, 211, 216, 222, 271, 287
volume. *See* Cell, size
wall, 58, 60, 63, 64, 66, 67, 108-110, 114, 127, 292, 297
Cell apposition
centrifugation, 54, 178, 179, 193, 196
chemical, 179, 197
electric, 2, 3, 176, 178, 181, 183-193, 197, 203, 267, 269-283
electro-mechanical, 181, 182
field-casting, 279-281, 306
magnetic, 182, 193
mechanical, 181, 182, 193, 291
natural, 177, 178. *See also* Spontaneous fusion
receptor-mediated, 179, 180, 203
sedimentation, 178, 182, 278. *See also* Levitation
ultrasonic, 182, 193, 201
Cell-cell contact. *See* Cell, contact; Cell apposition
Cell-cell fusion. *See* Electrofusion
Cell chains. *See* Cell apposition, electric, ultrasonic
Cell collection, 261, 269-273. *See also* Dielectrophoresis, negative, positive
Cell debris. *See* Cell, lysis
Cell electrophoresis. *See* Cell motion, electrophoresis
Cell engineering. *See* Bacteria transformation; Electrofusion; Electroinjection; Electrotransfection; Somatic hybridization
Cell motion. *See also* Dielectrophoresis, Electrorotation; Theories; Travelling-wave drives
dielectrophoresis, 176, 178, 183-193, 261, 266-283

electrophoresis, 186, 187, 190, 221, 269, 272, 332, 339
electrorotation, 261, 266-268, 283-299
travelling-wave-induced, 261, 266-268, 300-305
Cell-pair fusion. *See* Electrofusion, microfusion
Cell pellet. *See* Cell apposition, centrifugation
Cell polarization, 183, 185, 260-268, 270, 289-293, 304
Cell rotation. *See* Electrorotation
Cell suspension density, 14, 111, 112, 114, 118, 119, 178, 185, 190, 194, 195, 208, 214, 261, 267, 271, 285, 286, 366, 367, 369, 370, 372
Cell sorting. *See* Cell, separation
Cell-tissue electrofusion, 181, 182, 193
Cell volume regulation, 336-338. *See also* Osmotic swelling
Cellular dipole interaction pressures, 178, 187
Cellular oscillators, 284, 285
Centrifugation. *See* Cell apposition
Cetyl pyridinium chloride (CPCl), 158, 159, 161
Channels. *See* Ion channels
Chara sp., punch through effect, 344
Charge displacement, 11, 143, 145, 352. *See also* Cell polarization; Particle, polarization; Ponderomotive forces
Charge induction. *See* External electric field, membrane charging
Charge relaxation, 300, 301. *See also* Charge pulse technique
Charge pulse technique
cells, 4, 11, 294
charge decay, 157, 162-165
charging process, 157, 162-165
instrumentation, 155, 156
planar lipid bilayer membrane, 155-168
pulse duration, effect of, 164
supercritical field strength, 158, 163
Charge separation. *See* Particle, polarization
Charge-shift potentiometric fluorescent dyes, 5, 7, 21, 26, 339
Charging time. *See* External electric field, membrane charging
Chemical facilitators, electrofusion, 197-200, 209, 222
Chemically induced fusion, 175, 208, 214, 216, 227, 230. *See also* Polyethylene glycol
Chemiluminescence, 52, 53

Chemoattractants, 53
Chemotactic bacteria. *See* Bacteria,
    chemotactic
Chenopodium quinoa, electrotransfection, 59
Chimeric embryos, 178
Chinese hamster ovary cells, 11, 28, 39, 54,
    55, 75, 76, 216
Chlamydomonas sp.
    electroinjection, 42
    electrotransfection, 58
Chloride salts, electrofusion, 200
Chlorine. *See* Toxic electrolytic products
Chloroplast blebs, 18
Chloroplasts
    as biological probes, 204
    dielectrophoretic collection, 269
    electrofusion of, 196
    electroinjection of, 3
    electrorotation, 292
    membrane breakdown, 9, 17
    re-distribution, 207
    rotation spectra, 292, 293
    traits encoded in, 221-223
CHO. *See* Chinese hamster ovary cells
Cholesterol, 27, 139, 140, 141, 152, 160,
    351
    oxidized, 162-167
Chromaffine cells, electropermeabilization,
    44
Chromatin, 216
Chromosome damage, 24, 48, 50, 54, 55
Chromosomes, 174, 216, 220
Chrysanthemum sp., electrotransfection, 59
Circular dichroism, 353
Citrus sp., electrofusion, 222
Clastogenic agents, 216
Clavibacter sp., electrotransfection, 126
Cleavage plane, 178
Cloning, of embryos, 178, 223-227, 365
Co-cultures, electrofusion, 177, 178. *See
    also* Adherent cells
Co-field rotation. *See* Electrorotation, co-
    field
Cole-Cole plot, 293, 330
Colloid-osmotic pressure, 37, 38, 179, 192,
    214. *See also* Donnan-osmotic
    pressure
Colony-forming units (bacteria), 115, 117
Competence, 79, 109, 220
Complement, 227
Complex conductivity, 263. *See also*
    Permittivity
Complex dielectric constant. *See* Particle,
    polarization; Permittivity

Compressibility modulus, 140, 148
Compression dipole forces, 178, 187
Concentration gradients, current-induced, 23,
    24, 114, 343-346
Condenser energy, 150
Conductivity, external 14-16, 18, 22, 26, 27,
    33, 108, 113, 145, 146, 150, 156,
    183, 185, 201, 263, 273, 275, 283,
    286, 289-302, 304, 305, 327, 328,
    344, 346
Conductivity gradients, 300-302
Conductivity, internal. *See* Cell, internal
    conductivity
Conductometer, 286
Confluence. *See* Adherent cells; Cell,
    confluence
Confocal laser scanning microscope, 20
Conformational changes. *See* Field-induced,
    conformational changes
ß-Conglycinin, 61
Conifer cells, electrofusion, 219
Connective tissue cells
    electroinjection, 45
    electrotransfection, 69
Contact. *See* Cell, contact; Cell apposition
Contact-angle measurements, 28
Contact zone, 187, 188, 201, 204-206, 209-
    216. *See also* Cell apposition
Continuity equation, 145
Continuum approach, 138
Contra-rotating fields. *See* Electrorotation,
    anti-field
Convection, 191, 279. *See also* Joule heating
Corn oil, motion of, 300
Corynebacterium glutamicum,
    electrotransfection, 116, 126
Co-transfection, 41, 59-62, 65, 67, 68, 72,
    73
Coulter Counter. *See* Particle, analyzer
Cover slip, 194
Cracks, membrane, 36, 37, 127
Critical area strain, 139, 140
Critical field strength. *See* High intensity
    field pulses, critical strength
Critical pore radius, 150, 151, 153, 161
Critical stress, 139, 140, 147
Critical tension, 139, 140, 149
Critical thickness, 148
Critical voltage. *See* Membrane breakdown,
    voltage
Cross-linking. *See* Immobilization
Crossover frequencies, 268, 275
Cryopreservation, 111
Cucumis sp., electrofusion, 222

Culture conditions. *See* Bacteria
    transformation
Current
    capacitive, 263, 264
    circulating, 334
    through electropermeabilized membranes,
        5, 23-25, 31, 33, 163, 164, 191, 201
    ohmic, 264. *See also* Current-voltage
        characteristics
Current-induced concentration gradients, 23,
    24, 114, 343-346
Current-induced fields, 23, 24, 114, 215,
    343-346.
Current-induced volume flow, 23
Current-voltage characteristics, 4, 217, 294,
    343-346, 350
Curvature. *See* Membrane, curvature
Cut-off frequency, 32, 264
Cybrids, 174, 221, 223
Cylindrical electrodes. *See* Electrodes, wire
Cytochalasin, electroinjection of, 55
Cytoplasmic bridge, in electrofusion, 209-
    211
Cytoplasmic esterase activity, 8
Cytoplasmic inheritance, 223
Cytoplasts, 223
Cytoskeleton, 10, 28, 37, 39, 45, 50, 175,
    187, 198, 201, 208, 213-215, 339
Cytosol manipulations. *See* Bacteria
    transformation; Cell, cytosol;
    Electrofusion; Electroinjection;
    Electrotransfection; External electric
    field; Membrane breakdown
Cytotoxicity, 45, 46, 50, 51, 216

D

Dactylis glomerata, electrotransfection, 59
Damping forces, 152-154, 158, 159, 263,
    275, 278
Daucus carota
    electroinjection, 42
    electrotransfection, 59
Debye-Hückel ionic cloud, 330
Debye-Hückel length, 147, 185, 340, 341,
    349
Decay time. *See* External electric field,
    membrane discharging
Defects, membrane, 39, 40, 147-149, 152,
    153, 209. *See also* Cracks; Pores
Definitions. *See* Terminology
Deformation
    cells, 187, 192, 205, 284
    lipid vesicles, 18, 144, 147

Dehydrating agents. *See* Polyethylene glycol
Density. *See* Cell, density; Cell suspension
    density; Cell apposition,
    centrifugation; High-density media
DEP. *See* Dielectrophoresis
Depletion layer, 344, 345. *See also* Salt-shift
Desiccation-tolerant grass, electrofusion, 222
Destructive membrane agents, 289. *See also*
    Surfactants
Dextran, 37, 179, 197. *See also* FITC-
    dextran
Dialysis, 117
Diarachidonoyl phosphatidylcholine, 141
Dictyostelium sp.,
    electroinjection, 42, 43
    electrofusion, 178
    electrotransfection, 65
Dielectric breakdown. *See* External electric
    field; Membrane breakdown
Dielectric constant. *See also* Dielectric
    properties; Particle, polarization;
    Permittivity
    liquid, 145, 146, 150, 299, 340
    membrane, 145, 146, 148, 299, 342, 347
    particle, 269, 283, 284, 289
Dielectric field flow fractionation, 306
Dielectric interface, 263, 264, 291, 300, 301
Dielectric layers, 301
Dielectric loss, 283
Dielectric migration, definition of, 269
Dielectric motor, 266
Dielectric properties, 183, 187, 188, 261-
    266, 270, 283-285, 293, 301. *See
    also* Permittivity
Dielectric spectroscopy, 268, 270, 271, 287,
    306
Dielectrophoresis. *See also* Cell motion; Cell
    apposition; External electric field;
    Particle, polarization;
    Ponderomotive forces; Theories,
    dielectrophoresis
    in alternating fields, 2, 3, 37, 147, 183-
        193, 261-268, 370, 372
    in amplitude-modulated AC fields, 191,
        192, 370
    beneficial aspects, 178, 181, 187-190,
        270-283
    cages, 279, 280
    cell collection, 261, 268-270
    crossover frequencies, 268, 275
    in DC-fields, 14, 183, 185, 269
    definition of, 269
    factors controlling, 183-187
    field frequency, 183, 186, 187, 190, 211

field strength, 183, 187, 191, 275, 276
force, 183, 185, 261-265, 268, 270
  in high-conductivity media, 185, 273, 305
  instrumentation
    electrode assembly, 183, 193-196, 271
    power supply, 196, 365
  intracellular, 187, 204
  levitation. *See* Levitation
  in low-conductivity media, 183-193, 265,
    305, 336
  low-frequency, 186, 187, 272
  multiaggregate formation, 187, 272
  multi-body problem, 267, 270
  mutual, 14, 178, 185, 186
  negative, 183, 185, 186, 204, 266-268,
    271-283, 287, 299, 304, 305
  nuclear alignment, 187
  pitfalls, 190-193, 203
  positive, 183-185, 204, 205, 261-274,
    282, 284, 287, 394
  separation, 187
  spectroscopy, 268, 270, 271, 293, 298
  trajectories, 185, 264
  traps, 261, 268, 272
  turnover frequencies, 192, 267, 268
Dielectrophoretic response, definition of,
  273
Dielectrophoretic yield, 270, 273
Diffusion, non-Stokesian, 40
Diffusion layers, 343-346
Diffusion potential, 333-336
Digestive Enzymes. *See* Enzymes, and
  electrofusion
Dimethyl sulfoxide, 197, 218
Dimyristoyl phosphatidylcholine, 350
Diphytanoyl phosphatidylcholine, 156-160
Dipole. *See also* Particle, polarization;
    Ponderomotive forces
  concept, 265, 324, 325
  induction, 184, 185, 265, 285, 304, 328,
    352
  moment, 267, 269, 270, 289, 324, 351-
    353
    measurement of, 272
  orientation, 350, 352
  re-orientation, 352
Dipole-dipole interactions, 178, 187, 267,
  285, 326
Discharge chamber, 6, 32, 35, 112, 115,
  118, 365, 366
Discharging process. *See* External electric
    field, membrane discharging; High-
    intensity field pulses

Discocytes, 143
Disk electrodes. *See* Electrode arrays;
    Electrodes
Dispase, 197, 368
Dispersion, 263, 264
Displacement current. *See* Current,
    capacitive
Distearoyl phosphatidylethanolamine, 159
Divalent cations. *See* Ba^{2+} ions; Ca^{2+} ions;
    Mg^{2+} ions
Divergence. *See* External electric field,
    inhomogeneity
DNA. *See also* Plasmid
  breakage, 24, 120
  carrier. *See* Carrier DNA
  chromosomal, 116
  degradation, 39, 116
  dielectrophoretic collection of, 270
  electrophoresis of, 127
  injection of. *See* Bacteria transformation;
    Electrotransfection
  repair, 48, 50
  synthesis, 34, 44, 45, 49
Domain-induced budding, 36, 39
Domestic animals, embryo cloning, 178,
  223-227
Donnan-equilibrium, 330, 332, 333, 336,
  337, 344
Donnan-osmotic pressure, 336, 337. *See also*
    Colloid-osmotic pressure
Donnan-potential, 336
Donor embryos, 224-226
Double lattice charge membrane, 344, 345
Double layer. *See* Electric double layer
Driselase, 197
Drosophila sp., electrotransfection, 67
Drug-delivery systems, 47, 54, 56, 57, 77-79
Dryopteris paleacea, electroinjection, 42
Duration, electric field. *See* High-intensity
    field pulses
Dust precipitator, 261, 270
Dyes
  and electrofusion, 180, 191, 213
  exclusion. *See* Trypan blue
  field-induced uptake, 4-9, 19, 29, 30, 40,
    207. *See also* Electroinjection, of
    exogeneous compounds
  field-induced release, 4-9, 18, 23. *See
    also* Electroinjection, of exogeneous
    compounds
  membrane staining, 187, 213, 230, 276
  membrane voltage-sensitive, 5, 7, 21, 26,
    339
  retention, 8, 49-51

E

Ear cells, electroinjection, 45
Earthquakes, 2
EBV, 228
Echinocyte-stomatocyte transformation, 349, 351
Echinodermata, electroinjection, 43
Echinus esculentus, electroinjection, 43
E. coli. See Escherichia coli
Edge effects, in microstructures, 274
Edge energy, 147, 150, 151, 153, 154, 159, 161, 162, 168
EDTA, 34, 127, 366, 367, 369
Eggphosphatidylcholine, 152
Eggs, electrofusion, 178, 179, 181, 197, 204, 223, 225, 271. *See also* Blastomere; Cloning of embryos; Oocytes
Eggs, induction of polarity, 192
Egg-sperm electrofusion, 181
EGTA, 34, 217
Ehrlich ascites tumor cells, electrofusion, 188, 189, 216
Elastic
    compressive moduli, 140, 141, 148
    modulus of area stretching, 139, 140
    restoring forces, 148, 149, 167
    spring, 140
    theory. *See* Hook's law
Electric alignment. *See* Cell apposition; Dielectrophoresis
Electric breakdown. *See* External electric field; High-intensity field pulses; Membrane breakdown
Electric double layer, 330, 332, 339-343, 346, 350, 354. *See also* Zeta-potential
Electric field. *See* External electric field
Electro-acoustic fusion, 182, 193, 201
Electrocardiogram, 346
Electrochemical potential gradient, 331, 332, 334, 343, 350
Electrochemical trigger mechanisms, 331
Electrochemotherapy, 78, 79
Electrocompetence, definition of, 79, 109
Electrocompression, 36
Electroconformational coupling, 353
Electroconstriction, 353. *See also* Field-induced, membrane thinning
Electrode arrays. *See also* Microstructures
    cell motion, 2, 183, 271, 272-283
    electrofusion, 2, 183, 184, 193-196
    electropermeabilization, 2, 5, 6, 14
    electrorotation, 285, 286, 298, 299
    travelling-wave drives, 300-305

Electrode material, 6, 35, 120, 192, 193, 203, 204, 366
Electrode polarization, 186, 286
Electrodes
    cages, 279, 280
    disk, 272
    helical, 194, 195, 369, 370, 372
    interdigitated (castellated), 183, 271, 273, 275
    isomotive force, 271
    laminar flow system, 272
    meander, 194, 271
    microelectrodes. *See* Microelectrode techniques
    multi-electrode assemblies. *See* Microstructures
    pin-pin, 273
    pin-plate, 196
    planar arrays. *See* Microstructures
    plates, 6, 185, 195, 272, 285, 286, 370, 372
    ring, 272, 279
    strips, 6, 193, 194, 303, 304
    surface-to-volume ratio of, 274
    traps, 277, 278
    vibrating, 334, 346, 347
    wire, 6, 185, 193, 194, 202, 217, 273
Electrofusion. *See also* External electric field; Field-induced; High-intensity field pulses; Membrane breakdown
    adherent cells, 177, 178, 193
    albumin. *See* Albumin
    applications, 215-230
    benefits of, 176, 181
    cell alignment. *See* Cell apposition; Dielectrophoresis
    chemical facilitators, 197-200, 222
    co-culture, 177
    contact zone. *See* Contact zone
    in DC-fields, 176, 178
    field distribution, 184, 185, 193, 204-206
    field strength, 203-207, 212, 213, 215
    fusibility, 197, 198, 201, 208
    fusion index, 188, 189
    fusion yield, 178, 190-192, 194, 197, 198, 201, 203, 208, 214, 216, 222, 227, 371
    fusogenic state, 28, 208
    giant cell production, 39, 178, 190, 193, 198, 208, 216-219
    hemi-fusion, 213, 214
    high-conductivity media, 178, 182, 279, 282
    hybridization. *See* Somatic hybridization

hybrid yield, 182, 190, 192, 194, 197, 199-201, 203, 204, 214, 216, 222, 227, 228, 371
instrumentation, 177, 178, 193-196, 365
intergeneric fusion, 219, 222
interkingdom fusion, 218
intermingling process, 198, 199, 201, 203, 207-215
interspecies fusion, 174, 219, 222
intraspecies fusion, 174, 219, 222
kinetics, 187, 191, 199, 203, 207, 209, 211, 214
macrofusion, 194-196, 222
mechanism, 37, 187, 197-199, 205, 207-215
media ingredients, 36, 37, 199, 200, 369-371
membrane contact. *See* Cell, contact; Cell apposition
microfusion, 181, 193-196, 203, 220, 222, 223, 227, 261, 279, 282, 371, 372
under microgravity, 191
nuclear, 188, 189
organelles, 196, 204
osmolality of media, 36, 37, 39, 201-204, 369-371
partners, 182, 190, 191, 204, 205, 207, 208, 219-223
pitfalls, 190-193
post-alignment, 179, 181, 187, 191, 201, 203, 208, 215, 370, 372
pre-alignment, 176, 177, 179, 182, 191, 215
probability, 188-190
protocols, 369-372
pulse, 207, 208. *See also* High-intensity field pulses
in pulsed fields, 176, 186
in radio-frequency fields, 186
temperature, 203, 369, 370
vesicle formation. *See* Field-induced, vesicle formation
Electrogenic pumps, 296, 331, 334-336, 346. *See also* Ion pumps
Electrohydrodynamic fluid pumping, 300-303
Electroimplantation. *See* Electroinsertion
Electroinjection. *See also* External electric field; High-intensity field pulses; Membrane breakdown
algae, 42
applications, 42-57
asymmetric, 13, 17-24, 42, 43, 46, 52, 69
echinodermata, 43

of exogeneous compounds, 42-57, 203
field conditions, 5, 6
fungi, 42, 43
high-permeability state, 18, 27, 29, 33, 35, 37
instrumentation. *See* Discharge chamber; Electrode arrays; Electrodes
mammalian cells, 43-57
medium ingredients, 33-36, 38, 110, 111
of nucleic acids. *See* Electrotransfection; Bacteria transformation
osmolality of media, 35-37, 39, 69, 70, 367-369
of particles, 30, 31
pitfalls, 24, 25, 34, 35, 120
protocols, 366-369
single-cell level, 279
symmetric, 12, 17-24
temperature, 25, 27, 112, 120, 366-369
through cell wall, 25. *See also* Bacteria transformation
yeast, 43
Electroinsertion, 25, 41, 47, 78, 200
Electrointernalization, 3, 25, 26, 29-31, 38-41, 48, 51, 52, 54, 56, 58, 127, 128, 211, 212, 296
Electro-killing of cells. *See* Cell, lysis
Electrokinetic force spectra, 267, 268
Electroleaks, definition of, 40. *See also* Defects; Pores
Electrolysis, 35, 186, 187, 200, 273, 274
Electromagnetic braking, 283
Electromagnetic fields, 2, 215, 275, 353, 354
Electro-magneto-electrofusion, 182
Electromechanical compression, 36, 148, 149
Electromechanical fusion, 181, 182
Electromilking, definition of, 41
Electron beam lithography, 270, 273
Electroneutral pumps, 331, 334, 335. *See also* Ion pumps
Electron microscopy
blebs, 39, 127
charged colloids, 339, 346
electroinsertion, 30, 31
electropores, 36, 209
fusion, 179, 197, 198, 209
membrane openings, 36, 37
microvilli loss, 38, 39
sperm-egg fusion, 223
Electro-orientation, 146, 147, 192, 285
Electroosmosis, 23, 127, 346

Electropermeabilization; *See* External
electric field; Membrane breakdown
Electrophoresis. *See also* Free flow
electrophoresis, SDS-gel
electrophoresis
of cells, 186, 187, 190, 221, 269, 272,
332, 339
in field-induced solute release, 23, 127
of membrane components, 25, 36, 143,
346
Electrophoretic mobility, 198
Electrophoretic trapping of cells, 282
Electrophysiology. *See* Microelectrode
techniques
Electroporation. *See* Bacteria transformation;
Electroinjection; Electrotransfection;
Membrane breakdown
Electroporation, definition of, 3
Electropores, definition of, 3. *See also* Pores
Electrorelease, 4, 6, 7, 10, 17, 22, 114, 191.
See also Electroinjection, of
exogeneous compounds; Dyes
Electrorotation. *See also* External electric
field; Ponderomotive forces
in alternating fields, 192, 284, 285
anti-field, 266, 267, 283, 287-291, 294,
299, 304
applications, 289-299
cell properties, measurement of, 6, 18, 38,
110, 270, 290-299
cell walls, 292, 297
characteristic (maximum) frequencies,
268, 287, 288, 294-299
co-field, 262, 266, 267, 283, 287-291,
304
crossover frequencies, 268, 275
dielectric spectroscopy, 268, 270, 287-
299
field strength, 266, 286, 298
instrumentation, 285-288, 298
high-conductivity media, 290, 298, 299
historical background, 261, 283-285
low-conductivity media, 286, 289, 290,
294-298, 336
low-frequency, 283
membrane charging, 288, 289, 299
microvilli, contribution to, 38, 296, 297
multi-cell rotation, 285
null frequency method, 287, 288, 298
organelles, 292, 293
in rotating fields, 285-299
single-cell rotation, 285-299
spectra, 267, 268, 288-293, 296, 299
speed, 266, 288, 299

traps, 278
theoretical background, 261-268, 283-
285, 288-299, 301, 303
in weakly conductive solutions, 15, 16,
286, 287
zona pellucida, 292, 297
Electrostatic motors, 283, 284
Electrostatic repulsion, 175, 177, 340, 342,
351
Electrotransfection. *See also* Bacteria
transformation; External electric
field; Field-induced; High-intensity
field pulses; Membrane breakdown
algae, 58
bacteria
gram-negative, 121-123
gram-positive, 124, 125
fish, 67
fungi, 65-67
higher plants, 58-65
insects, 67
mammalian cells, 68-76
pitfalls, 34, 35, 39, 120
protocols, 366-369
protozoa, 67
stable transformants, 58-76
transient expression, 58-65, 67, 69-76
yeast, 28, 66, 67
Electrotransformation. *See* Bacteria
transformation; Electrotransfection
Electro-uptake, 3, 4, 6, 19, 29-32, 117, 203.
See also Electroinjection, of
exogeneous compounds; Dyes
Ellipsoidal cells, 111, 265, 293, 294. *See
also* Electro-orientation
Embryogenesis, 178, 223, 224
Embryos (Embryonic cells)
electrofusion, 193, 194, 201
electroinjection, 46
electrotransfection, 69
in nuclear transplantation, 223-227
in plant breeding, 223
preimplantation stage, 223, 224
storage, 226
Enamel organ cells, electrotransfection, 70
Encapsulation. *See* Immobilization
Endocytosis, 29-31, 45, 47, 209. *See also*
Electrointernalization
Endonucleases, 39, 117
Endoplasmic reticulum, 174
Endothelial cells, electroinjection, 47
Energy
bending, 147
Born charging, 36, 148

condenser, 150
edge, 147, 150, 151, 153, 154, 159, 161, 168
electrochemical, 331
free, 352
hydration layer, 28, 175, 177, 187, 200
kinetic, 154
line tension, 150
solvation, 150
Energy barrier. *See* Lipid bilayer membrane, energy barrier
Energy dissipation, 263
Entropy, 142, 332
Enucleation, 181, 223-225
Enzymes
in bacteria transformation, 110
contaminants, 198, 199
and electrofusion, 197-200, 209
electroinjection of, 40, 46, 48, 50, 51, 53-55, 77
and electropermeabilization, 10
heat inactivation, 199
isoenzyme analysis, 222
purification, 34
Epstein-Barr virus, 228
Eremosphaera viridis, 11
Erythrocyte ghosts, 18, 23, 28, 31, 47, 211, 214
Erythrocytes
as drug delivery systems, 47, 77
electrofusion, 176, 187, 198, 209, 213, 214
electroinjection, 47, 70
electrorotation, 264, 284, 285, 293
hemi-fusion, 213, 214
hemolysis, 34, 336
mechanical properties, 349
membrane breakdown, 5, 28, 39, 40, 47, 336
membrane composition, 339, 340
membrane potential, 336, 337
Na^+-K^+-ATPase, 331, 334, 335
size distribution, 5
shape transformation, 349, 351
volume changes, 336-338
Escherichia coli, 25, 108, 110-113, 116-119. *See also* Bacteria transformation
Esterase activity, 8
Ethanol, 117, 197, 301
Ethidium bromide, 19, 42, 46, 54, 55
Etioplasts, 9
Evacuolated protoplasts, 182, 221. *See also* Plant protoplasts

Evolution, 2, 3
Exchange, field-induced. *See* Electroinjection; Electro-release; Electro-uptake; Membrane breakdown, solute exchange
Exine, electrorotation, 292
Exocytosis, 30, 43, 44, 54, 209. *See also* Electrointernalization
Exogenous materials, injection of. *See* Electroinjection
Exonuclease, 116
Exponentially decaying field pulses, 31-33, 42-76, 112, 113, 121-126, 208, 366-369
External electric field. *See also* Cell motion; Charge pulse technique; Dielectrophoresis; Field-induced; Electrorotation; High-intensity field pulses; Membrane breakdown
alternating. *See* Dielectrophoresis
breakdown field strength, 5, 190, 203-207
cages, 265, 278-280
calculation of membrane voltage from, 6, 11-17, 204, 205, 212
casting, 279-281, 306
DC-shifted alternating, 33
electrode geometry. *See* Electrode arrays; Electrodes
electrolysis. *See* Electrolysis
electromagnetic, 2, 215, 275, 353, 354
electropermeabilization. *See* Membrane breakdown
exposure to, 4-11, 24-33, 112-114, 183-196, 261-305, 345, 346
discharge chambers, 6, 11, 32, 35, 41, 110, 112, 115, 118, 365
force calculations, 261-269, 278, 324-328. *See also* Ponderomotive forces
funnel, 278
growth in, 273, 275, 276, 306
heating effects. *See* Joule heating; Travelling-wave drives; Electrohydrodynamic fluid pumping
high-conductivity media. *See* High-conductivity media
high-frequency, 32, 33, 146, 182-190, 260-305
homogeneity, 6, 14, 196, 285, 286
induction by ion pumps, 334, 346, 347
inhomogeneity, 2, 6, 14, 143, 178, 183, 196, 204-206, 209, 261-283
interaction with membranes, 29-31, 38, 128, 143-152

line distribution, 144, 147, 184, 185, 193,
     204-206, 279, 281, 282, 286
  visualization of, 204-206, 271, 334
low-conductivity media. *See* Low-
     conductivity media
low-frequency, 28, 146, 186, 187
low-intensity, 353, 354
membrane charging, 7, 11, 12, 22, 26, 27,
     31, 32, 157, 162-165, 263, 288, 289,
     299
membrane discharging, 17, 113, 156, 157,
     162-165
microelectrodes. *See* Microelectrode
     techniques
microstructures. *See* Microstructures
permeabilization, 2-79, 111-116, 203-208,
     216
power supply. *See* Discharge chamber;
     Power supply
pulses. *See* High-intensity field pulses
punch through effect, 23, 24, 114, 215,
     344-346. *See also* Salt-shift
rotating. *See* Electrorotation
star-like, 279
and surface admittance, 14, 25, 204
toroidal, 279
trap, 261, 268, 272, 275-282
travelling-wave. *See* Travelling-wave
     drives

F

Fab fragment, 180
FACS. *See* Flow cytometry; Fluorescence
     activated flow cytometry
Fat particles, 269
Fatty acids, 351
Feeder cells, 372
Fertility, 220
Fertilization, 223-225, 331, 348
Fibroblasts, 199
  cultivation in electric field, 260, 275, 276
  electrofusion, 200
  electroinjection, 44, 46, 47, 51
  electrotransfection, 69, 74, 76
  levitation, 276
Ficoll, 37, 192, 195
Field
  electric. *See* External electric field
  electromagnetic, 2, 215, 275, 353, 354
  magnetic, 182, 265, 354
  ultrasonic, 182, 193, 201
Field-casting, 279-281, 306

Field exposure. *See* High-intensity field
     pulses, duration
Field-induced. *See also* Dielectrophoresis;
     Electrofusion; Electroinjection;
     External electric field
  Ca^{2+}-influx, 199, 209, 226
  cell deformation, 187, 192, 205, 284
  cell orientation, 111, 192, 199, 207
  cell motion. *See* Cell motion
  conformational changes, 28, 36
  cracks, 36, 37
  endocytosis. *See* Electrointernalization;
     Endocytosis
  exocytosis. *See* Electrointernalization;
     Exocytosis
  forces. *See* Ponderomotive forces
  fusion. *See* Electrofusion
  membrane loss, 29-31, 38, 209-213, 296
  membrane permeabilization. *See* External
     electric field; Membrane breakdown
  membrane thinning, 36, 148, 149, 353
  microaggregate formation, 279-281
  migration of molecules, 25, 36
  polarization, 183-187, 261-268, 289-293,
     300-305
  pores, definition of, 3. *See also* Pores
  redox reactions, 28
  rotation. *See* Electrorotation
  secondary processes, 33, 36, 37, 39, 164,
     191, 192, 332, 334, 336, 337. *See
     also* Electrofusion; Electroinjection;
     Osmotic swelling
  solute release. *See* Electro-release; Dyes
  solute uptake. *See* Electro-uptake; Dyes
  structural changes, 24, 25, 36, 37, 39,
     213. *See also* Electrointernalization
  surface conductance, 110, 183, 294, 297,
     346, 354
  vesicle formation, 29, 30, 39, 182, 199,
     209-213, 296. *See also* Bleb
     formation; Budding
Field-producing proteins, 347
Field propagation. *See* Travelling-wave
     drives
Field pulse. *See* High-intensity field pulses
Field-sensitive proteins, 347, 352. *See also*
     Ion channels
Film balance, 160
Finite difference methods, 264
Finite-element analysis, 264
Fish
  electrofusion, 226
  electrotransfection, 67

FITC-dextran, electroinjection of, 23, 42, 43, 46, 47, 49, 52
Fixation, 37
Fixed charges, 330, 332, 336, 339, 340, 344, 345, 349, 353
Flagellar movement, 348
Flip-flop, 28, 36, 211
Flory radius, 160
Flow cytometry, 6, 8, 9, 180, 220. *See also* Fluorescence activated flow cytometry
Fluid-fluid-interfaces, 300-302
Fluid-gas-interfaces, 300-302
Fluid polarization, 300
Fluid pumps, 300-303
Fluorescence activated flow cytometry, 17, 47, 48-53, 56, 71-73, 75, 76, 180
Fluorescence microscopy, 279
Fluorescent dyes. *See* Dyes
Forage grass, electrofusion, 222
Free energy, 352
Free flow electrophoresis, 17, 198, 203
Friction, 304. *See also* Damping forces; Medium viscosity
Friend cells, electrofusion, 218
Frozen-thawed donor embryos, 226
Fungi
    electrofusion, 219
    electroinjection, 42, 43
    electrotransfection, 65-67
Fusibility. *See* Electrofusion, fusibility
Fusion. *See* Electrofusion; Polyethylene glycol; Virus, fusogenic
Fusion media. *See* Media, electrofusion
Fusogenic additives, 179, 197. *See also* Polyethylene glycol; Virus, fusogenic
Fusogenic state, 28, 208

G

Gametes, electrofusion, 223
Gating currents, 352. *See also* Ion channels
Gel electrophoresis, 198
Genetic engineering. *See* Bacteria transformation; Electrofusion; Electrotransfection; Somatic hybridization
Ghost cells. *See* Erythrocyte ghosts
Giant algal cells, 11, 13, 23
Giant axon, 19, 333
Giant cell production, 39, 178, 190, 193, 198, 208, 216-219
Giardia lamblia, electrotransfection, 67

Glial tumor cells, electroinjection, 48
Gluconate, 35
Glucose, 367, 368
Glutamate, 35
Glutaraldehyde, 279
Glyceraldehyde-3-phosphatedehydrogenase, 34
Glycerol, 110, 111, 115
Glycidacrylate, 281
Glycine sp., electrotransfection, 59
Glycine, 110
Glycocalyx, 175, 197, 297, 339, 340, 342
Glycolipids, 339
Glycophorin, 339
    electroinsertion of, 78
Glycoproteins, 175, 339. *See also* Glycocalyx
Glycylglycine, 273
Goat, cloning of, 226
Gold electrodes, 6, 35, 120, 193, 203
Goldman equation, 350
Golgi vesicles, 174, 332
Gouy Chapman theory, 340, 341, 350
Gramicidin, 292, 297
Gram-negative bacteria. *See* Bacteria, gram-negative
Gram-positive bacteria. *See* Bacteria, gram-positive
Graphite electrodes, 204
Gravitational forces, 179
    levitation against, 270, 272-282
Growth. *See* Cell, growth; External electric field, growth in
Growth phase. *See* Cell, cycle
Guard cell, 217, 218
GUS gene, 58-61, 63

H

H^+-pump, 334, 335
Haemoglobin, 218, 336
Haemolysis, 34, 336
HAT medium, 372
Heat inactivation, of enzymes, 199
Heating. *See* Joule heating; Temperature
Heat shock, 59-61, 63, 65, 67, 114, 117
Heavy metals, contaminants, 35, 201
Helianthus sp.,
    electrofusion, 222
    trapping of, 277
Helical fusion chamber, 369, 370, 372
Helicobacter pylori, electrotransfection, 126
Helix, 352, 353
Helmholtz-equation, 340

Hemi-fusion, 213, 214
Hemisphere, cell. *See* Membrane
    breakdown, asymmetric, symmetric
Hemoglobin, 218, 336
Hemolysis, 34, 336
Hepatocytes
    electrofusion, 179
    electroinjection, 50
    electrotransfection, 73
Hepatoma cells
    electroinjection, 48
    electrotransfection, 70
HEPES buffer, 34, 110, 217
Heterokaryons, 174, 178, 182, 188, 216,
    220-222
Heteromyeloma cells, 202, 230, 371
Hexadecane, 159
High-conductivity media, 5, 6, 15, 16, 33-
    37, 120, 146, 178, 182, 273, 279,
    282, 290, 298, 299, 305
High-density media, 37, 192, 195
Higher plants. *See also* Plant protoplasts
    electrofusion, 178, 220-223
    electroinjection, 42, 220-223
    electrotransfection, 58-65
High-frequency fields. *See* Cell motion;
    External electric field, high-
    frequency
High-intensity field pulses. *See also* Charge
    pulse technique; Electrofusion;
    Electroinjection; External electric
    field; Membrane breakdown
    charging process, 7, 11, 12, 22, 26, 27,
        31, 32
    critical field strength, 5, 203-207
    discharging process, 31, 32, 112, 113,
        162-165
    duration, 10, 11, 24-27, 41, 108, 111-114,
        120, 127, 203, 207, 211, 345, 346,
        366-372
    instrumentation. *See* Capacitor discharge;
        Discharge chamber; Electrofusion,
        instrumentation
    number, 24, 36, 78, 113, 114, 200, 203,
        207, 208, 366-372
    pulse time interval, 24, 366-369
    radio-frequency pulses, 46, 48, 51, 54,
        58, 69, 71, 186
    shape, 31-33, 42-76, 112, 113, 203, 207,
        208, 211, 366-372
    supercritical field strength, 12, 23, 144,
        205, 206, 208, 212, 215
High-permeability state, 18, 27, 29, 33, 35,
    37

Histidine, 201, 368
HMAbs. *See* Monoclonal antibodies
Homogeneous electric field. *See* External
    electric field, homogeneity
Homokaryons, 174, 178, 188, 220
Hook's law, 140, 142
Hordeum vulgare, electrotransfection, 60
Hormone signaling, 331
Horseradish peroxidase, 44, 51
Human monoclonal antibodies. *See*
    Monoclonal antibodies
Human viruses, 176
Hyaluronic acid, 110
Hyaluronidase, 110
Hybridoma, trapping of, 277
Hybridoma production, 74, 78, 180, 195,
    202, 203, 227-230, 371, 372. *See*
    also B-cells; Electrofusion
Hybrid yield. *See* Electrofusion, hybrid
    yield; Somatic hybridization
Hydration forces, 175, 177
Hydration layer, 28, 175, 177, 187, 200
Hydraulic conductivity, 218
Hydrophilic pores. *See* Pores, hydrophilic
Hydrophobic interactions, 180
Hydrophobic pores. *See* Pores, hydrophobic
Hydrostatic pressure, 11, 23, 37, 108, 127,
    179, 214, 217. *See also* Colloid-
    osmotic pressure
Hyperimmunization, 228
Hypochlorite, 200
Hypo-osmolar media
    electrofusion, 36, 37, 39, 201-204, 208,
        212, 214, 370, 371
    electropermeabilization, 18, 20, 35-37,
        39, 69, 75, 369
    electrorotation, 294, 295, 297

I

Ice formation, 2
Image analysis, 271
Image force, 148
Immobilization, 216, 279, 371
Immortalization. *See* B-cells, hybridomas;
    Monoclonal antibodies; T-cells,
    hybridomas
Immunization, 228
Immunofluorescence, 230
Immunoglobulins, 203, 227-230
Immunomodulation, 227
Impedance spectroscopy, 267, 271, 286,
    293-295, 297, 298, 330
Indicator molecules. *See* Dyes

Inertia, 152, 153, 158, 161, 266
Infectivity, 127
Inflammation areas, 78
Inflammatory response, 53
Influenza virus, 175
Inhibitors, 40
Inhomogeneous electric field. *See* External
    electric field, inhomogeneity
Injection. *See* Bacteria transformation;
    Electroinjection; Electrotransfection
Inositol, in electromanipulation, 8, 18, 201,
    295, 369
Inositol cycle, 42
Inositol triphosphate, 35, 46, 49-52
Insects
    electrofusion, 219
    electrotransfection, 67
Insulin, 43, 52, 55, 56, 218
Insuloma cells, electroinjection, 48
Integrated circuits. *See* Microstructures
Interdigitated (castellated) electrodes, 183,
    271, 273, 275
Interfacial charge transport, 352
Intergeneric hybridization, 219, 222
Interkingdom fusion, 218
Intermingling process. *See* Electrofusion,
    intermingling process
Internal conductivity. *See* Cell, internal
    conductivity
Interspecific hybridization, 174, 219, 222
Intine, electrorotation, 292
Intracellular dielectrophoresis, 187, 204
Intracellular osmotic pressure, 11, 35, 37, 38
Intramembrane particle-free lipid domains,
    10, 197, 198, 205, 208, 209, 210,
    212
Intramembraneous particles. *See* Membrane,
    proteins
Intraspecific fusion, 174, 222
Intrinsic cellular oscillators, 284, 285
Intrinsic electric field, 22
Intrinsic membrane proteins. *See* Membrane,
    proteins
Inulin, 37, 39
Invagination process, 39. *See also*
    Electrointernalization
Iodoacetate, 221, 223
Ion carriers, 217, 331, 334, 335, 348
Ion channels, 7, 19, 22, 28, 36, 217, 218,
    331, 332, 343, 347, 352
Ion exchange beads, electrorotation, 285
Ion exchange membranes, 343
Ion gradients, collapse of cellular, 33, 34,
    337

Ion leakage, 298. *See also* Electro-release
Ion mobility, 284
Ion pumps, 7, 28, 330-335, 346, 348. *See
    also* ATPases; Electrogenic pumps;
    Electroneutral pumps
Ionic-strength modulation, 27
Ionophores, 292, 297, 334, 336, 351
Ion transporters. *See* Ion carriers
Irradiation, 216, 220-223, 228
Irreversible membrane breakdown. *See*
    Membrane breakdown, irreversible
Islets
    electrofusion, 218
    electroinjection, 55, 56
    immobilization, 217, 279
Isoelectric points, 180, 201, 336, 339
Isoenzyme analysis, 222
Isomotive force electrode, 271
Isonicotinic acid hydrazide, 110
Iso-osmolar media
    electrofusion, 36, 37, 39, 201-203, 207,
        208, 214
    electropermeabilization, 18, 20, 29, 36,
        37, 39, 369
Isotopes, 4

J

Joule heating, 24, 25, 114, 120, 143, 185,
    203, 261, 264, 273, 274, 284, 286

K

Kalanchoe daigremontiana, electrofusion,
    211
Karyoplast, 181, 223, 224
Kidney cells
    electroinjection, 48
    electrotransfection, 70
Killer activity, of yeast, 66
Killer toxin, 219
Kinetic energy term, 154
K^+ ion gradients, 33, 333, 336
K^+ ions
    and electromanipulation media, 18, 34,
        201, 367, 369
    and transition temperature, 350

L

La^{3+} ions, 200
Lactococcus lactis, electrotransfection, 108,
    118
Lactuca sp., electrofusion, 222

Laminar flow systems, 272
Langerhans's islets. *See* Islets
Langmuir adsorption isotherm, 341
Laplace equations, 11-17, 39, 111, 144, 145, 204
Large-scale fusion. *See* Electrofusion, macrofusion; Helical fusion chamber
Laser diffraction, 343
Laser traps, 282, 283
Latex particles, 30, 31, 47, 271, 272, 281
L-broth medium, 109, 112, 114, 115, 366
L-cells, 198, 296
    electroinjection, 18, 30
    electrophoretic mobility, 198
    electrotransfection, 69
    membrane potential oscillations, 348
    electrorotation, 38
Lead ions, contaminants, 35
LeChatelier's principle, 352
Lectins, 179
Leishmania sp., electrotransfection, 67
Lentil culinaris, electroinjection, 42
Leucine, 368
Leucocytes, electroinjection, 49
Leukemia cells
    electroinjection, 49, 50
    electrotransfection, 71-73
Levitation, 187, 261, 262, 268, 272, 275-283
    feedback-controlled, 270, 275, 282, 287
Li+, effect on transition temperature, 350
Light-scattering methods, 270, 271, 284
Lightning strike, 2
Lilium longiflorum, electroinjection, 42
Linearized DNA. *See* Plasmid, linear
Line tension, 150, 152
Lipid bilayer membrane. *See also* Membrane
    area stretching, 138-141, 147-149
    bending deformation, 39, 142, 143, 147
    breakdown, 4, 36, 40, 143-167, 209. *See also* Membrane breakdown
    capacitance, 155, 159
    charge pulse relaxation technique, 155-168
    cohesive properties, 141
    compressibility moduli, 139-141
    composition, 140, 152, 156-160
    conductance, 13, 145, 146, 150, 155, 156-159, 162-165
    curvature, 142, 143
    elastic moduli, 139-141, 148
    electric double layer, 340, 341
    energy barrier, 149, 151, 153

field effects on, 143-168
fluidity, 23, 27, 35, 200
line tension, 39, 150, 152
phase separation, 351
phase transition, 138, 350
planar
    electrofusion, 208
    irreversible breakdown, 152-161, 168, 208-213
    reversible breakdown, 36, 161-168, 208-213
resistance. *See* Lipid bilayer membrane, conductance
rupture, 39, 140, 147-161
shear deformation, 141
stability, 140, 142, 149, 151
surface tension. *See* Surface, tension
thickness, 148, 153, 159, 350. *See also* Field-induced, membrane thinning
trilaminar structure, 210, 212
vesicles. *See* Lipid vesicles
viscoelastic properties, 138-143, 152-161, 167
Lipid mobility, 23, 27, 35, 200
Lipid peroxidation, 28, 200
Lipid/protein junctions, 9
Lipid vesicles
    as drug delivery systems, 77
    electropermeabilization, 7, 18, 27, 29, 144, 147
    electrorotation, 285
    deformation 18, 144, 147
    field effects on, 143-152, 161, 162, 168
    field-induced formation, 29, 30, 39, 182, 199, 209-213, 296
    fusion, 147, 175, 198, 208
    as model systems, 138-143
    orientation, 146, 147
Lipopolysaccharides, 28, 110, 228, 371
Liposomes. *See* Lipid vesicles
Lipoteichoic acid, 110
Lipoxygenase activity, 42
Liquid chromatography, 303
Lithography, 271, 273
Liver cells
    electroinjection, 50
    electrotransfection, 73
L-mouse cells. *See* L-cells
Logarithmic growth phase. *See* Cell, cycle
Long-lived phenomena, 28, 208
Lossy dielectric particles, 185, 263, 284. *See also* Ponderomotive forces
Lotus sp., electrofusion, 222

Low-conductivity media, 15, 16, 110, 197,
200-203, 265, 273, 282, 286, 289,
290, 294, 295, 298, 305, 336
LPS. *See* Lipopolysaccharides
Lucifer Yellow, electroinjection of, 49, 51,
56, 57
Lung cells
electroinjection, 51
electrotransfection, 74
Lycopersicon sp.
electrofusion, 222
electrotransfection, 60
Lymph nodes, 228
Lymphoblasts. *See* B-cell
Lymphocytes. *See also* B-cells; T-cells;
Tumor-infiltrating lymphocytes
as drug delivery systems, 78
electrofusion, 179, 180, 202, 203, 227-
230, 371, 372
electroinjection, 51, 52
electrorotation, 295
electrotransfection, 74
transmembrane potential, 334
Lymphoma cells
electrofusion, 199
electroinjection, 49, 50
electrotransfection, 71-73
Lysis. *See* Cell, lysis
Lysolecithin, 168
Lysolipids, 197
Lysophosphatidylcholine, 152, 197
Lysosomes, 174
Lysostaphin, 110
Lysozyme, 110
Lytechinus pictus, electroinjection, 43

M

MAb production. *See* Monoclonal antibodies
Macrofusion, *See* Electrofusion, macrofusion
Macromolecules
dielectrophoretic collection of, 270, 271,
279, 283
injection of. *See* Bacteria transformation;
Electrotransfection; Electroinjection
Macrophages
electroinjection, 52
electrotransfection, 75
Magnesium ions, 34, 110, 127, 199, 200,
202, 350, 369, 371
Magnetically mediated cell alignment, 182
Magnetic fields, 182, 265
Magnetic particles, electroinjection of, 47,
78

Maize. *See Zea mays*
Mammalian cells. *See also* Electrofusion;
Electroinjection; Electrorotation;
Electrotransfection
electrofusion, 178, 188-192, 197-204
electroinjection, 43-57
electrorotation, 285, 292-299
electrotransfection, 68-76
Manganese ions, 200
Mannitol, 110
Marker molecules. *See* Dyes; Isotopes
Maxwell forces, 143-147, 149, 265, 300,
324
Maxwell stress tensor. *See* Maxwell forces
Meander fusion chambers, 194, 271. *See
also* Electrode arrays; Electrodes
Mechanical forces. *See* Hydrostatic pressure;
Turgor pressure
Mechanical membrane breakdown. *See*
Membrane breakdown, irreversible
Mechanical stress, 146-148. *See also*
Hydrostatic pressure; Turgor
pressure
Mechanical tension, 138-142, 148, 153, 161,
162
Mechanisms. *See* Membrane breakdown,
mechanisms of; Theories
Media
electrofusion, 15, 16, 178, 182, 199-203,
208, 282, 369-371
electropermeabilization, 5, 6, 15, 16, 18,
33-37, 108, 110, 120, 366-369
electrorotation, 15, 16, 286, 290, 294,
295, 298, 299, 336
high-conductivity. *See* High-conductivity
media
high-density. *See* High-density media
hypo-osmolar. *See* Hypo-osmolar media
iso-osmolar. *See* Iso-osmolar media
low-conductivity. *See* Low-conductivity
media
selection. *See* Selection media
Medicago sativa
electrofusion, 222
electrotransfection, 60
Medium polarizability, 273. *See also*
Dielectric constant; Dielectric
properties
Medium viscosity, 111, 263, 266, 275, 278
Melanoma cells, electroinjection, 52
Melittin, 336
Membrane. *See also* Lipid bilayer membrane
action potential, 332, 334, 346
area stretching, 138-141, 147-149

apposition. *See* Cell apposition
asymmetry, 27, 28, 142, 143
bending deformation, 142, 143, 147
blebs, 39, 127, 192
breakdown. *See* Membrane breakdown
capacitance, 18, 26, 38, 187, 270, 290, 294-297, 330, 343
charge displacement, 11, 143, 145, 352
charge separation. *See* Cell polarization
composition, 7, 9, 148, 263, 332, 339, 344, 346
conductance, 5, 13, 15, 21, 22, 36, 263, 270, 290, 294, 297, 298, 330, 343, 344
contact. *See* Cell, contact; Cell apposition
curvature, 36, 39, 143, 147, 177, 209, 342
diffusion potential, 333-336
electric double layer, 330, 332, 339-343, 350
electrofusion. *See* Electrofusion
fluidity, 10, 35, 110, 198, 200
glycocalyx, 175, 197, 297, 332, 339, 340, 342
hemi-fusion, 213, 214
hydration layer, 28, 175, 177, 187, 200
interfacial charge transport, 352
intermingling process. *See* Electrofusion, intermingling process
intramembranous particles. *See* Membrane, proteins
intrinsic electric field, 22, 332, 344, 347-350
ion carriers, 217, 331, 334, 335, 348
ion channels, 7, 19, 22, 28, 36, 217, 218, 331, 332, 343, 347, 352
ion permeability, 331-334, 336
ion pumps, 7, 28, 296, 330-336, 346, 348. *See also* ATPases
lipid domains, 10, 197, 198, 205, 208, 209, 210, 212
mitochondrial. *See* Mitochondria
mobile charges, 296, 297
nuclear. *See* Nucleus
permeabilization. *See* Electroinjection; External electric field; Field-induced; Membrane breakdown
plasmalemma, 9, 10, 182, 207, 263, 296
potential profiles
    lateral, 346, 347, 354
    perpendicular, 332-346
precompression of, 11
proteins, 7, 9, 10, 17, 28, 40, 179, 197, 198, 205, 209, 215, 331, 332, 339, 344, 351-353

punch-through effect, 23, 24, 114, 215, 344-346
repair processes. *See* Membrane breakdown, resealing
resistance. *See* Membrane, conductance
shear deformation, 141
skeleton. *See* Cytoskeleton
stability, 10, 34, 140, 142, 149, 198, 199, 216
staining, 187, 213, 230, 276
structure. *See* Membrane, composition
surface admittance, 14, 15
surface charge, 183, 186, 197, 198, 209, 262, 264, 300, 339, 340, 348-351
surface conductance, 110, 183, 294, 297, 346, 354
surface potential, 22, 209, 331-333, 339-343, 349, 350
tension, 19, 149, 153, 207, 212
thickness, 148, 153, 159
tonoplast, 9, 207
transbilayer mobility, 28, 36, 211
transmembrane potential, 5, 11-23, 179, 218, 263, 264, 331-338, 348-350
undulation. *See* Membrane, curvature
viscosity, 152, 154, 159, 161, 162
Membrane breakdown. *See also* Electrofusion; Electroinjection; External electric field; Field-induced; High-intensity field pulses
adherent cells, 5, 44, 177, 178
in alternating current fields, 32, 33, 113
asymmetric, 10, 12, 13, 17-24, 34, 42, 43, 46, 52, 69, 206, 207
chloroplasts, 9, 17
in DC-shifted alternating fields, 33
defects. *See* Cracks; Defects; Pores
dielectrophoretically aligned cells, 18, 32, 37, 188, 204-207
electrofusion, 176-231
electroinsertion. *See* Electroinsertion
electrointernalization. *See* Electrointernalization
electro-mechanical coupling, 36
eukaryotes, 2-79
field-induced thinning, 36, 148, 149
irreversible, 2, 14, 16, 17, 24, 25, 140, 148, 149, 151-161, 164, 203, 205, 206, 212-215, 284
lipid bilayer membranes, 4, 11, 36, 152-168
lipid vesicles, 7, 18, 27, 29, 144, 147
mechanical. *See* Membrane breakdown, irreversible

mechanisms of, 36-40, 147-152, 197, 198, 205-207
methodology, 4-7, 31-33
mitochondria, 17, 196
nuclei, 17
pores. *See* Cracks; Defects; Pores
prebreakdown state, 27
and pressure, 11, 36, 179
prokaryotes, 4-7, 10, 11, 107-128
pulsing in monolayers, 44-46, 48, 177, 178, 193
resealing, 4, 6, 7, 18, 24, 27-29, 35, 36, 39, 78, 112, 114, 115, 162-165, 168, 199, 203, 208, 209, 213
reversible, 2-7, 14, 16, 19, 23-26, 36-41, 108, 109, 127, 148, 149, 161-167, 205, 212-215
in rotating fields, 32
secondary processes, 36, 37, 39, 127, 164, 192, 332, 334, 336, 337
site of, 7, 9, 40
solute exchange, 3, 4, 6, 7, 17, 31, 33-35, 191, 203, 332, 334, 336
suspended cells, 4-7, 32, 39, 40
symmetric, 12, 17-24, 34
and temperature, 10, 17, 25, 29, 41, 78, 112, 114, 120, 167
time, 31, 37, 163, 346
transbilayer mobility, 28, 36, 211
vacuoles, 9, 17, 18
vesicle formation. *See* Field-induced, vesicle formation
voltage, 7, 10
    calculation of, 11-17, 111
    determination, 4-7, 164, 165
    factors affecting, 9-17, 23, 24, 111, 114, 148, 165-167, 215, 344-346
Membrane folding, 39, 296, 343
Membrane material, incorporation of, 297
Membrane, passive electric properties of. *See* Membrane, capacitance, conductance
Membrane proteins. *See* Membrane, proteins
Membrane stabilizing agents, 10, 34
ß-Mercaptoethanol, 367
MES, 34
Metabolites. *See* Membrane breakdown, solute exchange
Metal, for electrodes, 6, 35, 120
Metal ion contaminants, 35, 201
Methanococcus sp., electrotransfection, 110, 116, 126
Methotrexate, electroinjection of, 47, 51

Methylation, 117
S-Methyl-L-cysteine, 197
Mg^{2+} ions, 34, 110, 127, 199, 200, 202, 350, 369, 371
Microaggregates,
    field-induced formation of, 279-281
    field-induced motion of, 267
Microcarriers, 177, 279
Microchannels, 301-305. *See also* Microstructures
Microdroplets, electrofusion in. *See* Electrofusion, microfusion
Microelectrode techniques, 4, 7, 11, 23, 27, 217, 218, 332, 343. *See also* Charge pulse technique; Electrode arrays; Electrodes
    charge pulse relaxation, 4, 11, 155-168, 216, 294
    current clamp, 216
    current-voltage characteristics, 4, 217, 294, 343-346, 350
    microstructures. *See* Microstructures
    patch clamp, 5, 7, 27, 216-218
    voltage clamp, 5, 154, 155, 164, 168, 216
Microfilaments, 51, 55, 199
Microfluorimetry, 18, 19
Microfusion. *See* Electrofusion, microfusion
Microgravity electrofusion, 191
Micromanipulators, 220
Micromotors, 261
Microorganisms. *See* Bacteria
Microphotometry, 18. *See also* Photometry
Micropipettes, 138, 148, 151, 181, 220, 224, 349
Micropumps, 261, 301-303
Microstructures, 5, 183, 185, 193, 228, 260, 261, 265, 271, 273-283, 298, 301-306
Microtubuli, 214
Microvilli, 10, 38, 39, 175, 296, 297, 343
Microwaves, 215
Millipore filter, for electrofusion, 181
Mitochondria, 3, 9, 17, 216, 292, 332
    dielectrophoretic collection of, 269
    electrofusion of, 196
    membrane breakdown, 9
    microviscosity of membrane lipids, 351
    traits encoded in, 221-224
Mitogen, 228. *See also* Lipopolysaccharides
Mitosis-inducing protein, 216
Mn^{2+} ions, 200
Mobile charges, 296, 297
Modelling. *See* Theories

Model systems. *See* Lipid bilayer membrane; Lipid vesicles
Monoclonal antibodies, 74, 77, 78, 180, 195, 202, 203, 227-230, 371, 372
Monolayer cultures, pulsing of, 44-46, 48, 177, 178
Monolayer membranes, 351
Morphogenesis, 346
Morula stage, 224
Mosaic membrane structure. *See* Membrane, composition
Moss cells, electrofusion, 219
Mouse L-cells. *See* L-Cells
Mucin coat, 197
Multi-body problem. *See* Dielectrophoresis, multi-body problem
Multi-cell electrofusion. *See* Giant cell production
Multi-cell electrorotation, 285. *See also* Electrorotation
Multi-electrode assemblies. *See* Electrode arrays; Electrodes
Multinucleated cells. *See* Cell, multinucleated; Giant cell production
Multiple fusion. *See* Giant cell production
Multipoles, 267, 278-282
Multi-shelled particles. *See* Particle, multi-shelled
Multi-shell models, 146, 289-294
Muntjac cells, electrofusion, 188, 189, 216
Murine monoclonal antibodies. *See* Monoclonal antibodies
Muscle cells, electroinjection, 52
Muscle excitation, 331, 346
Mussels, electrofusion, 226
Mutagenic agents, 216
Mutual dielectrophoresis. *See* Dielectrophoresis, mutual
Mycobacterium sp., electrotransfection, 126
Myeloma cells
  asymmetric breakdown, 18, 20, 52
  electrofusion, 179, 180, 211, 214, 227-230, 371, 372
  electroinjection, 8, 9, 52
  electrorotation, 296, 297
  electrotransfection, 74, 75

N

Na$^+$-channels, 22
Na$^+$ ion gradients, 33, 333, 336

Na$^+$ ions
  and electromanipulation media, 34, 35, 201
  and transition temperature, 350
Na$^+$-K$^+$-ATPase, 331, 333-335, 353
Natural membrane contact, 175, 177, 178
Navier-Stokes Equation, 153
NBD-glucosamine, 23
Negative dielectrophoresis. *See* Dielectrophoresis, negative
Nernst equation, 336, 349
Nernst-Planck equation, 350
Nerve
  depolarization, 332
  excitation, 331, 334
Nervous cells, electrotransfection, 75
Neuraminic acids, 339, 340
Neuraminidase, 197, 339
Neuroblastoma cells, electroinjection, 52
Neuronal networks, 260
Neurosecretion, 16
Neurospora crassa, electrotransfection, 65
Neutral red, 190
Neutrophils
  as drug delivery systems, 54
  electroinjection, 52-54
Nicked DNA, 116
Nicotiana sp.
  electrofusion, 222
  electroinjection, 42
  electrotransfection, 60-62
Nitella sp., punch through effect, 344
NMR. *See* Nuclear magnetic resonance
Noble metal electrodes. *See* Electrode material
Non-equilibrium state, 332, 343, 346
Non-physiological solutions. *See* Low-conductivity media
Non-spherical cells, 111, 192, 264, 265, 293, 294
Non-Stokesian diffusion, 40
Non-uniform fields. *See* External electric field, inhomogeneity
Nose cells, electrotransfection, 75
Nuclear de-differentiation, 224
Nuclear magnetic resonance, 28, 79
Nuclear re-differentiation, 224
Nuclear transfer, 181, 223-227
Nuclear transplantation. *See* Nuclear transfer
Nucleolus, 216
Nucleosides, electroinjection of, 40, 49, 54
Nucleotides, electroinjection of, 44, 45, 46, 49, 51-57, 128

Nucleus,
   electrochemical potential, 332
   electrorotation, 292
   field-induced motion of, 187
   fusion, 174, 187-189, 214, 223
   membrane breakdown, 17
   organization, 216
   swelling, 39
   traits encoded in, 221-223
   transplantation, 223-227
Null frequency method, 287, 288, 298
Number of pulses. *See* High-Intensity field
      pulses
Numerical methods, 264, 265
Nyquist noise, 354

O

Oat. *See Avena sativa*
Octopole field cage, 279-281
Oocytes
   activation, 54, 224-226
   breakdown conditions, 26
   electrofusion, 193, 194, 197, 204, 207,
      271
   electroinjection, 54
   electrorotation, 292
   fertilization, 57
   nuclear transplantation, 181, 223-227
   pronuclear transplantation, 181, 224
Oogenesis, 178
Optical tweezers, 282, 283
Ore, veins of, 2
Organelles. *See also* Chloroplasts;
      Mitochondria; Nucleus
   dielectric spectroscopy, 306
   electrochemical potential, 332
   electrofusion, 196, 204
   electrointernalization, 3, 40
   electro-orientation, 192
   electrorotation, 292, 293
   high-frequency field exposure, 33
   membrane breakdown, 9, 24
   traits encoded in, 221-224
Organic compounds, in pulse medium, 34
Organogenesis, 223
Orientation. *See* Cell, orientation; Electro-
      orientation
Oryza sativa
   electrofusion, 222
   electrotransfection, 62
Oryzias latipes, electrotransfection, 67
Oscillators, cellular, 284, 285

Osmolality
   electrofusion media, 36, 37, 39, 201-203,
      208, 213, 214, 369-371
   pulse media, 18, 19, 35-37, 39, 69, 70,
      367-369
   rotation media, 286, 294, 296, 297
Osmotic swelling, 6, 11, 35, 37-39, 111,
      138, 179, 192, 201-203, 212, 214,
      215, 295, 297, 336, 337
Ouabain, 334, 348
Ovalbumin, electroinjection of, 45, 49-51,
      54, 56
Ovary cells. *See also* Chinese hamster ovary
      cells
   electroinjection, 54, 55
   electrotransfection, 75, 76
   membrane breakdown, 39
Oxidized cholesterol, 162-167
Oysters, electrofusion, 226

P

Pancreatic cells, electroinjection, 55
Pancreatic islets. *See* Islets
Panicum maximum, electrotransfection, 63
Paracentrotus lividus, electrofusion, 204
Partial fusion index, 188, 189
Particle
   analyzer, 4, 5, 15, 37, 180, 298
   alignment. *See* Cell apposition, electric
   caging, 261, 275, 279, 306
   chains. *See* Cell apposition, electric,
      ultrasonic
   collection, 261, 269-273. *See also*
      Dielectrophoresis, negative, positive
   conductivity, 267, 273, 289, 297
   dielectric properties; *See* Dielectric
      properties
   electroinjection of, 30, 31, 40, 47
   electrorotation. *See* Electrorotation
   field probes, 204-206, 271
   levitation. *See* Levitation
   loss-free, 185
   lossy, 185, 263, 284
   magnetic, 78, 182
   multi-shelled, 263-265, 289-294
   non-spherical, 264, 265, 294
   polarization, 183, 185, 260-269, 272, 301,
      304, 305
   precipitators, 261, 270
   single-shelled, 263, 265, 268, 289-294
   spinning. *See* Electrorotation
   trajectories, 185, 264
   trapping, 261, 275-283, 305

travelling-wave-induced motion, 261,
303-305
Passive electric membrane properties. *See*
Electrorotation; Membrane,
capacitance, conductance
Pasteurella multocida, electrotransfection,
109
Patch clamp technique, 5, 7, 216-218
Pb^{2+} ions, contaminants, 35
PBS. *See* Phosphate buffer solutions
PCC. *See* Premature chromosome
condensation
Pearl chaining. *See* Cell apposition, electric,
ultrasonic
PEG. *See* Polyethylene glycol
PEG-polymer, 160
Pellet. *See* Cell apposition, centrifugation
Penicillin, 110
Pennisetum purpureum, electrotransfection,
63
Peptide-mediated electrofusion, 180
Peptone, 367
Percoll gradient, 292
Periodate, 40
Peripheral blood, 203, 228, 229, 371
Perivitelline space, 181, 224
Permeabilization. *See* External electric field;
Membrane breakdown
Permittivity, 145, 183, 192, 263, 264, 267,
269, 273, 288, 300, 301, 327, 328.
See also Dielectric constant;
Dielectric properties
Permittivity of vacuum, 145, 150, 263, 284,
340
Petunia hybrida, electrotransfection, 63
pH, of electromanipulation media, 8, 34, 201
Phaeochromocytoma cells
electroinjection, 56
electrotransfection, 76
Phage RNA, 128
Phagocytes, as drug delivery systems, 78
Phagocytosis, 49
Phalloidin, electroinjection of, 46, 47, 48, 51
Phallotoxin, electroinjection of, 42
Pharmacological agents, 10, 77
Phaseolus sp.
electroinjection, 42
electrotransfection, 63
Phase contrast microscope, 147, 162
Phase separation, 351
Phase transition, 138, 350
Phenol red, 35, 38, 369, 372
Phenosafranine, 42
pH indicators, 35, 38

Phloxin B, 27, 34
Phorbol ester, electroinjection of, 44, 46, 55,
56
Phosphate buffer solutions, 33, 201
Phosphatidyl
choline, 139-141, 152, 156-160, 197
ethanolamine, 152, 159, 339
serine, 339, 342
Phospholipase, 49, 53, 198
Phospholipids
acidic, 28, 200
asymmetry, 27, 28, 40, 210, 339
Phospholipid translocase, 28, 339, 351
Phospholipid vesicles. *See* Lipid vesicles
Photoconduction, 284
Photo-lithography, 271
Photometry, 271. *See also* Microphotometry
Physarum polycephalum, electrotransfection,
65
Phytohemagglutinin, 228, 371
Picea glauca, electrotransfection, 63
Pigments, intracellular dielectrophoresis, 204
Pigs, cloning of, 226
Pinocytosis, 348
Pin-plate electrodes, 196. *See also* Electrode
arrays; Electrodes
Pisum sp., electrofusion, 222
Planar electrode arrays. *See* Microstructures
Planar lipid bilayer membranes. *See* Lipid
bilayer membrane, planar;
Membrane breakdown, lipid bilayer
membranes
Plant breeding, 58-65, 220-223
Plant (walled) cells
electrotransfection, 58, 60, 63, 64
turgor pressure. *See* Turgor pressure
Plant protoplasts,
electrofusion, 178, 181, 194, 197, 199-
201, 204-206, 209, 211, 217-223,
370, 371
electrokinetic force spectra, 267, 268
electropermeabilization, 10, 16-19, 27,
28, 42, 206, 368
electrorotation, 285, 291, 292
electrotransfection, 58-65, 368
enucleated, 223
evacuolated, 182, 221
guard cell, 217, 218
hybrid selection, 220, 221, 368
preparation, 368
regeneration, 34, 220, 223, 368, 371
Plasmacytoma cells
electroinjection, 52
electrotransfection, 74, 75

Plasmalemma. *See* Membrane, plasmalemma
Plasmid. *See also* Bacteria transformation;
    Carrier DNA; DNA; RNA; Virus,
    electroinjection of
  circular, 41, 59-61, 63, 65, 66, 68, 69, 71-
    73, 116
  concentration of, 116, 118, 367
  co-transformation, 41, 59-62, 65, 67, 68,
    72, 73
  curing, 119, 120
  degradation, 39
  exchange, 117, 119
  injection of, 58-76, 121-126. *See also*
    Electrotransfection
  linear, 41, 59-63, 65-76, 116
  nicked DNA, 116
  purification, 117
  relaxed DNA, 11, 116
  release, 117
  size, 116, 128
  supercoiled, 29, 41, 59, 60, 62-65, 69-76,
    116
Plate electrodes, 35, 112. *See also* Electrode
    arrays; Electrodes
Platelets
  aggregation of, 348
  as drug delivery systems, 56, 57, 78
  electroinjection, 6, 16, 56, 57
Platinum electrodes, 35, 120, 193, 203. *See*
    also Electrode arrays; Electrodes
Pluronic F-68, 159
Poisson equation, 342
Pokeweed, 228
  electroinjection of, 46
Polar body, 224
Polarizable particles. *See* Particle,
    polarization
Polarization. *See* Cell polarization; Electrode
    polarization; Particle, polarization
Polarization-induced motion. *See* Cell
    motion, Ponderomotive forces
Pollen
  electroinjection, 42
  electrorotation, 292
Polyacrylamide, 159
Polybrene, 217
Polyethylene glycol, 59-62, 64-67, 110-112,
    114, 115, 118, 159, 160, 175, 178,
    179, 197, 208, 230, 366
Polyimine, 217
Polykaryons. *See* Cell, multinucleated; Giant
    cell production
Poly(L)-lysine, 194

Polymers, 40, 192
Polypeptides, 37
Polyploidy, 219-221, 224. *See also* Cell,
    multinucleated
Polystyrene beads, 185, 186, 204, 206, 272
Ponderomotive forces, 182-187, 260-305
Pores. *See also* Defects; Cracks
  conductivity, 153, 154, 156, 158
  energy, 151, 153, 154. *See also* Energy
  expansion, 152-154, 157-161, 213, 214,
    215
  definition, 3, 40, 147
  field line distribution, 151
  formation, 18, 22, 36, 147-150, 153, 161,
    164, 209, 213
  fusion of, 161
  hydrophilic, 40, 150, 209, 210, 212
  hydrophobic, 150, 210, 212
  models, 12, 13, 22, 36, 37, 152-161, 209,
    210, 212
  permeability, 40, 127
  radius, 40, 44, 150, 151, 153, 157, 158,
    165
  resealing, 162, 164
  rim, 152, 154, 157-159
  single or multiple, 156-158
  stability, 37, 147-151, 162
Porous membranes, 177
Positive dielectrophoresis. *See*
    Dielectrophoresis, positive
Post-alignment. *See* Cell apposition, electric;
    Electrofusion, post-alignment
Post-breakdown state. *See* Electrofusion,
    post-alignment; Membrane
    breakdown, resealing
Potassium. *See* K$^+$ ions
Potato. *See Lycopersicon*
Potential. *See* Membrane breakdown;
    Surface, potential
Potential profiles, in membranes
  lateral, 346, 347, 354
  perpendicular, 332-346
Power supply
  electrofusion, 196, 365
  electropermeabilization, 6, 32, 365, 366.
    See also Discharge chamber
  electrorotation, 285-288
  travelling-wave drives, 300-304
Pre-alignment. *See* Cell apposition;
    Electrofusion, pre-alignment
Precipitators, 261, 270
Pre-implantation stage, 223, 224
Premature chromosome condensation, 216

Pressure
  cell turgor. *See* Turgor pressure
  mechanical, 11
  osmotic, 11, 37, 38
Progenitor cells, 223
Proliferation. *See* Cell, growth
Pronase, 10, 197-199, 209
Pronuclear transplantation, 179, 207, 224
Propidium iodide, electroinjection of, 8, 9,
      18, 20, 27, 45, 47, 51, 53-55
Propionibacterium jensenii,
      electrotransfection, 126
Prostaglandins, 197
Protease inhibitors, 199
Protein-free lipid domains, 10, 197, 198,
      205, 208, 209, 210, 212
Protein kinase, electroinjection of, 43
Proteins
  albumin. *See* Albumin
  band III, 198, 331
  degradation of, 10
  dipole moment, 352
  effect on membrane breakdown, 9, 198
  electroinsertion of, 78
  field-producing, 347
  field-sensitive, 347, 352. *See also* Ion
      channels
  intracellular. *See* Colloid-osmotic pressure
  membrane. *See* Membrane, proteins
  nuclear, 46
  surface, 28, 339, 342
  synthesis, 48
Proteolytic enzymes. *See* Enzymes, and
      electrofusion
Proton pumps, 334, 335
Protoplasmic streaming, 192
Protoplasts. *See* Bacteria protoplasts; Plant
      protoplasts; Yeast protoplasts
Protozoa
  electrofusion, 219
  electrotransfection, 67
Pseudomonas aeruginosa, 7, 114
Pulse-laser fluorescence microscopy, 7, 19,
      21, 25
Pulse medium. *See* Media,
      electropermeabilization
Pulse generator. *See* Capacitor discharge;
      Electrofusion, instrumentation;
      Discharge chamber
Pulses. *See* High-intensity field pulses
Pulse-spin-pulse technique, 118
Pumping of liquids, 300-303
Punch through effect, 23, 24, 114, 215, 344-
      346. *See also* Salt-shift

PVC particles, 272
Pyrene excimer formation, 351

Q

Quadroma cell lines, 75
Quadrupole field trap, 278, 282
Quincke rotation, 283
Quinine sulphate particles, 269

R

Rabbits, cloning of, 226
Radiation damage, 48, 216
Radiation pressure, 282
Radioactive isotopes, 4
Radio-frequency pulses, 46, 48, 51, 54, 58,
      69, 71
Raffinose, 110
Raphanus, electrofusion, 222
Receptor-mediated electrofusion, 179, 180,
      203
Receptors, 78, 228
Recombinant DNA technology. *See* Bacteria
      transformation; Electrotransfection
Rectangular field pulses
  electroinjection, 31, 42, 43, 46, 49, 54,
      113
  electrofusion, 196, 201, 203, 207, 208,
      370, 372
  electrotransfection, 58-62, 64, 66, 67, 70-
      72, 75, 76, 121, 122, 125, 126
Red blood cells. *See* Erythrocytes
Reflection coefficient, 37, 192
Refractive index, 282
Regeneration. *See* Plant protoplasts; Yeast
      protoplasts
Relaxed DNA, 116
Release. *See* Electrorelease
Renkin equation, 40
Resealing. *See* Membrane breakdown,
      resealing
Resistance. *See* Cell, internal conductivity;
      Membrane, conductance
Respiratory burst, 52, 53
Resting membrane potential. *See* Membrane,
      transmembrane potential
Restriction enzymes, electroinjection of, 46,
      48, 50, 54, 55
Retroviruses, 175
Reversible membrane breakdown. *See*
      Membrane breakdown, reversible
Rheogenic pumps, 330, 331, 334-336, 346

Rhodococcus fascians, electrotransfection, 116, 126

Rice. *See Oryza sativa*

Ring electrodes. *See* Electrodes

RNA, injection of, 41, 58-62, 64. *See also* Electrotransfection; Virus, electroinjection of

Rotating fields, 285-299. *See also* Electrorotation

Rotation spectra, 267, 268, 288-293, 296, 299

Rudbeckia sp., electrofusion, 222

Russel's viper venom, 55

S

Saccharomyces sp.
   electroinjection, 43
   electrotransfection, 29, 66, 67, 118

Saccharum sp., electrofusion, 222

Salmonella typhimurium, electrotransfection, 117

Salt-shift, current-induced, 23, 24, 114, 215, 343-346

Sandwich electrofusion method, 181

Schizosaccharomyces pombe
   electroinjection, 43
   electrotransfection, 67

SDS-gel electrophoresis, 198

Sea urchin eggs
   electrofusion, 204, 225
   electropermeabilization, 7, 18, 19, 21
   transmembrane potential, 21, 348

Secondary processes. *See* Membrane breakdown, secondary processes

Secretion, 348

Sedimentation, 178, 182, 191, 192, 272, 278. *See also* Levitation

Selection markers
   bacteria, 220
   mammalian cell, 220
   plant protoplasts, 220, 221
   yeast protoplasts, 219

Selection media, 109, 119, 190, 203, 208, 368, 372

Self-diffusion coefficient, 153, 161

Semi-axis. *See* Electro-orientation

Semiconductor-based pump-devices, 300-303

Semiconductor technology. *See* Microstructures

Sendai virus, 175

Sephadex beads, 277

Sequential pulsing technique, 119

Shape factor, 111. *See also* Laplace equations

Shape transformation, 142, 143, 147, 331, 349, 351

Shear deformation, 141

Sheep, cloning of, 226

Shoot regeneration, 59

Sialic acids, 339, 340

Signal transduction, 353

Silicon technology. *See* Microstructures

Single cell techniques. *See* Electrofusion, microfusion; Electroinjection, single-cell level; Levitation; Electrorotation, single-cell; Trapping of cells

Single-shell models, 146, 263-265, 268, 289-294

Sinusoidal electric field. *See* Dielectrophoresis, in alternating fields; Electrorotation, in alternating fields

Size distribution, of cells, 4, 5

Skin cells, electrotransfection, 76

Small-scale fusion. *See* Electrofusion, microfusion

Soap films, 152-154

Sodium. *See* Na$^+$ ions

Sodium deoxycholate, 197

Sodium oleate, 350

Solanum sp.
   electrofusion, 222
   electrotransfection, 63

Solute exchange. *See* Electro-release; Electro-uptake; Membrane breakdown, solute exchange

Solutions. *See* Media

Solvation energy, 150

Somatic hybridization, 174, 187, 219. *See also* Electrofusion
   basic research, 215-219
   cloning of animals, 178, 223-227
   hybridoma technology, 202, 203, 227-230
   plant breeding, 220-223
   yeast biotechnology, 219

Sorbitol, 201, 202, 367, 368, 371

Sorghum bicolor, electrotransfection, 63

Sound velocity, 182

Sound wavelength, 182

Southern hybridization, 222

Soybean. *See Glycine*

Space charge effects, 14, 15, 204

Spectra. *See* Dielectrophoresis; Electrorotation

Spectrin, 213, 214

Sperm-egg electrofusion, 181, 223
Spermidine
  and electrofusion, 179, 197
  electroinjection of, 43
Spermine
  and electrofusion, 179, 197
  electroinjection of, 43
Spherulation. *See* Electrofusion, kinetics
Sphingomyelin, 141
Spin label, 339, 340
Spin resonance, 284, 285
Spleen, 228
Splenocytes. *See* B-cells; Hybridoma
      production; Lymphocytes; T-cells
Spodoptera frugiperda, electrotransfection,
      67
Spontaneous fusion, 174, 175
Square pulses. *See* Rectangular field pulses
Squid axon, 19, 333
Sr^{2+} ions, 127
Stability, of membranes. *See* Membrane
Stable transformants. *See also* Bacteria
      transformation; Electrotransfection
  eukaryotes, 58-76
Stainless steel electrodes, 193. *See also*
      Electrode material
Standing ultrasonic wave, 182
Staphylococci, 120
Stationary growth phase. *See* Cell, cycle
1-Stearoyl, 2-oleoyl phosphatidylcholine,
      139-141, 152
Stern theory, 340, 341
Stoke's friction, 304
Stomatocytes, 143, 349, 351
Streaming
  convective, 264. *See also* Thermal field
      effects
  electro-osmotic, 23, 127, 346
  protoplasmic, 192
  travelling-wave-induced, 300-303
Streptavidin, 180
Streptococci sp., electrotransfection, 109,
      110
Streptomyces lividans, electrotransfection,
      108, 126
Stress tensor. *See* Maxwell forces
Strongylocentrotus purpuratus,
      electroinjection, 43
Strontium ions, 127
Submicrosecond imaging, 7, 19, 21
Sucrose, 110, 179, 201, 337, 338
Sulfolobus sp., electrotransfection, 108
Supercoiled DNA. *See* Plasmid, supercoiled

Supercritical field pulses. *See* High-intensity
      field pulses
Supercritical field strength, 12, 23, 144, 158,
      163. *See also* Charge pulse
      technique; High-intensity field
      pulses
Surface (Cell)
  acidic phospholipids, 28
  admittance, 14, 15, 204
  antigens, 28
  charge, 183, 186, 197, 198, 209, 262,
      264, 300, 339, 340, 348-351
  charge density, 145, 350
  charges, binding to, 35
  conductance, 110, 183, 294, 297, 346,
      354
  double layer. *See* Electric double layer;
      Zeta-potential
  glycocalyx, 175, 197, 297, 339, 340, 342
  Gouy-Chapman-Stern theory, 340, 341,
      350
  hydration layer, 28, 175, 177, 187
  lipopolysaccharides, 110
  pH, 349
  potential, 22, 209, 331-333, 339-343, 349,
      350
  pressure, 351
  proteins, 28, 339, 342
  reduction, 212
  structure, 296
  tension, 150-154, 160-162, 168, 351
  waves, 36
Surface coat. *See* Surface, glycocalyx
Surface tension of fluids, 196
Surface-to-volume ratio, 6
Surfactants, 110, 168. *See also* Cetyl
      pyridinium chloride
Suspension density. *See* Cell suspension
      density
Sycamore tissue cells, electrofusion, 209
Symbiotic theory, 2
Symmetric membrane breakdown, 12, 17-
      24, 34
Synchronized cells, 38, 45, 69, 76, 296
Synkaryons, 174

T

Tannin, 279
T-cells
  electrorotation, 295
  hybridomas, 227, 228

Temperature and
 breakdown voltage, 10, 17, 167
 electrofusion, 203, 213, 214, 369, 370
 electroinjection, 25, 27
 electroinsertion, 78
 electrotransfection, 29, 41, 112, 120, 366-369
 heat inactivation of enzymes, 199
 Joule heating. *See* Joule heating
 resealing, 27, 203
 and travelling-wave drives, 300, 301
Temperature gradients, 279, 300-303, 306
Temperature-sensitive mutants, 120, 220
Tension. *See* Line tension; Membrane, tension; Surface, tension
Terminology
 defects, 147
 dielectrophoresis, 269
 dielectrophoretic response, 273
 dielectrophoretic yield, 273
 electrocompetence, 79, 109
 electroinjection, 3
 electrointernalization, 3, 29-31
 electromilking, 41
 electropermeabilization, 3
 electroporation, 3
 electropore, 3, 40
Tetrahymena thermophila
 electrotransfection, 67
Theories
 defect model, 153
 dielectrophoresis, 185, 261-268
 dipole concept, 265, 324, 325
 electrofusion, 197, 205-215
 electrorotation, 261-268, 283, 284, 288-299, 301, 302
 Gouy-Chapman-Stern theory, 340, 341, 350
 membrane breakdown, 11-17, 36-40, 147-152
 multiple electrodes, 264, 265, 325-328
 multishelled particles, 264, 325-328
 travelling-wave drive, 261-268, 300-305
 viscoelastic models, 138-143, 152-155, 167
Thermal field effects, 24, 25, 114, 120, 143, 185, 203, 261, 264, 273, 274, 284, 286
Thermal fluctuations, 149, 153
Thermal gradients, 279, 300-303, 306
Thermal motion, 146, 269, 354
Thermal noise, 353, 354
Thermionic valves, 283
Thermodiffusion, 306

Thermodynamic equilibrium, 332, 336
Thermophoresis, 306
Threonine, 110
Thylakoids, contribution to electrorotation, 292
Thymidin kinase, 46
Time constant. *See* High-intensity field pulses
Tissue-cell electrofusion, 181, 182
Tobacco. *See Nicotiana*
Tomato. *See Lycopersicon*
Tonoplast, 9, 207. *See also* Vacuoles
Toroidal fields, 279
Torque, 262, 264, 268, 275, 283, 284, 289-293, 324
Toxic electrode material, 35, 120, 366
Toxic electrolytic products, 35, 186, 187, 200, 273, 274
Toxins, 331, 333, 336
Trajectories, 185, 264
Transbilayer mobility, 28, 36, 211
Transfection. *See also* Plasmid
 efficiency, 58-76, 110, 112, 114, 116, 117, 127
 electric, *See* Electrotransfection
 frequency, 58-76, 117
 other methods, 59-62, 64-72, 74-76, 107, 110, 116, 127
Transformation. *See* Transfection
Transgenic plants, 59, 62. *See also* Somatic hybridization
Transient expression, 58-65, 67, 69-76. *See also* Bacteria transformation; Electrotransfection
Transition temperature, 350
Transmembrane potential, 5, 11-23, 179, 218, 263, 264, 331-338, 348-350. *See also* Ion Pumps; Laplace Equations
Transport numbers, 23
Trapping of cells, 261, 268, 272, 275-283, 305
Travelling-wave drives
 anti-field motion, 262, 266, 267, 304
 characteristic frequency, 301, 305
 co-field motion, 266, 267, 304
 electrohydrodynamic fluid pumping, 300-303
 microelectrode arrays. *See* Electrode arrays; Electrodes
 particle motion, 266, 275, 303-305
 spectra, 268
 speed, 266, 304, 305
 theoretical background, 261-268

Trilaminar bilayer structures, 210, 212
TRIS buffer, 34, 217, 366, 367
Triticum sp., electrotransfection, 63, 64
Trypan blue, 27, 34, 44, 45, 48, 54, 55, 192, 369
Trypsin, 197
Turbidity, 366
Tumor cells. *See* Carcinoma cells
  transmembrane potential, 334
Tumor-infiltrating lymphocytes, 228. *See also* B-cells
Tungsten complex, 296
Turgor pressure, 11, 108, 127, 217
Turnover frequencies, 192, 267, 268
Tween-80, 110
Two-shell model, 293

U

Ultrasound, for cell alignment, 182
Undulations, membrane surface. *See* Membrane, curvature
Uniform electric field. *See* External electric field, homogeneity
Unstirred layers, 343-346
Uptake, field-induced. *See* Electro-uptake
UV-absorbing material, 111

V

Vacuoles
  asymmetric breakdown of, 18
  breakdown voltage of, 9, 17, 207
  electrofusion, 190, 207
  electrorotation spectrum, 292
Valinomycin, 334, 351
van der Waals forces, 177
Vaseline, 194
Veins of ore, 2
Velocity, of pore opening. *See* Pores, expansion, rim
Vertebrates
  electrofusion, 223-230
  electroinjection, 43-57
  electrotransfection, 67-76
Vesicles. *See also* Bleb formation; Budding
  as drug delivery systems, 77
  as model systems, 138-143
  deformation, 18, 144, 147
  electropermeabilization, 7, 18, 27, 29, 144, 147
  electrorotation, 285
  field effects on, 143-152, 161, 162, 168

field-induced formation, 29, 30, 39, 54, 182, 199, 209-213, 296
  fusion, 147, 175, 198, 208
  orientation, 146, 147
Viability. *See* Cell, viability
Vibrating electrode technique, 334, 346, 347. *See also* Electrode arrays; Electrodes
Vicia faba, electrofusion, 217, 218
Vigna sp., electrotransfection, 64
Vigor analysis, 220
Villiation. *See* Microvilli
Vimentine, 47
Vinca rosea, electrotransfection, 64
Virus, 331, 333, 339
  electroinjection of, 47, 61, 62
  fusogenic, 175, 176, 208, 214, 216
Viscoelastic model, 138-143, 152-161, 167
Viscosity. *See* Medium viscosity; Membrane, viscosity
Viscous effects, 263, 266. *See also* Damping forces
Visualization of electric field distribution, 204-206, 271
Vitellina, electrorotation, 292
Voltage. *See* Membrane breakdown, voltage
Voltage clamp technique, 5, 154, 155, 164, 168, 216
Voltage-current characteristics, 4, 217, 294, 343-346, 350
Voltage-gated channels. *See* Ion channels
Voltage-sensitive fluorescent dyes, 5, 7, 21, 26, 339
Volume regulation, 336-338

W

Water flow, field/pressure-induced, 11, 23, 37, 39, 175, 179
Water structure, 201
Wave-form, of field pulse. *See* High-intensity field pulses, shape
Wavelength, 182
Weakly conductive solutions. *See* Low-conductivity media
Wheat. *See Triticum*
Whole-cell capacitance, 38, 294, 296, 297
Wire electrodes. *See* Electrodes, wire

X

Xenomolecules, injection of. *See* Electroinjection, of exogeneous compounds

Xenoparticles, injection of. *See*
    Electroinjection, of particles;
    Particle
Xenopus eggs, electrofusion, 225
X-ray, 215, 220, 350. *See also* Irradiation

Y

Yarrowia lipolytica, electrotransfection, 67
Yeast
    cultivation in electric field, 276
    culture conditions, 367
    electrorotation, 285, 292, 297, 298
    electrotransfection, 66, 67
    industrial strains, 219
    killer activity, 66
Yeast extract, 367
Yeast nitrogen base, 367
Yeast protoplasts
    electrofusion, 197, 201, 217, 219
    electropermeabilization, 6, 29
    electrotransfection, 29, 66, 67, 367, 368
    preparation, 367
    wall regeneration, 368

Z

Zea mays
    electrofusion, 223
    electrotransfection, 64, 65
Zeta-potential, 332, 339, 346
Zn^{2+} ions, 34, 200
Zona pellucida, 178, 179, 181, 197, 207,
        292, 297
Zygotes, 223
Zymolyase, 367

DATE DUE

AUG 0 9 1996			
AUG 1 6 REC'D			